ANCIENT
CHINESE
WARFARE

ANCIENT
CHINESE
WARFARE

RALPH D. SAWYER
WITH THE BIBLIOGRAPHIC COLLABORATION
OF MEI-CHÜN LEE SAWYER

BASIC BOOKS
A Member of the Perseus Books Group
New York

Published by Basic Books,
A Member of the Perseus Books Group

Books published by Basic Books are available at special discounts for bulk purchases in the United States by corporations, institutions, and other organizations. For more information, please contact the Special Markets Department at the Perseus Books Group, 2300 Chestnut Street, Suite 200, Philadelphia, PA 19103, or call (800) 810-4145, ext. 5000, or e-mail special.markets@perseusbooks.com.

Designed by Pauline Brown
Text set in 12-point Dante

Chinese calligraphy by Lee Ting-rong

The Library of Congress has cataloged the printed edition as follows:

Sawyer, Ralph D.
  Ancient Chinese warfare / by Ralph D. Sawyer ; with the bibliographic collaboration of Mei-chun Lee Sawyer.
      p. cm.
  Includes bibliographical references and index.
  ISBN 978-0-465-02145-1 (hardcover : alk. paper) — ISBN 978-0-465-02334-9 (e-book) 1. Warfare, Prehistoric—China. 2. Military art and science—China—History. 3. Weapons, Ancient—China—History. 4. China—History, Military—To 221 B.C. 5. China—History—Shang dynasty, 1766–1122 B.C. I. Sawyer, Mei-chün. II. Title.
  U43.C6S288 2011
  355.020931—dc22

                                                                    2010051391

10 9 8 7 6 5 4 3 2 1

FOR
LEE MEI-CHÜN

✦ CONTENTS ✦

✦ Preface ✦

Ancient Chinese Warfare and its companion, *Western Chou Warfare*, were started more than thirty years ago but were soon de-emphasized, though never abandoned, to investigate more accessible topics because insufficient archaeological material was available for assessing many aspects of ancient Chinese military history. Even though dramatic new finds such as San-hsing-tui (Sanxingdui) can still provoke astonishment and significantly affect historical understanding, the accumulation of thousands of discoveries and hundreds of highly relevant reports over the intervening decades has not only resulted in something akin to a minimal critical mass, but also considerably diminished the impact of archaeology's accidental nature. To cite just one example, reports on Wang-ch'eng-kang in the early 1980s indicated the existence of a fortress consisting of two small but conjoined square citadels roughly 100 meters on a side that immediately prompted heated arguments about its possible identity as an ancient Hsia (Xia) capital. However, a partial excavation of the greater site in the early twenty-first century has now revealed that the "King's City" once enclosed a massive 300,000 square meters within its substantial outer fortifications, considerably buttressing claims for an imperial role.

Although my efforts over the last few years, whether in the cold of Korean winters or heat of interminable Indonesian summers, have been focused on this volume, many more could easily be spent. No one has ever been granted indefinite longevity, yet it is difficult to escape the persistent feeling that only now, after nearly a half century of pondering Chinese topics, am I approaching some requisite level of understanding upon which the entire topic should be restudied. This is particularly true with respect to the ancient period because of the inescapable necessity of relying on innumerable interpretive archaeological reports and scholarly explications of oracular and bronze inscriptional materials, the core of this book.

Despite the convenience of the Internet and the growth of extensive (but not yet fully accessible or comprehensive) databases, exhaustive examination of all relevant articles on any single aspect of ancient Chinese military history, even something as focused as arrowheads, remains impossible. Paradoxically, numerous materials that were once relatively available through interlibrary loan, especially Japanese books and articles, have become even more difficult to acquire due to declining library holdings, an unwillingness to relinquish physical possession, and insufficient staff to provide the photocopies previously enjoyed. Nevertheless, despite the elusiveness of a few known titles and no doubt ignorance of many more, articles through the end of 2008 from the major Chinese historical and archaeological journals, as well as numerous minor ones assembled in collated volumes over the past few decades, and various books and site reports published in the last half century or more provide the basis for this study.

Now that belief in objective history has been discarded, it need hardly be mentioned that all works of this type are necessarily highly individualized creations that are guided by particular views and interpretations, however eclectic. Thus, for example, although increased coverage of Northern zone knives might well be merited, their study has been foregone for examinations of more focal or directly relevant topics such as the role of the *yüeh* (large battle-axe) in solidifying and apportioning martial authority. Selectivity has been particularly severe in the area of contextual history despite the temptation to pursue many threads of Hungshan, Liang-chu, and other cultural ascensions with military implications or to examine the fortifications at a number of additional sites. Fortunately *The Cambridge History of Ancient China*, despite being silent on the Neolithic and earlier periods, provides extensive background and analysis that may be taken as fundamental, even though I sometimes disagree with their exposition and conclusions, just as I have never assumed that the latest scholarship, no matter how enthusiastically embraced by the scholarly community, necessarily represents an advance or correctness.

A number of minor issues and decisions might well be noted. To take the last first, the concluding chapter, "Musings and Imponderables," although somewhat of a summation, rather than striving for an unattainable "conclusion," is intended to raise basic questions and indicate significant topics, a few of which will reappear in a future study. The Shang's extinction, being more a tale of conquest than simple collapse, is similarly deferred to the next volume on the rise and dominance of the Chou.

The discussion essentially proceeds in two streams, the textual narrative and a collateral expansion of many aspects and subtopics in the endnotes where matters of purely Sinological import and more general points of military history are explored. To ensure that *Ancient Chinese Warfare*, which is intended for the broadest possible audience of interested readers rather than just Sinologists, would not only be accessible but also published at a reasonable price rather than an exorbitant one appropriate to a research tome, certain decisions were made that will no doubt be bemoaned by reviewers.

As originally conceived, *Ancient Chinese Warfare* was to have a number of maps, many of them quite basic, for the convenience of the reader. However, this volume is little concerned with campaigns, obviating any need for phase and tactical depictions, and superior multicolor historical and topographical maps are available on the Internet, making the time and expense of producing them unnecessary. Similarly, many more illustrations of ancient weapons were originally planned, but preliminary readers found them less useful than the relatively few generic versions now included, a fortuitous development since requests made to various archaeological museums and publications in China for permission to reproduce many interesting specimens (and maps) were, on the whole, never even acknowledged.

In part to minimize cost and because it was felt that entries in Chinese characters would be more useful to those with competence in the language, no effort has been made to provide a bibliography of

Romanized titles for Chinese and Japanese books and articles. A highly compressed form of reference has also been employed in the endnotes, though shortened titles appear when an author has two or more publications in a single year or similarly titled papers. Rather than encompassing hundreds of books of contextual importance or even the many archaeological reports that individually describe one or two weapons, the expanded basis of this study, the bibliography is also confined to works cited in the endnotes. However, additional titles will appear in subsequent publications, and if interest merits, a complete bibliography of both Western and Chinese works will be made available on Sinostrategist.org.

As in our previous works, although I am responsible for the historical content, writing, military texts, theorizing, and conclusions, Mei-chün has contributed immeasurably to this volume through her management of the vast hoard of research materials underlying this work and the preparation of the bibliography. In addition, because she has willingly, if not always enthusiastically, suffered the tedium of many military discussions, contributed valuable insights, and tramped around the world prowling ancient fortifications, investigating military museums, and attending military conferences while continuing to participate in our ongoing intelligence and corporate consulting, I take great pleasure in dedicating this book to her.

Ralph D. Sawyer
Spring 2009

✦ A Note on Pronunciation ✦

As I have repeatedly commented, neither of the two commonly employed orthographies facilitates the pronunciation of Romanized Chinese characters for the uninitiated. Each system has its stumbling blocks, and I cannot imagine that *qi* in *pinyin* is inherently more comprehensible to unpracticed readers than the older, increasingly discarded Wade-Giles *ch'i* for a sound similar to the initial part of "chicken" or *x* for the simple "she," although they are certainly no less comprehensible than *j* for *r* or even *t* for *d* in Wade-Giles. However, as this work is intended for a broad audience, many of whom will have little experience with Romanized Chinese words apart from a few occurrences in the news and my other works, in which Wade-Giles has exclusively been used, the hyphenated break between syllables facilitates pronunciation, and specialists should have equal facility in either system—we have continued to employ Wade-Giles here with the exception of an idiosyncratic *yi* for *i* and contemporary provincial names.

As a crude guide to pronunciation we offer the following notes on the significant exceptions to normally expected sounds:

t, as in *Tao*: without an apostrophe, pronounced like d (*pinyin* d); otherwise t

p, as in *ping*: without an apostrophe, pronounced like b (*pinyin* b), otherwise p ch, as in *chuang*: without an apostrophe, pronounced like j (*pinyin* j and zh), otherwise ch

k, as in *kuang*: without an apostrophe, pronounced like English g (*pinyin* g), otherwise k

hs, as in *hsi*: pronounced like English sh (*pinyin* x)

j, as in *jen*: pronounced like r (*pinyin* r)

Thus, the name of the famous Chou (or Zhou in *pinyin*) dynasty is pronounced as if written "jou" and sounds just like the English name "Joe."

1.

PRELIMINARY ORIENTATIONS AND LEGENDARY CONFLICTS

◆‧◆ ─────────────────────────── ◆‧◆

When warriors battle over territory, slaughtering each other until they fill the fields or fight over a city until the battlements are filled with the dead, it should be termed devouring human flesh for the sake of terrain. Death is an inadequate punishment for such crimes. Those who excel in warfare should suffer the most extreme punishment, those who entangle states in combative alliances the next greatest. If the ruler of a state loves benevolence, he will be without enemies everywhere under Heaven.

—MENCIUS

FOR TWENTY-FIVE HUNDRED YEARS China has viewed the late prehistoric era as an ideal age marked by commonality of interest within clans and external harmony among peoples. This vision of a golden era, nurtured by the sagacious legendary rulers known as the Yellow Emperor, Yao, Shun, and Yü, was fervently embraced by the intellectual persuasions that came to be known as Taoist and Confucian, although from radically different perspectives and with rather contradictory objectives. Confucian literati-officials in Imperial China did not merely believe that Virtue alone had subjugated the recalcitrant, but also vociferously promoted its efficacy to thwart military solutions to external threats. An essentially pacifistic yearning, it would shape much of China's military heritage and frequently preclude aggressive action and adequate preparation, however dire the need.[1] Yet it was a severely distorted image that conveniently ignored the Yellow Emperor's storied military activity and

1

the great feats wrought by Kings T'ang of the Shang and Wu of the Chou, unquestioned paragons of righteousness who still had to strive mightily to suppress the wicked.

The numerous viewpoints and diverse conceptions formulated over the centuries included a less optimistic, more realistic understanding that posited warfare as innate and deemed turmoil and conflict inescapable, although it would never dominate the intellectual terrain or prevail in court discussions.[2] Despite encompassing highly disparate materials and a few contradictions, the classic military writings compiled in the Warring States period perceive the "golden age of antiquity" rather differently. The recently recovered *Sun Pin Ping-fa* characterizes the legendary era as a time when warfare, not virtue, wrought peace:[3]

At the time when Yao possessed All under Heaven there were seven tribes who dishonored the king's edicts and did not put them into effect. There were the two Yi (in the east) and four others in the central states. It was not possible for Yao to be at ease and realize the advantages of governing All under Heaven. Only after he was victorious in battle and his strength was established did All under Heaven submit.

In antiquity Shen Nung did battle with the Fu and Sui; the Yellow Emperor did battle with Ch'ih Yu at Shu-lü; Yao attacked Kung Kung; Shun attacked Ch'e and drove off the Three Miao; T'ang of the Shang deposed Chieh of the Hsia; King Wu of the Chou attacked Emperor Hsin of the Shang; and the Duke of Chou obliterated the remnant state of Shang-yen when it rebelled.

Immersed in an age of unremitting warfare that saw untold combatants slain and numerous states extinguished, Sun Pin concluded that virtue had not only proven insufficient in the past, but also remained fundamentally unattainable:

If someone's virtue is not like that of the Five Emperors, his ability does not reach that of the Three Kings, nor his wisdom match that of the Duke of Chou, yet he says, "I want to accumulate benevolence and righteousness, practice the rites and music, and wear flowing robes and thereby prevent conflict and seizure," it is not that Yao and

Shun did not want this, but that they could not attain it. Therefore, they mobilized the military to constrain the evil.[4]

Sun Pin deemed conflict to be innate and warfare inescapable: "Now being endowed with teeth and mounting horns, having claws in front and spurs in back, coming together when happy, fighting when angry, this is the Tao of Heaven, it cannot be stopped."[5] Despite its ostensibly Taoist perspective, the eclectic *Huai-nan Tzu* essentially seconded his belief:

> Now as for the beasts of blood and *ch'i*, who have teeth and mount horns, or have claws in front and spurs in back: those with horns butt, those with teeth bite, those with poison sting, and those with hooves kick. When happy, they play with each other; when angry, they harm each other. This is Heavenly nature.
>
> Men have a desire for food and clothes, but things are insufficient to supply them. Thus they group together in diverse places. When the division of things is not equitable, they fervently seek them and conflict arises. When there is conflict, the strong will coerce the weak and the courageous will encroach upon the fearful. Since men do not have the strength of sinews and bone, the sharpness of claws and teeth, they cut leather to make armor, and smelt iron to make blades.[6]

Hsün-tzu, a late Warring States period philosopher simplistically remembered for his assertion that human nature is inherently evil, identified human desire as the root cause of conflict: "Men are born with desires. When their desires are unsatisfied they cannot but seek to fulfill them. When they seek without measure or bound, they cannot but be in conflict. When conflict arises, there is chaos; with chaos, there is poverty."[7] Conversely, the authors of another late Warring States eclectic work believed that individual weakness in the face of natural and human threats constituted the very basis for social order:

> Human nature is such that nails and teeth are inadequate for protection, flesh and skin inadequate to ward off the cold and heat, sinews and bones inadequate to pursue profit and avoid harm, courage and

daring inadequate to repulse the fierce and stop the violent. Yet men still regulate the myriad things, control the birds and beasts, and overcome the wild cats while cold and heat, dryness and dampness cannot harm them. Isn't it only because they first make preparations and group together?

When groups assemble they can profit each other. When profit (advantage) derives from the group, the Tao of the ruler has been established. Thus when the Tao of the ruler has been established, advantage proceeds from groups and all human preparations can be completed.[8]

Social order is thus envisioned as having been forcefully imposed by conscientious men of wisdom, the legendary Sage emperors, rather than engendered by radiant Virtue. As Hsün-tzu notes, constraints had to be formulated:

The former kings hated their chaos, so they regulated the *li* (rites and forms of social behavior) and music in order to divide them, nourish the people's desires, supply what the people seek, and ensure that desire does not become exhausted in things, nor things bent under desire.[9]

Even the somewhat esoteric *Huai-nan Tzu* conceded that the existence of evil compelled the primal leaders to resort to harsh measures:

In antiquity, men who were greedy, obtuse, and avaricious destroyed and pillaged all under Heaven. The myriad people were disturbed and moved, none could be at peace in their place. Sages suddenly arose to punish the strong and brutal and pacify the chaotic age. They eliminated danger and got rid of the corrupt, turning the muddy into the clear and danger into peace.[10]

Their actions assumed an outwardly directed martial form but were not undertaken for personal profit:

When the ancients employed the military it was not to profit from broadening their lands or coveting the acquisition of gold and jade.

It was to preserve those about to perish, continue the severed, pacify the chaotic under Heaven, and eliminate the harm affecting the myriad people.[11]

With slight variation, most of the classic military writings justify undertaking military campaigns solely for the purpose of protecting the state from aggression and rescuing the people from any suffering that might be inflicted by brutal oppressors:

Taking benevolence as the foundation and employing righteousness to govern constituted uprightness in antiquity. However, when uprightness failed to attain the desired objectives, they resorted to authority. Authority comes from warfare, not from harmony among men.

For this reason, if one must kill people to give peace to the people, then killing is permissible. If one must attack a state out of love for their people, then attacking it is permissible. If one must stop war with war, although it is war, it is permissible.[12]

The ancient sages did not just rectify the disorder about them, but also created the very means for waging war:

Those who lacked Heavenly weapons provided them themselves. This was an affair of extraordinary men. The Yellow Emperor created swords and imagized military formations upon them. Yi created bows and crossbows and imagized strategic power on them. Yü created boats and carts and imagized tactical changes on them. T'ang and Wu made long weapons and imagized the strategic imbalance of power on them.[13]

The legendary cultural heroes had thus been compelled to decisively thwart chaos and quell disorder to preserve the populace. However, as the military writings emphasize, their approach equally entailed the pursuit of righteousness, cultivation of virtue, and implementation of measures intended to mitigate the people's suffering and improve their welfare. This devolution from a tranquil, ideal age prompted the authors of *Huang-shih Kung's Three Strategies* to assert: "The Sage King does not take any pleasure in using the army. He mobilizes it to execute the violently perverse and to rectify the rebellious. The army is an inauspicious implement and the Tao of Heaven abhors it. However, when its use is unavoidable it accords with the Tao of Heaven."[14] This is a highly complex, essentially contradictory situation, because "the Tao of Heaven abhors it," yet conflict similarly expresses "the Tao of Heaven" and "cannot be stopped." Warfare is thus paradoxically inescapable and, in many views including that of Confucius himself, a crucial human endeavor for which training and preparation are required.[15]

THE SEMILEGENDARY PERIOD

Archaeological discoveries over the past several decades have suddenly infused life into previously shadowy remnants of ancient Chinese civilization, validating many early assertions about the Shang and nominally substantiating, with appropriate allowance for interpretative frameworks and the effects of millennia, vague images of the Hsia and the legendary period. In addition, many traditional battle tales that attained a life of their own within popular culture deserve recounting irrespective of their historical inaccuracy. Scholarly audiences apart, countless generations across the ages, even emperors and generals, accepted their historicity, as does much of the Chinese populace today.[16] Moreover, despite the concept and portraits of the Five Emperors being generally acknowledged as having been radically shaped, if not actually created, in the Warring States period, it has still been argued that mythical tales embody events and reflect significant developments in the course of Chinese civilization, including warfare, and can be parsed and scrutinized for clues and insights. Texts considered to be late fabrications, such as the *Shang Shu's* "Canon of Yao," are similarly seen as valuable repositories of vestigial memory and therefore well worth detailed—synonymous with "imaginative"—pondering.[17]

According to early writings and traditional belief, the most famous legendary battles arose between the great progenitor known as the Yellow Emperor and two powerful opponents: first Yen Ti, the Red Emperor, and then Ch'ih Yu, a tribal leader thought to have served as one of the Red Emperor's officials before he rebelled. As depicted in the monumental *Shih Chi*, China's first synthetic history, the Yellow Emperor was a judicious commander as well as a cultural paragon:

> The Yellow Emperor, a descendant of the Shao-tien clan, was surnamed Kung-sun and named Hsüan-yüan. When he was born his spirit was already penetrating; while an infant he could speak; as a child he could reply intelligently; and when growing up he was substantial and acute, as brilliant as an adult.
>
> Shen Nung's clan was in decline in Hsüan-yüan's time. The various lords encroached upon each other and acted brutally and perversely toward the hundred surnames.[18] Shen Nung's clan was unable to chastise them. Thereupon Hsüan-yüan practiced employing shields and halberds in order to conduct punitive expeditions against those who would not offer their fealty. The various lords all came to submit, as if they were his guests. However, no one was able to attack Ch'ih Yu, the most brutal of all.
>
> The Red Emperor encroached upon the various clan leaders, so they all gave their allegiance to Hsüan-yüan. Accordingly, Hsüan-yüan cultivated his Virtue and put his weapons in order; regulated the five *ch'i*;[19] cultivated the five grains; was solicitous toward the myriad peoples; took the measure of the four quarters; and trained the bears, leopards, and tigers[20] in order to engage in battle with the Red Emperor in the wastes of Pan-ch'üan. Only after three engagements did he realize his objective.
>
> Ch'ih Yu revolted and did not follow the Yellow Emperor's edicts. The Yellow Emperor summoned the armies of the clan chiefs and engaged Ch'ih Yu in battle in the wilds of Chuo-li, capturing and slaying him. The clan chiefs all honored Hsüan-yüan as the Son of Heaven, and he replaced Shen Nung, becoming the Yellow Emperor. The Yellow Emperor then pursued and rectified all those under Heaven who failed to submit, but left the tranquil alone.[21]

Although the *Shih Chi's* account has traditionally provided the basis for popular portrayals, several other texts from the late Warring States and early Han preserve fragments that are often employed to amplify the depiction. For example, *Chuang-tzu* states: "Being unable to attain to complete Virtue (and thereby persuade him to submit), the Yellow Emperor engaged Ch'ih Yu in battle in the wilds of Chuo-li. The blood flowed for a hundred *li*."[22] The *Hsin Shu* graphically asserts that "the Yellow Emperor implemented the Tao but Yen Ti did not obey, so they engaged in battle in the wilds of Chuo-li. The blood spilt was great enough to float a pestle."[23] A late T'ang dynasty work paints an even more melodramatic portrait: "The Yellow Emperor and Ch'ih Yu engaged in battle in the wilds of Chuo-li. Ch'ih Yu created a great fog so that the armies were all confused. The Yellow Emperor then ordered Feng-hou to fashion a needle instrument in order to discriminate the four quarters and subsequently captured Ch'ih Yu."[24]

Another version of the battle appears in the *Canon of Mountains and Rivers*, a former Han dynasty compilation of late Warring States material.[25] In discussing a "woman wearing blue clothes" who was sometimes sighted in the Ta-huang-pei area, the narrative notes: "Ch'ih Yu fabricated weapons and attacked the Yellow Emperor. The Yellow Emperor then ordered the winged dragon Ying to assault him in the wilds of Chi-chou.[26] Ying Lung gathered up the waters, whereupon Ch'ih Yu asked Feng Po (Wind Duke) and Yü Shih (Rain Commander) to unleash fierce winds and rain. The Yellow Emperor then had the Heavenly female deity Pa (who wore blue clothes) sent down and the rain ceased. The Yellow Emperor subsequently slew Ch'ih Yu. However, Pa was unable to re-ascend to Heaven and wherever she dwelt it never rained."

This version clearly reflects a clash of two regional cultures, the central Lungshan and Tung Yi. Each of the combatants strove to re-create familiar topographical conditions for which their long-accustomed tactics would be more suitable. If, as traditionally believed, the Yellow Emperor originally inhabited the comparatively dry central plains area where mobility was relatively unlimited, his forces would have found Ch'ih Yu's damp, marshy (southeastern) environment inconvenient, if not fatal. The winged dragon was called upon to evaporate the water, which then formed clouds, but when Ch'ih Yu counteracted that measure, the Yellow Emperor called down celestial forces, essentially prefiguring his great Heavenly power in later Huang-Lao and Taoist religious thought. Rather than some ethereal figure, this Heavenly woman (also known as the "drought demon") was so powerful that any area in which she stayed would invariably dry out. To prevent future calamities among the people, the Yellow Emperor reportedly shifted her dwelling place north of the Red River.

A number of other threads in the Ch'ih Yu tradition have implications for both ancient and traditional military history in China. The conflict supposedly occurred about 2600 BCE, the incipient period of Chinese bronze metallurgy, and Ch'ih Yu is closely associated with discovering, or at least exploiting, copper's ductility to fabricate new metal weapons, justifying his inclusion as one of eight Chou era spirits under the designation of Weapons Master.[27] A paragraph in the *Kuan-tzu* on conserving natural resources for the state's benefit reflects the underlying premise that multiplying the number of weapons had led to increased carnage:[28]

After the Yellow Emperor had practiced these (conservation) measures for some ten years the Ko Lu mountains split asunder and poured forth water and metal. Ch'ih Yu gathered and smelted the metal to make swords,[29] armor, spears, and halberds that he employed to subjugate nine feudal lords that year. Similarly, when the Yung Hu mountains cracked and water came forth followed by metal, he gathered and smelted it to fabricate Yung-hu halberds and Hu-fu dagger-axes. That year he subjugated twelve feudal lords. Throughout the realm rulers wielding shields and halberds arose in anger and the wilds were filled with prostrate corpses. This was the beginning of manifesting dagger-axes.

The chapter actually begins with Kuan-tzu, Duke Huan's famous minister, asserting that copper is found in 467 of China's 5,371 named mountains and iron in 3,609. Although the major deposits that would eventually be exploited by the Shang are located in the southeast, the provinces of Shandong and Shanxi (in the east and west respectively) both had limited reserves and would have been appropriate sites for the development of metallic weapons. Somewhat more fancifully, Ch'ih Yu is sometimes depicted as having a bronze head, and the drums in his army are similarly said to have been fashioned from bronze, vivid embodiments of his metallic nature as well as the metaphysical basis that allowed the Yellow Emperor to employ the magical properties of ox-hide drums to vanquish the metallic ones.[30]

Many legendary accounts that emphasize Ch'ih Yu's fierceness and brutality point out that he coerced dissident groups into participating in his revolt through the creation of shackles and widespread use of five harsh punishments. His rebelliousness and mythical aspects were further exaggerated over the centuries. Imperial era tales thus describe him as extremely ugly, even abnormal, in various ways ranging from being marked by a dark red or black complexion to having the horns and feet of an ox or possessing six arms. Ch'ih Yu was supposedly so fierce that his name alone was sufficient to terrify the general populace, and the Yellow Emperor supposedly pretended he wasn't dead in order to employ his image to frighten the perverse.

Paradoxically, Ch'ih Yu's fierceness outweighed his brutality—or perhaps the two have a certain martial appeal in combination—because even today the Miao of southern China continue to worship Ch'ih Yu as their ancestor.[31] Furthermore, across history he has not only been highly esteemed but also inexplicably seen as one of the chief progenitors of the Hua-Hsia culture and the true embodiment of courage.[32] Liu Pang, founder of the Han, venerated both the Yellow Emperor and Ch'ih Yu as war spirits. He continued to be revered in the Han, recent People's Republic of China (PRC) theoretical publications have advocated emulating him, and some Koreans also continue to recognize him as an ancestor.[33]

Finally, in a ritual expression of magical thinking, Ch'ih Yu was supposedly dismembered after being defeated by the Yellow Emperor, perhaps even rendered into a meat paste and eaten, thereby symbolically vanquishing him and absorbing his courageous spirit. Two recently re-covered bamboo texts dating to the late Warring States period include accounts of this primeval conflict.[34] In one called "Five Rectifications" ("Wu Cheng"), the Yellow Emperor is depicted as trying to impose order on the world and eliminate warfare but being advised that anger and the impulses of blood and ch'i must first be eliminated. When his measures and (more important) Virtue fail to bring about the desired order, he is compelled to accept the painful lesson that even though warfare is baleful, failing to resort to it when necessary equally results in failure. He therefore bestirs himself, engages Ch'ih Yu in battle, and captures him.

A second account, found in "Rectifying Chaos" ("Cheng Luan"), describes the ritualized punishments he inflicted upon Ch'ih Yu. The Yellow Emperor "peeled off his skin, made an archery target, and had his men shoot at it. Those who hit it the most received rewards. He cut off his hair and arrayed it below Heaven calling it Ch'ih Yu's banner. He stuffed his stomach to make a football and had the men kick it, the most successful again being rewarded. He turned his body into mincemeat, mixed it with bitter salts, and had all under Heaven consume it." Although the ostensible intent of these brutal punishments was frightening the realm into accepting his prohibitions and vision of order, such behavior dramatically contradicts the vaunted image of a paragon of Virtue and legendary Sage who held sway over the realm through righteous charisma alone.

Premised upon the existence of the Yellow Emperor, Red Emperor, and Ch'ih Yu, traditionally oriented historians continue to synthesize cohesive views of the Yellow Emperor's battles. Before considering the archaeological data and demythologizing these tales as the product of Warring States thought, it may prove informative to summarize one influential twentieth-century interpretation that asserts historical evidence supports the veracity of these traditional versions despite a cascade of prominent works that firmly deny these titanic figures ever existed except as tribal totems.[35]

First, it is assumed that Ch'ih Yu and Yen Ti were the same person and that the earliest accounts portray a single series of battles rather than sequential struggles between the Yellow Emperor and two different antagonists. Second, these clashes apparently arose out of a fundamental conflict between the "agriculturalists," who were descended from Shen Nung, honored even today in most temples as the progenitor of agricultural and medicinal knowledge, and members of the Yellow Emperor's clan, who were distinguished by their ability to exploit such technological innovations as the chariot and bow. Because of the inherently disproportionate strengths of these fundamentally dissimilar orientations, the Yellow Emperor's victory was inevitable. Bows and arrows would have enabled his forces to direct missile fire onto their enemies before engaging, whereas Ch'ih Yu had to rely upon shock weapons despite being a skilled weapons fabricator. In addition, the Yellow Emperor may have exploited pipes and drums, two of his other inventions, to stimulate the troops' morale before the engagement and direct them during the battle.[36]

Tactically, the Yellow Emperor was active and aggressive and learned as he proceeded. Having been cast in the role of a defender, someone protecting his home territory against brutal invaders, he also enjoyed the psychological advantage derived from "moral superiority."[37] Conversely, the blatantly aggressive Ch'ih Yu exhausted himself in assaulting numerous other states before undertaking the final expeditionary campaign, allowing the Yellow Emperor to fashion a preemptive defense by moving to establish himself at Chuo-li and intercept Ch'ih Yu. (Whether the Yellow Emperor's troops arrived first and had an opportunity to rest and improve their position or were thrown pell-mell into the conflict remains unknown.) Given his much-praised sagacity, they should have been fighting upon chosen ground and therefore enjoyed some sort of positional advantage, just as Sun-tzu later advocated.

It is also believed that the Yellow Emperor's clan benefited from improved transport because he had created the *ch'e*, a general term for wheeled vehicles. (Even without chariots, another invention often credited to the Yellow Emperor—boats and oars—would have been useful in crossing the Yellow River.)[38] Moreover, he reputedly established a rudimentary bureaucracy that included an office for military affairs that was initially held by Li Mu, traditionally regarded as China's first minister of war. But perhaps the most interesting aspect of this reconstruction is the contention that the battle arose not simply over regional hegemony, but over natural resources, apparently an inland salt production area near Chieh-ch'ih in the western province of Shanxi. Ch'ih Yu is therefore identified as the charismatic leader of the Chiu Li (Nine Li), who were reportedly centered in the area of modern Nanyang in Henan province, and thus as Yen Ti's descendant as well as the earliest ancestor of the famous Chiang clan that would figure prominently in Chou military history.

Many other theories have been formulated to account for the scope and nature of what has traditionally been deemed China's first war, though always on the assumption that the extant traditional accounts preserve an essential kernel of facts about events that have been verbally transmitted, often in highly circumscribed geographic regions, from antiquity.[39] The Yellow Emperor's traditional association with central China is not unchallenged, and arguments over their origins and respective loci of activity sometimes result in a virtual refighting of the original clash by modern partisans. For example, one analyst has argued that the Red Emperor's clan arose in the central plains area but the Yellow Emperor's clan in the Shandong and identifies them with the Tung Yi,[40] whereas another envisions the Yellow and Red Emperors as both having emerged during the P'ei-li-kang and Yangshao cultures and being active in the central Yellow river valley.[41] A variant of the Eastern interpretation sees the Yellow Emperor's clan as arising in the east and defeating the Red Emperor's clan to their west before assimilating them and moving into the central area themselves.[42]

Ch'ih Yu is sometimes seen as the leader of the Eastern Yi centered about Shandong rather than of the Miao in the south, and the conflict is understood as having arisen over grazing and agricultural lands in the central region of China, one in a series over rights and leadership among the Yellow Emperor, Agricultural Emperor (Shen Nung), Ch'ih Yu, and Fu Sui.[43] Ch'ih Yu is also thought to have headed an alliance of eighty-one or other similarly large number of tribes that seem to have devoted themselves to martial activities and developed new weapons, possibly even of primitive bronze, under his clan's leadership.[44] They were therefore able to supplant the Yellow Emperor throughout the "nine corners" of his domain—the eight directions and the center—the number nine being an indefinite reference for "everywhere" in ancient China. In contrast, because the Yellow Emperor could only muster his own clan and five others, his forces would have been far fewer, accounting for his initial defeats.

The Yellow Emperor is said to have dammed a river to prevent Ch'ih Yu from crossing, but adverse weather conditions, which favored those more accustomed to the wet weather indigenous to the east and south, may have resulted in the dikes breaking and Ch'ih Yu being able to exploit the conditions to wrest additional victories.[45] However, the Yellow Emperor was able to prevail when the weather dried out, possibly because

Ch'ih Yu had exhausted his army en route to the final confrontation, an error against which most of the classic military writers would subsequently warn.

Ultimately the Yellow Emperor prepared well, trained his troops, and organized them into some sort of cohesive, responsive force. If the battle occurred at Chi-chou in Hebei, his ability to exploit the dust clouds created by the strong winds blowing across the parched yellow soil may have been a critical factor. With the terrain obscured and the four directions unclear, his forces would have enjoyed a unilateral opportunity to advance, because he not only chose the battlefield but also reputedly possessed the superior technology of the southern-pointing chariot.[46]

However, traditional claims that the Yellow Emperor, as one of the first ancestors, grand progenitors, and magical creators, contrived numerous artifacts essential to civilization, including two vital to military technology, are not sustained by archaeological evidence. His name in the *Shih Chi*, Hsüan-yüan, has long been cited as evidence that he invented the chariot, because the individual terms *hsüan* and *yüan* refer to the horizontal axle pole and the shaft(s) extending to the front of a chariot. Although this interpretation became an article of faith to the Han, it is completely unfounded because the term did not appear until the Warring States period.

Furthermore, even though vestiges of narrow vehicle tracks have been found at the presumed Hsia capital at Erh-li-t'ou, and traditional accounts assert that the Shang employed chariots to vanquish the Hsia in a decisive battle that would date to about 1600 BCE, there is no archaeological evidence that either the Hsia or the early Shang had horse-drawn, functional war chariots. Therefore the Yellow Emperor, who would have been active in the Lungshan period predating the Hsia itself, might only have fashioned some sort of cart or wheelbarrow.[47] If they could have been fabricated in quantity and proved reasonably durable, these carts certainly would have facilitated the transport of essential supplies at a time when the only alternative was human portage.

A famous sentence in the *Yi Ching* states that the Yellow Emperor also invented bows and arrows by "stringing a branch to make a curved bow and shaving branches to fashion arrows" and that "the advantage of bows and arrows was to overawe all under Heaven."[48] Although archaeological excavations have uncovered numerous bone and stone arrowheads dating to the Lungshan period with which the Yellow Emperor should be identified, the preceding Yangshao culture already shows extensive evidence of social stratification and the widespread use of bows and arrows beyond what would have been reasonably necessary for hunting purposes. Moreover, arrowheads of great antiquity have been found in Shanxi, proving that they not only existed but were also employed some twenty millennia prior to the Yellow Emperor's era.[49]

Nevertheless, being a martial chieftain in a period probably just beginning to extensively exploit the bow's potential in warfare, the Yellow Emperor may have emphasized archery training and employed missile fire more vigorously and systematically than other leaders, perhaps even initiated the technique of massed fire. Although totally speculative, these measures could account for the bow and arrow's close identification with the Yellow Emperor and perhaps indicate a critical factor in vanquishing enemies wielding shock weapons alone.

Two other aspects that have particularly attracted attention are the "fog" that reportedly covered the battlefield and the "southern-pointing chariot." A naturalistic explanation for the former would be that some sort of temperature shift had occurred after heavy rain, creating an inversion that trapped the moisture at low levels and thus obscured the battlefield, but a simpler one just requires the presence of fog or low-level clouds. However, neither interpretation can be sustained in the face of statements that a strong wind had arisen, presumably at the Yellow Emperor's behest since his transcendent cosmic attunement should have allowed him to magically modify the climatic conditions.[50]

Perhaps Ch'ih Yu deliberately created a smoke screen, possibly one infused with chemical irritants such as would be witnessed in the Warring States period and used in the Spring and Autumn as well.[51] If so, it would certainly constitute the earliest known incident of deliberately employing a smoke screen, even though it might have proven of limited utility unless the winds, which would have quickly dispersed any artificial clouds, were blowing to his advantage. Within this incapacitating miasma the Yellow Emperor could only have resorted to his southern-pointing chariot, an ingenious device that must have taxed even the Sage Emperor's inventiveness. Remarkably, this "southern-pointing chariot," much romanticized in the traditional literature, is no myth but instead a small chariot with a mounted figure whose outstretched hand, once initially set, always points south. However, it was probably invented in the later Han,[52] explaining the lack of references to it and the fog until somewhat later.

Considering all the possible factors, interpretations, archaeological evidence, and perspectives of the scholars who have extensively studied these ancient materials, a reasonable conclusion might be that at some indeterminate time in the middle of the third millennium BCE a few significant battles resulted from two tribal alliances, each seeking to control increasingly greater territory. The first battle marked the ascendancy of the clan subsequently identified as the "imperial" clan, that of the bear, led by an individual termed the "August" one, who eventually (in later Chou literature) was accorded the title of Yellow Emperor and became the focal nexus for many cultural inventions and achievements.

No doubt his and the Red Emperor's clans were closely related, perhaps even derived from a common ancestor such as the great agricultural deity Shen Nung, as some highly credible scholars have imaginatively suggested.[53] The clans therefore clashed not just over the central plains area, but also over who would monopolize the power and leadership of the tribal confederation, whatever the actual character of their alliance. If Marxist analysis has any validity, the process of shifting from a matriarchal to a patriarchal age may have provided the stimulus for this struggle or was perhaps synonymous with it, with the conflict actually stretching over generations between related clans. All the archaeological evidence—the plethora of recovered arrowheads, existence of major moats and walled fortifications, and burial patterns that display increased class distinction—indicates the rise of powerful clan leaders accompanied by a shift from universal military obligation to a dominant martial clan that consolidated power within the tribe and undertook responsibility for waging war outside it.

Viewed from the dynamics of tribal conflict, the Yellow Emperor may therefore have been the truculent leader of an obstreperous group that challenged the Red Emperor's established authority and ultimately wrested control of the collective organization.[54] Whether the Yellow Emperor's clan was economically more prosperous or simply more martial and self-disciplined than Shen Nung's, whom power and privilege had perhaps made soft, as would happen to virtually all of China's ruling clans, cannot be known. If the *Shih Chi* account even minimally reflects actual events, the issue of supremacy wasn't decided with the sort of single clash that supposedly characterizes primitive warfare, but through at least three battles. Preliminary skirmishes between members of the opposing alliances probably preceded any command decision to mobilize their forces and act more decisively. Although the Yellow Emperor's cause has traditionally been identified with morality and the forcible restoration (or imposition) of order was his avowed aim, whether he actually enjoyed some sort of superior moral claim remains doubtful.

Bows and arrows would have initially been employed, followed by spears, clubs, hand axes, and other shock weapons, even agricultural tools, upon closing for combat. However, neither swords nor the ubiquitous dagger-axes of later times had yet come into existence, and chariots were equally absent, traditional accounts to the contrary. Depending upon the effectiveness of their training and degree of rudimentary organization, the battles probably resolved into general melees marked by hundreds of individualized fights of the type thought to have marked so-called primitive warfare.[55] At their conclusion, the vanquished either fled or submitted and became an integral part of the Yellow Emperor's alliance, evidence that this was more of an intratribal conflict for dominance than a war of extermination wherein the defeated would either be slain or enslaved, as in the Shang and later times.[56]

The extant literature similarly portrays the Yellow Emperor's subsequent conflict with Ch'ih Yu as a struggle between good and evil, morality and licentiousness, as well as between a benevolent despot who had garnered the people's allegiance and a brutal leader who forcefully coerced it. This sort of archetypical depiction would be repeatedly encountered across the millennia as dynasties fell and their successors wrote historical tales to justify their own violence and excesses to posterity. The laconic *Shih Chi* account actually records little beyond a conflict between an alliance leader and an obstreperous subleader, but later versions significantly embellish the story.

This battle with Ch'ih Yu should perhaps also be accepted as simply a struggle for supremacy between two subgroups with different totems seeking to dominate a newly forged, extended alliance. Although the Yellow Emperor may have profited from his experience to create a more formidable or cohesive fighting organization, their weapons would have been unchanged. Although both groups were obviously mobile, the Yellow Emperor essentially responded to Ch'ih Yu, since the latter played the role of an invader, later termed a "guest" by military theorists, yet managed to shape the battlefield. However, whether Ch'ih Yu's movements constituted a strategic initiative is fruitless to discuss at this level of warfare.

Overall, the archaeological records in fact indicate a transitional period of rising conflict, increasing class distinction, and the evolution of political and concomitant military power or, depending upon theoretical emphasis, military strength and concomitant political power. The details have been mythologized, but these traditional accounts may still reflect actual tribal conflicts and preserve the names of the most impressive martial leaders. Apart from whatever truths may be embedded within them, whatever ancient realities they may reflect, these legendary versions also display the subsequent historical mindset, one not irrelevant to understanding the views and motives of commanders and emperors across the centuries. However, asserting more than this leads only into the realm of even more rampant speculation.

2.

ANCIENT FORTIFICATIONS, I

◆—◆—◆————————————————————◆—◆—◆

L ONG VIRTUALLY DEFINED by the mythic aspects of its Great Wall,
China's tradition of wall building far exceeds the most exaggerated
claims spawned by its famous icon. As early as 7000 BCE, protective ditches
had already appeared in scattered settlements along the two great river
systems and their tributaries. However, rather than being employed to
construct defensive fortifications, the excavated soil furnished the raw
material for structural foundations and for raising the entire settlement
above the surrounding terrain, thereby preventing inundation from
pooling rainwater and overflowing streams and providing a slight tactical advantage.

In response to escalating threats, the concepts and technology of
defensive fortifications fitfully but continually evolved over the next
five millennia until what has been claimed to be the distinctive form
of the double-walled Chinese city protected by an external moat was
finally realized. Several stages can be discerned: shallow perimeter
ditches; simple ditches augmented by mounded earthen walls, the latter
simply the by-product of ditch and moat excavation; earthen walls deliberately constructed with early pounding techniques, generally augmented by perimeter ditches or early moats; massive, rigorously
constructed *hang-t'u* (stamped or pounded earth) walls coupled with
expansive protective moats; and the culmination of Chinese martial
engineering, massive rammed earth walls faced with stone or brick,
buttressed with waist walls, and systematically augmented by conjoined
exterior and interior moats.

Although low stone walls had been erected in conducive, semiarid
regions such as Inner Mongolia as early as the Neolithic period, solid

walls of stone, brick, and even marble were not constructed in China until recent centuries, and then only in limited areas. Moreover, regional practices and localized disparities would persist across the centuries despite consistent progress in understanding the engineering principles required for fortification construction, the development of work methods, and the evolution of administrative measures. Some towns in the late Shang and early Chou continued to employ nothing more than ditches long after massive fortifications had become commonplace.[1] Conversely, peripheral states such as Chao that faced external threats began constructing lengthy defensive structures, appropriately termed "long walls," along their proclaimed borders late in the Warring States period, a practice that would eventually culminate in the colossal Ming dynasty edifice.

Ditches and walls have always been the first defensive measures adopted in response to environmental threats and human violence. In consonance with the Marxist dictum that civilization evolved from a matriarchal, egalitarian society through patriarchal-based structures and an inherent tendency to warfare, Chinese scholars have long claimed that the earliest Neolithic ditches were intended simply to prevent domesticated livestock from venturing forth and thwart incursions by wild animals. However, although they may have been minimally effective in achieving the former, the ditches would never have deterred agile predators from entering a settlement. Therefore, they must have performed a defensive function, one that was enabled by the raised profile of most early villages.

Many premodern civilizations fabricated highly formidable walls from readily available rocks and laboriously quarried stone blocks, but others resorted to more easily worked, though perishable, materials such as wood to erect functional barriers, ranging from hedgehogs and simple palisades through complex log fortresses. Those dwelling in environments bereft of readily accessible trees and stone were compelled to construct primitive earthworks, employ sun-dried mud blocks, or exploit the evolving technology of kiln-fired bricks. Although China was still heavily forested throughout the Neolithic period, the soil deposited in the alluvial plains, even the famous sticky "yellow earth" washed down by the Yellow River, lent itself far more easily to digging, carving, and shaping with the basic tools of the time—axes, scrapers, and short shovels cobbled together by affixing appropriately shaped pieces of stone or bone to a wooden handle—than did hewing trees or chiseling stone from local outcroppings. Concerted pounding could compress and harden the soil to an almost concretelike substance roughly equivalent to sedimentary rock that would, with minimal maintenance, endure for millennia.

A number of strategically advantageous locations near rivers or amid mountains were continuously occupied from Neolithic times. As these settlements evolved into fortified towns in the Lungshan period and labor became available or the degree of threat increased, their massive walls were enlarged. Preexisting structures were generally reshaped and significantly augmented rather than simply rebuilt, often deliberately expanded out over the very ditches that had protected them. Many Shang sites subsequently served as the nucleus for resplendent cities when the early Chou established the fortified power centers that eventually evolved into heavily fortified Spring and Autumn and Warring States capitals.[2]

Unfortunately, despite their critical role in civilization's evolution and impact on warfare, the history and technology of Chinese fortified settlements have yet to be systematically studied.[3] Nevertheless, discoveries over the last several decades have pushed the defensive horizon well back into the Neolithic period, permitting a tentative recounting of the salient features within the context of China's military history.[4] However, contemplation of more profound questions, including whether increasingly urbanized, fortified sites stimulated warfare or simply reflected warfare's increased scope and intensity, must be foregone.

Particularly in the earliest periods, when the populace was comparatively small and competition for land minimal, burgeoning groups invariably sought to exploit the terrain's strategic advantages. Later ages might be able to employ vast labor resources to overcome natural obstacles and reshape the terrain's configuration, but Neolithic groups had to choose their sites for their environmental benefits and protective features. Except for mountainside villages constructed in semiarid locations, migrant settlers seeking well-watered areas generally established their communities alongside marshes, lakes, ponds, streams, and rivers.

The many early towns that were not abandoned despite suffering extensive flood damage show the overwhelming importance of immediate proximity to active water sources.

Still bodies of water may have presented little flood risk, but the danger of being inundated by a sudden increase in the volume of an adjacent river or stream prompted the implementation of basic protective measures. Insofar as embankment building and preventative channeling did not begin until well after the technology of rammed earth had been reasonably perfected, the sole options were choosing somewhat raised terrain, especially natural terraces, or artificially increasing the village's overall height to produce a sort of mound settlement, the precursor of the platform city.

Cognizant of water's effectiveness as a defensive barrier, settlement planners—there is considerable evidence that even early settlements were built according to design rather than haphazardly evolved—frequently chose sites amid multiple bodies of water. Many early towns were surprisingly positioned within the confluence of two or three rivers or between a river and nearby lake or marsh rather than simply above or along the outer bank of a single river, despite the high flood risk and the inconvenience of having to cross these watercourses on a daily basis.

CHARACTER AND FUNCTION OF EARLY WALLS

Even in the Lungshan period (3,000 to 2,000 BCE) immediately prior to the Hsia dynasty, Chinese walls had already reached astonishing dimensions and sometimes exceeded twenty-five meters (eighty-one feet) in width. Constructing these fortifications must have required a Herculean expenditure of energy that inflicted intense misery on the workers, yet the motive for their massiveness remains unknown. Later periods would exploit the resulting top surfaces to deploy contingents of soldiers, set up countersiege engines or large trebuchets, and erect low protective walls or crenellations for archers and crossbowmen. However, combatants in the Yangshao, Lungshan, and Shang lacked large defensive devices, employed very limited infantry forces, and didn't otherwise require a large platform to repel attackers because city assaults were rarely undertaken, making the ranks to the rear largely ineffective.

Because these ancient peoples never built watchtowers or constructed bow emplacements on top of these ancient walls except possibly above a few gates, another justification for their inordinately great width must be sought.[5] One commonly advanced explanation claims that this massiveness was required to withstand the scouring effects and relentless pressure that might be exerted by nearby creeks and rivers, that only substantial walls and deliberately emplaced dikes could prevent the disastrous inundations that plagued settlements along the Yellow and Yangtze rivers, as well as their more vigorous tributaries, throughout Chinese history.

Despite the assertion frequently seen in the later military writings that "in walls, massiveness is best," this explanation remains inadequate for several reasons. First, many cities protected by imposing walls were located far from problematic rivers and streams. Second, height rather than width has always proven the most formidable psychological deterrent and greatest physical obstacle to external forces, yet China's walls were almost invariably more expansive than necessary to sustain their height. Third, the problem of erosion could be solved in part by facing the exterior portion with smaller rocks, overlaying exposed surfaces with hardened clay, constructing stone knee walls, or encapsulating the earthen core in stone or brick, as witnessed from the Han onward. Truly thick walls would only become necessary in the Spring and Autumn period, when aquatic warfare commenced and powerful assaults effected by damming rivers and diverting mountain streams could not be withstood without this massiveness.[6]

Nevertheless, indestructible bulk creates a sense of awesomeness, and insurmountable fortifications foster a sense of security. Apart from functioning as a military bastion and serving as a refuge against such natural onslaughts as typhoons, walls define communities and nurture a sense of uniqueness by physically isolating the members of the settlement. Western psychoanalytic literature has long pondered the significance of this inclusiveness and the symbolic sense of sanctuary that walls can create, and popular analysis continues to attribute China's ethnocentric tendencies (or supposed "citadel mentality") to their deliberate employment over the millennia, including to segregate the much-denigrated "barbarians." Compelling the populace to engage in organized wall building would have enhanced this sense of common identity while presumably solidifying the chief's authority and absorbing excess disruptive energy.

Whatever the validity of these insights, it should be remembered that geopolitically defined states did not appear prior to the late Western Chou. Walls were first constructed around villages and towns, then tentatively between states to demark borders as much as thwart incursions, and only late in the Warring States period between so-called sedentary China and the nomadic steppe peoples. Internal compartmentalization slowly developed in late Lungshan cities, then accelerated dramatically in the Shang and thereafter, increasingly segregating the royal quarters from the privileged populace, defining the artisanal sectors, and segmenting the residential and industrial areas into smaller and smaller units, ostensibly for defense but more often to prevent people from moving about freely and to thwart evildoers. Apart from impeding external intrusions, the walls came to be deemed an essential measure for regulating the populace, controlling their activities, and managing intercourse with the outer world, functions that they inherently performed from inception even in the absence of explicit theorizing.

More important, rather than just preserving the inhabitants and their property, walls provided the means to protect military forces, project power, and control the countryside. Prior to the advent of cannon and explosives, fortified walls constituted tremendous combat multipliers in early China because they allowed small but determined garrisons to successfully withstand virtual hordes of aggressors. Accordingly, in a much misunderstood and frequently misquoted statement, Sun-tzu condemned wasteful assaults on fortresses as the lowest possible tactic and incisively postulated that even minimal forces ensconced in a defensive posture would prove more than adequate. Fortifications could thus be exploited to maximize the value of limited power rather than simply provide a place of refuge in times of weakness and necessity.[7]

The growth of settlements, largely synonymous with the evolution of civilization in ancient China, fundamentally affected the history of warfare. As populations increased, class differentiation arose, and ruling elites emerged, a few developed into sizable military and administrative centers. Material wealth began to accumulate in them; ceramic, bronze, and other early craft industries developed in their immediate environs;

significant surpluses produced by the burgeoning agricultural sector were stored within them; and they became centers for consumption. Even if (as vociferously claimed) their growth was power based rather than economically derived, rudimentary trading practices also appeared that further stimulated the evolution of certain fortified towns into regional centers as well as highly desirable military targets.

The nature of warfare in the Neolithic remains unknown, but based on Shang tactical practices and the lack of destructive evidence at the sites so far investigated, it can be concluded that these massive fortifications proved effective in deterring attacks. Rather than seeking to conquer and annex them, chieftains in search of the spoils necessary for self-aggrandizement and motivating their followers targeted contiguous areas and unprotected settlements. Extensive grain pits discovered at Wu-an Ts'u-shan dating to the second half of the sixth millennium BCE, which still held the equivalent of some fifty metric tons of millet, provide incontrovertible evidence of agriculture's ever-increasing productivity and that Neolithic settlements were already generating astonishing grain surpluses, one of the foundations of power and vitality in ancient China.[8]

Although ditches, then walls, developed in accord with the rise of agriculture and animal husbandry, other factors apparently stimulated their comparatively rapid evolution, including environmental degradation, climatic changes, population increases, and the ever-intensifying struggle for edible natural resources and wild animals in the less fertile and more arid areas. A significant shift to a threat-dominated context is generally envisioned as having occurred during the transitional early Lungshan period. However, the recent discovery of astonishingly early Neolithic defensive measures and numerous studies of ancient weapons and skeletons that predate this demarcation indicate China had a long heritage of violence that may well have commenced with its famous Beijing Man.

The oldest presently known Chinese settlements, mostly scattered individual hamlets, date back prior to the Neolithic Age that commenced in parts of China by approximately 10,000 BCE.[9] For the next 3,000 years primitive yet constantly evolving agricultural practices spawned slowly growing settlements that were generally concentrated near rivers and accessible marshes. Unsurprisingly, so-called Hua-Hsia or Chinese civilization thus developed primarily (but not exclusively) in the middle and lower reaches of the Yellow and Yangtze rivers, both well-watered areas with soft, alluvial soil.

Although barter and other forms of minimal interaction within a day's walking distance seem to have occurred with the inception of these settlements, life in that era of "ten thousand villages" probably differed little from the *Tao Te Ching's* idealized depiction:[10]

> *Although neighboring states look across at each other,*
> *And the sounds of roosters and dogs be mutually heard,*
> *Unto old age and death, the people do not travel back and forth.*

The earliest among the several sites that have been at least minimally excavated and radiocarbon dated is **Nan-kuang-t'ou** at Hsü-shui in Hebei, which flourished in the ninth or eighth millennium BCE.[11] Next in succession appears to be the **P'eng-t'ou-shan** cultural site on the plains just north of the Li River in Hunan, essentially located in the middle reaches of the Yangtze river basin. Although dates stretching from 7800 to 5795 BCE have been suggested for this site, it was probably occupied from 6900 to 6300 BCE.[12] This settlement is particularly important for the history of Chinese warfare because it was already protected by a circular ditch, and the dwellings, which average some thirty square meters and are much larger than formerly, were probably intended for extended family use.[13] Moreover, the benefits of rice cultivation were already evident, satisfying one of the oft-stated conditions for true warfare.

The Yellow river basin area, marked by several cultural clusters including the P'ei-li-kang, has also yielded numerous sites with dates ranging from 6100 to 5000 BCE, as has the middle Yangtze River area. One of the oldest ditched settlements from this era, the now well-known Yangshao village of **Pa-shih-tang** in Hunan, reportedly dates back to about 6000 BCE and was continuously occupied for about a thousand years before finally being abandoned for another two thousand.[14] Located near a marsh, Pa-shih-tang was defined by an encircling protective ditch that had been augmented for defensive purposes by a low interior wall and therefore marks the earliest stage of conjoining walls with ditches. Despite meeting a river that flows in the north and west (which would have provided additional protection at a distance), from the terrain's contour and evidence that the ditch was periodically cleared of debris, it never functioned as a moat but may have provided drainage for the settlement.

The site, which exploited a slight natural rise in the terrain, spans some 200 meters north to south and 170 meters east to west. The low interior wall was constructed by simply mounding up soil excavated from the ditch, none of the pounding or other work that characterizes tamped earth walls having been employed. Two stages are visible. The initial effort apparently created a slight wall some 0.5 to 1.0 meter high and 6 meters wide that was characterized by a gradual pitch of 20–30 degrees. The accompanying ditch had a width of 1 to 2 meters but a very shallow depth of less than 0.5 meter, barely enough to impede aggressors standing at the bottom as they confronted the low wall. However, without any increase in height the base of the wall was subsequently widened to about 7.5 meters and the mouth of the ditch opened to 4 meters (with a bottom expanse of 1.5 meters), as well as deepened to a functional 1.5 to 2.0 meters, sufficient to impede aggressors and thus clearly defensive in intent, despite opinions to the contrary.[15]

A terrace settlement at **Ta-ti-wan** in Gansu's eastern portion, reportedly the earliest Yangshao site yet discovered in the province, eventually evolved to cover about a million square meters. Continuously occupied for some 3,000 years starting about 5800 BCE, it too was protected by a circular ditch (or possibly moat) that ranged from 5.5 to 8 meters wide at top and from 2.5 to 3.5 meters at the bottom, and had a depth of 3.5 to 3.8 meters.[16]

The period from 5000 to 3500 BCE is generally considered the late Neolithic. Although discussions of copper's discovery and bronze's development remain highly speculative, their incipient beginnings are said to fall within this era, even though the products of bronze technology do not become truly noticeable until the appearance of the Hsia. Defensive ditches continue to feature prominently in such famous, well-defined sites as **Pan-p'o** and **Lin-t'ung Chiang-chai**.[17] Both were continuously occupied for several centuries in the Lungshan period, but based on structural remains and numerous artifacts including arrowheads, Pan-p'o is now considered one of the defining sites for the preceding early Yangshao culture.[18]

Located approximately 800 meters from the Ch'an River, which has now shifted from one side of the village to the other, and in the immediate vicinity of the strategically critical city Hsi-an (Xian), Pan-p'o stands about 9 meters above the nearby river plain. A typical prehistoric village, its earliest stages have been variously dated to between 5000 and 4000 BCE.[19] A massive protective ditch some 6 to 8 meters wide at the top and 1 to 3 meters at the bottom, with an average original depth of 5 to 6 meters and a full circumference of 600 meters, surrounded the site, delimiting a total area of about 50,000 square meters. This ditch was marked by a strongly defensive contour: the outer wall has a nearly vertical drop that would have made controlled descent difficult, but the inner wall near the village gradually slopes away, preventing blind spots below the interior rim and fully exposing aggressors to bow shots from above.

Remnants of a meter-high wall formed by simply piling excavated dirt into a continuous mound have been detected just inside the town's perimeter. These somewhat extemporaneous fortifications seem to mark an intermediate stage between simply employing a protective ditch and deliberately conjoining ditches and moats with carefully erected, rammed earth walls. Moreover, insofar as the soil excavated from protective ditches had previously been used for building platforms and raising a settlement's overall height (and Pan-p'o seems to have been constructed on a 0.5-meter platform),[20] the wall seems to have been intentionally constructed and was apparently augmented late in Pan-p'o's period of occupation, contrary to assertions that it lacks a deliberate character.[21]

A second, semicircular interior ditch that defines and protects about a third of the settlement apparently was constructed about the same time. Much reduced in scale, its remnants vary between 1.4 and 2.9 meters in width at the top and 0.45 and 0.84 meter at the base, and it has a depth of 1.5 meters, or about a man's height. It may have been a functional precursor to the walled-off royal quarters that would become visible in Shang and other double-walled cities. At least two guardhouses seem to have controlled access to the interior realm and the outer compound, presumably additional evidence of class differentiation having emerged.[22] Wooden palisades appear to have been erected on both the interior and exterior of the main ditch, implying great concern with security problems. These remnants reinforce the idea that Pan-p'o's defenses mark a transitional stage between simply relying on ditches or palisades and erecting the rammed earth walls typical of fortified towns and cities.

Another early, fully excavated Yangshao site identified with the Pan-p'o culture that has attracted considerable attention is Chiang-chai, located near Hsi-an.[23] Sitting atop a slight plateau defined by the two rivers that flow around three of its sides, it was backed on the fourth by Mt. Li to the southeast. The well-defined dwelling area, roughly 18,500 square meters, consisted of a central square surrounded by houses that were inwardly oriented, probably for defensive as much as psychological purposes. Nearly circular at some 160 meters east to west and 150 meters north to south, it is partially defined by four ditch segments that run from the east to the south, presumably the remnants of a continuous perimeter system that encircled the entire core area. Its population has been estimated at about 400 to 450.

Although it is generally claimed that these ditch remnants were intended for both drainage and protection, their defensive function probably would have been primary, because the site seems to have been chosen for the enhanced security offered by the Wei River some four kilometers to the north (but apparently somewhat closer when running in its old course in antiquity) and the Lin River to the southwest. The southernmost ditch segment appears to have once been connected with the Lin River, but its height above the river's level would have prevented water from filling it. (Ditches become moats when they penetrate the level of the local groundwater or are lowered sufficiently to derive water from a nearby source, whether a river or lake.) The dimensions of the ditches vary considerably, but they would have been sufficient to impede aggressors: 1.5 to 3.2 meters in width at the top, tapering down to between 0.5 and 1.3 meters at the bottom, and about 1 to 2.4 meters deep.

Defensive ditches dating back to the middle Neolithic period, or 7000 to perhaps 5000 BCE, have also been found throughout southern Inner Mongolia, western Liaoning, and northeast Hebei, including the general area of Beijing. Somewhat oversimplified, it can be stated that ditches constituted the primary defensive measures in the Hsing-lung-wa (variously dated as 6200/6000 to 5600/5500 BCE), Hung-shan (3500–3000 BCE), and Hsia-chia-tien (2000–1500 BCE).[24] However, because these were hunting cultures, the ditches need not have been particularly formidable to effectively demark the settlement's bounds and sufficiently retard enemies.

At the definitive site of **Hsing-lung-wa** the ditch had a radius of 80 to 100 meters, was about 2 meters wide and 1 meter deep, and was broken only by a single entrance on the northwest. Another site, **Pei-ch'eng-tzu**, is bordered by a river on the western side and therefore required ditches just on the remaining three.[25] Among the other small, circular sites composed of well-arrayed dwellings found in Inner Mongolia and western Liao-tung that have been identified as Hsing-lung-wa, one at **Ao-han-chi** in Inner Mongolia stands out. The dwellings, which were concentrated into eight well-ordered rows of ten, encompassed a surprising 50 to 80 square meters each (with two even attaining 140 square meters), much in contrast to contemporary Yellow River dwellings of a mere 4 square meters.

Ditches continued to be employed as the sole defensive measure at many sites even after wall building began to emerge. For example, an immense ditch varying between 15 and 20 meters in width and marked by depths of 2.5 to 3.8 meters has recently been discovered in Hubei near Sui-chou. Its somewhat oval shape of 316 meters north to south and 235 east to west enclosed a site of 57,000 square meters. Constructed in the Ch'ü-chia-ling around 3000 BCE, it was used well into the Shih-chia-ho, even though the ditches were allowed to deteriorate relatively early. A well-established settlement with a wet rice agricultural basis, it is one of at least ten such settlements in the middle Yangtze River area known to have had comparable protective ditches.[26]

EVOLUTION OF THE FORTIFIED TOWN

With allowance for regional variation, the idealized form of the Chinese city—an inner, generally segmented, and fortified sector containing the royal quarters, palaces, and ritual complex; outer walls encompassing the important inhabitants; and an external area for the general populace, workshops, and livestock—basically evolved during the fourth and third millennia from precursors that had previously been encircled by just ditches or moats.[27] A number of important representatives are described below, albeit in somewhat tedious detail, to facilitate pondering the technical nature of these developments and provide a sense of the increasingly complex, extensive, and well-engineered fortifications that were erected to defend Neolithic villages and towns.

Ch'eng-t'ou-shan, at Li-hsien near the Yangtze River, although originally a Ta-hsi cultural settlement, has come to be regarded as a paradigmatic Ch'ü-chia-ling cultural manifestation. Because it was located slightly north of the Tan River, south of the Ts'en River, and northwest of Tung-t'ing Lake in a historically wet, rice-growing area, the town's inner platform had to be raised above the countryside, resulting in a so-called platform city (t'ai ch'eng) of approximately 80,000 square meters.[28] Although demarked by a circular wall some 325 meters in diameter, the town's defenses were further augmented by a very expansive 35- to 50-meter-wide moat with a challenging depth of 3 to 4 meters. The moat exploited an old riverbed for part of its course and was connected to a nearby river, ensuring a constant supply of water for defense, drinking, and transport.

In contrast to a sharp exterior declination of 50 to 85 degrees, the interior of Ch'eng-t'ou-shan's walls displays a very slight slope of 15 to 25 degrees. Comparatively primitive rammed earth techniques that employed 20-centimeter clumps of earth were used to build the walls, any gaps being filled with river pebbles. Although composed by ten layers of clearly differentiated soil stacked in a triangular cross-section, the entire edifice seems to have been quickly mounded up on the terrain's surface. (The individual layers are readily distinguishable by their distinctive soil colors and the sand that was dusted on their surfaces to facilitate pounding and drying.)

A nearly perfect circle punctured by prominent gates in all four directions and an additional water gate, the site has been dated to approximately 2800 BCE or the middle of the Ch'ü-chia-ling culture, and may be taken as representative of the early stage of compound fortifications. However, before this wall-and-moat combination was realized,

Ch'eng-t'ou-shan's defenses had already passed through three stages that illustrate how early ditch-protected settlements evolved into strongly fortified towns.[29]

About 4500 BCE, Ch'eng-t'ou-shan was already outlined by a 15.3-meter-wide, 0.5-meter-deep moat buttressed by a slightly mounded interior wall about 0.75 meter in height. Between 4000 and 3500 BCE the enclosed area was laboriously expanded and the fortifications significantly augmented by rebuilding the inner wall on the outer portion of the old ditch, resulting in a new width of 8 to 10 meters and functional height of 1.6 to 2.0 meters. Although the newly excavated moat was somewhat narrower—only 12 meters at the top and 5.5 meters at the bottom—it now had a useful depth of 2.2 meters.

A second outward thrust of some 40 meters added 20,000 square meters within the site's new 318-meter diameter and resulted in a circular wall with a base width of 8.9 to 15 meters, a remnant height of 1.65 to 2.6 meters, and a moat that was still about 12 meters wide at top but had been dredged to the formidable depth of 5 meters. The final renovation and expansion apparently occurred between 3000 and 2700 BCE in the early Ch'ü-chia-ling when the preexisting moat was filled; the walls were widened to 20 meters and heightened to at least 4 meters; and a new, immense, 35- to 50-meter-wide moat with a highly functional depth of 4 meters was excavated.

Being located about four kilometers from the Yellow River and twenty-three kilometers northwest of Cheng-chou, the early walled site at **Hsi-shan** lies in the core cultural area.[30] Occupied throughout the Yangshao period, it flourished from 3300 to 2800 BCE but was eventually destroyed through a Ta-wen-k'ou cultural intrusion. A heavily rounded square approximately 180 meters across, it perhaps stands somewhat between purely circular settlements and the square design of fortified Lungshan towns.[31] Originally protected by a simple ditch that encompassed a total area of about 25,000 square meters, it was subsequently expanded by walls, a partial inner ditch, and a moat to cover some 34,500 square meters.

Having been constructed with three different techniques, the walls at Hsi-shan are unusually complex. They consist of a tapered original core, a second outer wall added for additional strength, and finally a top layer that extends across both of them and increased the overall height. (This top layer is said to be composed of looser sand, no doubt evidence of an urgent need to improve the fortifications in the face of external threats.) The core wall's augmented portion encroached upon the original ditch, requiring that it be filled and another one cut further out, a highly laborious double procedure.

The inner wall, which was constructed atop a slight excavated foundation of perhaps 0.5 meter, stands about 1.6 meters high; the extended outer wall was built to a matching height; and the 4.35-meter-wide augmentation at the top raised the overall profile 0.74 meter, though current remnants vary between 1.75 and 2.5 meters. The fortifications average 3 to 5 meters in width, though the heavily strengthened corners expand to 8 meters and protrude slightly higher.

The walls were constructed from layers of pounded soil that had ash and fibrous plants intermixed in it for strength. The center section, which varies somewhat in width, and the other two portions used a total of three techniques. Rather than being continuously laid down within forms that were shifted after completing each section, the core basically employed a sectional block method. Although tamping had long been employed, Hsi-shan is said to be the first site to use forms to constrain the soil being compacted.[32] Evidence of holes from the posts that retained the boards can still be discerned, but it is limited, and some parts of the wall were constructed by simply piling the dirt up and pounding it.

The boards also differed in size, resulting in a lack of uniformity in the extent and height of the individual layers. (They varied between 1.5 and 2 meters for the longitudinal portion and 1.2 meters for the cross-sectional portion, three blocks being required to create a core width of 5 meters.) Boards of 5-centimeter thickness and about 0.5 meter in width were apparently employed to create the core wall's slightly laddered shape, but the layers within the individual sections vary from 4–5 centimeters thick in the well-pounded areas to 8–10 centimeters in the looser areas and outer wall.

Gaps for entrances punctuate the wall remnants on the north and east, a bridge originally spanned the moat on the west, and there was also a gap in the moat opposite the north gate. Although the opening was about 10 meters wide, this potential weak point was shielded by a 7-meter-long protective wall centered in the gap but erected about 5 meters to the exterior, resulting in an advanced defensive device. Although only 1.5 meters wide, it had been pounded hard for permanence. The external ditch, generally described as a moat, was continuous except for a single break opposite the north gate. Reports of its extent vary, but it was at least 3.5 meters wide and had a minimally functional depth of 1.5 meters, though portions seem to have reached 7 meters in width and perhaps 4 meters deep.

Contrary to the rarely disputed view that large Chinese walled sites were deliberately created as military and administrative centers rather than the product of economic activity, Hsi-shan seems to have been based on trade.[33] Several reasons have been offered for this conclusion, the most important being the town's location near the Yellow River just where the mountains and plains intersect. Clearly advantageous for defense, the topography was equally conducive to communication, transport, and interaction with the numerous nearby settlements that it apparently dominated to some degree. Highly favorable to agriculture, the area was still marked by enough localized differences to ensure some diversity in indigenous products, a requisite for trade.

The presence of numerous well-constructed storage pits indicates that in addition to functioning as a trading center, Hsi-shan was a production site for ceramics. Both the size and number of the storage pits increased over time, coincident with the town's walls being improved and refurbished, suggesting that prosperity was the driving force behind the augmented defenses. The several strongpoints visible around Hsi-shan provide further evidence of an ongoing need to thwart raiders and brigands.

Numerous Lungshan sites that have been identified in the middle and lower Yellow River and Yangtze basins similarly provide impressive evidence of fortification activity on a massive scale.[34] **Meng-chuang**, located on the plains just south of the T'ai-hang Mountains and somewhat north of the Yellow River despite being in Henan, was established on a low terrace in an extremely wet area that continues to be marked by numerous springs, rivers, and very high groundwater levels even today. Inhabited from the P'ei-li-kang cultural strata onward, it was initially fortified about 2400 BCE and then continuously occupied until about 2100 BCE, the end of the Lungshan period, after which it lay deserted for perhaps two centuries.[35] Meng-chuang's abandonment has generally been attributed to severe flood damage and its demise seen as evidence that great floods occurred toward the end of the twenty-second century BCE due to a brief climatic shift.[36]

Essentially a rectangle that extended 375 meters north to south and 340 east to west, the 14-meter-wide fortifications were constructed by excavating the soil to the interior and exterior of the walls and then pounding it between framed boards, traces of which have been found on the interior side. This produced an inner ditch some 6 to 8 meters wide and 2 to 3 meters deep and a 20-meter-wide moat marked by depths of 3.8 to 4.8 meters. The wall attained a maximum height of 4 meters in the east but apparently varied considerably, perhaps having been only half a meter high in the west at one time. Ten meters of open ground, sufficient for defensive purposes, interceded between the wall's exterior and the moat.

The so-called Hai-tai Shandong Lungshan culture (2600 to 2000 BCE) was early on marked by an important innovation, the excavation and careful buildup of foundation trenches under the walls.[37] Important sites include Pien-hsien-wang, Ting-kung, and T'ien-wang, but perhaps the most significant is **Ch'eng tzu-yai** on the lower reaches of the Yellow River, which dates to about 2500 BCE. The somewhat rectangular town was surrounded by hard-packed earthen walls that ran 450 meters from north to south and 390 meters east to west.[38] Erected on the banks of an old river that also served as the moat's water source, the walls once rose about 6 meters above ground level. Being just 13.8 meters wide at the base and 7 to 9 meters at the top, they are strikingly narrow for such height, especially as the interior had a gradual slope and only the outer face was nearly vertical.[39] Moreover, the walls were constructed on top of an extensive foundation trench up to 3 meters deep and 13.8 meters wide that was composed of layers of dirt strengthened with intermixed stones. In addition, they were fronted by a sharply defined ditch around the outside from which much of the soil was taken, whereas the area outside the walls seems to have been given a slight downward gradient before construction began.

At Ch'eng tzu-yai the foundation consists of well-compacted, generally uniform layers approximately 12–14 centimeters thick. The walls all display identical construction techniques, each ascending layer being roughly 3 centimeters narrower, no doubt the result of placing the boards at the top of the previous layer, creating a ladderlike effect. Despite the somewhat greater irregularity in the density and thickness of the layers found in the outer portions of the wall, the technique of tamping earth between well-constructed forms was already somewhat advanced. Evidence that pounding tools 3 centimeters in diameter were employed remains visible on the major layer surfaces, but not on the multiple thin layers that almost indistinguishably comprise them.

In contrast with earlier plateau settlements (chü-ch'iu) that minimally improved the topography's natural advantages and relied on relatively primitive barriers, Ch'eng-tzu-yai represents the new stage of so-called platform cities or t'ai-ch'eng, towns well fortified with significant walls. Unfortunately, Ch'eng tzu-yai's dating is marred by discrepancies that affect its evaluation and detract from possibly identifying it as an early Hsia capital.[40] However, the walls were probably first constructed around 2600 BCE and repaired and rebuilt numerous times thereafter.[41] Furthermore, an older, somewhat larger Lungshan town underlies it, suggesting that the site had been occupied for centuries before it eventually reached some 200,000 square meters and was populated by tens of thousands. Some forty minor Lungshan sites have been discovered within a twenty-five-kilometer radius, most averaging about 20,000 square meters even though six fall between 30,000 and 60,000, evidence that Ch'eng-tzu-yai functioned as a significant focus of power.

Also located in Henan but along the middle reaches of the Yellow River is **P'ing-liang-t'ai**, a typical Lungshan fortified town (dated to about 4355 BP) marked by an interior that had been deliberately raised some 3 to 5 meters above the surrounding countryside.[42] A slightly distorted square with rounded corners, its four 185-meter-long walls enclose a dwelling area of 34,500 square meters. The wall's remnants show a height of between 3 and 3.5 meters and a width tapering from 13 meters at the base to between 8 and 10 meters at the top. Two gate openings have been discovered in the north and south walls along with evidence of small guardhouses built from unfired brick on either side of the south gate. A stone road runs directly between the two gates, and ceramic drainpipes ensured the inner grounds would remain dry. A broad moat some 30 meters in width and with an original depth of approximately 3 meters (whose excavated soil was incorporated into the walls) provided a formidable initial line of defense.[43]

P'ing-liang-t'ai's walls were apparently erected in an unusual two-stage effort by first constructing a thin core wall 0.8 to 0.85 meter wide and 1.2 meters high from preformed blocks of brown clay interspersed with burnt clay fragments. The 15- and 20-centimeter-thick layers of yellow and grayish soil were pounded hard within frames built from small boards, marking the first site to use small rather than wide planking. (Surface impressions indicate that the round and oval pounding tools were bundled in groups of four.) An outer wall was then raised against this inner core with similar techniques until the top was reached, whereupon the conventional rammed earth process was employed to increase the height of the continuous wall now formed by the original core and outer-facing portion. Although somewhat more advanced, the technique is similar to that employed at Ch'eng-tzu-yai and therefore considered comparatively primitive.

Ching-yang-kang lies about four kilometers from the Yellow River's present course at a sort of cultural crossroads in Shandong. Encompassing roughly 380,000 square meters within a slightly rounded but essentially rectangular enclosure, it constitutes one of the largest sites yet found in the Yellow River plains area. The perimeter totals roughly 1,150 meters, but the rectangle varies in width between 300 and 400 meters. The walls, which are about 19 to 20.5 meters at the base and taper to between 10.5 and 12.5 meters at the top, still retain a remnant height of 2 to 3 meters. (The site was continuously occupied right through the Ch'ing dynasty, the original fortifications being repaired and reconstructed about ten times over the intervening 4,000 years.) The walls, all erected on excavated foundations, consist of multiple layers made up of distinctive though repeated types of earth that range between 5 and 20 centimeters in thickness, the greatest variation being found at the top and bottom.[44]

Vestiges of concentric inner and outer walls define the important bastion of **Pien-hsien-wang**, strategically located on a terrace between two old rivers in northern Shandong.[45] However, the walls were neither constructed nor employed at the same time, the outer one having replaced the inner and even used portions of it to protect a greatly expanded area. Roughly 240 meters on a side, they encompassed 57,000 square meters within a somewhat irregular square marked by rounded corners and 10-meter-wide gate openings on all four sides. The fortifications were constructed from unusually thin layers a mere 4 to 10 centimeters thick and erected on excavated foundations that range between 7 and 8 meters wide.

The original, square inner compound of just 10,000 square meters was protected by 100-meter-long walls with a base width of 4 to 6 meters that were erected on a deep foundation trench. The pounding tools seem to have varied from oval to circular and been distinctively larger than at other sites, often 10 centimeters in diameter. The outer wall dates to about 1800 BCE, the inner one approximately a century earlier.[46] The destruction of much of the latter during the outer's construction provides an unusual glimpse into the dynamics of settlement expansion and the growth of power centers.

Located in Hubei near Ching-chou on the river plain formed by the Chiang and Han rivers, **Yin-hsiang-ch'eng** consists of a significant wall system that at one time underwent a major reconstruction.[47] Although both of the wall's phases date to early in the Ch'ü-chia-ling cultural phase, the site must have been strategically advantageous, because the fortifications were erected over a Ta-hsi Wen-hua cultural layer and the town was continuously occupied down through the Shang and Chou dynasties. Constructed on massive foundations that attain an expansive 46 meters on the eastern side, the walls generally vary between 10 and 25 meters wide. Even though major portions are missing, the 900-meter remnants suggest that the total length may have been 1,500 to 1,600 meters. Although presently standing only 1 to 2 meters over the raised interior platform, they tower 5 to 6 meters over the outer moat, forcing anyone who successfully negotiated the 30- to 40-meter watery span to confront a very formidable height.

The initial step-shaped wall that subsequently acted as the core of the expanded fortifications had a top width of 6.5 meters and was constructed of alternating layers of 5- to 20-centimeter-thick grayish white and grayish yellow soil, with each of the layers being well pounded and smoothed. It was flanked with an interior protective knee wall and an outer wall of yellow earth, both of which were smoothed and hardened. The final wall was constructed by heightening and broadening the original by repairing the top and expanding the base with about 4 meters of soil excavated from the nearby moat, thereby covering the original inner knee wall as well. Although never thereafter expanded, the fortifications were repeatedly repaired during the Shang and Chou periods.

Several other important Ch'ü-chia-ling sites dating from 3000 to 2600 BCE, despite really being "fortified towns," are generally termed cities because of their size and comparatively advanced stage of development: Tsou-ma-ling, Yang-hsiang-ch'eng, Ma-chia-yüan, Chi-ming-ch'eng, and Shih-chia-ho.[48] The ancient city at **Tsou-ma-ling** in Hubei is located about four kilometers from the Yangtze River next to Shang-chin Lake, its water source in antiquity. Six cultural layers ranging from the late Ta-hsi through the Ch'ü-chia-ling and middle Shih-chia-ho are discernible, impressive evidence for the site's strategic and environmental desirability.[49] The fortifications, which were probably erected in the early Ch'ü-chia-ling period, defended a town that flourished throughout the era. The walled enclosure delineates an irregular circle marked by various indentations that obviously resulted from attempts to conform to the terrain's characteristics. Extending roughly 370 meters east to west and 300 meters north to south, Tsou-ma-ling's approximately 1,200-meter rectangular circumference encompassed an area of 78,000 square meters.

The fortifications stand a substantial 4 to 5 meters high over the interior platform but a formidable 7 to 8 meters above the surrounding countryside and vary between 25 and 37 meters in width at the bottom and 10 and 20 at the top. The 13-meter-wide core wall was constructed from tamped layers composed of alternating yellow and gray soils that range from 10 to 30 centimeters in thickness. Most of the soil was apparently excavated from the 30- to 35-meter-wide, 2-meter-deep moat that exploited preexisting depressions in some places. Defensive platforms constructed near the town's gates controlled access to the interior.

The ancient fortified city at **Ma-chia-yüan** in Hubei mainly reflects the Ch'ü-chia-ling cultural phase but also shows remnants of the later Shih-chia-ho. It was built on a slight tumulus to exploit the terrain's configuration and protected by massive walls up to 32 meters wide that were steeply angled on the outside but gradually sloped on the inside. More trapezoidal than square in outline, the eastern and western walls were 640 and 740 meters long respectively, with the southern slightly exceeding 400 meters but the northern much shorter at 250 meters, creating an enclosure of some 240,000 square meters.[50]

Because the site's interior stands about a meter above the outside terrain, the walls loomed five meters over the interior and some six meters over the exterior. Foundation remnants atop the fortifications indicate elementary defensive works once existed, and there are openings for gates on all four sides, as well as for additional water gates in the western and lower eastern walls. A protective moat encircled the town, the western portion of which integrated a swiftly flowing river with steep banks that must have furnished the requisite water supply.

Because **Chi-ming-ch'eng** was located in the middle of the Yangtze watershed, an area with multiple rivers as well as numerous lakes and marshes, flooding had to have been a primary concern.[51] Although the Yangtze flows to the north and northeast, where it makes a sharp downward bend, the closest river is the western branch of the Sung-tzu. Within a greater site that extends some 500 meters north to south and 400 meters east to west, the 1,100 meters of fortifications essentially defined a square with rounded corners that encompassed 150,000 square meters.

The massive walls taper from 30 meters at the base to 15 meters at the top, with the remnants varying in height from 2 to 3 meters. Cross-sectional analysis reveals at least seven layers, and the interior contains a tamped earth platform typical of ancient sites. A protective moat some 20 to 30 meters wide with an average depth of 1 to 2 meters and a total circumference of 1,300 meters completely surrounded the town. Being a Ch'ü-chia-ling cultural site, it should date somewhere between 3000 and 2600 BCE.

Yang-hsiang-ch'eng in Hubei reportedly dates to the end of the Ch'ü-chia-ling cultural phase. Its rectangular walls run about 580 meters from east to west and 350 meters from north to south, encompassing about 120,000 square meters within the interior of its 10- to 20-meter-wide bulwark. Apart from the usual city gate openings, a water gate apparently was located on the north side. The walls themselves are composed of alternating layers of yellow and gray soil of varying thickness and are surrounded by a 30- to 45-meter-wide protective moat with a slight depth of 1 meter.[52]

Shih-chia-ho in Hubei, the defining site for the Shih-chia-ho cultural phase, encompasses an enormous 1.2 million square meters within a rectangular enclosure marked by slightly rounded corners of 1,200 meters from north to south and 1,100 meters east to west.[53] The walls, which towered 6 to 8 meters high and were 8 to 10 meters wide at the top, were constructed in mounded, minimally pounded layers between 10 and 20 centimeters thick. The entire city was protected by an enormous moat some 4,800 meters in circumference that reportedly varied between 60 and 100 meters wide. The Eastern Ho River flowed quite close on the eastern side; mounds and low hills in the northwest, west, and south would have impeded aggressors in antiquity. An enclosure in the northeastern part created by walls that were 4 to 6 meters wide at the top and 10 to 12 at the base defined a separate section measuring roughly 510 by 280 meters, of unknown importance. The total population has been estimated at between 30,000 and 50,000.

It has been suggested that in comparison with other early fortified towns, the walls of the six Ch'ü-chia-ling sites—Shih-chia-ho, Tsou-ma-ling, Ch'eng-t'ou-shan, Yang-hsiang-ch'eng, Ma-chia-yüan, and Chi-ming-ch'eng—are much longer, generally being 100 to 200 meters per side, and their defensive needs different from those of the so-called core Lungshan area in Hubei and Shandong.[54] Obviously this is not the only possible conclusion, because greater site sizes and larger walls may simply reflect increased population and material wealth and therefore greater capacity for such projects or even military dominance of the surrounding countryside. Nevertheless, their increasingly formidable character is undeniable. Moreover, based on the walls apparently having been constructed solely for protection against external enemies and other contextual evidence, despite arguments to the contrary it appears that Ch'ü-chia-ling culture was already dominated by martial rulers and that conflict between groups of settlements seeking to increase their wealth and power had already spawned violent clashes. Only significant threats could have made fortifications essential for survival, justifying the diversion of enormous manpower to wall building.[55]

Finally, down in the southeast a remarkable site in Zhejiang with complex conjoined fortification measures, possibly the remnants of the Liang-chu capital, has recently been investigated. Although reports remain sketchy, this southeastern Yangtze River area city was undoubtedly surrounded by a slightly rectangular enclosure marked by somewhat rounded corners, with rough dimensions of 1,800–1,900 meters north to south and 1,500–1,700 east to west. The walls were erected on 40- to 60-meter-wide foundations created from a ditch that included stone pieces systematically layered on top of a 20-centimeter-thick clay bed, and they were protected by an exterior moat roughly 45 meters wide and 1.5 meters deep. The active city covered a total area of about 2.9 square kilometers, and the area within the walls encompassed five small hillocks in all, one each in the northeast and southwest corners and three more in the center. The yellow earth comprising the walls was brought from the nearby mountains rather than simply sourced around the perimeter, vivid evidence of the inhabitants' ability to mobilize a massive labor force to execute a clearly predesigned city.[56]

3.

ANCIENT
FORTIFICATIONS, II

✦•✦━━━━━━━━━━━━━━━━━━━━━━━━━━━━✦•✦

CULTURES IDENTIFIED WITH the semiarid Northern zone were traditionally deemed semicivilized and disparaged as "barbarian" by
the denizens of Imperial China because they were perceived as lagging
far behind Hua-Hsia material and intellectual levels. However they might
be interpreted, the interminable steppe–sedentary clashes that would
plague both realms throughout Chinese history commenced in the
Shang, if not earlier. However, vestigial evidence for martial threats
having troubled the Northern zone itself exists in defensive works that
date back prior to the currently known conflict horizon.

Ditches constituted the primary defensive measures in the Hsing-
lung-wa (6200/6000 to 5600/5500 BCE), Hung-shan (3500–3000), and
Hsia-chia-tien (2000–1500) cultures.[1] However, recent decades have witnessed the discovery of numerous Lungshan villages where the inhabitants chose to erect walls instead of relying on ditches, including a
group of twelve sites in central and southern Inner Mongolia characterized by protective walls and dwellings constructed from stone rather
than earth.[2] Originally occupied between 3000 and 2300 BCE, an interval during which localized ecological constraints prompted the
initiation of new settlements by splinter groups, they were abandoned
by 1500 BCE because the climate cooled below the point of sustainable
agricultural yields.[3]

The segregated quarters, variation in building size, large sacrificial
altars, well-developed pottery, and handful of bronze artifacts discovered
in these twelve towns are interpreted as evidence of growing class differentiation and the emergence of localized chiefs. The sites themselves
vary in size from a minimal 4,000 square meters to a very substantial

130,000 but are primarily small and must have been inhabited by limited populations of perhaps a thousand.

Although the sites may be geographically divided into three groups, they all appear to have functioned as military citadels because their walls are not only constructed of stone but also display significant defensive characteristics. For example, gate openings are generally protected by screening walls, and the sites with the fewest natural advantages of terrain often employ parallel doubled walls augmented by external moats or ditches. Such extensive, determined measures, undertaken in an era of simple tools, can only be interpreted as evidence of a pervasive fear of aggressors.

These fortified settlements differ in their exploitation of the terrain's configuration. The five sites in the Pao-t'ou area, despite being situated on comparatively raised ground, are all located on the lower slope of Mt. Ta-ch'ing in the region south of Mt. Yang and thus are marked by north-to-south declinations.[4] They are also fully walled except where precipitous cliffs obviate any need for further defensive works and generally comprise aggregates of smaller citadels positioned in close proximity. For example, Wei-chün has three distinct walled enclaves, whereas A-shan, the largest in the group, and Sha-mu-chia each have two. (Spaced some 250 meters apart, the square one at A-shan measures 260 meters by 120 meters for a total 31,200 square meters; the second tapers from 120 meters to 50 meters along a length of 240 meters.)

In contrast, the four towns studied in the Tai-hai area essentially assume the shape of a traditional Chinese bamboo basket, three of the sides being high and the remaining one low, the equivalent of the basket's opening.[5] Rather than multiple bastions and semisubterranean dwellings, they are internally marked by smaller, segregated walled areas and vary in size from 20,000 square meters to Lao-hu's surprising 130,000 square meters.

Insofar as they average roughly 4,000 square meters each, terrain considerations must have heavily constrained the six small sites clustered along the southern flow of the Yellow River.[6] Although they all exploit local heights and take full advantage of the Yellow River's confluence with nearby valleys and ravines, they also augment their defensive posture with doubled exterior walls, external ditches and moats, and protective galleys on both the interior and exterior of entrances.

Another recently excavated western Liaoning hillside site marked by integrated defensive measures provides further evidence that fear of attack had become a crucial factor in village design and location even in the north during the first half of the second millennium BCE. Dating to the lower Hsia-chia-tien culture (2000 to 1500 BCE or roughly contemporaneous with the Hsia), the 40,000-square-meter village is defined by extensive ditches on three sides and a steep cliff on the remaining or southern border. The ditches on the east and west, both of utilized preexisting ravines, are an astonishing 10 meters deep, and a double stone wall marked by a few crenellations and 17-meter internal spacing runs down the western side. Additional walls isolate a section in the north, and other internal walls provide interior barriers in the remaining portion, completing the formidable bastion.[7]

Walls constructed from larger stones stabilized with smaller stones and plugged with pebbles wedged into the gaps define these sites.[8] Since remnants of tamped earth walls may still be found in this area after thousands of years, and the soil in the vicinity of the Yellow River is particularly conducive to rammed earth construction, the inhabitants must have deliberately opted to employ stone. Insofar as stone walls more readily resist scouring by floodwaters than does pounded earth, the inhabitants could have positioned their settlements along the local rivers, which would have been far more convenient for drawing water. Defense against human threat rather than flood avoidance therefore must have been the paramount factor in selecting somewhat distant, moderate heights for their villages.

Despite the highly conservative interpretations offered by archaeologists, these sites in the semisteppe region of southern Inner Mongolia that date to central China's Lungshan and Hsia periods thus show a strong martial orientation. The solidity of the stone walls coupled with a pattern of dispersed towns inhabited by single extended clans implies a high, if localized, ongoing threat level. Whether they posed a threat to each other, or the entire culture was reacting to external dangers, is unclear. However, it appears that neither larger aggregates nor great chiefs who could unify them against foreign marauders or organized forces from China's interior had yet emerged.

Somewhat surprisingly, just as the Shang ascended to power, these settlements were abandoned when the inhabitants apparently adopted a more nomadic form of existence based primarily on pastoral rather than agricultural practices. This suggests it was not just climatic degradation but also military pressure from the southeast that ended the viability of their sedentary lifestyle.[9] As a result they apparently evolved into the peoples of the so-called Northern complex, who not only came to dominate the region but may also have acted as a conduit for the chariots and distinctive weapons that suddenly appeared in the thirteenth-century Shang.

One other important northern site has been discovered in Liaoning province.[10] Numbering among those that comprised a virtual defensive line along the 42nd degree of latitude, where the state of Yen would construct its border wall in the Warring States period, it dates to the late Shang or very early Chou and typifies settlements that fully exploit riverside locations and natural ravines while being oriented to open vistas. Its incomplete, badly crumbling protective walls outline an elongated, somewhat axe-shaped compound that tapers down at the top or north. Situated between two rivers that sweep protectively from the northwest to the northeast, it is bordered by a 300-meter-long, 20-meter-deep steep ravine off to the northeast.

The compound extends 430 meters on the eastern and western sides, 150 meters across the south, but only 80 meters in the north. The walls have a width of about 3.2 meters and remnant heights that vary between 1.2 and 1.8 meters. They were constructed on a 1-meter-deep foundation by erecting two facing portions from large, indigenous stones and plugging any gaps with soil and small stones before filling the interior with a composite mixture of rocks and earth and leveling the top, a highly localized technique. A 4.5-meter-wide wall about 0.8 meter high, gradual 15-degree slope, and hardened surface protected the interior. However, the steep cliff on the eastern side seems to have been deemed adequately formidable for defensive purposes, obviating any need for further fortifications.

SICHUAN PRECURSORS

Impressive but comparatively late walled towns have also been found in the semitropical hinterlands of Sichuan, far to the southwest of the

Hua-Hsia core. Here, at a crossroads for trade and transport in the upper Yangtze River basin, lies the fertile plains area that eventually evolved into the famous states of Shu and Pa. As might be expected from the fiercely independent clans having retained localized power well into imperial times, this relatively isolated area was early on marked by chieftains capable of organizing and commanding massive work projects. A number of significant sites associated with the Pao-tun culture around Ch'eng-tu or otherwise generally dating to about 2500–1800 BCE, roughly contemporaneous with the late Lungshan and the Hsia's projected dates, have been excavated in recent decades.[11] Although not considered immediate ancestors to the now famous San-hsing-tui culture, the Pao-tun manifestations are deemed precursors.[12]

Even though tamped earth methods and a finely laddered profile were often employed, the techniques used to enclose these sites tended to lag behind those along the Yellow River. However, the terrain's characteristics, particularly the nearby rivers, were fully exploited to create substantial defensive fortifications that assumed the usual basic configurations of squares, rectangles, and circles.[13] Representative Sichuan sites include Pao-tun itself; Mang-ch'eng at Tu-chiang-yen, where the Ch'in would undertake its famous irrigation project; P'i-hsien, Yü-fu-t'un; Ch'ung-chou Shuang-ho; and Tzu-chu.

These sites are all marked by double concentric walls, whether constructed simultaneously or in different eras as part of a town's expansion, with Mang-ch'eng and Ch'ung-chou Shuang-ho even being said to be the first double-walled cities in China; the employment of river pebbles on exterior wall faces to improve weathering and retard flood erosion; and deliberately planned but often rapidly executed construction, including at Pao-tun. Brief characterizations of a few sites will provide a sense of the dimensions and variations that existed within this moist, semitropical location.

The now definitive site of **Pao-tun**, basically a rectangle that runs 1,000 meters from north to south and 600 east to west, was constructed on the edge of a natural terrace on the Ch'eng-tu plains along the Mien River. Occupied from about 2600 to 2300 BCE, it was protected by walls ranging from 29 to 31 meters at the base to 7.3 to 8.8 meters at the top that still display remnant heights as great as 4 meters.

Yü-fu-t'un, located on a plain 1,700 feet above sea level, lies about twenty kilometers southwest of Ch'eng-tu, two kilometers from the Chiang-an River, and seven kilometers from the Min-chiang River. Constructed about 2000 BCE, the now badly damaged fortifications originally protected some 320,000 square meters within an irregularly shaped enclosure closely configured to the terrain, whose walls varied between 400 and 500 meters in extent.[14] Most of the perimeter's 2,110 meters of walls stand between 2 and 3.7 meters high and are characterized by the usual trapezoidal shape as they expand from 19 meters at the top to nearly 29 meters at the base.

The walls were constructed on unimproved ground from discrete layers that generally range from 10 to 35 centimeters, some of the thicker ones actually consisting of numerous unpounded, thin layers. River pebbles were also intermixed to form an external protective layer. The core wall was further inured against deterioration and weathering by two half walls on the interior and exterior. A lengthy depression found to the north may be the remains of a defensive ditch that gradually filled in over time.

Essentially a rectangle of approximately 620 by 490 meters that enclosed 300,000 square meters, the walls at P'i-hsien vary between 8 and 40 meters in width and display remnant heights of 1 to a surprising 5 meters.[15] Although generally dating to the Pao-tun culture and employing the same technology, they were apparently built in two stages, the original walls having been thickened by overlaying. (For example, a section that was originally 1.9 meters at the top, 10 meters at the base, and 2.4 meters high was expanded to 7.1 meters at the top, 20 meters at the base, and raised to a height of 3 meters.)[16]

Yen-mang-ch'eng, Ch'ung-chou Shuang-ho, and Tzu-chu all exploited the defensive advantages of two concentric walls that coincidentally created an isolated killing zone that would entrap anyone who managed to penetrate the outer perimeter. At Yen-mang-ch'eng an intervening ditch further augmented this formidable defense. The inner rectangle, approximately 290 to 300 meters north to south and 240 to 270 meters east to west, consisted of walls between 5 and 20 meters in width that retain a remnant height of 1 to 3 meters. Those of the outer compound, which vary between 7 and 15 meters in width and still protrude 1 to 2.5

meters above the terrain, extended 350 meters from north to south and at least 300 from east to west. Two moats were incorporated for additional protection, and the entire edifice was apparently erected in a single effort, though the inner walls show evidence of subsequent reconstruction.

Although now partially destroyed by the nearby P'ang-hsieh River, **Ch'ung-chou Shuang-ho**'s rectangular inner citadel was about 450 by 200 meters and originally had walls ranging between 5 and 30 meters wide that still display heights of 2 to 5 meters.[17] The badly damaged exterior wall varied in width between 3 and 10 meters but retains a height of 0.5 to 2 meters. Additional protection was provided by an unusual moat interspersed between the walls that varied in width between 12 and 20 meters.

The interior and exterior walls at **Tzu-chu Ku-ch'eng** were similarly separated by an intervening ditch. The 400-meter interior wall varies between 5 and 25 meters in width and stands 1 to 2 meters high. Those on the exterior, although displaying the same height, are a relatively narrow 3 to 10 meters in width. However, they were constructed by simply mounding the soil.

In recent decades two major cities less than thirty miles apart, dating back to the late Hsia or early Shang dynasty, have been partially excavated in Sichuan: San-hsing-tui at Kuang-han, site of the later state of Shu, and Ch'eng-tu, tentatively identified with Pa.[18] Somewhat different in character, they clearly interacted with the Shang but were neither submissive nor externally controlled, even though the Shang may have viewed them as at least nominally integrated into their sphere of influence. However, excavations at **San-hsing-tui** have produced a number of astonishing finds, primarily large bronze figures, trees, and enormous masks unlike anything previously recovered in the Shang and Chou cultural areas. Shang casting techniques were employed, but the dramatic stylistic and thematic differences marking these creations attest to the inhabitants' indigenous might, cultural power, and an ability to withstand Shang military and political power.[19]

San-hsing-tui was probably a theocratic center that developed coincident with the emergence of a new local ruling clan or tribe. Although thought to be a precursor of Shu culture, tremendous controversy revolves around almost every question concerning its origins and nature.

The very identity of San-hsing-tui's progenitors also remains a matter of debate, some analysts claiming that they were descended from the Yellow Emperor's clan and originally either were coterminal with the Hsia or contributed to Erh-li-t'ou culture, others attributing crucial formative influences to the Hsia and Shang.[20] However, for the purposes of military history it is the existence of a flourishing, fortified city providing additional evidence of a powerful alternative culture that is paramount.

Unlike P'an-lung-ch'eng (discussed in the next chapter), San-hsing-tui was not a Shang bastion constructed by deploying military power or dispatching Shang clan members to a border area. Rather, it was a locally constructed, walled city that employed common *hang-t'u* fortification techniques that may have been acquired from the Shang, whether directly or indirectly. The walls, which retain a remnant height of up to 6 meters in some places, are particularly impressive. Those on the east and west are 1,600 meters and 2,100 meters in length; the north and south walls are both about 1,400 meters; the total core area is estimated at 2,600,000 square meters; and the overall cultural site occupies an expansive 12 square kilometers. Evidence has been found of royal quarters, numerous building foundations ranging from ordinary structures covering some 10 square meters to larger ones roughly 60 square meters in extent, and segregated production and living areas, as well as very extensive ritual or religious activities.

The walls were apparently erected at the same time as the palace complex, confirming their defensive function even though aggressors might have easily scaled the gradually sloping exterior. However, squads of soldiers could also have been stationed on top to repel intrepid attackers, and the massiveness of the walls alone would have had a deterrent value in addition to reinforcing the site against recurrent flooding. The deep trench created by excavating the earth for the walls was converted into a moat that augmented their defenses and continues to provide highly visible evidence of their sophisticated grasp of contemporary fortifications technology.

Few weapons have been found at San-hsing-tui, with those so far recovered being fabricated from jade and thus more symbolic than functional, highly suggestive of a lack of external challenges.[21] Despite the massiveness of the walls, the buildings and associated artifacts are said

to betray a pervasive spirituality, suggesting that the city was a commercial and ritual center rather than an administrative and military enclave. Based on one commonly employed measure that allots roughly
155 square meters per household, nearly 82,000 people could have
dwelled within the walled enclosure and perhaps some 250,000 in the
immediate vicinity.[22] The population seems to have been highly intermixed not only in terms of economic classes and occupations, including
bronze workers, but also in ethnicity, because several tribes apparently
migrated in from the surrounding region.

Being located on a major trade intersection, San-hsing-tui was perfectly situated to benefit from diverse external stimuli even as a powerful
ruling hierarchy capable of ordering a disparate population evolved.
Because most of the artifacts display unique characteristics, the clan
that ultimately wrested control may have originated elsewhere and
ruled by force of conquest, just as the generalized distribution of the
few recovered weapons would tend to suggest. Its period of fluorescence seems to have extended from the dynastic Hsia through the
late Shang.[23]

In contrast to San-hsing-tui, the commercial center of **Ch'eng-tu**
apparently lacked walls despite having once deployed a wooden pal-
isade. However, extensive effort must have been required to construct
the twelve bridges required by their location in a basically wet area criss-
crossed by rivers. Parts of the site have been radiocarbon dated to
roughly 4010 BP or well within the putative Hsia period, but the first
significant cultural layer apparently corresponds to the early Shang, suggesting a date closer to 1600 BCE. Excavated pits attest to Ch'eng-tu's
occupation down through the Shang period, just as at San-hsing-tui.

The population in the fifteen-square-kilometer area of Ch'eng-tu
has been estimated at a robust 280,000, making it another powerhouse
capable of fielding a massive army and therefore likely to have been totally independent of, if not actively opposed to, the Shang.[24] However,
Ch'eng-tu clearly enjoyed some sort of commercial relationship with
the Shang, because Shang bronze vessels have been recovered for which
the city lacked production facilities. Moreover, unlike many fortified
settlements in the central plains that evolved into political centers before
becoming commercial centers, Ch'eng-tu and San-hsing-tui seem to

have been economically robust from the outset and only subsequently developed the requisite political and military apparatus.[25]

Finally, an early Bronze Age site about four kilometers square has been discovered at **Chin-sha**, some thirty-eight kilometers west of San-hsing-tui.[26] Although the numerous artifacts and divinatory practices indicate strong Shang influence, Chin-sha has been interpreted as the center of another independent, peripheral people sufficiently powerful to challenge the Shang. Just as at Ch'eng-tu, the large site lacks defensive walls. However, the Mo-ti River flowed through its midst, and the city was situated between two rivers that flowed to the north and south that would have significantly impeded aggressors. Evidence for a considerable level of martial strife is seen in a small statue of a kneeling figure who displays a contorted facial expression expressive of fear or horror and has his hands tied behind his back. Archaeologists have tended to interpret this figurine as evidence of class distinctions and internal conflict, but he seems more likely to have been a doomed prisoner of war, especially because sacrifices of this type were already prevalent in the Shang at this time.

FORTIFICATION TECHNOLOGY AND METHODS

Most historians believe that the erection of expansive walls became possible only with material prosperity, the emergence of class differentiation, and the rise of political power. Without doubt, raiding and pillaging would not have been worthwhile until people and places began accumulating goods and wealth or, as the *Tao Te Ching* notes, "When gold and silver fill the hall, robbers and thieves will come." While counter-arguments need not be raised, tribal societies apparently clashed for reasons other than plunder, including hatred, revenge (the most often mentioned), threat reduction, domain expansion and resource control, and seizing prisoners for enslavement. Nevertheless, even in early agricultural settlements or villages where hunting remained primary, theft and forceful seizure may still have represented a comparatively efficient expenditure of time, threatening the undefended.[27]

Whenever troops were inadequate or slaves unavailable, the requisite labor had to be co-opted from the local settlement or within the greater sphere of political dominance. Diverting numerous workers from the crucial tasks of farming, hunting, and fishing implies that a minimal surplus in foodstuffs such as millet had been achieved, some sort of need perceived, and a basic plan formulated. A chieftain or village council with sufficient power to coerce the inhabitants into undertaking the project also had to exist. If successful, the effort would certainly enhance the chieftain's authority and contribute to consolidating the ruling clan's power.

Because great strength was not necessary, merely determined effort on a vast scale, nearly anyone could perform at least some of the tedious labor required for excavating ditches and constructing walls. Warring States writings indicate that people were expected to participate in accord with their physical abilities, women as well as men, a practice that probably dated back to antiquity. For example, the *Mo-tzu* states that those who can build should do so and that those who can carry earth or measure should also be appropriately employed.[28] The *Huai-nan-tzu*, although composed fifteen hundred years after the event, notes that, "in employing the people to construct earthworks, Yi Yin had the lanky dig, those with strong backs carry earth, those with discriminating eyes determine level, and the bent-over work on plastering."[29] Brief statements in theoretical military writings such as the *Wei Liao-tzu* indicate that women participated in Warring States siege defense by constructing and repairing the walls.

Although constructing earthen walls might seem to be simplicity itself—merely pile up immense amounts of soil and rely on gravity to keep it in place—reality proves otherwise. Massive mounds constitute formidable obstacles, but if basic engineering and construction principles are not observed, the work will lack integrity and may collapse under its own weight. Moreover, to function effectively a wall must have sufficient height to preclude jumping over it, be nearly vertical to make climbing it difficult, and have well-compacted soil so as to provide a solid footing for any soldiers engaged in mounting an active defense atop the ramparts.

Comprehensive planning and thoroughgoing organization are therefore required to construct conjoined walls and moats. General dimensions have to be determined, topographical features analyzed, and the terrain's load-bearing capability estimated. The construction has to proceed in sequential steps; the laborers must be organized; and the diverse tasks of excavating the soil at designated sites, transporting it in reed baskets, depositing it in appropriate locations, watering as necessary, and then pounding for the lengthy period required to attain concrete-like hardness within prepositioned wooden forms must be systematically assigned.

An entry in the *Tso Chuan* for 510 BCE, five centuries after the demise of the Shang but timeless in its essentials, describes the process of wall building:[30]

> First they assessed the length as the number of *chang*,[31] estimated the height [of the ground] as high or low, measured the thickness and thinness, plumbed the depths of the moat and ditch, graded the soil, determined [where to acquire the required materials], estimated the time required for the work, calculated the number of soldiers and ordinary workers to be employed, pondered the materials that would be used, and recorded the amount of provisions in order to mandate the amount of work for each of the feudal lords.

Apart from the earliest walls created by simply piling up dirt and loosely packing it down without the use of retaining forms, multiple discrete layers were always employed that sometimes interspersed dry sand, pebbles, organic materials, pottery shards, and even straw, twigs, and branches in arid areas. After the wooden components were fabricated, the forms had to be correctly positioned and temporarily secured at each stage before being repeatedly shifted upward or sideways as the filling and pounding proceeded layer by layer and section by section. As the walls became more massive over the centuries, experience taught that a solid foundation that penetrated well below the surface and was capable of sustaining the immense weight had to be rigorously prepared. Deep trenches were therefore excavated and the foundation systematically built up with layers of well-pounded soil in a process identical to erecting the wall itself.

Although later practices suggest that workers were probably responsible for furnishing their own provisions and perhaps finding local shelter, whether numbering just a few hundred or soaring to thousands, they had to be fed and sometimes temporarily housed. Compelled to exert themselves from dawn to dusk irrespective of the weather, to endure dusty and muddy conditions, the conscripts must have found the work exhausting. Some sense of their misery may be gleaned from an ode preserved in the *Book of Odes*, a compilation that includes authentic material from the early Chou dynasty as well as compositions no doubt dating to the end of the Western Chou period. Despite having been composed long after the Neolithic ended, the depiction retains its validity because the basic techniques had remained unchanged:[32]

> *Crowds brought the earth in baskets;*
> *They threw it with shouts into the frames;*
> *They beat it with responsive blows;*
> *They pared the walls repeatedly, and they sounded strong.*
> *Five thousand cubits of them arose together,*
> *So that the roll of the great drum did not overpower (the noise of the*
> *builders).*

Of particular interest is that the work effort involved in building the Chou capital was not only supervised, but also controlled by the beat of a drum.

Anyone who has labored with a shovel or pick, whether digging foxholes or landscaping a backyard, can easily envision the physical stress involved in constructing the simplest tamped earthen walls. Even when they were exploiting naturally occurring veins of loose soil or digging along riverbeds, ancient workers wielding cumbersome bone and stone shovels with less than ideal shapes and minimum cubic capacity must have found the effort extremely taxing. Highly compressed soil, clay, intermixed pebbles and stones, even surface vegetation and entangling roots that had to be cut and cleared away, would have immeasurably lengthened the process.

Although most walls were built with soil excavated from adjoining ditches, technological acumen early on mandated the interspersed employment of different types of soil, some of which had to be sought out, excavated, and then transported back over considerable distances and possibly difficult terrain. The simplest encirclement would have required thousands of worker-days, evidence of the magnitude of their dedication, as the population base in the average Neolithic settlement would rarely have exceeded several hundred to perhaps a few thousand. (In comparison with the work's requirements, such low numbers immediately suggest that slaves, prisoners, or nominally subjugated peoples were coercively employed on larger projects, prima facie evidence that localized warfare was already proliferating.)

Defensive fortifications require regular maintenance and occasionally extensive repairs to prevent deterioration. The exposed surfaces seem to have frequently been finished with an exterior coating of sticky clay that congealed in the sun or was occasionally directly fired, as well as normally rendered as smooth as possible to prevent the enemy from ascending.[33] Nevertheless, earthen walls unenclosed by brick or stone were always susceptible to weathering. Rain wreaked the greatest havoc from direct impact and runoff, but the drying effects of wind and the sun, coupled with repeated cycles of heat and cold, expansion and contraction, could rapidly degrade the integrity of exterior portions. Minor cracks might rapidly expand, causing the wall to quickly erode under unrelenting seasonal moisture or the occasional torrents unleashed by typhoons in coastal regions.

Vegetation (other than slippery grass) that might have provided a measure of surface protection had to be removed due to the invasive nature of roots and because bushes and shrubs might conceal stealthy assailants or provide a hand grasp for midnight enemies. Moats had to be similarly cleared of rapidly spreading vegetation and ditches of newly sprouting shrubs that might momentarily shelter enemy troops from archers shooting down upon them. Keeping the requisite angle of fire clear and preventing blind spots created by vertical precipices certainly account for the gradual slope that marks most ditch walls that abut fortifications. In contrast, the walls on the far side were normally rendered as perpendicular as possible to impede enemy descent, and most early ditches were characterized by wide mouths and narrow bases. Debris from the ever-eroding walls and accumulating leaves that could reduce a ditch's effectiveness over time also had to be periodically cleared.

A number of historians have pondered the amount of time required to construct Neolithic and later fortifications in China. Although the relatively small site of Pao-tun Ku-ch'eng required moving 250,000 cubic meters of dirt, the walls at Cheng-chou are estimated as having required somewhere between 870,000 and 1,439,000 cubic meters of soil, depending on assumptions about the wall's average width.[34] However, the size of the city's interior space allows some crude population and labor estimates to be conjured that coincidentally suggest the site's immensity and the Shang's astonishing power.

In an early experiment it was found that the surprisingly meager amount of 0.03 cubic meter of soil per hour—roughly a cubic foot—could be excavated from the surrounding terrain using bronze tools appropriate to the early Shang and only 0.02 cubic meter per hour with stone implements, drastically less than anyone experienced in digging with modern, relatively sharp shovels might have expected.[35] Assuming that 10,000 laborers working 330 ten-hour days annually were impressed for the task and organized into 3,000 assigned to digging, 3,000 to transporting, and 4,000 to pounding, roughly eighteen years would have been required.[36]

One of the questions besetting this estimate is whether the tasks are properly apportioned, whether the number assigned to the pounding work is too low or high. Thinness being the secret to compacting the layers and realizing the extraordinary hardness required, very small-diameter branches or dowels, often bundled, were employed. Moreover, modern reconstructions have repeatedly shown that each layer requires many hours of consistent work rather than quick tamping with heavy stone weights to achieve true, virtually sedimentary solidity.[37]

A larger labor force would naturally accomplish the task more quickly, while fewer men and more women would take correspondingly longer. Whether slaves, prisoners of war, soldiers, lower-class members of the Shang or its subordinate proto-states, or all of these were impressed for the task, they must have required a large number of support personnel to provide the food, water, and clothing necessary for the builders and themselves, suggesting that up to 13,000 people might have been engaged in the project.

Based on the oft-suggested estimate that the average five-person household occupied roughly 155 square meters of space,[38] 20,000 families may have lived within the confines of Cheng-chou's approximately 3 million square meters. Even excluding the large numbers of slaves, artisans attached to the workshops, and farmers tending small suburban gardens who would have been housed outside the walls, this translates into 100,000 people. If each family furnished one person for labor service, voluntarily or otherwise, the requisite number could easily have been assembled.

Without further replication efforts, these projections must be deemed highly conjectural. A number of other calculations have resulted in estimates that range from 5 through 8 to 12.5 years for the work at Cheng-chou.[39] Even though still requiring massive amounts of labor and great organizational effort, the more reasonable assumption of 0.5 cubic meter per day per worker would significantly cut the likely time from 18 years, depending on how many were actually employed in digging.[40]

Several complicating factors and technical complexities that stem from the nature of the soil, the design and shape of the wall, and the wall's exterior surface could also adversely affect the construction rate. Soft, loose soil facilitates excavation work but makes pounding more difficult and complicates the preparation of a sufficiently strong and stable foundation. (Deep excavations for ditches and moats also pose significant risks because of the danger of collapse.) Core walls were sometimes first prepared; many older walls were expanded more than once, often right out over previously excavated defensive ditches or moats; and exterior knee walls were added for strength and protection.

Neither stone, which would have required enormous preparatory work including quarrying, dimensioning, and transporting but constituted an essentially impervious surface (including to enemy diggers), nor mud blocks or bricks were employed for wall facings in the Neolithic period and Shang dynasty. However, sometimes pebbles gathered from nearby rivers and embedded in the face provided minimal additional protection against water's scouring action, and readily collected stones were employed by peoples in the Northern complex.

In response to these issues, external threats, increasing experience, rising population, and greater wealth, the defensive measures implemented at every level not only became more extensive and complex over time, but also more regularized. The first ditches, whose dirt was often used to raise the entire community or provide early building foundations, evolved into ditches conjoined with internal ramparts created by simply mounding up the excavated soil. Eventually it was discovered that pounding the earth as it was piled up would compact and harden it, improving its defensive characteristics while simultaneously raising its resilience.

The need for excavated foundations was quickly realized and the practice of layering quickly emerged, possibly because pounding the soil as it is laid down proved the only effective method. Experience taught that intermixing layers of soil with different characteristics could produce a stronger, more stable wall, a practice that was applied to the excavated foundations and the massive building foundations then being constructed. Thereafter, Chinese walls and platforms consistently show improvement in the uniformity and number of their layers right through Erh-li-t'ou and Erh-li-kang, when thin layers of ten to eleven centimeters were routinely pounded to uniform consistency with bundles of small rounded tools. The placement of additional dry ditches and moats on the exterior and occasionally interior ultimately produced the integrated defensive systems visible in the late Hsia and early Shang at such sites as Yen-shih and Cheng-chou.

The earliest fortified towns generally assumed two basic shapes, circular and rectangular, though portions of a wall's course might be modified to resolve or exploit abnormalities in the terrain, resulting in trapezoids, crushed corners, and odd indentations. In addition, ancient fortified sites invariably took advantage of heights provided by tumuli and mounds and exploited natural depressions, lakes, marshes, and flowing water. In fact, one of the primary conceptualizations of Chinese military science—that configurations of terrain convey strategic advantages—had already been recognized and was being actualized, at least in rudimentary form, in the Lungshan period.[41] Moreover, gate openings were fixed neither in number nor in position, but located largely in accord with exigency despite a preference for centering on each of the walls. With the addition of inner citadels or royal quarters demarked by internal walls, the final form of the Chinese capital, one theoretically distinguished by inner and outer walls rather than the thickness of the fortifications or integrity of the moat-encircled defensive compound, would be realized.

4.

THE HSIA

·•·•·——————————————————·•·•·

T HE EXISTENCE OF THE HSIA, traditionally regarded as the first of
China's three great founding dynasties and one of the chief progenitors of Chinese civilization, has not only long been questioned but
also vehemently rejected by scholars caught up in the skeptical spirit of
the past century. Although it is commonly asserted that the Hsia was a
literate culture that kept records on bamboo slips, a crucial step in creating a rudimentary administrative system, none have been found, and
little other evidence for writing has been recovered apart from a number
of recurring symbols that appear to be forerunners of certain common
Chinese characters.[1] Moreover, despite the evidence for an obviously
vibrant civilization having existed at Erh-li-t'ou, no written references
to the Hsia appear prior to the Chou, including in Shang oracular inscriptions, where some boastful hints might be expected.[2] When not
rejected outright, the Hsia has therefore been generally ignored, with
even the comprehensive *Cambridge History of Ancient China* commencing
its study with the Shang, the earliest dynasty for which literate materials
such as oracle bones and bronze inscriptions have been recovered.

However, other historians have concluded that the increasingly
massive and detailed archaeological evidence uncovered in recent decades incontrovertibly indicates that the Hsia not only constituted a
powerful political entity, but also controlled or otherwise influenced a
substantial area in central China from an original administrative center
around Mt. Sung in Chung-yüeh.[3] Once they emerged as a dynastic
state, their core domain ranged from Yü-hsi in Henan, especially the
Luo-yang plains and Yi, Luo, and (the upper reach of the) Ying river
areas, through Chin-nan in Shanxi and westward into the eastern part
of Kuan-chung, including the Fen, Hui, and Su river basins, as well as

northern parts of Hubei and southern parts of Hebei.[4] Moreover, by identifying the Hsia with the late Lungshan, Hsin-chai, and Erh-li-t'ou cultural layers, these historians believe an essentially reliable history can be compiled that coherently integrates Warring States written accounts of the thirteen generations and sixteen kings with site reports and recovered artifacts.[5]

The resulting portrait depicts a transition from scattered Neolithic settlements to a few dominant fortified towns accompanied by social stratification, economic differentiation, and gradual immersion in warfare of unspecified character. Thus, given the current startling discoveries and plethora of references to the Hsia throughout Chinese history, it seems more reasonable to assume that a proto-state known as the Hsia emerged through vigorous, aggressive action than to dogmatically assert its nonexistence and then examine the era's military history. Moreover, despite being dubious or perhaps even worthless, it is also necessary to scrutinize traditional historical accounts and conceptions because of their impact on military and political thought in subsequent ages.

Interesting myths and fascinating legends about Yü and the origins of the Hsia, largely irrelevant to the study of China's military history, abound. However, Yü has always been accorded universal recognition as the Hsia's first monarch, founder of the line that ruled the state and controlled its domain until the Shang finally ousted the tyrant Chieh. Some accounts identify him as a fourth-generation descendant of the Yellow Emperor, thereby implying a sanctified authority rivaling the other two Sage emperors, Yao and Shun, but most assert that Shun voluntarily ceded the emperorship to him because he was the most virtuous and qualified person in the realm.[6]

Yü, however, apparently contravened this virtuous precedent by passing the throne to his son and thus established clan rule marked by lineal descent, an act for which his detractors have condemned him. In response, his admirers have rationalized, if not justified, this blatantly selfish lapse by claiming that the people voluntarily flocked to his son rather than to Yi, the righteous figure to whom he purportedly yielded, or that the mandate had been decreed by Heaven and therefore no man, not even Yü, could contravene it.[7]

No matter how disparate and steeped in myth these depictions may be, the Hsia's founding figures have always been defined by their achievements in water management and China's fabled bureaucracy envisioned as originating in the quest to master the associated administrative difficulties.[8] Undertaken at Shun's behest, Yü's ultimate, perhaps sole, achievement was taming the floods that periodically inundated the Yellow River basin by planning and overseeing the construction of ditches and canals to disperse them. His method differed radically from that of his father, Kun, whom Emperor Yao, Shun's predecessor, had similarly saddled with taming the rampant waters but who had failed because his levees and dams ultimately impeded their flow, resulting in disaster whenever seasonal surges from rain or melting snow burst through them. The dikes also caused silt to be deposited in the river's channel rather than allowing it to refertilize the fields, blocking the flow and raising the riverbed.[9]

According to the classic account found in *Mencius*:

In Yao's era, when the world was not yet tranquil, the rampant waters flowed uncontrollably, inundating the terrain everywhere under Heaven. Grasses and trees grew luxuriantly, birds and beasts proliferated and prospered. The five grains did not flourish, birds and beasts encroached upon the people. Trails from hooves and tracks from birds transected the Middle States. Yao, alone being troubled by this, entrusted Shun with responsibility for administering corrective measures.

Shun therefore had Yi employ incendiary techniques, so Yi ignited and burned the mountains and marshes, forcing the birds and beasts to flee for refuge. Yü deepened the Nine Rivers and dredged the Chi and T'a, facilitating their flow toward the sea. He cleared out the obstacles in the Ju and Han Rivers and arrayed the Huai and Ssu, facilitating their flow into the Yangtze. Only thereafter could the Middle States feed themselves. During this time Yü spent eight years outside his home, and even though he passed by his gate three times, never entered it.[10]

Since water has always been crucial for irrigation and daily life, as well as exploitable for defense and occasionally offense, the Hsia's entwinement with hydraulic engineering may have many unexplored implications for military history.

Even though the techniques for constructing earthen fortifications were already well advanced in the Yangshao and Lungshan periods, because he tried to constrain the rampant waters by "walling" them with dikes, Kun has traditionally been credited with building the first walls and sometimes condemned for thereby antagonizing the people. For example, in a chapter devoted to warning about the perils entailed in abandoning the Great Tao (synonymous with the true, natural Way) and instead relying on the "minor techniques" that governments and administrations invariably employ to their own detriment, the *Huai-nan Tzu* observed:

> In antiquity, when Kun of the Hsia erected a wall three *jen* [twenty-four feet] high, the feudal lords turned their backs upon it and those beyond the seas developed crafty hearts. When Yü learned that All under Heaven were in revolt, he destroyed the walls and leveled the moats, dispersed goods and wealth to the people, buried their armor and weapons, and overspread them with Virtue. Those beyond the seas then came as submissive guests and the Four Yi offered tribute. When he assembled the feudal lords at Mount T'u, a myriad states came bearing symbolic jade pendants and silks.[11]

In the same spirit the passage concludes: "When armor is made solid, weapons become sharp; when walls are erected, assaults are born." This sort of thinking represents a divergent but crucial trend in Chinese political thought that believed artifice provokes retaliation and warfare and therefore envisioned a return to primary Virtues implemented in selfless fashion as the only solution.[12]

Although legends clearly speak of three Sage emperors, Yü may simply have been an emblematic clan figure who symbolized the tribe's devotion to water management, a focus that resulted in ameliorating dramatic river fluctuations, reducing the catastrophes that apparently plagued the realm from 4000 to 3000 BCE while increasing agricultural productivity.[13] Although a millennium or more before Yü's purported era, this would be particularly appropriate considering that the Hsia's progenitor has always been said to be Hou Chi, Lord of the Millet, and the Hsia itself is thought to have emerged through agricultural strength, including perhaps irrigation measures that ensured surpluses sufficient to support diverting vital manpower to military tasks.

Yü's achievements were naturally magnified as these legends evolved in the Warring States and beyond, resulting in the evolution of alternative interpretations. For example, Yü's father had supposedly been banished, imprisoned, or even executed for failing in this same task despite laboring for nine exhausting years, but his fate may have stemmed from other causes, such as irreverently criticizing Yao for ceding the throne to Shun, or may simply represent the result of two clans, both descended from the Yellow Emperor, clashing. Remarkably, Yü still accepted Shun's mandate to undertake the onerous task, thereby submitting to the newly established emperor and acting to reduce the people's misery:[14]

In the time of Emperor Yao the overflowing waters filled Heaven and expansively embraced the mountains and heights, causing misery to the people below. When Yao sought for someone who could control the waters, his ministers and the Four Chiefs all said that Kun could. Yao said, "Kun is someone who rebels against orders and injures his clan. He cannot." The Four Chiefs all responded: "If you rank everyone, no one is more worthy than Kun. We would like you to test him." Thereupon Yao listened to the Four Chiefs and employed Kun to control the waters. Nine years passed but the waters had not yet been stilled nor had his efforts proven successful.

Yao then searched for someone to undertake the task and again got Kun. Thereafter Shun ascended to employment, personally assumed the Son of Heaven's administrative tasks, and undertook an imperial tour of inspection. When his travels revealed that Kun's efforts to tame the waters lacked visible achievement, he imprisoned him at Mount Yü to await death. Without exception, All under Heaven viewed Shun's imposition of punishment as correct. Shun then raised up Kun's son Yü to continue Kun's responsibility.

When Yao died, Emperor Shun asked the Four Chiefs, "Who may fruitfully complete Yao's task and serve as chief official?" They all replied, "If Duke Yü is made Director of Works, he can successfully complete Yao's achievements." Shun said, "Let it be so" and then ordered Yü, "Exert yourself in leveling the waters and the land!" Yü bowed his head to the ground in obeisance, but yielded to Hsieh, Hou Chi, and Kao-yao. Shun said, "Go and oversee your work."

Yü was quick, perceptive, capable, and resilient. He never contravened Virtue, his benevolence was approachable, his words were credible, his utterances became legal statutes, his body exemplified measure, and his actions were weighed. Relentless and industrious, he was a standard and model. Yü accordingly undertook to fulfill the emperor's mandate together with Yi and Hou Chi. He ordered the feudal lords and hundred surnames to mobilize labor forces in order to shift the earth. He traveled among the mountains marking out the trees [for roads] and defined the high mountains and great rivers.

Yü, perturbed that his father Kun's efforts had not been successful and that he had been executed, labored his body and troubled his mind. He dwelled outside for thirteen years without ever daring to go inside his gate even when passing by his house. He kept his clothes sparse but was respectfully filial to ghosts and spirits. He kept his palaces meager but expended all his funds for ditches and channels. When traveling on land he mounted a chariot, when traveling on water he employed a boat, when traveling on mud he used wood plank shoes, and when traveling in the mountains he employed spiked shoes. In his left hand he held a level and cord, in his right a compass and square.

He recorded the [effects of] the four seasons in order to open the Nine Provinces, connect the nine river ways, bank up the nine marshes, and measure the nine mountains. He ordered Yi to give rice to the common people for planting in the low-lying wetlands. He ordered Hou Chi to provide the people with food difficult to obtain. When their foodstuffs were few, he balanced them with excesses from areas of plenty in order to equalize the feudal lords. Yü then implemented the system of tribute based upon what was appropriate to each area and thus facilitated the profits of the mountains and rivers.[15]

Several odes in the *Shih Ching* also stress Yü's work ethic and self-sacrifice, creating a much-admired persona that would be cited whenever later bureaucrats wanted to inspire the people or indirectly rebuke a profligate ruler.

ORIGINS AND PREHISTORY

Various dates, derived in part from early written sources but significantly modified to reflect radiocarbon techniques, have been assigned to the Hsia dynasty, with 2200 to 1750 BCE and 2200 to 1600 BCE having previously been the most common. However, 2100 to 1521 BCE is now deemed orthodox despite considerable criticism, acrimonious counterarguments, and a probable Shang conquest date of 1600. The preceding era—the mid- to late third millennium BCE, which witnessed a sudden proliferation in weapons, the expansion of defensive fortifications, and initial utilization of bronze in warfare—has long been revered as the age of heroes. Reflecting a thrust toward demythologizing antique legends, the vaunted cultural icons have been apportioned to the middle and late centuries, though not without ongoing disagreements about specifics.[16]

It is also possible to envision these icons' reigns not as singular events but instead as sequences of ten or twelve generations, broad indicators of various stages of civilization symbolized by certain heroic characteristics, obviating the need for unattainable chronological precision.[17] However, it should not be forgotten that they are all noted for their military accomplishments as much as for their contributions to cultural and material life.[18] Whatever the actual dates of their reigns, the traditional accounts clearly reflect the emergence of great chieftains who were increasingly glorified with the passage of time, Yü thus becoming the first "global" ruler.

Generally speaking, the Yellow Emperor has traditionally been seen as active about 2700 or 2600 BCE;[19] Yao dominated the stage somewhere around 2300 or 2200; Shun ascended to power about 2200 or 2100; and Yü, identified as the Hsia's first monarch and the dynasty's progenitor, arose sometime in the twenty-first century BCE.[20] A number of reign periods and key events, including Yü's ascension, have recently been computed from important eclipses and other astronomical observations, such as a rare five-planet conjunction, embedded in the *Bamboo Annals* and other Warring States compilations, many well argued but others wildly speculative. Among the possibilities suggested for these legendary totemic figures are 3709–2221 BCE for the Yellow Emperor, 2397–2275 for Yao, and 2274–2222 for Shun,[21] while 1953 BCE seems to be the best possibility for Yü's first year as ruler.[22]

The question of Hsia precursors is too complex to pursue in detail even if, as here, Erh-li-t'ou culture is considered synonymous with the historical entity known as the Hsia and late Lungshan manifestations are deemed predynastic forms. Nevertheless, certain aspects of their ascension to power deserve contemplation because the Hsia undoubtedly emerged through conflict. Unfortunately, frequent shifts of their early capital and the contentious nature of origination theories, including that they evolved from Henan Lungshan culture through early Hsin-chai manifestations into early Erh-li-t'ou,[23] considerably complicate the effort. The larger question as to whether or not the Hsia, Shang, and Chou were ethnically homogenous, whether they had a single or multiple origins, also looms large.[24]

The Hsia might be understood as a chiefdom that began as a localized power, perhaps one marked by a rudimentary administrative apparatus, but evolved into a despotic form of overarching rulership through struggle and coercion rather than acclamation.[25] Emperor Yü emblematizes this transition from an alliance chieftain to an incipient despot, from the stage of loosely grouped settlements to a somewhat integrated domain. Moreover, in contrast to legends that extol the virtuous Sage rulers voluntarily yielding to the most worthy, traditional accounts indicate a highly lethal clash developed over Yü's successor.[26] Whatever its inception, in some sense the proto-dynastic Hsia state can therefore be understood as commencing with Yü.[27]

Any attempt to chronicle Hsia history can perhaps be reduced to two mutually entangled questions: Where did they originate, and when did a minimal Hsia identity emerge? The quest for origins necessarily begins in the area that they presumably controlled in their final embodiment at Erh-li-t'ou: much of Henan and Shanxi, a possible early administrative center around Mt. Sung in Chung-yüeh, the upper reaches of the Ying and Ju rivers, the Fen and Hui river areas, and other areas previously noted, including strongpoints at the perimeter such as P'an-lung-ch'eng. However, without doubt, the Hsia's core domain migrated from a somewhat peripheral, still-disputed origination point to a focal area around the Yi and Luo rivers, including Cheng-chou and Yen-shih.

The question of precursors is complicated by the several distinctive cultures then populating greater China, not just constantly evolving but interacting in every imaginable way, ranging from mutual infusion through displacement and armed conflict. Contrary to the long-held but now-discredited traditional view that all cultural developments radiated outward, the direction of influence constantly shifted throughout the Yangshao and Lungshan periods, no single group or culture always predominating. Rather than inventions, practices, and beliefs simply flowing outward, attributes from peripheral cultures, especially those evolving in the east and southeast, significantly affected the core.[28]

The widely accepted assumption that the Hsia evolved directly out of the Lungshan cultural stage or a particular variant such as Henan or Shandong Lungshan has recently been challenged and even rejected in favor of other possibilities.[29] However, because of the incontrovertible presence of Lungshan elements in the core Hsia domain immediately before its appearance, the basic question tends to be whether late Lungshan is synonymous with early Erh-li-t'ou, Erh-li-t'ou culture is Lungshan's direct successor, or some other culture or intermediate stage intercedes. Somewhat less orthodox but still a possibility is that Erh-li-t'ou culture represents a derivation from some other culture or combination of cultures, including Tung Yi, Liang-chu, and Yüeh-shih.[30]

The idea that pre-Hsia culture evolved in the east along the middle and lower reaches of the Yellow River, either in Shandong or Henan, has numerous proponents, including those who posit an eastern origin for the Yellow Emperor.[31] Perhaps the most interesting formulation stresses that multiple shifts in the Yellow River's course in Shandong compelled a greater intermixing of peoples (and presumably conflict over territory and resources),[32] especially in the contentious region between the Chi and Yellow rivers where the Hsia would eventually clash with the incipient Shang.[33] In antiquity the Hai-tai region (which essentially encompasses the area in Shandong between the Yellow River and the Yellow Sea and up to the Pohai gulf but not the Shandong peninsula) was a particularly fertile area due to the lowest reaches of the Yellow River constantly shifting and flooding. The cultures that arose there, including Pei-hsin, Ta-wen-k'ou, and Lungshan, are known to have dynamically interacted with central China at various stages. Well-fortified towns developed in the Lungshan period, and regional centers such as Ch'eng-tzu-yai and Pien-shien-wang, situated about 100 *li* apart, emerged.[34]

In contrast, the perceived continuity of cultures from Lungshan to Erh-li-t'ou at Teng-feng and Yü-chou on the upper reaches of the Ying River have caused the area to be proposed as a possible western origination point for Hsia culture.[35] The inhospitable nature of much of the general area would have favored the unimpeded evolution of isolated groups, but also thwarted expansion and amalgamation. The lower reaches of the Fen River and its tributaries, site of the T'ao-ssu Lungshan culture, have also been suggested as probable sources, as well as equally rejected.[36]

Finally, based on the (perhaps dubious) perception that many of the Hsia's important cultural elements, including covenants, marriage customs, esteem for jade,[37] large axes, military expeditions, agricultural practices, sericulture, and sacrifice, all originated in the southeast, a southern inception theory has also been proposed. Furthermore, the disappearance of Liang-chu culture and the migration of their populace into the central region coincident with the Hsia's ascension raises questions about the nature of their interaction.[38]

Irrespective of their point of origination or whether they were not instead an amalgamation of different peoples,[39] the era immediately preceding the Hsia clearly witnessed a fairly rapid transition from somewhat isolated settlements with rudimentary defenses to well-protected towns and regional centers marked by defensive systems consisting of massive walls and conjoined moats. Because of the incredible manpower required in an age of stone tools, complex fortifications would never have evolved without the compulsive effects of fear or pervasive concern about raids and destructive assaults.

Without doubt the increasing population, closer proximity of the settlements, reduced viability of external resources acquired through hunting and fishing, and a greater emphasis on agricultural practices that were polluting and exhausting the land and compelled occasional shifts of the populace caused an escalation in the frequency and lethality of conflict in the centuries preceding the Hsia's appearance. Whether through coping with these threats or other challenges, more powerful clan chiefs emerged who acquired the power of life and death over others, as well as the ability to compel participation in massive civil projects, including the construction of palace foundations, levees, and walls, thereby reinforcing their own authority.[40] New weapons evolved and society acquired a much more martial character, with military values being esteemed and deceased warriors increasingly being honored by the presence of weapons, especially battle axes, in their graves, particularly in late Erh-li-t'ou culture.

Finally, although the Hsia's administrative structure and their agricultural growth are often attributed to their success in mitigating the damage caused by flooding during the comparatively wet predynastic period and controlling the waters themselves, from the early Hsia to its extinction the climate became significantly drier.[41] More wells had to be dug, and conflict objectives probably changed somewhat from struggling to occupy dry, relatively secure mounds and other high points near vital water resources to battling for control of dwindling wetland areas.[42] The diminished rainfall would also have reduced the land and aquatic animal populations, causing significantly increased competition for their acquisition.

EARLY SITES AND CAPITALS

Aided by radiocarbon and other dating techniques, scholars have assiduously sought to identify archaeologically excavated sites with the Hsia and Shang capitals discussed in such late written materials as the *Chu-shu Chi-nien* and *Shih Chi*, thereby chronicling their respective histories and attesting to the antiquity and continuity of Chinese civilization.[43] According to these traditional writings, the Hsia and the Shang shifted their capitals a number of times for unspecified but presumably strategic and environmental reasons.[44] Purported Hsia capitals include Hsia itself, Yang-ch'eng, Yüan, Lao-ch'iu, Hsi-ho, and Chen-hsün, most of which have been speculatively matched with various sites, including two that have already been discussed, Pien-hsien-wang and P'ing-liang-t'ai.

At present only Hsiang-fen T'ao-ssu, Teng-feng Wang-ch'eng-kang, P'ing-liang-t'ai, and Erh-li-t'ou have been even minimally justified by archaeological evidence, though Erh-li-t'ou is now officially, if not universally, recognized as the last capital of Chen-hsün. However, the fortified cities of T'ao-ssu, Wang-ch'eng-kang, Hsin-chai, Ta-shih-ku, and Tung-hsia-feng, all of which fall within the Hsia's projected dates of 2070 to 1600 BCE, seem to provide somewhat concrete form to the previously elusive dynasty. Being located in peripheral areas in proximity to antagonists such as the Yüeh-shih and Tung Yi, several of these sites may have been secondary capitals or major strongpoints rather than true Hsia capitals,[45] and even Pien-hsien-wang probably functioned as a Hsia bastion, then suddenly declined, perhaps coincident with the Hsia's extinction.

Artifacts excavated from **P'ing-liang-t'ai's** five main layers are variously dated, but the walls themselves may have been constructed as early as 2370 ($\pm$ 175) BCE, placing them in the putative transitional period for Yao and Shun.[46] However, insofar as the enclosed area amounts to only a third of many earlier Yangshao sites and probably had a population of fewer than a thousand, despite claims that it was a predynastic or early dynastic Hsia capital, at best it probably was merely an early military fortress.

T'ao-ssu at Hsiang-fen in Shanxi, the defining site for the increasingly discussed T'ao-ssu culture, has sometimes been suggested as the locus for Yao, Shun, and Yü,[47] as well as the first Lungshan city where emblematic evidence of Hsia culture can be discerned.[48] Located in an area that has long been identified as core Hsia domain as well as Emperor T'ang's legendary bastion, the overall site encompasses some three million square meters. The densely populated society was already marked by clear class distinctions, and the graves of esteemed males are distinguished by the inclusion of weapons and other symbols of martial status or achievement. For example, 111 stone arrowheads (out of 178 objects) were interred in one grave; others include not only a small number of arrowheads, but also stone axes.[49] The presence of some thirty bodies whose appearance indicates they died violently, coupled with the essential abandonment of the walls even while the site flourished, has prompted the suggestion that Hsiang-fen was the site of Ch'i's capital after their internal conflict.[50]

Another early fortified town that has been tentatively identified with the Hsia dynasty, the aptly named **Wang-ch'eng-kang**—"king's (*wang*) city (*ch'eng*)"—was strategically situated on a well-chosen hillock (*kang*) just a couple of hundred meters to the west of the Wu-tu River and about 600 meters north of the Ying River. Excavated in two stages approximately two decades apart, it consists of a small citadel in the northeast corner of a much larger, roughly 348,000-square-meter enclosure. The nature of this double enclosure has prompted assertions that Wang-ch'eng-kang represents an early, even the earliest, manifestation of the Chinese royal city.[51]

The inner or original fortified enclosure of roughly 20,000 square meters consisted of two essentially square citadels of approximately 90 to 100 meters per side, arrayed on an east–west line with a common center wall. The corners were square, and there seems to have been a gate at the eastern corner of the western section's southern wall, abutting the common center wall. Unfortunately, extensive flood damage, leveling, and agricultural exploitation over the millennia have almost obliterated the walls, preventing an accurate estimation of their original height.[52]

The walls themselves must have been fairly narrow, perhaps only 5 meters wide, because the carefully prepared, 2-meter-deep foundations have a top opening of 4.4 meters but taper down to a 2.54-meter-wide base. The fairly well-leveled tamped earth layers, generally 10 to 20 centimeters thick but as thin as 6 centimeters, are clearly demarked by an intervening 1-centimeter layer of coarse sand that was intended to prevent the tamping tools from adhering to the soil being compacted. Considerable variations in the size, shape, and depth of the pounding marks indicate that river stones may have been employed, consistent with early practice. The site's prolific middle Lungshan artifacts, coupled with this tamping technique, suggest that the walls were built in the mid- to late Lungshan period, before the emergence of the Erh-li-t'ou phase.

The walls of the greater Lungshan enclosure, one of the largest yet discovered in Henan, nearly define a square that runs some 600 meters east to west and 580 meters north to south. The northern wall, which displays the laddered cross-section typical of the period, was constructed on an excavated foundation using the same technique as the inner citadel. From a top width of 6.8 meters it expands to 12.4 meters at the base, retains a remnant height of 1.12 meters, and consists of ten layers of tamped yellowish sand that vary from 4 to 28 centimeters in thickness after having been compacted with 5- to 7-millimeter-diameter tools.

Moat segments of approximately 620 and 600 meters just 4 meters outside the walls protected the northern and western sides, thereby completely encircling it with a water barrier in conjunction with the Wu-tu and Ying rivers. These external moats were refurbished at least twice during the city's occupation. As originally constructed in the Lungshan period, they were about 15 meters wide and tapered down to 10.4 meters at the very significant depth of about 5.2 meters. During the Erh-li-t'ou phase they were about 15.8 meters wide at the top and an augmented 7 meters deep, but only 2 meters wide at the bottom; and in the Spring and Autumn they had an expanse of 16 meters, a base width of 10.8 meters, but an average depth of only 3.6 meters.

Radiocarbon dating shows that although the area was intermittently populated from at least the P'ei-li-kang, the earliest wall-building activity dates roughly to the first half of the twenty-first century BCE, slightly later than T'ao-ssu and the era frequently identified with the Hsia's inception.[53] The smaller compound predates the larger, the latter having destroyed a portion of the former during its construction, but both probably fall into the late Lungshan era. Recovered artifacts and alterations to the moat in both the Erh-li-t'ou and Spring and Autumn periods indicate that Wang-ch'eng-kang was occupied almost continuously until the Warring States period due to its strategic location and the natural fertility of the greater surroundings.

As later historical writings associate the semilegendary progenitors of the Hsia—Yao, Shun, Yü, and Yü's father Kun—with this area, and the site has historically been known as the "king's city," it has been suggested that the smaller citadel was constructed by Kun, the "first wall builder." The greater city, which would befit the Hsia's first ruler and have required a massive effort to construct and thus reflect the rise of centralized authority, is therefore envisioned as having been Yü's first capital of Yang-ch'eng.[54] The segmented double quarters would have delimited the royal quarters, the greater site would have encompassed a population of perhaps 50,000 people, and the fortifications coupled with the exploitation of the two nearby rivers would have provided considerable protection.

If, however nebulously, Wang-ch'eng-kang can be identified with Yü, the fortified town located at **Hsin-chai** in Hsin-mi, Henan merits possible consideration as his successor's capital of Huang-t'ai.[55] Recent radiocarbon dating confirms that Hsin-chai stands at a transition point between the late Lungshan and Erh-li-t'ou and might therefore be considered incipient Erh-li-t'ou or Hsia.[56] Located on the north bank of the Wei River (now the Shuang-chi) and flanked by the Wu-ting and Ch'ih-chien (or Sheng-shou) rivers on the west and east respectively, it was nearly surrounded and thus naturally protected by three rivers as well as a preexisting ditch.

Nevertheless, commencing in the late Lungshan period, Hsin-chai underwent several stages of complex fortification efforts when the first walls were constructed and a moat excavated to the exterior, the size of the former being about 3 meters in height and 4.5 meters in width, with a narrow moat of about 2.7 meters that tapered to 1.3 meters at the bottom and ranged in depth from a very shallow 0.34 to a relatively functional 2.3 meters.[57] The walls were improved to some degree early in the Hsin-chai period, then rebuilt with rammed earth construction techniques further to the exterior following floods that apparently filled the previous moat. These new walls reached the more significant height of just under 5 meters in some sections, though others dipped as low as perhaps 2 meters. A new moat with roughly double the width and depth of the previous one was excavated for augmented protection, and evidence of an interior moat has also been reported. However, after some minor revisions and improvements in the late period, all the walls and even the moat were inexplicably destroyed in the Erh-li-t'ou period, being replaced by a new 1,500-meter protective waterway said to average 6 to 14 meters wide and 3 to 4 meters deep.

Hsin-chai is an intriguing site that well illustrates not just the continued occupation of a strategic location, but also the deliberate process of expanding a fortified town by laboriously filling previous moats and ditches before erecting walls on them with soil excavated for a new exterior moat. The ruling authorities must have had considerable power and a strong impetus to compel the people to undertake what can only be considered a double process instead of simply rerouting a portion of the external moat and unilaterally expanding one side or adding a second section rather than uniformly reworking the entire encirclement in all directions. Excluding the area of the walls and moat, the interior at the end of the Hsin-chai phase encompassed about 700,000 square meters, the largest late Lungshan/Hsin-chai site known and certainly a reasonable candidate for a pre-Erh-li-t'ou Hsia capital.

In addition, an unusual site attributed to the Hsin-chai phase of early Hsia culture has been uncovered somewhat outside the historically identified Hsia focal area to the north above the Sung-shan Wan-an Mountains.[58] Its unique defensive technology—four square, concentric ditches marked by rounded corners of somewhat varying width but fairly uniform spacing—provides incontrovertible evidence that the occupants intended to project power into the hinterland and protect the approach to the Hsia heartland. The outermost ditch spans some twelve meters, ranges from a startling five to six meters deep, and is characterized by laddered sides. A single defensible roadway penetrates the series of ditches on the southwest, away from the prime direction of threat.

Although considerable skepticism, not to mention vehement opposition and rejection, continues to beset the identification of all the previous sites with pre-Hsia culture, **Erh-li-t'ou** at Yen-shih has been more generally acclaimed as the Hsia's last capital and thus presumably Chenhsün, where T'ai Kang, Yi, and Chieh all dwelled.[59] Numerous problems remain, including the apparent discontinuity between the cultural complex visible at Erh-li-t'ou and the characteristics of late Lunghsan culture, as well as the apparently heterogeneous nature of the populace, an admixture of clans and peoples (as would be expected from integrating conquered groups) rather than a single Hsia entity.[60]

Although backed by a small range of hills, Erh-li-t'ou was situated in the eastern part of the Loyang plains between the eastward-flowing Luo and Yi, two rivers that could provide ample water and constituted natural defensive barriers. Some 2.5 kilometers from east to west and 1.5 kilometers from north to south, the site encompassed structures that ranged from palatial to ordinary and dedicated areas such as workshops and burial grounds, all of which preserve evidence of a sophisticated culture. Surprisingly, the site lacks the massive earthen fortifications that define Lungshan and previous Erh-li-t'ou manifestations, only the royal city having been protected by a limited, double-walled compound. However, Erh-li-t'ou was located in the midst of a strategically advantageous area populated by late Lungshan, Yangshao, and pre-Erh-li-kang towns. Several capitals and otherwise important fortified cities would also be constructed over the millennia within 25 kilometers, including Han-Wei Ku-ch'eng, Ch'eng-chou, and Luo-yang.

Ongoing investigations continue to compel modifications of Erh-li-t'ou's envisioned nature, role, and significance. From the artifacts already recovered it is clear that the inhabitants pursued all the highly skilled occupations appropriate to a royal capital and that trade, tribute, and conquest resulted in the acquisition of diverse, nonindigenous products. Agriculture flourished in the surrounding, fecund earth; domesticated animals including cattle, sheep, pigs, and dogs were all raised; and fishing augmented the basic food supply. The specialized handicraft industries fabricated sophisticated items from a wide variety of materials, including bone, stone, clay, jade, metal, lacquer, and plant fibers. Weaving was well advanced, and wheel-thrown ceramics had appeared. Liquor, whose liberal use would ironically be condemned by the Shang before reputedly coming to plague its own rulers, was also fermented and enjoyed, as attested by bronze drinking vessels.

An early bronze foundry containing copper fragments and molds has been uncovered, and numerous other metal artifacts that provide evidence of significant productive capacity have been found. Among the items produced from copper and high copper-bronze alloys were fishhooks, jewelry, knives, axes, and comparatively simple cauldrons intended for both everyday and ritual use. Weapons must have been highly valued, because many were fashioned from bronze and numerous bronze arrowheads have been recovered. (The use of bronze rather than stone for such single-use, irretrievably lost items has been interpreted as evidence of access to an ample copper supply and extensive mining activity.) Finely worked symbolic jade weapons such as axes, knives, and halberds, which must have been intended to mark the military prowess and authority of tomb and grave occupants, indicate the importance attached to martial values in the Hsia.

The segmented, clearly defined royal quarters of 108,000 square meters were not only well planned and carefully executed, but also protected by a minimal fortified enclosure in the shape of a slightly distorted rectangle.[61] As projected from existing remnants, the eastern wall was originally 378 meters long, the western 359, the southern 295, and the northern 292. The eastern and northern walls were erected on untreated ground, the western and southern on shallow foundation trenches. Being a maximum of 3 meters at the base, 1.8 to 2.3 meters at the top, and 0.75 meter high, the walls were more symbolic than defensive, apparently intended to minimally demark and thus conceptually segregate the royal domain rather than protect it against incursion. Composed simply of reddish brown soil tamped into layers 4 to 12 centimeters thick, they were erected in Erh-li-t'ou's second period and continuously maintained thereafter.[62]

Another similarly constructed, 2-meter-wide, tamped earth wall has recently been discovered about 18 to 19 meters beyond the southern side. Running parallel to the south wall, the 200 meters so far uncovered were constructed on a slightly wider foundation trench and may well have been part of a fully encompassing second enclosure. (A few wall fragments have recently been reported on the western side as well.) Somewhat surprisingly, the wall apparently dates to the middle or later fourth period and thus presumably to after the Shang conquest. Apparently the Shang simply occupied the old royal city and reinforced the perimeter with a typical double enclosure, creating a limited bulwark against the indigenous populace and external inhabitants.

Because capitals have traditionally been understood as defined by massive walls, segmented royal quarters, palatial structures, and ancestral temples, the absence of perimeter walls at Erh-li-t'ou has long prompted questions about the city's role.[63] However, apart from questions about the applicability of this assumption, it was not impossible to have a secure site without them, provided that the area benefited from natural strategic defenses and external strongpoints could be erected to control access routes. The recently discovered citadel at Ta-shih-ku in Hsing-yang, which was probably intended to project Hsia power at the limits of its domain, no doubt falls into this category of forward defenses.[64]

Located about 22 kilometers northwest of Cheng-chou and 13 kilometers south of the confluence of the Chi and Yellow rivers, Ta-shih-ku is roughly 70 kilometers east of Erh-li-t'ou. Mt. Mang lies just to the north, numerous small rivers and streams are found in the general area, and the Luo River now passes through the western part of the enclosure's remnants, but sand deposits and the 30-degree declination of the northwest corner wall suggest it probably once flowed somewhat outside the first walls. In the late Hsia, Ta-shih-ku would have occupied a forward position along the historically projected interaction zone with the nascent Shang cultures in the region between the Yellow and Chi rivers.[65] However, it was too close to Erh-li-t'ou to take seriously current claims that it was King T'ang's preconquest capital of Po.

Excavations have found extensive evidence of four Erh-li-t'ou cultural layers at Ta-shih-ku. The second and third layers show a flourishing site but diminished occupation thereafter, as well as the sudden intrusion of lower Erk-li-kang elements. This suggests that Shang forces probably occupied the city between Erh-li-t'ou's third and fourth periods, coincident with the widespread eclipse of Erh-li-t'ou culture in other statelets and at Teng-hsia-feng and Meng-chuang. Ta-shih-ku's loss must have been the result of a penetrating incursion, one that would have enabled the Shang to subsequently mount a direct strike on Erh-li-t'ou by passing between the Hsia allies of Ku and K'un-wu, one to the southeast, the other somewhat to the northwest.[66] The interval between the city's conquest and the Hsia's final subjugation would therefore have been quite short.

The strongly fortified, rectangular city encompassed roughly 510,000 square meters, about one-eighth the size of Erh-li-t'ou. It was protected by a continuous wall that was further augmented by an external moat that apparently exploited a part of the Suo River's course. (The river flowed along the lower southwestern portion of the southern wall before turning up through the city and then wrapping around well to the outside of the city complex.) Although somewhat varying dimensions have been reported, the overall site appears to have been just under 1,000 meters from east to west and 600 north to south. (A ditch in the north presently extends about 980 meters, and the wall on the south is about 950 meters. However, the last third or so of the western part tapers down in trapezoidal fashion to just 300 meters in extent.)

Even though the walls were apparently undamaged, postconquest the Shang embarked on an extensive building program intended to ensure the bastion's security amid enemy terrain. Key measures included a deep ditch on the interior edge of the original moat that considerably widened it and added the surprise of sudden depth. Although these renovations have obscured the exact width in many places, the original

Hsia moat probably presented a watery expanse some 5 to 9 meters wide, marked by a functional depth of about 3 meters and a very uneven bottom contour about 3.25 meters wide. Subsequent Shang excavations then expanded it right up to the wall's new foundation, for a width of between 13 and 14.55 meters, including a sharply defined innermost portion of 1.5 meters that had a formidable depth of 4 to 6.8 meters.

The original Hsia walls consisted of well-defined layers that had been carefully erected on leveled ground. Though the expanded moat cut into their exterior, overall the Shang's layered additions still augmented them. The remnant walls presently protrude slightly less than a meter above the terrain but probably reached at least 3 meters in antiquity. Marked by the usual trapezoidal shape, they are about 7 meters wide at the top and 13 at the base. Some parts of the walls lie on alluvial sand deposits, others span refilled areas, and the entire configuration conforms to the terrain's features.

Generally composed from seven discernible layers of fairly uniform tamped earth ranging from 4 to 10 centimeters in thickness, the walls utilized a complex mix of soils apparently excavated from different layers of the moat.[67] The interior wall is somewhat unusual in having been erected by overlayering a sharp upward protuberance with a wider section that overcapped it and even extended out onto a new inner foundation in some areas before three outer sections were added. All the walls display a sloping contour indicative of a laddered construction created by staggering small boards with height increases.

RESOURCE CONTROL POINTS

The general thesis that the earliest Chinese states directly emerged as political entities rather than evolving out of suddenly burgeoning economic centers remains generally unchallenged despite the evidence from Hsi Shan. Nevertheless, the work of archaeologists more sensitive to questions of resource acquisition and control is gradually adding a new dimension to the otherwise fairly cohesive picture of how localized Lungshan chiefdoms developed into integrated, incipient states with well-defined economic and ritual centers capable of exercising broad administrative control.[68]

The Hsia and Shang were both agricultural-based economies in which animal rearing played a vital role, but the discovery of metals and development of metallurgical techniques, coupled with the difficulties of maintaining power and the demands imposed by externally oriented warfare, soon saw great emphasis placed on securing such crucial natural resources as copper, tin, and salt. Expansive productive activities had to be undertaken in order for the state's workers to fabricate the requisite luxury, ritual, and military goods in sufficient quantity. Mines, smelting sites, transportation corridors, and fabrication facilities also had to be protected from local brigands and external marauders.

Evidence that the Hsia deliberately subjugated or otherwise colonized peripheral areas along the perimeter of the Yi-Luo river valley to ensure vital mineral production becomes prominent in Erh-li-t'ou's third period, when copper-based metallurgy began to advance in the flourishing capital. The settlements vary in their extent and sophistication, but they clearly display evidence of an imposed presence. Most lay at the confluence of rivers or the mouths of valleys that provide access into nearby mountains and, apart from serving as transportation foci, operated as accumulation points and warehousing centers.

Even though the most extensive and sophisticated production facilities, especially those for ritual objects, were confined to the capital, large-scale craft facilities dedicated to manufacturing objects that fully utilized the indigenous resources are often found. Tung-hsia-feng and P'an-lung-ch'eng, two well-known sites that would have enhanced importance in the Shang, are prime examples. Both served as transport and production centers during the Hsia, though P'an-lung-ch'eng lacked walls in the Hsia.

The fortified town of **Tung-hsia-feng** in southern Shanxi controlled access to the somewhat dispersed mineral deposits found in the Chung-t'iao mountain area, primarily copper with important trace metals intermixed, including tin and zinc.[69] Coincident with Erh-li-t'ou's initial fluorescence, local Tung-hsia-feng graves suddenly begin to show the presence of definitive Erh-li-t'ou products, especially ceramics. Molds for identically styled arrowheads and axes have also been recovered, and a shift in the divination media from pig to oxen bones mirrored a change that occurred in Erh-li-t'ou's second period. Although these items are insufficient evidence in themselves, the swift, no doubt militarily based imposition of Hsia forces is confirmed by a sudden flourish of defensive activity.

Two basically concentric ditches 130 meters and 150 meters in diameter, separated by 5.5–12.3 meters, have been excavated. The interior ditch is approximately 5 to 6 meters wide, but the exterior only 2.8 to 4 meters. Precisely constructed laddered sides and depths of about 3 meters characterize both ditches. A dramatic increase is seen in the number of arrowheads attributable to this period and a concomitant shift to their being fabricated mainly from bronze rather than bone. Taken together, the evidence suggests that Tung-hsia-feng had been deliberately transformed into a military center designed to ensure the acquisition and production of the metals increasingly deemed vital to the ruling clan's prosperity and dominance.[70]

5.

WARFARE
IN THE
HSIA

◆‑◆‑━━━━━━━━━━━━━━━━━━━━━━━━━━━━━━━◆‑◆

D ESPITE THE DISCOVERY of numerous Lungshan and Erh-li-t'ou sites, the rise of the Hsia—variously labeled China's earliest "monarchy" and its first slave society—remains an enigma. How one clan group, part of a widely disseminated culture, evolved until it could eclipse other cultures and politically dominate the many clans, tribes, and proto-states about it to a significant extent remains the subject of divisive speculation. A comparative examination of the cultural artifacts so far recovered within the context of China's traditional historical records suggests that several important military events mark the Hsia's history: the early subjugation of nearby peoples, especially the San Miao, in protracted conflict; infighting over the royal succession, resulting in a civil war; conflict with the Tung or Eastern Yi and Han Chuo's subsequent seizure of power; Shao-k'ang's dramatic restoration of the ruling house following a lengthy period of preparation; inexorable decline through decades of luxuriant peace; and finally immersion into the much-excoriated royal excesses that presumably allowed, as well as supposedly justified, the Shang to overthrow the "tyrannical" and "perverse" Hsia.

Unfortunately, these episodes have to be reconstructed by winnowing out and harmonizing numerous fragmentary glimpses preserved in such traditional records as the *Chu-shu Chi-nien* (*Bamboo Annals*), *Tso Chuan*, and *Shih Chi*, just as scholars have done over untold centuries. Even though many passages are dubious, others spurious, and some outright forgeries, they retain important secondary value in revealing how China has traditionally understood its history, as well as in grounding concepts that remain an ineradicable part of contemporary general consciousness.

Nevertheless, obviously mythical events such as Yi shooting down nine extra suns and claims that lack any archaeological basis, such as the chariot's much ballyhooed employment in Hsia warfare, though necessarily noted, must ultimately be rejected. Succinctly stated, combat in this period was conducted by men on foot, in loosely organized forces of limited strength, almost entirely with bows and arrows and crushing weapons such as axes, clubs, dagger-axes, and a few spears (but not swords) primarily fabricated from stone rather than metal.

Contrary to his image as a selfless individual who labored unceasingly on behalf of all the people, destroyed harmful weapons, and tore down his father's walls to stifle rebellious discontent, Yü also quashed the San Miao and consolidated the clan's power by reputedly executing the lord of Fang-feng for arriving late at a conclave.[1] Moreover, he is traditionally credited with wielding bronze weapons and closely associated with metallurgy in general, the inception of the famous nine great cauldrons (for the Nine Provinces that he supposedly demarked) symbolic of legitimate rule sometimes being attributed to him. However, although metallurgy certainly developed in the Hsia, whether indigenously or through the importation of techniques via the steppe, in his era the nascent capabilities would have been limited to hammering out small knives.[2]

The Hsia's conflict with the San Miao, essentially a century-long process even though it is usually identified with Yü's reign, is well attested by the pattern of recovered artifacts.[3] The San Miao were not just defeated but virtually extinguished; the few survivors scattered and their culture in its manifestation as Shih-chia-ho during the era of Yao, Hsun, and Yü in the Tung-t'ing and P'o-yang lake areas simply vanished. Rather than a gradual amalgamation, their definitive ceramic styles and totems were abruptly replaced in the late twenty-third century BCE in southern Hubei and Henan by late Lungshan artifacts. In fact, rather than simple tribal animosity, a quest for material goods, or a concerted effort to seize prisoners to impress as slaves or employ as sacrificial victims, their strongly dissimilar religious practices and totems may have caused their ongoing clash.[4] Furthermore, unlike the Tung Yi, who displayed an increasing affinity with Hsia culture, after the conquest the fragmented and relatively isolated San Miao groups that survived retained their distinctiveness,[5] another indication of a fundamental clash in customs and views, not to mention political domains.

Animosity had marked relations between the Hsia and San Miao as early as their precursor cultures, late Yangshao and early Lungshan for the nascent Hsia and Ch'ü-chia-ling for the San Miao. The San Miao had been closely allied with the Tung Yi (Eastern Yi) of the Ta-wen-k'o phase, with whom they shared a number of cultural traits, including bird totems, and had cooperated with them in fielding coalition-type forces against late Lungshan antagonists, preventing the latter from dominating either of them.[6] It has even been suggested that the San Miao may have originally numbered among the Tung Yi before they were expelled because of their identification with Ch'ih Yu.[7] However, Tung Yi culture was itself gradually displaced by, or evolved into, late Shandong Lungshan and thus came to differ significantly from the San Miao's later manifestation as Shih-chia-ho, inclining the former to identify more with the Hsia even though they had to forcefully oppose Hsia domination attempts. Furthermore, although the Tung Yi no longer supported the San Miao, they remained a significant power throughout the Hsia dynasty, and even the Shang felt compelled not just to neutralize them before attacking Chieh, the last Hsia monarch, but also to gain their support.

Materially and economically, the San Miao and pre-Hsia had also diverged with the passage of time. Perhaps due to their increased mastery of the rivers and early use of irrigation, the pre-Hsia had significantly expanded their agricultural yields over the centuries encompassed by the late Yangshao and third Wang-wan cultural phases, whereas the San Miao reportedly stagnated. This no doubt allowed them to accumulate the surpluses vital to developing specialized industries, diverting men to military activities, and to nurture sufficient royal clan power to dominate their own people, displacing communal or tribal leadership. (As the classic Chinese military writings would subsequently stress, economic prosperity underpins the possibility of military power.) The early Hsia's increasing dominance from the twenty-third century onward culminated in what has traditionally been encapsulated as Yü's legendary conquest of the San Miao, even though the scope of his actual achievement may be open to question.[8]

Proponents of the pedantically held view that the virtuous, if their "Virtue" is sufficiently perfected, will always prevail over the evil and perverse envision Yü's symbolic conquest of a Miao chieftain as depicted in the *Shang Shu* as irrefutable proof. Despite being a late fabrication, the importance of the section now titled "Great Yü's Plans" to political and military thought from the Wei-Chin period onward, particularly the reasons cited to justify the "punitive" attack, cannot be overemphasized:[9]

Emperor Shun said, "Alas, Yü, only this Miao chieftain refuses to be submissive. Go and mount a campaign of rectification against him." Yü assembled the many lords and then pronounced his intention to the army: "All you masses in firm and martial array, heed my commands. Irascible is this Miao chieftain. Obfuscated and confused, he is not respectful. Insulting and rude, he takes himself to be sagacious but contravenes the Tao and overthrows Virtue. Superior men remain in the wilds, menial men hold office. The people have abandoned him and will not preserve him. Heaven is sending down calamities. Consequently, in company with you many warriors, in accord with my instructions I will attack him for his offenses. Now if you are united in mind and strength, his conquest will be achieved."

When the Miao still contravened his mandate after thirty days, Yi Yih advised Yü: "Only Virtue moves Heaven, there is no distance it does not reach. Arrogance brings about loss but humility receives increase. This is the Tao of Heaven. In the early days when Emperor Shun was living at Mount Li,[10] he went out into the fields and daily called out and cried before compassionate Heaven, and also to his father and mother, accepting responsibility for all his offenses and taking the guilt for unseen evils upon himself. He then respectfully commenced serving Ku Su, his father, appearing grave and fearful until Ku also was fully in accord with him. Since perfect sincerity moves the spirits, how much more will it affect the Miao chieftain!"

Yü bowed in respect of his superlative words and voiced his assent. He then put the army in order and returned it in proper array.[11] Emperor Shun then augmented and bespread his civil Virtue and had dances with shields and feathers performed between the two staircases (in the court). Seventy days later the Miao chieftain came to court.

The idea that cultivating virtue and making a highly visible, righteous display would "magically" affect even a distant Miao chieftain some 2,500 *li* away would subsequently be much quoted in imperial court discussions and memorials whenever nonmartial courses of action were advocated. However, traditional commentators offer the interesting observation that Shun, despite being venerated as one of antiquity's paragons, failed to properly balance the requirements of the civil (*wen*) and martial (*wu*) when he precipitously deputed Yü on this punitive mission against the Miao without first formally announcing their offenses. Since the practice of the civil required adherence to certain idealized forms of behavior and protocols, his premature employment of conspicuous martial power was doomed from the outset.

Yü's inability to cow the Miao chieftain into submitting prompted Yi Yih's recommendation that he rededicate himself to perfecting his sincerity, synonymous with initiating diplomatic measures designed to persuade him to acknowledge the constraints of Hsia political domination. The lack of any reference to combat having arisen suggests that Yü's efforts consisted of a martial display, nothing more. The elaborate dances performed in Shun's own court, presumably under his direction rather than Yü's, have accordingly been interpreted as expressing the emperor's desire to suppress any tendency toward the martial alone, thereby restoring the balance of *wen* or the civil (symbolized by the feathers) and *wu*, here symbolized by a shield rather than an axe, halberd, or bow, the era's primary weapons. However, irrespective of the validity of these creative interpretations, for historical purposes these passages might simply be seen as a remnant memory, one reflecting the intensity and longevity of the clash between various Miao tribes and the Hsia's founders.

Mo-tzu, who exploited accounts of these events for his own persuasive purposes, concocted a dramatically enhanced version:

In antiquity, when the San Miao were causing great disorder, Heaven mandated that they be put to death. Strange apparitions of the sun came out at night, it rained blood for three mornings, dragons spawned in the ancestral temple, and dogs cried in the market. Ice formed in the summer, the earth cracked, springs bubbled up, and the five grains were transformed. The people were greatly shaken.

Emperor Kao Yang issued an edict in the Hsüan Palace and Yü, personally bearing the Heavenly jade pendant of authority, mounted a punitive expedition against the Miao. Lightning and thunder suddenly arose, a spirit with a human face and the body of a bird held a jade tablet to repress the Miao's auspicious power. The Miao forces were thrown into chaos and thereafter fell apart. After Yü conquered the San Miao, he charted the mountains and rivers, discriminated things by upper and lower, and respectfully administered the extremities. Neither spirits nor men contravened his rule, All under Heaven were tranquil. This is the way in which Yü waged his campaign against the Miao.[12]

Other writings suggest that Yao not only deputed Yü to smite the San Miao but had earlier personally led a punitive expedition against the rebellious San Miao dwelling among the Southern Man, defeating them in a clash on the banks of the Chou River and compelling them to withdraw and change their customs.[13] Similarly, the *Shang Shu* states that Yao "expelled the San Miao out to San-wei," the latter synonymous with the westernmost point conceivable, one subsequently identified with a location south of the now famous Tunhuang,[14] and deputed his son Chu to control the region around Chou-shui.[15]

Archaeologists have recently suggested that evidence for these earlier clashes with the San Miao may also be seen in the sudden extinction of the Shih-chia-ho variant known as the Ch'ing-long-ch'üan in the region of Yao's supposed activity, the Tan-chiang and T'ang-pai-ho river basins.[16] Throughout its area of distribution in the upper and middle Han River, middle and lower Tan-chiang, the T'ang-pai-ho river basin, and upper Huai River area, this phase—considered the northernmost expression of the Shih-chia-ho—was displaced by localized cultures, presumably as a result of Yao's conquest and ensuing forced emigration. The radiocarbon-based dates obtained for this displacement fall between 2600 and 2150 BCE, roughly in the centuries ascribed to Yao and Yü, suggesting that although the campaign was attributed to Yao, he merely personifies an evolving cultural conflict. Thereafter, Yü's activities would necessarily have been directed to the south, to the plains of Hsi-hu, the other regions having already been cleared of a tenacious enemy.

According to accounts that contradict the peaceful transfer of authority from Yü to his son Ch'i, the dynasty's first significant military conflict was actually a battle over succession. In these versions Ch'i eventually vanquishes Yi, a meritorious official who had become famous for his efforts to reduce the people's misery during Shun's reign and been personally designated by Yü to succeed him. Ironically (since he similarly embarked on the path of indulgence shortly after securing the throne), Ch'i's attack was supposedly justified by Yi's dissolute and negligent behavior. These events apparently engendered the idea that Yü never designated Ch'i as his successor, but instead forced his son to fight for the throne by somehow predestining the conflict, even though the *Bamboo Annals* note that Yi withdrew to his own state while Ch'i went about convening the lords and was much honored by Ch'i when he finally died in the latter's sixth year of reign.

Irrespective of its lack of veracity, this account may be understood as symptomatic of unremitting strife within the ruling clan, not to mention among the various extended clans and other tribes or peoples populating the area at the time. Moreover, it apparently presaged centuries of forthcoming strife between the Hsia and eastern powers, since it has been suggested that Duke Yi's clan originated in the east and was numbered among the Tung Yi.[17]

The Hsia's next battle quickly arose when Ch'i, presumably in command of core clan forces, confronted the Yu-hu—theoretically members of Shun's own clan—because they had either rebelled or simply refused to acknowledge his sovereignty. Hsia forces prevailed (at an engagement made famous by the *Shang Shu* chapter "The Oath at Kan") and the Yu-hu submitted:[18]

> You men who have charge of the six affairs, I solemnly proclaim that Yu-hu-shih has forcefully contumed the five elements (phases) and neglectfully abandoned the three (principles of) uprightness.[19] Heaven is about to employ force to end his mandate.[20] Today I am just respectfully implementing Heaven's punishment.
>
> If the left does not attack on the left you will disrespect my edict; if the right does not attack on the right, you will disrespect my edict. If you charioteers do not keep your horses under control, you will disrespect my edict.

Those who follow my edict will be rewarded at the ancestral temple; those who do not follow my edict will be exterminated at the altar to earth, and I will extinguish your entire family.

The oath falls within the martial tradition of boldly announcing the enemy's offenses to justify imminent military action, thereby inciting fervor in the troops. However, it is somewhat at variance with comments in the *Ssu-ma Fa*: "The rulers of the Hsia dynasty administered their oaths amidst the army for they wanted the people to first complete their thoughts. They rectified their Virtue, never employed the sharp blades of their weapons, and granted rewards but did not impose punishments."[21]

It has long been claimed that Ch'i's oath proves the existence of chariots manned by a driver, an archer, and a shock-weapons bearer at this battle, confirming implications derived from a listing of the *ch'e cheng* or "chariot commandant" among the Hsia's officials. Given that Hsi Chung, known to history as Yü's chariot driver, is credited with inventing the chariot (or at least excelling at constructing chariots), if the oath were at all authentic this would be plausible. However, in the absence of chariot remnants or traces of rotting wood in any excavated late Lungshan, Erh-li-t'ou, or even early Shang site, such claims can only be cast aside as unwarranted.[22]

According to the *Bamboo Annals*, Yü had previously faced dissension within his immediate family, being compelled in his eleventh year to banish his youngest son, Wu Kuan (whose name ironically means "martial observations"), to the Hsi Ho (Western Ho) region, an area west of the Yellow River often noted in later history for its natural strategic advantages. Wu Kuan apparently flourished in this conducive area; he revolted four years later and the king, rather than commanding in person, deputed Duke Shou of Ts'ai to lead the punitive expedition that eventually forced Wu Kuan to submit.[23]

Not unexpectedly, Ch'i's death precipitated internal chaos. His five sons battled for the throne, and T'ai-k'ang, who eventually succeeded him, apparently followed his dissolute ways. T'ai-k'ang's excesses in turn provided an opportunity for the Yu-ch'iung of the Eastern Yi, under Hou Yi, to move westward out of the lower Yellow and Huai river valleys, attack the capital, and eventually occupy it. Little is known about the battle, but probably as the result of conflating myth and reality, tradition identifies Hou Yi as a great archer who followed in the footsteps of his ancestors, chief archery officials under Yao and Shun, or as Duke Yi, just noted for his confrontation with Ch'i, who was also credited with shooting down the nine extra suns incinerating the earth, as well as taming the poisonous animals.

Discounting the mythic aspects, Hou Yi's pronounced skill in archery probably symbolized the Eastern Yi's greater dependence on hunting and fishing in contrast to the more agriculturally oriented Hsia. Thus the Yu-ch'iung's deadly proficiency reputedly forced Shao-k'ang's son Chu, who eventually completed the lineage's restoration, to fabricate China's first body armor so as to survive an onslaught of arrows before engaging in close combat. (He is also credited with inventing the spear, a rather startling and unlikely claim given that spears and javelins are normally among the first weapons developed in any culture, but possibly indicative of Chu being the first warrior to bind a bronze, rather than stone, spearhead to a wooden shaft.)

Remarkably, the victorious Huo Yi is also portrayed as having been quickly seduced by royal pleasures and thus lost his own head when Han Chuo, whom he had foolishly entrusted with the actual administration, gradually developed a power base sufficient to usurp the throne:

In antiquity when the Hsia had begun to decline, Huo Yi moved from Hsü to Ch'iung-shih[24] after which he displaced the Hsia government by exploiting [the disaffection of] its people. Thereafter, relying upon his skill in archery, he did not attend to the people's affairs but dissolutely immersed himself in hunting out on the plains. He dismissed [his old meritorious officials] Wu Luo, Po Yin, Hsiung K'un, and Mang Yü, and instead employed Han Chuo.

Han Chuo, the vituperative heir of the Po-ming clan, had been exiled by its leader, the duke of Han. However, Duke Yi of the Yi peoples brought him in, trusted, and employed him as his chief minister. Chuo ingratiated himself in the inner palace and bribed those (in the government) outside it. He kept the people ignorant and manipulated them while encouraging Yi's indulgence in hunting. He planted his deceptions and evil in order to seize the government, and those both inside and outside all submitted. As Yi still did not awaken, when he

WARFARE IN THE HSIA

was about to return from his hunt, his family troops[25] killed, cooked, and fed him to his sons. Since they could not bear to eat him, his sons all died before the gates of Ch'iung. Thereupon (the high official) Mi fled to the Yu-ko people.

Thereafter Chuo had two sons by Yi's wives, Ao[26] and Yi. He relied upon slander and evil, prevarication and artifice, displayed no virtue to the people, and had Ao employ the army to extinguish the Chen-kuan and Chen-hsün (who had earlier provided refuge to Hsiang). He had Ao occupy Kuo and Yi occupy Ko. However, Mi came back from the Yu-ko to assemble the remnants of the Chen tribes, exterminate Chuo, and establish Shao-k'ang as ruler. Shao-k'ang exterminated Ao at Kuo and his son Chu exterminated Yi at Ko. Thereafter the Yu-ch'iung clan vanished.[27]

The severity and barbarous consequences of Han Chuo's subversive actions, uncommonly attributed to antique periods, seem to reflect the excesses to which early ethnic or totemic clashes were apparently prone, as well as prefiguring those seen in later millenarian clashes and witnessed in the Cultural Revolution. Moreover, the ready adoption of Yi's wives, an action symbolic of conquest that is further attested by the birth of two sons, suggests that his policy of currying favor within the inner palace may have penetrated far into Yi's personal quarters, affecting his consorts as well as the personal retainers who murdered him.

Han Chuo's attempts to exterminate the royal line's remnants set the stage for the marvelous tale of Shao-k'ang's survival, cleverness, and endurance:

In antiquity Ao slew the Chen-kuan and then attacked the Chen-hsün in order to exterminate Emperor Hsiang of the Hsia. The emperor's pregnant consort Min escaped through a hole in the wall and returned to her clan, the Yu-jen, where she eventually gave birth to Shao-k'ang. Shao-k'ang became chief shepherd among the Yu-jen and, fearing Ao's capability, made preparations against him. Ao had Chiao search for him, so Shao-k'ang then fled to the Yu-yü where he became chief cook in order to eliminate this threat. The chief of the Yü gave him two of his daughters surnamed Yao as wives and a fortified town at

Lun together with one *ch'eng* of fields and one *lü* of troops so that he might manifest his Virtue and initiate his plans. He then gathered in the Hsia's masses, expanded his officials, and deputed Nü Ai to spy on Ao and his youngest son Chu to entice Yi. Subsequently he vanquished Ao and Yi and restored Yü's heritage.[28]

Because of the differences in Hsia and Yi customs, the lengthy eight-decade interval during which the Hsia lost much, if not all, control of its territory may also be envisioned as a clash of peoples and cultures. Relations between the Yi tribal groups and the Hsia obviously waxed and waned over the dynasty's four centuries depending on their relative power, just as those between the Shang and its contemporary states did in the next period. Peace, however coercive, essentially prevailed when the Hsia was strong and the Yi fragmented and thus coerced or forced into submission, but conflict arose when both were strong or the Hsia wavered, allowing incursions and rebellion.

The Yi, also called the Tung Yi and Nine Yi, encompassed nine main tribes reportedly composed of eighty-one clans: the Ch'üan,[29] Yu, Fang (Square), Huang (Yellow), Pai (White), Ch'ih (Red), Hsüan (Dark), Feng (Wind), and Yang (Sun). There are also frequent references to the Huai Yi, a term that probably encompassed populations from two or more of the Nine Yi who happened to dwell in the fertile Huai river valley. Although they may have been individually incapable of withstanding Hsia supremacy, when allied they constituted the only power capable of challenging Hsia dominance and, in aggregate, certainly exceeded the Hsia in strength and extent.

Hou Yi, who initiated the revolt, was affiliated with a Yi clan group known as the Yu-ch'iung, and later traditions assert that "four of the Yi rebelled against the Hsia when T'ai-k'ang lost the state." The *Bamboo Annals* seemingly confirm this by chronicling three additional encounters during Emperor Hsiang's tortured reign. In his initial year, despite his much compromised and largely powerless position as a refugee in Shang-ch'iu (which intermittently served as the site of the incipient Shang state), he still attacked the Huai Yi with the prince of Shang's support, no doubt to damage Hou Yi's base of support. Remarkably, the next year he conducted "punitive expeditions" against the Feng Yi and Huang Yi, no doubt to the east and southeast.

Surprisingly, some five years later the Yu Yi visited as submissive guests, apparently signifying their temporary elimination as an immediate threat. However, only after Shao-k'ang's son Chu conducted major expeditions in his eighth year as far as the Eastern Sea and San-shou did the Nine Yi essentially return to the fold.[30] Thereafter, they are not mentioned until the end of the dynasty, when the next-to-last ruler, Emperor Fa, wisely strengthened the capital's walls after various Yi reportedly came as submissive guests. Finally, in Emperor Chieh's third year, the Ch'üan Yi entered Ch'i (where the Chou would later originate) and rebelled. However, Mo-tzu, who esteemed the Hsia, notes in addition that Emperor Yü died on the road after mobilizing his forces and venturing east to attack the Nine Yi, suggesting the dynasty commenced and concluded in violent entanglements with the Yi.[31]

Relative peace punctuated by occasional campaigns against peoples such as the Huai, Ch'üan, Huang, San Shou, and Chiu Yüan, all reportedly members of the Yi, prevailed for some ten generations after Shao-k'ang and Chu restored royal power.[32] However, in the Hsia's final moments Chieh apparently pursued a policy of strong military activity that harmed the people, including a campaign intended to subjugate the recalcitrant Min-shan, a Yi proto-state, which ultimately undermined the state's ability to defeat foreign incursions as well as further antagonizing the Tung Yi.[33] Eventually the Shang revolted, destroyed the ruling house, occupied the state, and dispersed the populace, who then may have become the ancestors of the Chou and various steppe peoples, including eventually the Hsiung-nu.[34]

POLITICAL ORGANIZATION AND MILITARY STRUCTURE

Shang Shu and *Shih Chi* chapters describe a Hsia organizational structure that presumably integrated various Hsia peoples and submissive entities, whether tribal groups or proto-states, in a series of expanding rectangles within which the acknowledged degree of submission correspondingly declined.[35] Naturally this schematized characterization has long been justifiably disparaged as a late, highly idealized portrait, just as the conflicts already discussed would require. However, it certainly expresses the underlying reality that those entities closest to the royal capital, whether clan members or subordinate proto-states, formally enfeoffed or not, would have been the most susceptible to military coercion.

Hsia power projection capabilities would have correspondingly decreased with increasing distance from the core, inevitably allowing greater independence and self-assertion. Moreover, because the degree of dominance, though not presently ascertainable, would have depended on factors such as accessibility and logistical support, it could not have been uniform even throughout the generally acknowledged realm of maximum influence in the Yi and Luo river area or the irregular projections out to the west and down to Pan-lung-ch'eng.

The territory within the immediate vicinity of the capital, said to have encompassed 500 *li* in all directions, forming a square 1,000 *li* on each edge, was probably under the king's direct control, since it is termed *tien* and was primarily responsible for providing the foodstuffs.[36] The next 500 *li*, depicted as the region of the feudal lords, has been traditionally understood as including the ministerial families within the first hundred, key Hsia supporters in the next hundred, and some larger states with their capitals in the remaining 300 *li*.

The martial purposes of the next or third region, theoretically running from 1,000 to 1,500 *li* out, merit note. Divided into just two zones, the first 300 *li* were conceived as a transitional zone, the last civilized bulwark against the uncultured challenges threatening from without, where *wen* should be emphasized. The remaining 200 *li* in this schema were to be devoted to "flourishing martial values (*wu*) and defensive activities" and must have comprised a border that differentiated Hsia terrain from hostile peoples in the two outermost regions designated as "*yao-fu*" and "*huang-fu*," meaning something like "essential submission" and "barren submission," or more realistically "barely submitting" and "the wilds," respectively.

Particularly interesting within the context of Hsia–Yi conflict is the demarcation of the first 300 *li* of the *yao-fu* as the Yi's abode, thereby locating them within a ring 1,500 to 1,800 *li* from the Hsia capital. The outer 200 *li* of this fourth region were reserved for exiles, but the wastes of the immediately outer region, the first 300 *li* of the fifth region or *huang-fu*, were supposedly populated by the very uncivilized Man barbarians. The most distant 200 *li* were then designated as a repository for serious offenders banished beyond the pale of all society, whether

Hsia or foreign. Although this highly idealized delineation reflects later conceptualizations of harmonized order, it not only informed the traditional understanding of Hsia political organization, but was also an often advocated theoretical model for Imperial China's foreign relations from the Han onward.[37]

It has long been asserted that the incipient origins of many bureaucratic organs can be traced back to the Hsia and that a sort of proto-bureaucracy is already discernible.[38] Although important individuals could simply have been deputed to undertake specific responsibilities as needed, administering the realm would no doubt have required minimally defined bureaucratic positions with at least rudimentary authority. The early rulers presumably established a limited but basically effective core of officials under their direct control, a sort of working staff, with responsibility for crucial activities being broadly construed. Titles such as *ssu t'u* ("minister of agriculture") and "chief archer" then presumably evolved from repeatedly assigned tasks.

The *Li Chi*, a late Warring States ritual text, claims that the Hsia had a hundred officials (*pai-kuan*), obviously a nominal figure for a group of proto-officials whose responsibilities were probably not sharply differentiated. Among those identified are the *ssu fu* ("four supporters"), *liu ch'ing* ("six ministers"), *ssu cheng* ("four rectifiers"), *ssu t'u* ("minister for labor"), and *t'a-shih*, a functionary who supported, advised, and supposedly criticized the ruler.[39] The Hsia also reputedly had an official to oversee the chariots, a post initially held by Hsi Chung, but the absence of any evidence for chariots, carts, or even simple wagons makes its existence highly unlikely. Finally, the *chou shih* (who clearly was not a military functionary despite the implications of the term *shih*, which eventually came to signify "commander" or "general") was responsible for the segmented administrative area known as a *chou*.

It has generally been claimed that the Hsia and its immediate predecessors were predominantly civil-oriented societies whose members acted as warriors whenever necessary, but reverted to their roles as farmers or artisans with the conflict's termination.[40] Insofar as this interpretation is based on the absence of symbols of power and martial achievement in graves from the period, any sudden efflorescence of weapons, especially when coupled with clear evidence of violence such as crushing blows and arrowheads embedded in bones, must mark a transition to esteeming military prowess. These indicators began to appear in the mid- to late Lungshan, great axes symbolic of both punitive and military power suddenly being found in ever-increasing numbers.

Assuming power at the end of the Lungshan, the Hsia may be expected to have confronted a variety of martial challenges on an irregular basis. If it was ruled by a chieftain and the ruler's clan dominated, the extended tribe would have been compelled to participate in combat.[41] According to later historical reports, disgruntled clans occasionally defected and even mounted physical challenges to the leadership, compounding the martial challenges. Rather than being philosopher kings in a peaceful land, Hsia rulers almost certainly relied on charisma, personal prowess, clan connections, and martial skill to prosper amid tribal infighting and external challenges. Even though most military historians confidently assert that the Hsia did not maintain a standing army,[42] it would be highly unlikely for the ruler not to have been protected by a body of men with pronounced martial abilities who would form the core of any broader combat effort. However, the degree to which the martial dominated the civil remains unknown, though suggestions that the Hsia emerged because it embraced warrior culture seem more likely to accurately characterize the historical situation.

Whatever the size and organization of the Hsia's military forces, some sort of training in the era's weapons and coordinated action would have been required to field an effective contingent composed of highly individualistic warriors. Although hunting provided an opportunity to practice group action, and weapons training probably was undertaken within the family as part of normal upbringing, it has been suggested that the Hsia established organized schools (*hsiao*) dedicated to archery instruction, initiating an institution that would continue into the Shang (with the *hsü*) and Chou (with the *hsiang*).[43] Even though archery may not have been any more difficult to master than effectively wielding a dagger-axe, this thrust would accord with traditional China's esteem for archery's complexity. These early "schools" or training academies need not have been particularly formalized to have played a significant role, one perhaps understandable in the context of the traditional view that they initiated education in China.

If the proto-bureaucracy cannot be known, the primarily ad hoc military structure is even more uncertain. However, it is generally believed that all administrative positions, whether preassigned or delegated in exigencies, were dual in nature: anyone participating in clan power, already being a person of significance, would be expected to perform military functions. The initial words of the "Pronouncement at Kan"—"you men who have charge of the six affairs"—have prompted claims that the Hsia already had six armies because the term translated as "six affairs" is understood by commentators as "six armies." However, this interpretation lacks justification, and even if they existed, probably being responsible for the six main administrative affairs of government, they would have simply been co-opted to act as field officers or commanders.

Ch'i's harangue also indicates the existence of left and right forces: "If the left does not attack on the left you will disrespect my edict; if the right does not attack on the right, you will disrespect my edict." Despite assertions to the contrary, rather than evidence for well-formed, integrated units such as regiments or armies, this more likely indicates an operational segmentation into left and right flanks. However, the edict certainly shows a presumption of the power to enforce military discipline and a willingness to execute the disobedient in untold numbers.

One other early passage has given rise to claims that the Hsia already had a structured military organization. The famous *Tso Chuan* account previously cited contains a frequently noted sentence indicating that the chief of the Yu-yü gave Shao-k'ang two of his daughters as wives, a town at Lun, one *ch'eng* of land, and one *lü* (regiment) from the masses. Depending on how the original text is construed, *lü* is understood as referring to a military unit that (according to very unreliable statements in the *Chou Li*) traditionally numbered 500 men. However, even if the Hsia's military organization were founded upon the pyramid of fives frequently adopted in the Spring and Autumn and Warring States periods, given the smaller population base and more limited resources, the *lü* may have only amounted to 125 men. Moreover, if based on the decade system as subsequently employed by several peripheral states during the Chou's decline, it may have been a mere 100 men, the equivalent of a Warring States company, particularly if the squad/platoon/company structure had not yet evolved and it was a clan-based aggregation. However, this sentence is frequently cited as evidence that non-clan military forces already existed whose members were presumably drawn from the farmers or other workers who could be summoned for military service as necessary.[44]

6.

THE
SHANG
DYNASTY

◆‧◆——————————————————◆‧◆

W HEN ATTEMPTING TO reconstruct the history of the Shang one
question looms large: how to evaluate and employ the traditional accounts and seemingly precise geographical statements scattered throughout various Spring and Autumn and Warring States texts,
which, if actually based on now-lost records, may preserve vital information about the Shang. Many modern scholars simply reject all non-
archaeological material, but centuries of incisive reading have produced
a detailed portrait of the Shang well worth pondering, a starting point
to be proven or disproven by the many artifacts and numerous bamboo
strips that continue to be recovered. This traditional account has not
only influenced numerous generations of Chinese, but also continues
to furnish core material for discussions in contemporary PRC media
and China's ongoing search for uniqueness. Whatever its reliability,
cultural inertia will probably ensure that this account persists for several more decades, making the original materials worth pondering
in themselves.

Despite unbroken reverence for the Hsia and the recent discovery
of probable Hsia sites, most scholars continue to recognize the Shang
as the first Chinese dynastic state because it remains the earliest to be
documented by archaeologically recovered textual materials, many of
which attest to the veracity of fundamental elements in Ssu-ma Ch'ien's
Shih Chi account. Slightly more than a century ago the now-famous
"dragon" (oracle) bones were discovered; Anyang, site of the Shang's
last capital, was first explored just prior to World War II; and several

major Shang enclaves, not to mention numerous smaller sites among the many hundreds already identified, have been partially excavated over the past five decades.

Apart from excavation reports that have been torturously slow in being published due to PRC academic and political complexities, oracle bones provide the chief resource for reconstructing the life and nature of the Shang. Constantly increasing but currently numbering about 200,000, a quarter of them have been deemed reasonably informative. Unfortunately, they only cover the last nine kings from Wu Ting to Hsin (best known as the tyrant Chou), who ruled from Anyang during the dynasty's final two centuries.[1] Cryptic notes on inquiries made to Heaven and invocations to the ancestors on a wide variety of state and royal matters ranging from military campaigns to prospects for the harvest, these prognostications inevitably express the ruler's perspective, are naturally selective, and were probably never intended to be preserved. Nevertheless, the extensive work, ranging from the necessarily tedious to the brilliantly insightful, completed since their discovery has now provided sufficient material to construct a tentative picture of the Shang.[2] In addition, the weapons, cauldrons, ritual jades, ceramics, and other items that have been recovered from tombs, hoards, and storage pits vivify cultural and psychological dimensions merely hinted at in the oracle bones as well as concretely documenting the process of technological evolution.

Although the exact nature of the Shang, even their very name,[3] remains a matter for debate, characterizations range from a large chiefdom of heterogeneous composition, through an increasingly bureaucratic territorial state, to a despotic monarchy.[4] Within China it has traditionally been regarded as a dynasty because the rulers succeeded each other and the clan maintained its authority, but it apparently began as a strong tribal chiefdom or self-contained clan state in a circumscribed location. Moreover, as the oracle bones reveal, it remained one entity among many throughout its rule, never able to completely dominate the ancient world in the fashion traditionally depicted. However, the Shang early on evolved a still imperfectly glimpsed, ad hoc administrative structure staffed by royal clan members and close subordinates of the king, who were entrusted with particular tasks for specific times that proved capable of dealing with such focal issues as accepting tribute, opening up lands, organizing the hunt, and directing military campaigns.

The Shang reputedly moved their capital five times after conquering the Hsia. Aided by radiocarbon and other dating techniques, scholars have assiduously sought to identify archaeologically excavated sites with the Shang capitals discussed in such late written materials as the *Bamboo Annals* and *Shih Chi*, coincidentally attesting to the antiquity and continuity of Chinese civilization. After numerous articles and hundreds of pages of detailed and often highly prejudiced argumentation, a probable sequence can perhaps be hazarded for the various capitals that shows the evolving geostrategic situation as the Shang expanded and contracted, flourished and declined.

The earliest bastion was Yen-shih, just six kilometers from Erh-li-t'ou, the final capital the sprawling ritual and administrative complex at Anyang, with the most important intermediate site being Cheng-chou. These massive, expensive, and no doubt organizationally difficult shifts were probably undertaken to impose stronger political control or escape military pressure, but may also have stemmed from material causes, including the diminishing productivity of nearby fields, depletion of conveniently located copper veins, or religious reasons.[5] However, the Shang maintained its identity throughout the period and interacted with the proto-states and tribal peoples about it in a variety of ways as the dynasty continued to evolve, subjugating some, forming alliances with others, and treating the rest indifferently or antagonistically.

Throughout the Shang's existence the king was a powerful, essentially theocratic chief whose personal charisma and authority underlay and sustained the ruling structure.[6] He apparently devoted great effort to performing divination and conducting sacrificial rituals to the ancestors and Ti, a still imperfectly understood deity on high, accounting for the tremendous number of ritual bronzes employed in court and religious life and the astonishing number of human victims—prisoners of war as well as subordinate groups—slain in their performance.[7] In fact, contrary to the idealized depictions concocted by later Confucian literati emphasizing that the Shang cultivated Virtue and implemented humanitarian policies among the populace, whatever their quasi-religious justifications, the royal elite were extremely brutal in asserting and maintaining power. Their sites are littered with the tortured, often decapitated bodies of victims, sometimes singly, others in large groups, apparently intended to guarantee the auspiciousness of a building, solidity of a wall, performance of some rite, or protection of a tomb. Martial aspects often dominated in the latter, with at least some of the figures being expected to eternally stand guard even though family members, retainers, servants, and other unfortunates also accompanied the deceased in the afterlife.[8]

TRADITIONAL ACCOUNT OF THE SHANG'S RISE

The Shang has traditionally been depicted as compelled to vanquish the Hsia because the last ruler's perversity adversely impacted the people, the only acceptable justification for overthrowing an acknowledged ruler. In contrast to the peaceful transitions exemplified by Yao yielding to Shun and Shun to Yü, the succession was achieved through military conquest. However, apart from any questions about material culture, two imponderables persist: How could an unknown clan or tribal group develop the military prowess and organizational structure necessary to gain victory over a presumably well-entrenched entity, however circum-scribed its power, and what were the military dimensions of the process? Two focal issues in turn significantly impact any understanding of the conquest: the location of the Shang administrative center when they commenced whittling away the Hsia's domain and the degree to which the eight early capitals might have provided residual support after having been abandoned.

Despite a myriad of vestiges in literary and archaeological records, perhaps because of a dearth of reliable material and the sometimes prevalent perception that they are of little consequence, the Shang's in-cipient beginnings remain obscure and its early history nebulous. More-over, in comparison with the Chou conquest of the Shang, an event that spawned centuries of intense speculation and continues to stimulate academic debate, the Hsia's overthrow has aroused but minor interest despite important questions having been raised about the Shang's origins; their relationship with the Hsia, Yi, and Chou cultures; the evolution of their metalworking techniques; and the administrative structure and methods that permitted them to dominate a considerable area.[9]

In the absence of adequate archaeological materials, the main events in the Shang's ascension have long been gleaned from historical writings of varying antiquity, ranging from the *Shang Shu* (*Shang Documents*) and *Shih Ching* (*Classic of Odes*) through the *Yi Chou-shu* (*Lost Chou Documents*).[10] Extrapolated, they yield a sequence of indeterminate reliability that has historically been accepted in China and therefore retains inestimable value for understanding Chinese self-perception as well as subsequent military thought. Moreover, despite contemporary skepticism and a thoroughgoing emphasis on archaeological materials, creative analysts continue to derive imaginative theories of origin solely from these traditional written materials and by demythologizing various ancestral legends.[11]

The *Shih Chi*'s "Yin Pen-chi" ("Basic Annals of Yin"), which was unquestionably accepted as providing a true account of the dynasty for nearly two millennia, indicates that the clan's progenitor Ch'i assisted Yü in successfully taming the unruly rivers. For his efforts Emperor Shun appointed him Minister of Works and enfeoffed him at Shang, from where the clan presumably derived its name. The final shift of the ritual and administrative center to Anyang where Yin-hsü—the "wastes" or remnants of Yin—is located gave rise to the Shang being referred to as "Yin" from the Chou onward. However, as attested by oracle bones and contemporary inscriptions, members of the Shang never referred to themselves by this name.[12]

Speculation about the Shang's namesake has long centered on a location in northern Henan near the Shandong border now known as Shang-ch'iu, the "mound" or "hillock" of Shang.[13] As confirmation, however dubious, it is often noted that this area was once the site of the ancient state of Sung, where the vanquished Shang populace were allowed to maintain a vestigial state.[14] Although the reverse—that Shang-ch'iu may have derived its name from the proto-Shang people having moved there—seems equally possible, it also claimed that they retained Shang-ch'iu as their capital even after King T'ang's conquest, wherever their administrative and military centers were. However, apart from being nebulous and unsubstantiated, given Yen-shih and Cheng-chou's siting, advanced development, and opulence, it seems extremely unlikely that it would have somehow retained a role as either the functional or ritual capital during the state's initial period of fluorescence.

It is variously theorized that the Shang, as manifest as the Erh-li-kang phase, developed out of Henan or Shandong Lungshan culture through various intermediates; originated among the Eastern Yi or Northern Jung; arose in the west or even south; or evolved out of the same cultural background as the Hsia, just as claims that they were descended from Yao basically assert.[15] However, the most frequently espoused view asserts that the Shang were primarily an eastern people whose name derived from an early location, whether simply a place-name or some sort of recognized fief.[16] Formulated by Wang Kuo-wei, Tung Tso-pin, Kuo Mo-jo, and other giants of ancient studies in the past century on the basis of traditional literary sources, this widely accepted theory still lacks archaeological substantiation. Moreover, none may be forthcoming, because any artifacts remaining from the most commonly proposed site, Shang-ch'iu in eastern Henan, are buried under twenty feet of Yellow River silt and thus essentially irrecoverable despite ongoing efforts.[17]

According to variants of this theory, whether through the dramatic shifts of the capital summarized in the literary records or simply through gradual emigration, the early Shang embarked on a sort of northwestern progression from an initial site in Shandong into middle Hebei, proceeded as far as Yi-shui, then moved to the western part of Hebei and finally back to the western part of Shandong.[18] Their final assault on the Hsia would therefore have originated in the east just as literary accounts that refer to them attacking a capital in the west demand, but required King T'ang to first send troops around to the south and ultimately strike from the west in order to actualize Chieh's prophetic dream.

In contrast, theories of northern origin stress that several of the numerous groups that emerged over the ages among the peoples identified with the north and the northern complex developed sufficient power to aggressively challenge sedentary China's rulers.[19] Whether the Shang are seen as originating in the Po-hai area, around Yu-yen, or in Tung-pei, possibly out of Hungshan or Lungshan cultures, this thesis ultimately assumes a downward movement along the eastern side of the T'ai-hang Mountains to the site from which they mounted their final assault. If accurate, it is an explanation that entails considerable impact because the environmental stimuli and their allies in Hebei, a strategically advantageous area, would have differed considerably. Moreover, in envisioning the Shang's final preconquest capital being in the east, the northern origin theory can equally accommodate any reconstructed conquest sequence that entails first neutralizing the smaller coalition states arrayed in an arc running from east to south.

The third view revitalizes an ancient theory by asserting that the Shang arose in the west, one expression even claiming that T'ang's final, prestrike capital of Po is the recently discovered fortress at Yüan-ch'ü in Shanxi.[20] Although intriguing, this theory suffers from two glaring defects: the radiocarbon dates do not cohere with the probable conquest period, and several of the minor states that were systematically conquered prior to assaulting the Hsia itself would have to be relocated in a western-to-southern arc around the Hsia capital. However, this deficiency is rectified if, as proposed, the Shang emigrated eastward into the Shandong area before ultimately descending along the eastern side of the T'ai-hang Mountains.

A variant of the theory that assumes the proto-Shang peoples primarily populated a natural corridor delimited by the T'ai-hang's foothills to the west and the old course of the Yellow River to the east, both of which would have constituted significant impediments for aggressors, suggests that global cooling precipitated their movement southward.[21] Originally a comparatively wet area populated by numerous ponds, lakes, marshes, and streams that had formed over the previous two millennia, the terrain in this 120-kilometer-wide swath apparently suffered considerable drying, loss of small wildlife, and changes in forest composition commencing about 2000 BCE after the temperature had dropped about 3 degrees C and rainfall amounts had substantially declined over the previous 500 years.[22]

Areas farther to the north were more drastically affected, the reduction in rainfall making agriculture infeasible outside of the line along which the Great Wall would eventually be constructed. Confronted by gradually deteriorating conditions, migrating to more viable terrain must have been an obvious option. Even though the Shang were thus motivated by enhanced prospects for survival rather than any grandiose design for conquest, frequent clashes with well-entrenched indigenous groups must have been inescapable. Ironically, postconquest, a slight warming trend prevailed virtually throughout the Shang era.

Although eastern China was an area much in flux as clans with various cultural affiliations expanded, contracted, intermixed, and moved about, and the eastern thesis currently predominates, the issue not only remains unresolved but is essentially bereft of the requisite evidence for realistic assessment. The military implications of the Shang's origin therefore continue to be unframed and unexplained, as does whether these reputed capital movements were simply seminomadic shifts of focal locale; expressions of a will to power; compelled by environmental degradation or internal strife; or undertaken to expand their basic domain, consolidate control, and project power. Numerous articles have attributed them to every possible cause, including a strategic determination to move the population center away from looming threats and, contradictorily, deliberately positioning the capital closer to contiguous enemies to blunt and contain them. If the Shang were stimulated by population pressures, each displacement may have cumulatively expanded their territory, the excess populace allowing them to extend their control over additional, rather than alternative, areas.

Some historians have argued that all the predynastic shifts were made en masse, no residual forces being left behind, but others assert that a significant portion of the population remained to retain control over the area, implying that it continued to be secured. However, archaeological evidence indicates that several of the later capitals were actually abandoned despite being in threat-free areas. Moreover, despite numerous archaeological reports and extensive speculation on Erh-li-t'ou, Yen-shih, Cheng-chou, Huan-pei, Hsiao-shuang-ch'iao, and An-yang, the rationale for the commonly recognized postconquest movements of the combined administrative and ritual capital also remains equally unknown, together with which locations correspond to the traditionally enumerated names.

Due to lacking the clarity of the Kanmu Emperor's movement of the capital from Nara to Kyoto to avoid Buddhist monastic power in 794 CE or Peter the Great's decision to construct a new political and cultural center in the marshes of St. Petersburg, in the context of the Shang's rising power these shifts probably reflect a transition from loosely organized settlements to cohesive states. Accordingly, even though primarily intended to reorient or increase the Shang's domain of influence, they must have also played a role in elevating the leader's status and solidifying the power of the ruling clan.

CONQUEST OF THE HSIA

According to popular recounting, King T'ang established the Shang as China's preeminent power fourteen generations and eight shifts of their capital after their putative founding. Despite the momentousness of their conquest, the *Shih Chi* characterized the engagement in a single sentence, the Grand Historian opting not to offer any explanatory passages despite the materials presumably at hand. Fortunately the *Shang Shu*, *Mencius*, and various other Warring States works, all composed at least twelve to fifteen centuries after the Hsia's overthrow, contain extensive, albeit scattered and fragmentary, material on their clash. However imaginative or inaccurate they may appear today, because there was little reason to doubt their veracity in antiquity, the traditional accounts derived from them and the *Shih Chi* were fervently embraced from the Han onward. As political and military thinkers integrated them into their theoretical constructs, they therefore came to affect court discussions, the image of Virtue, and the very conceptions of revolution and warfare.

For virtually everyone the essential facts were few but dramatic. When the Hsia became oppressive and debauched, the Shang cultivated its own Virtue, developed key alliances, and reluctantly assumed leadership of the disaffected lords before engaging in the decisive confrontation in which they easily vanquished King Chieh, the tyrannical last ruler. The *Shang Shu*, subsequently deemed one of the original five Confucian classics, came to provide the fundamental impetus for interpreting King T'ang's assault not just as a righteous action mounted on behalf of all the people, but also as a punitive effort undertaken with Heaven's sanction, even its mandate.

The chief document cited in justification is "T'ang's Oath," essentially a proclamation of purpose issued by King T'ang just before the campaign to elicit (or otherwise coerce) the requisite support:[23]

> Come you multitudes of people, all of you listen to my words. It is
> not I, the little child, who dares practice what can be termed rebellion.

Hsia is guilty of numerous offenses, so Heaven has ordered that he be destroyed.

Now you multitudes are saying, "Instead of having compassion on us, our lord makes us abandon the work of harvest to cut off and rectify Hsia." I have indeed heard the words of you multitudes, but the chief of the Hsia has committed offenses. Because I fear Shang Ti, I dare not but punish him.

Now there are those among you who say, "What do Hsia's offenses have to do with us?" The king of the Hsia has utterly exhausted the strength of his masses, everywhere wearing them down in their towns. His masses have all become lazy and will not cooperate with him. They are saying, "When will this sun expire? We will all perish with you." The Hsia's virtue being so, I must now go there.

I pray you will assist me, the one man, to bring about Heaven's punishment. I will greatly reward you. Do not let there be any disbelief among you for I will not eat my words. If you do not obey the words of my pronouncement, I will exterminate your entire families, there will be no pardon.

Although the "T'ang Shih" was probably composed no earlier than the Western Chou, prompted by traditional belief that King T'ang gave a stirring invocation before the campaign, it has assumed immense importance throughout imperial Chinese history.[24] From the contents, vestigial or not, it appears that T'ang's own people not only lacked enthusiasm for undertaking the campaign but also even murmured against it, implying that even "virtuous" rulers had to motivate the people before they would disrupt their lives and engage in military activities.[25]

Perhaps because other *Shang Shu* authors in the fourth century CE felt it necessary to buttress the king's claim—and similarly that of King Wu of the Chou, equally guilty of regicide unless Heaven had sanctioned his revolt—of having acted righteously, another section of the *Shang Shu* depicts King T'ang as personally questioning the validity of his actions by ruefully saying, "I am afraid that in future generations, I will fill everyone's mouth."[26] This device allows one of his righteous ministers, Chung Hui, to then justify his actions as both virtuous and necessary:[27]

Truly, when Heaven gives birth to the people they have desires. Without a ruler they would certainly become chaotic. Truly, Heaven spawns wisdom and intelligence in the father for the era.

The ruler of the Hsia had confused virtue, the people had fallen into mud and ashes. Heaven made a gift of courage and wisdom to you, our king, so that you might visibly rectify the myriad States and continue Yü's ancient responsibilities. You just acted in accord with its methods, upholding and executing the mandates of Heaven.

The king of the Hsia was guilty of offenses, contorting and distorting the intentions of Heaven in order to disseminate his commands among the people. Ti accordingly condemned him and employed the Shang to receive its mandates and you to enlighten its armies.

King T'ang again condemns the Hsia in another late section, entitled "T'ang Kao," which purportedly preserves his pronouncements upon returning to the capital. However, the lines most frequently cited in subsequent centuries, whether to simply recount the glorious Shang conquest or justify some divergent course of action, are again attributed to Chung Hui:[28]

When the earl of Ko treated those bringing provisions as enemies, King T'ang's punitive campaigns began with Ko. When he campaigned in the east, the Yi in the west complained. When he campaigned in the south, the Ti in the north complained, "Why does he make us, alone, last?" Wherever he ventured, the people congratulated each other in their households, saying, "We have been waiting for our lord. Our lord has come and we have respite."

In the absence of oracle bone or other contemporary records, the ancient spin masters simply fleshed out the rumors and legends transmitted from antiquity. The tale then assumed a life of its own as a true portrait of events, with the exemplary King T'ang reluctantly assuming the awesome yet baleful responsibility of executing Heaven's will and restoring the people's welfare. Thus the *Shang Shu* portrays King T'ang as a paragon of virtue: he is devoted to his task, oblivious to licentious distractions, effective in action, and focused in deed. Unfortunately, the

Shang proved to be a brutal theocratic state throughout its reign, long before perversity and licentiousness purportedly doomed it to extinction.

By drawing on numerous fragmentary records, traditionally oriented scholars have pieced together a fairly comprehensive picture of the conquest that merits consideration in itself, as well as a potential composite from which the historical outlines might be gleaned.[29] In summary, the Shang began to flourish after removing to a strategically advantageous, semiremote position somewhat north and east of the Hsia capital, where they would be reasonably free from immediate interference though still nominally submissive;[30] established their own network of allies and client states, presumably through Virtue, but more likely by emphasizing their evolving power while championing the idea of benign and beneficent government (which certainly would have translated into policies toward the elite and chiefs more than the common people, whose interests would be critical only in a military crisis); coercively expanded their power base through local conquests, some (according to the *Shang Shu*) ironically undertaken on behalf of the Hsia (such as against the Yu Yi), just as the Chou would centuries later on behalf of the Shang; maintained close relationships with the Tung Yi to gain their support and preclude their enmity; probed the strength of the Hsia and fidelity of its subject states; held a preliminary assembly of its own allies to determine the intensity of their commitment; defeated and otherwise neutralized the Hsia's allies;[31] gathered general intelligence and exploited the ruler's paranoia by attacking the capital from an auspicious direction, forcing his flight; and finally defeated the K'un-wu, the Hsia, and even the San-tsung, where Chieh had sought refuge. Throughout these efforts King T'ang reputedly employed insightful advisors such as Yi Yin, a figure to whom sacrifices were still being offered during the dynasty's last centuries at Anyang, attesting to his perceived importance.[32]

Effecting this sort of revolution would have required a lengthy period to accumulate essential materials, forge alliances, and gradually overcome nearby recalcitrant states. At what point King T'ang first became enamored of regal power is unknown, although *Mencius* suggests that Yi Yin may have been instrumental in persuading him to undertake an armed rebellion because Chieh's odiousness was growing and the people were becoming increasingly disaffected.[33] The first entry of importance in the *Bamboo Annals*, for Chieh's fifteenth year, indicates that King T'ang moved his people to a site called Po, a designation that has been interpreted as basically signifying any site the Shang central government occupied rather than a particular place-name. Presumably this was to position himself away from the Hsia's immediate power and pernicious influence, contrary to postulates that Po was Cheng-chou or Chiao-tsuo, both within easy marching distance of the capital at Erh-li-t'ou. At this early stage the move must have been sanctioned by Chieh, insofar as he apparently had appointed the Shang as a *fang-po* or ducal state entrusted with military authority to act on behalf of the Hsia.

Thereafter, according to the *Bamboo Annals*, in Chieh's seventeenth year the Shang dispatched Yi Yin to the Hsia, where he remained for three years before returning. Perhaps with an eye toward ultimately attacking the Hsia, commencing with Chieh's twenty-first year the Shang began to exert its military power by conquering the Luo before going on to mount a repressive expedition against Ching in the south, subjugating them as well, possibly on behalf of the Hsia if traditional accounts are to be believed. Ta-shih-ku may also have been seized about this time,[34] causing Chieh to become suspicious.

When King T'ang foolishly ventured to Chieh's court in the twenty-third year he was imprisoned in the Summer Tower, where he remained for a year before finally being freed. Why Chieh shortsightedly failed to execute him is unknown, but immediately after his release many of the so-called lords ventured to T'ang's court, where they were welcomed as honored guests.[35] The king must have cultivated these and similar relationships with such eastern groups as the Yu-shih, Yu-jeng, Yu-hsin, and Yu-pi for some three years before initiating the campaign that exterminated the Wen.[36]

In Chieh's twenty-eighth year the K'un-wu, no doubt perturbed by the specter of Shang power and, as another *fang-po*, perhaps acting as Chieh's alter ego, attacked the Shang, with indeterminate results. After convening his allies, King T'ang went on to seize Wei and mount an unsuccessful assault on Ku, which proved unconquerable until the next year, presumably in another campaign rather than as the result of a protracted siege. (About this time, perhaps foreseeing imminent doom, the Hsia's Chief Historian fled to the Shang, in itself a highly baleful omen.)

The Shang then mounted punitive expeditions against the K'un-wu in each of the next two years, finally defeating them after attacking the Chen-hsün by way of Erh in Chieh's thirty-first year.

The final clash between the Shang and Hsia melodramatically unfolded at Ming-t'iao amid severe thunder and rain, well befitting the overthrow of one dynasty and establishment of another. However, King Chieh somehow managed to escape and found refuge among the San-tsung, compelling T'ang to march against the San-tsung and defeat them at Ch'en, eventually capturing Chieh at Chiao-men. Remarkably, according to the *Bamboo Annals* and many other accounts, King T'ang exiled Chieh to Nan-ch'ao in Anhui rather than executing him. Ironically, the last ruler of the Shang—historically portrayed as a perverse clone of King Chieh—would suffer a far more violent fate in essentially the same circumstances.

Certain aspects of these events as chronicled by the *Bamboo Annals* may be usefully amplified by other sources, insofar as they seem to have preserved vestiges of the events or became historically important thereafter. First, in particular, although Mencius frequently emphasized the power of Virtue attracting and conquering, one of his passages notes that a total of eleven different military actions were required before the Shang could subdue the Hsia and its allies, hardly the sweeping victory of Virtue unopposed! Irrespective of any hostile intent, some of these states probably had to be preemptively eliminated to preclude their exploiting the power vacuum created by the absence of Shang military forces.

Second, in a passage describing why the Shang attacked Ko that no doubt reflects the new theoretical concerns with ritual practice prevalent at the time of its composition in the middle Warring States period, the "welfare of the people," thereafter the watchword of many military actions and theoretical writings, came to be emphasized:

Wan Chang inquired, "If Sung, which is a small state, were to implement benevolent governmental practices, Ch'i and Ch'u would detest it and attack it. What can be done?"

Mencius replied: "When T'ang dwelt in Po the duke of the neighboring state of Ko had discontinued offering sacrifice. When T'ang dispatched an envoy to inquire why he did not offer sacrifice the duke replied, 'I lack the means to provide sacrificial animals.' T'ang accordingly had oxen and sheep sent to him but the duke ate them and still did not offer sacrifice.

"T'ang again dispatched an envoy to inquire why he was not offering sacrifice, to which he replied: 'I lack the means to supply the ritual vessels of millet.' T'ang thereupon dispatched the masses from Po to go and plow the fields on his behalf and had the young and old convey provisions. The earl of Ko led the populace to intercept them and seize the wine, edibles, millet, and rice, slaying anyone who wouldn't turn them over. They even killed a youth and seized his provisions of millet and meat. When the *Shu* states, 'The earl of Ko treated those who would supply him with food as enemies,' this is what it means."

"Since King T'ang mounted his punitive expedition because the earl had killed this youth, all within the four seas said: 'It isn't because he seeks the riches found under Heaven, but to avenge an ordinary man and ordinary woman.'[37] [As the *Shang Shu* states,] 'When T'ang initiated his punitive efforts, he commenced with Ko.' After eleven campaigns he no longer had any enemies under Heaven. When he faced east and went forth, the Yi peoples in the west murmured; when he faced south and went forth, the Ti in the north complained, 'Why does he put us later?' The people looked to him just as if longing for rain during a severe drought. Those venturing to market didn't stop, those working in the fields continued their weeding unchanged. In punishing the ruler but consoling the people he was like a seasonal rain falling. The people were greatly pleased. As the *Shu Ching* states, 'We have been waiting for our lord. When our lord comes, will we not be free from punishment!'"[38]

This passage came to exert a formidable influence on court debates about military action, and the concept of not disturbing the people by campaigning solely against the chief miscreants, though rarely observed in practice, proved to be particularly important. It was also increasingly emphasized in such classic military writings as the *Ssu-ma Fa* as the carnage escalated dramatically in the Warring States period and was frequently raised during the chaos of subsequent millenarian revolts and dynastic overthrows.[39]

As to other aspects, the *Lü-shih Ch'un-ch'iu*, a late third-century BCE eclectic text, contains an expanded discussion of Yi Yin's career that focuses on his activities as a spy rather than virtuous adviser. In essence, Sun-tzu's assertion that, "in antiquity, when the Shang arose, they had Yi Yin in the Hsia. When the Chou arose, they had Lü Ya in the Shang," canonized him as one of China's first two spies.[40] Nevertheless, writings such as the *Shang Shu* not only fail to sustain Sun-tzu's opinion, but instead portray him as a recluse who deliberately sought out King T'ang and persuaded him to embark on the righteous course that could sweep away the Hsia.[41]

The *Lü-shih Ch'un-ch'iu* preserves a dramatic passage that not only integrates the virtues of a moral paragon with the deviousness of a seasoned spy, but also shows that King T'ang created a painful cover story for Yi Yin:

Because Emperor Chieh of the Hsia dynasty was immoral, brutal, perverse, stupid, and greedy, All under Heaven quaked in fear and were greatly troubled. However, reports of his deeds were contradictory and confused, so it was difficult for anyone to fathom the true situation. Availing himself of the emperor's awesomeness, prime minister Kan Hsin was brusque and insulting to both the feudal lords and the common people. Because the worthy and outstanding ministers were all anxious and resentful, the emperor killed Kuan Lung-feng to stifle rebellious stirrings.

The common masses, in consternation and confusion, all wished to emigrate elsewhere and, being afraid to speak directly about the situation, lived in constant terror. Although the great ministers shared the same worries, they dared not cooperate to mount a revolt. Meanwhile Emperor Chieh increasingly regarded himself as a great worthy, boasted of his transgressions, and even viewed wrong as right. The imperial way was blocked and stopped, the populace in total collapse.

Terrified and troubled that the realm was not at peace, T'ang wanted Yi Yin to go and observe the mighty Hsia. However, he feared that Yi Yin would not be trusted so he personally shot him with an arrow. Yi Yin then fled to the Hsia. After three years he reported back to T'ang at his enclave at Po: "Earlier Chieh was befuddled and deluded by his consort Mo Hsi and now loves his two concubines from Min-shan so he doesn't have any concern for the common people. The will of the people does not support him, upper and lower ranks detest each other. The hearts of the people are filled with resentment. They all say that Heaven fails to have pity on them, so the Hsia's mandate is over."

T'ang exclaimed: "What you have just said is much like a prophetic lament!" They then swore an oath together to evidence their deter-mination to extinguish Hsia.

Yi Yin went again to observe the mighty Hsia where he learned from Mo Hsi that "the previous night the emperor had dreamt there was a sun in the west and a sun in the east. When the two suns en-gaged in combat, the western sun emerged victorious and the eastern sun was vanquished." Yi Yin reported this to T'ang. Although the area was suffering from a severe drought, T'ang still mobilized their forces in order to keep faith with the oath he had sworn with Yi Yin. There-after he ordered the army, after proceeding out from the east, to make their advance into Hsia territory from the west. Chieh fled even before their blades had clashed, but was pursued out to Ta-sha where he was finally killed.[42]

This episode already incorporates many crucial elements of spycraft. First, from the outset Yi Yin was deliberately employed as a covert agent to gather critical intelligence about the Hsia and its ruler. Second, he was given a plausible pretext for seeking refuge with Chieh, an arrow wound emblematic of T'ang's hatred. Third, Chieh had not only suc-cumbed to licentiousness but also committed the fatal mistake of dis-placing his consort, who then acted as an archetypal slighted beauty by secretly revealing the king's nightmare to Yi Yin. T'ang's astonished re-action to Yi Yin's initial report similarly indicates the importance of prophetic events, dreams, sayings, and phenomena in this era.

Encouraged to undertake otherwise dubious campaign measures, King T'ang scripted a scenario that would actualize Chieh's secret fears, including an assault from the west. Although the choice of direction was dictated by the nightmare, it may have been tactically advantageous because the Hsia probably left the west comparatively undefended on the assumption that any attack would originate in the east. Chieh's downfall was also undoubtedly hastened because the hostile state of Min-shan had wisely exploited the debilitating power of women and sex by presenting two new concubines, intended to fascinate him and thereby deter further assaults on their small state.

A millennium after these events military theorists such as Wu-tzu would emphasize the need to probe the enemy with minor stimuli to accurately gauge personality and intent. An incident preserved in the *Shuo Yüan*, a first-century BCE thematic compilation, envisions Yi Yin as having advocated just this sort of probe:[43]

> T'ang wanted to attack Chieh but Yi Yin said, "Let's reduce our tribute and observe his actions." Enraged, Chieh mobilized the armies of the Nine Yi in order to attack the Shang. Yi Yin said, "It is not yet possible. If he is still able to mobilize the Nine Yi armies, the offense lies with us." T'ang then acknowledged his offense and asked to be allowed to submit and again forward the requisite tribute. However, the next year, when he again did not forward the required tribute and Chieh, enraged, tried to mobilize the armies of the Nine Yi, they did not respond.[44] Yi Yin then said, "Now it is possible." T'ang mobilized their armies, attacked and destroyed the Hsia, and removed Chieh to Nan-ch'ao.

Whether the Hsia had degenerated to the degree attributed to Chieh, the Shang simply excelled as history's first propagandists, or Chou and later spin doctors concocted the whole tale may never be known. However, it seems likely that Chieh's behavior had so antagonized the Hsia clan states and other peoples that they welcomed an alternative under which they might enjoy greater freedom—or at least fewer burdens—and by being early supporters, perhaps more prestige and therefore independence.

King T'ang is portrayed as having visibly cultivated and consciously manifested an image of solicitude and righteousness. This would have attracted more distant peoples, especially those suffering under the yoke of Hsia tributary demands, and cemented nearby allegiances, creating the requisite foundation for attacking King Chieh. When the Shang

finally struck, their motivation was ostensibly righteous, and they apparently claimed that as Heaven (or Ti on High) sanctioned the change,
the Hsia had forfeited its legitimacy. (At this moment the late concept
of a "mandate of Heaven" had not yet been formulated, but the idea
that the rulers could lose the blessing of Ti, to whom the Shang would
constantly appeal in their divinations, and thus the right to rule, well
coheres with subsequent Shang religious thought.)

If these traditional accounts hold any validity, the Shang's strategy
was complex, its incubation lengthy, and its final measures brilliantly
executed as they consciously exploited a newly created tactical imbalance
of power and Chieh's psychological weakness through covertly gathered
intelligence. King T'ang can therefore be seen as having exemplified
Sun-tzu's subsequent concept of first creating the circumstances in
which one cannot be defeated (by neutralizing potential enemies, especially the Eastern Yi, who could have easily exploited the power vacuum created by the conflict) and then those for achieving victory (by
forming alliances, including with the Yüeh-shih, and acquiring true
knowledge of the enemy), thereby gaining victory essentially unopposed
over a crumbling foe. Even though this is a somewhat romanticized
portrait, one remarkably and no doubt deliberately similar to the Chou's
conquest of the Shang centuries later, to the extent that Chieh was
flawed and terrified and King T'ang had to confront certain military realities to gain victory, it may contain some degree of truth.

7.

SHANG CAPITALS, CITADELS, AND FORTIFICATIONS

◆·◆————————————————◆·◆

DEFINITIVE EVIDENCE FOR the Shang's military and economic power may be seen in several important archaeological sites, including Yen-shih, Cheng-chou, Anyang, and P'an-lung-ch'eng. (The first two are critical for understanding the early history of the Shang, the last for its projection of power southward, while Anyang was the capital over the final two centuries.) Although much of Yen-shih and Cheng-chou still remains buried despite several probing excavations, enough has been uncovered to identify them both as major sites dating back to the Shang's initial period. However, their relative sequence and importance have become matters of acrimonious contention. Proponents argue over whether Yen-shih was the initial capital of Po or at least a secondary or western capital known as Hsi Po, or Cheng-chou was Hsi Po or perhaps Ao, the site to which Chung Ting moved the Shang administrative head-quarters in the middle period, as well as other possibilities.[1]

The most reasonable explanation—which will be argued below after examining their basic features but need not be accepted to understand the military character and significance of Yen-shih—is that the multi-layered site at Cheng-chou, with its vast dimensions and numerous, impressive artifacts, was a fully developed royal capital constructed at a continuously occupied Shang site, but from the outset Yen-shih was a military bastion. It was probably erected shortly after the conquest in the heart of enemy territory to serve as a fortress even though Erh-li-t'ou itself had been occupied because, contrary to traditional accounts, many Hsia groups remained unsubmissive, as might be expected if the Shang campaign had focused on eliminating the ruling clan.

In addition to these critical cities, several hundred Shang era sites have now been located, suggesting that the total number of substantial fortified towns or cities, including those of other peoples, may have numbered about seventy. In consonance with the historical materials furnished by the oracle bones from Anyang, a few such as Wu-ch'eng and Tung-hsia-feng have greatly contributed to a much revised understanding of the dynasty's power, extent, and dynamics.

All the Shang cities bear witness to increasingly sharp class distinctions and the evolution of royal groups with immense power, including life and death, over the general populace; the development of massive, sophisticated buildings; segmented inner quarters for the highly privileged; the slow emergence of burgeoning economies based on increased agricultural yields, animal husbandry, handicraft production, conspicuous consumption, and limited trade; the further perfection of bronze technology, resulting in large workshops capable of casting massive cauldrons, fine symbolic items, and weapons, the latter in multiple molds; and clear differentiation in the dwellings within these citadels and across their expanded sites.[2]

The Shang was also capable of mobilizing the large labor forces required to undertake ever greater fortification work, vast communal projects, and deeper, more complex tombs. Although Yen-shih and Cheng-chou embody important Lungshan characteristics, including expansive wall systems, they are undoubtedly early Shang fortified capitals. However, just as at Erh-li-t'ou, no defensive fortifications apart from a single moat have yet been discovered amid the opulent remains at Anyang, immediately raising the question of whether those rulers were too immersed in the pursuit of pleasure to undertake them or simply felt that surpassing military power rendered them unnecessary. If so, the absence of defensive walls proved a fatal conceit, because the last emperor lacked a defensible refuge after his forces were vanquished at Mu-yeh.

Although traditional scholars have long acrimoniously but futilely argued over the identity and location of the preconquest capitals, recent discoveries showing the preconquest forcible displacement of Hsia cultural elements apparently confirm that the incipient Shang seized terrain early on and erected citadels to block its nemesis at peripheral points of confrontation. In particular, the remnant structures at Chiao-tsuo in Fu-ch'eng, Henan, appear to comprise a preconquest bastion that was constructed in a crucial area where the Hsia had been exerting, possibly even expanding, its influence and thus clashed with the Shang, then an increasingly dynamic post-tribal entity intent upon projecting power and broadening its domain.[3]

Abutting the T'ai-hang Mountains in the Shen River area, Chiao-tsuo's roughly 300- by 310-meter-long walls form a virtually perfect square that encompassed somewhat more than 90,000 square meters, impressive but too limited for the site to have served as anything more than a detached city or secondary bastion despite suggestions that it might have been the predynastic capital of Yung. Based on remnants, the original walls apparently averaged 15 meters in width and stood at least several meters high. Although marked by the usual 8- to 12-centimeter-thick layered construction typical of the period—nineteen layers being required for the 2.1-meter-high eastern wall—a somewhat unusual method of pounding with bundles of fifteen to twenty comparatively small sticks was employed. All the walls were erected on ample excavated foundations, ensuring their stability.[4] Several major structures have been discovered, and the citadel's interior was divided into two compounds.

Apparently maintained as a militarized city until somewhat after Cheng-chou was abandoned, postconquest Chiao-tsuo presumably continued to function as a control point over the Hsia populace and then a bulwark against the Eastern Yi. By the middle of the dynastic Shang period, after the Yi had been defeated and Hsia opposition had evaporated, it may have become an unnecessary expenditure[5] or a casualty of the internal turmoil supposedly responsible for Cheng-chou's demise.

Another obvious military bastion, the Shang-fortified city at **Yen-shih** is located some ten kilometers east of Luoyang in the Luo River watershed just above the present-day river, close on a marsh that encroaches from the southeast. First discovered and excavated in 1983, it continues to be the subject of ongoing investigations in the quest to determine its historical identity.[6] Several possibilities continue to be argued despite the findings of the Xia-Shang-Zhou dynastic project: Chen Hsün, the last Hsia capital; Hsi Po, the first Shang capital; an early Shang secondary capital, perhaps paired with Po (which is then said to be Chengchou); an early Shang military bastion; T'ai Wu's new capital of Hsi; T'ai Chia's T'ung Palace;[7] or even P'an Keng's new capital, known as Po Yin.

As the diagram shows, the city eventually assumed the shape of a contorted rectangle that tapers slightly downward in the upper northeast corner and is severely squashed by the marsh in the lower southeast. With a remnant circumference of 5,330 meters, the outer walls span some 1,700 meters from north to south and 1,215 meters from east to west, although the southernmost portion extends only 740 meters before turning upward. A total of seven gates have been identified; eleven large roads oriented along the cardinal directions crisscross the interior; and a formidable ditch some 20 meters wide and 6 meters deep encircles the walls at a distance of approximately 12 meters. Clearly functional in intent, its near side gradually slopes downward, facilitating archery fire while precluding protective concealment, but the far side is a daunting, nearly vertical, 19 feet. When filled with water and functioning as a moat, its 20-meter expanse, although easily negotiated by swimmers, would have required rafts or boats to convey any sort of siege or assault equipment.

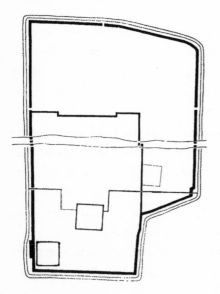

The remnant walls of the greater enclosure presently stand 2.9 meters over the outside terrain and 1.5 to 1.8 meters above the interior, but were certainly higher when first erected. They are 17 to 18 meters wide at the base and taper to just under 14 meters at the top and have a perpendicular outer face to preclude easy ascension. Constructed by the usual pounded earth technique that produced layer thicknesses varying between 8 and 13 centimeters, they were underpinned by a similarly compressed, multi-layer foundation excavated down between 1.3 and 1.7 meters, with an expanse slightly narrower than the ground-level wall base.[8] However, the consistent use of a single soil type rather than intermixed layers of distinctively different soil is interpreted as evidence of either less sophisticated engineering practices or haste in constructing the defenses. Nevertheless, the layers are described as hard and dense, as they must have been for the walls to have still protruded some 2 meters above the terrain as late as the Han and Wei periods,[9] with the outer walls being erected upon a carefully leveled and formed inner core.

More recent excavations have uncovered the interior enclosure shown on the diagram as well as a self-contained palace area slightly south of its midpoint and two other segregated areas. The royal quarters were set off in a roughly square enclosure measuring about 210 meters east to west and 230 meters north to south, creating a segmented area of somewhat over 40,000 square meters that contains evidence of palace foundations.[10] As the wall averages only 3 meters in width, it clearly was more symbolic than functional in intent.

The royal quarters were initially surrounded by fortifications that assumed a fundamentally rectangular shape that extended 1,100 meters from north to south and 740 meters from east to west. Far narrower than the walls of the greater enclosure, they average 6 to 7 meters in width and were erected upon a shallow, meter-deep foundation. The freestanding portions retain a remnant height of only 0.2 to 0.6 meter, but they were certainly higher in antiquity. Constructed before the outer fortifications subsumed them, they demark the approximately 800,000 square meters that constituted the first Shang enclave at this site. However, within decades the highly dynamic Shang more than doubled the city's bounded area to 1,900,000 square meters, even building right out over their bronze workshops. The final configuration thus assumed the palace, inner city (ch'eng), and outer confines or suburbs (kuo) widely discussed in late Chou and later literature as the highly idealized form of traditional Chinese capitals. As a result of this evolutionary expansion, Yen-shih is sometimes said to be the first to manifest it.[11]

Two walled enclosures, each with dimensions somewhat more than half the royal city, have also been discovered within the greater fortifications. Evidence suggests they may have been military barracks or storage areas for weapons and provisions. An 800-meter-long underground ditch roughly 2 meters wide, ingeniously constructed from wood and stone and linked to all the buildings in the palace compound, runs from the royal quarters out below the city's walls. Originally characterized as a drainage ditch, its connection to a large artificial pool some 128 meters in length, 20 meters wide, and 2 meters deep suggests it was designed to provide water rather than simply drain it away. (This pool is the earliest artificially constructed urban reservoir yet discovered in China.)[12]

The inner walls, roads, gates, and artificial pond, being highly regularized and expertly constructed out of tamped earth, clearly show the systematic execution of a well-crafted strategic design that emphasized structural organization and active defense. Additional features such as two gatehouses that could control access to the entire settlement and the interior quarters, coupled with the narrowness of the gates, emphasize Yen-shih's martial character, one highly appropriate for a new power center that had to be forcefully imposed in the enemy's very core.

The citadel's outer walls, which were erected within a few decades of the inner enclosure, and most of its functional remains are said to date to the fourth Erh-li-t'ou and lower Erh-li-kang phases, evidence that Yen-shih was not an ordinary site that gradually evolved over time, but was deliberately erected.[13] Conversely, although austere, the site is too palatial to have simply been a military bastion, suggesting it must have been employed for some decades as the first capital (known as Hsi Po) before royal power was shifted back to the new capital at Cheng-chou.[14] Key radiocarbon dates for artifacts from the site fall just about the time traditionally claimed for the Shang's ascension of 1600 and the astronomically derived year of 1588.[15] Late geographical and eclectic writings such as the *Lü-shih Ch'un-ch'iu* and *Ch'un-ch'iu Fan-lu* preserve abundant, albeit much later, textual evidence that suggests King T'ang established his first capital of Hsi Po in this area, perhaps even roughly at this site.[16]

Apart from being located in the Hsia heartland, the site could also control vital passes to the east and west and project power to the troublesome northwest, where the Shang would continually experience severe challenges. Armies going east out through the Hei-shih and Hu-chien passes would move directly into the Hua-pei plains; marching westward through Han-ku Pass and T'ung Pass, they would cross into the critical Kuan-chung area; advancing northward, they would enter Chin and Shaanxi after fording the Yellow River; and going south on campaign through Yin-ch'üeh, they would move into the Ying and Ju river basins. From this single location all the military campaigns spoken of in *Mencius* and other writings, as well as any to the west and north-west, would have been feasible.[17]

Thus, the identification of Yen-shih as T'ang's first capital, in conjunction with Erh-li-t'ou being the last capital of the Hsia, seems well founded, particularly if Ta-shih-ku is considered to have been Po after being wrested from the Hsia or predynastic Cheng-chou fulfilled that role. Moreover, the discovery of wheel tracks on an interior road, though no longer evidence of the very first use, shows that by 1600 BCE the Shang was already employing primitive vehicles that would have provided considerably greater earthmoving capability than laboriously toted baskets. Molds, bronze fragments, charcoal, and crucibles indicate that a significant bronze workshop was operated on the site even though vessel production continued at Erh-li-t'ou. However, the bronzes and other items lack inscriptions and are generally simpler than those subsequently identified with the Shang, befitting an earlier, more martial site. The few tools and agricultural implements recovered indicate that rather than being a major production or economic center, Yen-shih depended upon the greater Shang domain for provisions and logistical support.

The now famous Shang capital at **Cheng-chou**, a city near the Luo River in Henan province in the middle lower course of the Yellow River area, has also been the subject of numerous excavations since its discovery.[18] Before Yen-shih was identified and explored, many scholars vociferously argued that Cheng-chou had to be the first Shang capital of Po, others that it was Ao or even a Hsia capital.[19] However, in contrast to Yen-shih's comparative austerity, the extensiveness of its bronze and other production facilities, numerous opulent bronze cauldrons recovered, the stage of cultural and artistic development, and similar factors clearly indicate a more politically and economically developed stage.

Based on radiocarbon datings and an analysis of the underlying layers, the walls at Yen-shih were probably constructed some sixty to eighty years before those at Cheng-chou, vital evidence that the former was probably the Shang's initial postconquest capital.[20] Thus, even though many analysts still argue that both were initial capitals and should be identified with King T'ang, the defining Erh-li-kang manifestation of Cheng-chou may well have been erected as a new palatial and ritual center and assumed the functions of the Shang capital under Ta Keng.[21] Claims that Cheng-chou was most likely the capital of Ao, mentioned in numerous traditional texts such as the *Shih Chi*, *Bamboo Annals*, and *Shang Shu* associated with Chung Ting, are certainly balked by Cheng-chou's lengthy occupation, the hundred years far exceeding the brief reigns associated with Ao.[22]

Exploratory excavations indicate that Cheng-chou was a vast site whose city walls encompassed some 25 square kilometers within their total length of 6,960 meters. Situated roughly along the compass directions on a height between two minor rivers, the city obviously functioned as a royal administrative center as well as the dwelling area for important members of the populace. Evidence of extensive royal quarters in the northeast portion of the city, apparently protected in part by an internal moat, has been uncovered, and several large production sites for bronze, bone, and ceramic products have been identified in virtually every direction outside the city.[23] In both quantity and quality the artifacts unearthed at each site greatly exceed the limited amount recovered from Yen-shih and include rather specialized items such as wine vessels. The inscribed bronzes preserve evidence of the development of written language, and the ritual cauldrons that continue to be recovered, some as large as a meter high and weighing hundreds of pounds, are so magnificent as to be identifiable only with the royal household. They provide incontrovertible evidence that the markedly advanced cultural site of Cheng-chou, which obviously traded with other areas, must have been the Shang capital during its occupation.[24]

The originally massive and unusually high walls apparently towered some 10 meters over the outside terrain and had a formidable width of 20 meters at the base but just 5 meters at the top. (Excavated portions actually run from 1 to 9 meters high and 4.8 to 22.4 meters wide, with the corners being about 30 meters thick.) More specifically, the eastern wall is 1,700 meters long, the western 1,870 meters, the southern 1,700 meters, and the northern 1,790 meters, with openings for gates in all but the eastern wall. Generally constructed over an unusually narrow excavated foundation of only 2.5 meters in width, but with the usual downward trapezoidal shape and pounded, leveled ground, the fortifications actually consist of a 10-meter-wide core wall to which protective waist walls of some 5 meters in width were appended on either side.

Constructed in 3.8-meter-long sections framed by wooden boards between 2.5 and 3.3 meters long and 0.15 to 0.30 meter high, the layers are clearly defined, well leveled, and generally uniform, being between 8 and 10 centimeters thick but occasionally dipping to as little as 3 centimeters and bulging up to 20. However, little variation is evident in their composition, all the layers being composed largely from a mixture of sticky red soil and grayish red sand pounded to maximum hardness. The tools employed, bundled rods roughly 2 to 4 centimeters in diameter, left permanent impressions up to 2 centimeters deep on the surface of each layer. The two external protective walls were similarly pounded, and the outer one was coated with a layer of protective pebbles, presumably to forestall erosion by falling rain and perhaps buttress it against floodwaters.[25]

More recent excavations have discovered remnants of another 5,000 meters of substantial walls that externally encircled the southern portion and an external ditch or moat. Discernible wall heights presently range from 1.2 to 2.3 meters, with widths in the various sections reported of 12, 17, and an expansive 25 meters, all constructed on an excavated foundation trench.[26] In the northeast corner of the city, evidence of an even earlier 100-meter-long wall some 8 meters in width has also been reported. Together with artifacts from predynastic Shang and other cultures, it indicates that Cheng-chou was an important location even before it became a Shang capital, perhaps King T'ang's initial city of Po.[27]

STRATEGIC ASSESSMENT AND HISTORICAL IMPLICATIONS

Assessments of late Hsia and early Shang sites generally differ in how they identify the fortified cities of Yen-shih and Cheng-chou rather than in how they interpret the basic archaeological data. Whether Yen-shih was just a secondary capital or the first, primary capital of the Shang, though historically important, is somewhat less significant for military history than its definitive martial character. Moreover, when coupled with the evidence that Erh-li-t'ou is almost certainly Chen-hsün, the last capital of the Hsia, Yen-shih's strategic function becomes obvious and virtually requires that Cheng-chou be viewed as the second Shang capital. This does not deny a martial function to Cheng-chou as well, especially since its role as a weapons production center has been clearly established by the discovery of numerous molds for casting bronze axes, halberds, and arrowheads in workshop ruins.

Being just six kilometers away from Erh-li-t'ou, Yen-shih must have been constructed to project power and consolidate control, to nullify the influence of the old Hsia center at Chen-hsün, which continued to flourish. In contrast, although Cheng-chou certainly retained an important military role in controlling people to the east and south, it has been characterized as a relatively open site that lacked the internal controls and defensive measures found at Yen-shih, and thus more of a ritual and administrative center.[28] The walls, though massive, are also marked by gradually sloping protective waist walls, hardly a desirable feature against assaults.[29]

A simple comparison of the relative might of each of these sites will aid in differentiating their character and functions. As already noted, Yen-shih lacked the productive basis for independent existence and had only minimal workshops and agricultural fields. In contrast, Cheng-chou included not only numerous large, specialized workshops, but also loosely integrated agricultural fields in nearby areas, accounting for the recovery of farming tools. The area enclosed by Cheng-chou's walls totals some 3 million square meters, whereas the original compound at Yen-shih, a citadel of 740 meters by 1,100 meters before expansion, totaled about 800,000 or slightly more than a quarter of Cheng-chou's size.

Whatever the dating of their walls, both sites were occupied over long periods, suggesting that Cheng-chou, much as Ta-shih-ku, may have been a pre-Erh-li-kang site initially occupied by the predynastic Shang as they expanded out of their core area in Chi-chung and undertook to subjugate the peoples in the nearby Luo-tung area, neutralize the Eastern Yi, and vanquish the Hsia's last ally, the K'un-wu, over the four years chronicled by the *Bamboo Annals*.[30] After overcoming the Hsia, the Shang required a new military citadel at Yen-shih to consolidate and impose its power in the Hsia heartland even as it continued to maintain its presence and project power in the east through some eleven additional campaigns.[31] Having completely subjugated the various Hsia groups, ensured the area's security with a strong bastion, and expropriated the Hsia's wealth, Cheng-chou was apparently fortified on a grand scale to become an opulent administrative and ritual center for the newly aggrandized rulers, Yen-shih functioning as secondary capital and strongpoint.[32]

Until recently many analysts believed that all the archaeological evidence pointed to a major disjuncture between the third and fourth periods of Erh-li-t'ou's occupation, justifying the traditional accounts of the Shang emerging through military conquest rather than simple evolutionary displacement.[33] The four distinct layers clearly visible at Erh-li-t'ou have generally been correlated with four stages of activity: establishment, expansion, flourishing, and decline. The first three display evolutionary continuity, but the fourth, in addition to the apparent abandonment of palace buildings and other evidence of contraction, suddenly shows an admixture of lower Ch'i-t'an Shang culture, very similar to that characterizing the first of Yen-shih's three periods.[34]

However, further excavations and the publication of additional reports indicate not only that Erh-li-t'ou continued to flourish in phase 4 (now dated as 1564 to 1521), but also that a new palace surrounded by tamped earth walls and other buildings were constructed.[35] In addition, the site seems to have continued as a major bronze production center until Erh-li-t'ou suddenly declined to the level of an ordinary village late in the sixteenth century, when much of the populace may have been forced to migrate to Cheng-chou.[36]

No evidence of a massive disaster or destruction by fire has yet been uncovered at Erh-li-t'ou. This coheres well with traditional, highly idealized accounts that the Shang targeted the chief miscreant, the one man, and that the actual fighting occurred away from the capital. Somewhat surprisingly, despite the importance of fortifications in Lungshan cultures and their derivatives, rather than enclosing the city with protective walls, the Hsia had opted to rely on a fortified palace at Erh-li-t'ou.[37] Perhaps the founders thought that the sheltering hills surrounding them and their location in the center of the Loyang basin would provide an adequate defense. In marked contrast, Yen-shih's initial occupants immediately opted to construct the massive walls typical of Shang fortified cities even though Yen-shih equally enjoyed significant advantages of terrain, including Mt. Mang to the north and the Luo River to the south.

The unprotected city, with most of its population intact, was probably compelled to capitulate. Moreover, having refrained from vengefully destroying it, the Shang could then exploit Erh-li-t'ou's productive capabilities with, as attested by the paucity of Erh-li-kang artifacts, only a minimal presence. However, they opted to impose significant control from the security of the nearby bastion of Yen-shih, perhaps because they felt Erh-li-t'ou to be indefensible given its lack of walls. Yen-shih's initial period thus not only coincides with Erh-li-t'ou's fourth period, but also displays some Hsia elements intermixed among predominantly Shang artifacts.[38]

A number of aspects clearly indicate an ongoing evolution in the capabilities and practices of material culture, particularly in the technology and styles of bronze casting, one of the hallmarks of Shang civilization. Although some magnificent ritual bronze vessels have been recovered, many of the artifacts discovered at Erh-li-t'ou are comparatively simple and, having been produced from an extremely high copper alloy, thin walled. Even allowing for the continuity that should be expected given that Erh-li-t'ou continued as a ritual production center for decades, the bronze vessels and implements found at Yen-shih early on show evidence of an advanced mixture that mingled tin and a large lead component while reducing the copper to approximately 80 percent, allowing larger, heavier, thicker objects (including weapons), marked by better curves and more refined surfaces, to be cast from multiple-part molds.

Significant differences in burial practices and ceramics can also be discerned at the two sites, even though ceramics were already being deemphasized with a shift in focus to bronze objects. A second definitive cultural aspect, the use of symbols or early writing, also advanced, if only in terms of the number of symbols appearing on the artifacts. However, despite claims of essential continuity the numbers remain minuscule, hardly the language of the oracle bones identified with Anyang.

The palace quarters similarly show distinctive Hsia and Shang placements. The essentially self-fortified compound at Erh-li-t'ou, which extends some 100 meters on a side, contains a highly systematized central hall with a surrounding portico built in the third period of fluorescence that faces south and, apart from being slightly offset toward the east, is fundamentally centered and thus surrounded by industries and living quarters. However, the palace compound at Yen-shih, which apparently also faced south, was located in the southern portion of the fortified city, slightly offset to the west, between two other walled compounds that seem to have been reserved for defensive functions.[39]

Yen-shih thus resembles Cheng-chou and other well-known Shang cities and fortified towns in which the markets, living quarters, and industries are usually found "behind" or to the north of the inner compound or administrative center, within the enclosure created by the outer walls. The construction techniques employed at Yen-shih, including the incorporation of a drainage ditch with a very slight but sufficient pitch, the solidity of the walls and foundations, and the overall systematic organization of the city thus all manifest perceptible differences and incremental advances over Erh-li-t'ou practices.[40]

SHANG EXTENT AND FLUCTUATIONS

For more than two millennia traditional Chinese historians have dogmatically asserted that China's three founding dynasties—the Hsia, Shang, and Chou—completely dominated the minor states and relatively insignificant, uncouth tribes about them. Even twentieth-century topical analyses of Shang events tended to view any deviation from this idealized depiction as an aberration or chimera, the consequence of flawed interpretation. Thus, for example, the Kuei-fang's presumptuous challenge of Shang authority, incontrovertibly recorded in the *Yi Ching*, has been understood as exceptional not because of the infrequency of such affronts, but because the Kuei-fang obdurately shunned Shang wisdom, culture, and sagacity. However, the innumerable archaeological artifacts and divinatory records recovered over the past century depict a remarkably different situation. Thus, most scholars now agree that the Shang's political power and material achievements affected an extensive area in central China, allowing it to impose its rule and cultural norms to a significant degree, but not to the exclusion of stimulating the development of local cultures and indigenous styles.[41]

Although Shang clan states within the core domain were basically homogenous, in more remote regions localized focal cultures such as those ensconced in the future state of Ch'u, the astounding enclave at San-shing-tui, and the remote southeast continued to evolve their own identities, habits, and material characteristics. Moreover, cultural and technological stimuli did not just radiate outward from the Shang, but also inward from the diverse tribes and distinctive peoples about them, prodding Shang developments.[42] For example, the Shang not only adopted images and practices from the Tung Yi, but also absorbed bronze decorative elements from the earlier Liang-chu culture (3200–2100 BCE).[43] Conversely, many peripheral peoples found the Shang's opulence, magnificent bronzes, intricate jade carvings, silks, wine, and profusion of foods difficult to resist, especially when ostentatiously displayed in the royal court and offered as potential recompense in tributary relationships.

However idealized, the Shang might be envisioned in terms of concentric realms marked by varying degrees of dominance and influence. The personal lands of the royal house, close clan members, and a few privileged "others" were located in the core; somewhat farther out were the more distant clan members and subjects who had been integrated into the Shang hierarchy.[44] However, the degree of control diminished as the distance increased, with the truly outer peoples or proto-states generally termed "*fang*"—a somewhat problematic appellation that may have been reserved for enemy states—vacillating between allegiance and relative independence, sometimes even aggressive opposition and outright rebellion.

Because Shang rulers certainly thought in these terms, the submissive were expected to furnish tribute (including the all-important divinatory media) and participate in military campaigns.[45] The king controlled the disbursement of core lands[46] and probably much of the terrain in the extended subjugated realm and also awarded a number of titles such as *po* (duke) and *nan* (earl) to the privileged, no doubt in recognition of actual power rather than as an attempt to create a systematically enfeoffed hierarchy such as would be implemented by the Chou, even though there seems to have been a basic distinction of *nei-fu* and *wai-fu*, submissive clan members and external peoples, respectively.[47]

Shang power and culture were at one time believed to have been confined to the middle and lower Yellow River. However, although they never reached the upper Yangtze, extensive Shang sites have been discovered in most of Shandong, the middle and southern parts of Hebei, all of Henan, eastern and southwestern Shanxi, middle and southern Shaanxi, most of Hubei, the northern part of Hunan, northwest Jiangxi, and western Anhui. Based on archaeological evidence, especially bronzes, weapons, and the exploitation of natural resources, it can be concluded that the Shang completely dominated in some areas, but indigenous local cultures remained vigorous in others. This immediately raises questions about the nature of the Shang presence: Was it a purely military occupation, a colonization carried out by a collateral member of the Shang ruling household, or merely the result of recognizing a powerful local chieftain who had nominally submitted and thus retained some independence even though accepting the main trappings of Shang culture, voluntarily or otherwise?

Traditional accounts state that 10,000 states all paid homage to King T'ang immediately after the conquest, obviously an exaggeration even if "ten thousand" is understood simply as "myriad" or some vague but large number, as in most military contexts. Nevertheless, this depiction well characterizes the evolving dynamics as contiguous groups realized the need to at least nominally acknowledge the suddenly ascendant Shang. However, over the centuries the degree to which clan members, allies, subjugated peoples, and external groups were actually subservient would fluctuate with the vibrancy of Shang central power. The number of clan leaders and peripheral lords paying homage at the court probably varied between much of the known, or at least readily accessible, realm to just Shang members resident in the core domain and a few close allies.[48] Even if, as claimed, relationships existed with some 500 peoples and states, only a fraction of them were at all noteworthy.

Recent archaeological overviews, coupled with hundreds of detailed reports, allow the vectors of Shang power to be sketchily plotted. Extensive evidence indicates that the Shang conquest of the Hsia quickly impacted not just the Yen-shih to Cheng-chou region, but also a widespread area where lower Erh-li-kang manifestations suddenly displaced or overlaid Erh-li't'ou (Hsia) cultural artifacts.[49] Several fortified towns constructed about the same time as or slightly later than Yen-shih and Cheng-chou illustrate the Shang's ability to project power and establish defensive strongpoints against incursion, as well as the thoroughness of early efforts to subjugate the recently conquered populace until they became somewhat assimilated, coincidentally attesting to the fictional nature of claims that the Shang dominated through Virtue alone.

However, as predicted by the *Tao Te Ching's* explication of extremes being unstable, apparently because of internal strife the Shang experienced a generalized contraction of its power and influence after a century or so. Once the capital was finally shifted to Anyang, Wu Ting managed to initiate a vibrant resurgence whose promise was subsequently subverted by the pleasure-oriented final rulers. Nevertheless, even under their reputedly debauched leadership, right through its waning days the Shang continued to vigorously pursue external campaigns into the east and southeast.

During the initial phase of their rule from Yen-shih and Cheng-chou the Shang aggressively expanded north to Chi-pei, south to P'an-lung-ch'eng, along the eastern bank of the Han River, down into the lower Han River and T'ung-t'ing Lake areas, and even into parts of Shandong and Anhui.[50] In the west, Shang forces penetrated to the farthest parts of Kuan-chung and up to Ch'i-shan (Mt. Ch'i) and Chou-yüan, establishing a presence in the western part of Shanxi, Shaanxi's northern plains, along the northern Wei River, and eventually Han-chung, though their influence rapidly diminished west of a demarcation line that runs down along the lower reaches of the Ching River in Yao-hsien.[51] In addition to quickly occupying former Hsia fortresses and enclaves in the west, including Tung-hsia-feng, Shang forces established bastions such as Yüan-ch'ü and a number of smaller citadels to control the perimeter, resist incursions, and provide staging points for military expeditions. As in the east, members of powerful clans also seem to have been deputed to occupy the hinterlands in Shaanxi but were supplemented by local lords.[52]

Several sites within Kuan-chung where settlements rapidly multiplied in the quest to exploit mineral resources represent the westernmost extent of Shang culture. However, despite an initially strong presence that saw mining expanded and foundries established at the outset, this formerly Lungshan/Hsia (Erh-li-t'ou) cultural area retained many indigenous traits[53] and managed to reassert its independence when the Shang contracted and turned eastward late in the middle period with the shift to Anyang.[54] Among the more interesting representatives are Lao-niu-p'o and Chin-hsi, the former located in the fertile Kuan-chung plains area near Hsi-an (Xian). Typical Shang manifestations found at this highly militarized site capable of controlling access in several directions include the palace, graves, and a horse and chariot burial.[55] The early Shang **Chin-hsi** site is located in the southern quarter of Chin-hsi, an area where the T'u-fang would become active. Being another military bastion erected in former Hsia territory, the tamped earth walls were constructed with techniques identical to those employed at Cheng-chou, and all the recovered artifacts are similar to those associated with the Shang capital.[56]

Sometimes suggested as having been a secondary capital, the bastion at **Yüan-ch'ü** was probably the most formidable fortified city constructed by the Shang out in the western region. Located roughly one hundred miles west of Cheng-chou along the course of the Yellow River in Shanxi, the fortifications fully exploited the terrain's strategic advantages.[57] Erected on a semiplateau that projects out from a western ridge of hills and towers some fifty to sixty meters above the surrounding countryside, it relied for its initial defense on steep cliffs that drop away on the north, south, and east. Although an assault might still be attempted from the west, it was protected from the north around to the south by encircling rivers. (The minor Po-ch'ing River crosses to the north as it descends to the southeast, until it merges with the Yellow River flowing eastward across the south.) In addition, the Yen River, located slightly to the northeast, flows from north to south, merging with the Yellow just to the east of the Po-ch'ing.

The presence of massive walls well integrated into a comprehensive defensive system defines Yüan-ch'ü as a bastion rather than simple administrative center. Further evidence for the site's definitive military character may be seen in the disorganized nature of the burials, the presence of dismembered victims in more than 50 percent of the graves, and a few corpses that still have arrowheads embedded in their limbs.[58] Located in the northeast corner of the plateau so as to more easily command the plains, the compound's walls extend 338 meters on the north, 395 in the west, 336 down the east, and about 400 across the south, creating a slightly distorted rectangle that enclosed 130,000 square meters. None of the remnant walls exceed 2 meters in height even with their excavated bases included, but they range in width from 6 to 15 meters and probably once reached at least 7 or 8 meters in height.

Lacking the deterrence of cliffs in the west, a partial double-wall system—the first known construction of this type in China—was employed on the western and southern sides to thwart attacks emanating from the hills. The northern portion of the western wall diverged to comprise a screening rampart for an eight-meter-wide interior gate opening that was thus well insulated from direct assault. Although this outer wall was considerably thinner, it created a passageway some seven to ten meters wide that not only allowed for protected entrance and egress, but also provided a staging area for troops. Remnants of a double-wall system and vestiges of a similarly shielded gate show it was replicated on the south side as well. A smaller gate was located at the northeastern corner and another presumably on the eastern side, though the latter is no longer visible.

As a further reinforcement and thus clear evidence that threats were expected to emerge from the west, a formidable 446-meter-long ditch roughly 8 to 9 meters wide paralleled the entire length of the western wall about 6 to 8 meters from the footing. Although it extended right down to the plateau's southern edge, the lack of water resources in this elevated area would have precluded its use as a moat, but the steep edges and impressive depth would have been sufficient to impede aggressors.

As might be expected from its location in the midst of a highly fertile, well-watered alluvial basin bordered by mountains on three sides and the Yellow River to the south, the site was continuously occupied from the Neolithic period. Cultural layers ranging from the Yangshao (when the town was protected with ditches) through Lungshan and Erh-li-t'ou are all visible beneath suddenly intrusive Erh-li-kang artifacts deposited when the walls were erected by utilizing earlier ditches to create the foundations and the town was turned into a fortress. The artifacts and various layers so far uncovered have stimulated the usual divergent conclusions on the site's likely date and exact political nature, but without doubt it was constructed shortly after Cheng-chou itself. Moreover, the sharp transition from Erh-li-t'ou to Erh-li-kang culture conclusively shows that this was a deliberate Shang effort that coincided with its westward expansion into Chi-nan and Yü-hsi. Whether it was focused on controlling the former Hsia populace, securing local mineral deposits, warding off threats and incursions from more distant peoples, or some combination of these remains uncertain.[59]

Eastward, what may be interpreted as Shang colonization and pacification efforts that probably commenced in the Cheng-chou era during Chung Ting's reign proceeded along the T'ai-hang mountain range out into Shandong and Hebei and the plains area. Hsing-t'ai, Han-tan, and Shih-chia-chuang are thus among the numerous Shang sites marked by heavy upper Erh-li-kang cultural manifestations. Artifacts recovered from Ta-hsin-chuang near Chi-nan City also indicate it was directly controlled by the Shang and probably had become a Shang city through the sudden displacement of Yüeh-shih (Tung Yi) cultural elements, though the latter retain prominence and were apparently well integrated.[60]

The well-known site of T'ai-hsi at Kao-ch'eng in Hebei also shows the sudden intrusion of Erh-li-kang culture into what had previously been intermixed Lungshan and early Shang strata with pronounced northern cultural (semipastoral) characteristics.[61] Situated in the corridor between the T'ai-hang foothills and the Yellow River, the fortified town occupied a crucial location from which movement in all directions, including access down into the Anyang area, could be controlled, explaining its evolution into a powerful regional center by the early Anyang phase.[62]

The discovery of various symbols of power, including great axes, shows that these were fundamentally Shang military regions under the control of a commander or ya. For example, a large bronze yüeh with an iron blade (!) and another bronze yüeh decorated with an animal motif have been recovered at T'ai-hsi.[63] Inscribed yüeh recovered from Su-pu-t'un, a site that has been tentatively identified with the former state of P'u-ku made famous by the Shang conquest, indicate that a "commander Ch'ou" (Ya Ch'ou) governed as ruler of P'u-ku.[64] Small bastions or strongpoints have also been found scattered around the southern and western periphery of Shang power, such as near Mt. Wang-chia in Yün-meng, T'ung-ku-shan in the Pa area in the upper Yangtze River, and Ching-nan-ssu, also in the ancient Pa area. Although marked by fewer decidedly Shang characteristics, they must have functioned as military emplacements that ensured the security of trade and trans-portation routes.[65]

Down in the Chiang-nan plains region, a complex area marked by strife between Lungshan and local cultures from the late Neolithic on-ward,[66] lies the heavily fortified town of **P'an-lung-ch'eng**. Evidently constructed about 1500 BCE in the heart of the territory that would sub-sequently be known as Ching, it seems to have been completely aban-doned during the interval between Kings P'an Keng and Wu Ting coincident with the rise of Wu-ch'eng as a major power center, evidence of ongoing Shang retrenchment after the initial period at Cheng-chou.[67] Situated some 350 miles south of the latter in the middle Yangtze river valley close to modern Wu-han at the highest point on the Han River, this bastion essentially replicated Cheng-chou both physically and culturally but on a considerably smaller scale. Although the greater site exceeds a million square meters, the actual walls total a mere 1,000 meters, in comparison with Cheng-chou's 7,100, with the basic fortified enclosure being just 290 meters north to south and 260 meters east to west. Having been constrained by the terrain's contour—the entire fortification lies on one of several contiguous fingers that project out into Lake P'an-lung—the walls outline a slightly distorted rectangle and encompass about 75,000 square meters.

The stamped earth techniques used for the foundation and walls are virtually identical to those employed at Cheng-chou, clear evidence not only of their erection slightly after the latter, but also that *hang-t'u* fortification methods had become highly systematized. The uniform layers average 8 to 10 centimeters in thickness, while the excavated walls retain a remnant height of 1 to 3 meters despite erosion having claimed at least a meter or two. However, their widths vary dramatically, ranging from 21 to 38 meters on the north and 18 to 45 on the west, the two sides with comparatively greater threat exposure, to between 21 and 28 meters on the south.

A 10- to 12-meter-wide exterior moat with a depth ranging between 3.9 and 4.6 meters that exploited a preexisting ditch and achieved its deepest point in the north further augmented the defenses. In addition, the lake effectively isolated and thus protected P'an-lung-ch'eng on three of its four sides, while the numerous rivers and lakes in the immediate region would have further impeded ill-prepared aggressors.

Without doubt P'an-lung-ch'eng was constructed in an inherently hostile area. The transition from Erh-li-t'ou to Erh-li-kang artifacts indicates the forcible imposition of externally based authority on a local community that had previously been engaged in acquiring and forwarding mineral resources to the Hsia capital.[68] Located at a transportation crossroads that allowed access to virtually every corner of China, including via the Yangtze up into Sichuan and down to the sea, and at the gateway to the vital copper and other mineral resources located in the middle and lower Yangtze that were being increasingly consumed by burgeoning Shang ritual culture, P'an-lung-ch'eng constituted a stand-alone, fortified city in the southern hinterland.[69]

The richness of the grave goods indicates the city certainly served as the headquarters of a local ruler, perhaps someone awarded ducal status within the Shang complex of dependent and annexed states. Although it has been argued he was a strong local chieftain who enjoyed generous Shang recognition, he may well have been a Shang clan member and the enclave a highly militarized Shang community.[70] The unusually large percentage of weapons found among the excavated grave goods offers further confirmation of the settlement's martial character. Apart from a unique, great symbolic axe of power, other weapons recovered from the ruler's tomb include three axes, five halberds, two spearheads, seven knives, and eighteen arrowheads. Whatever the commander's identity and status, P'an-lung-ch'eng proves that the Shang was capable of projecting power and consolidating control in a remote area, as well as levying the labor component necessary to construct a major fortified town.

The Chiang-nan plains region also saw the establishment of **Wu-ch'eng** in Jiangxi, another fortified city heavily imbued with Shang characteristics.[71] Far from the center of Shang civilization, Wu-ch'eng is located well south of the Yangtze at approximately the same latitude as Ch'ang-sha in the west and Wen-chou in the east. Situated on a low ridge, it rises above the Hsiao River, which flows from west to east just above it. Additional streams and bodies of water are found among the hillocks to the north and another stream runs along the south, all no doubt potentially effective in retarding enemy incursions, particularly because the Hsiao's sixty-meter-wide sand bed indicates it had a much greater flow in antiquity. Numerous Shang sites are scattered about the surrounding terrain, no doubt because of the area's fertility.

The walls enclose a relatively spacious 610,000 square meters, sufficient for a population of about 20,000. Three cultural layers roughly contemporaneous with Cheng-chou, early Anyang, and late Anyang or early Western Chou phases overlie the late Neolithic foundations. A few of the more than 500 items excavated bear primitive characters, some judged to be earlier forms of Shang oracle bone script, others deemed highly localized. Bronze artifacts include a few weapons such as a short knife with a protruding handle from the first period, a halberd and an axe from the second, and arrowheads from the third.

The stone and ceramic items are dominated by tools employed in handicraft industries and daily life, cooking utensils, and wine vessels, some of which were also cast in bronze. A number of stone molds designed to produce items in the style of artifacts from P'an-lung-ch'eng and the early Anyang phase have also been recovered. Many were created to produce arrowheads and similar small weapons, others allowed the casting of larger pieces such as axes, and a few even have inscribed characters. A small foundry has been discovered, but it would not have been capable of producing the full range of recovered items by itself.

Critical questions have arisen about the city's significance and interpretation because the ceramics and bronzes display a markedly indigenous character despite their basic similarity to Shang styles dating to Cheng-chou and thereafter. Some analysts have suggested Wu-ch'eng is a derivative culture from P'an-lung-ch'eng rather than a direct Shang enclave, but none have ventured any bolder evaluation than that the culture "was the product of an indigenous Geometric environment" that early on developed extensive metallurgical industries.[72] However, it is located at a strategic point that could have repelled threats from the south, west, and east and controlled the trade passing through from highly disparate regions. Even Wu Ch'i referred to this area when attempting to enlighten his ruler about the error of relying on advantageous locations when he said, "Formerly the Three Miao had Tung-t'ing Lake on the left and P'eng-li Lake on the right, but they didn't cultivate Virtue and righteousness and Yü obliterated them."[73]

The sudden appearance of Shang-style ceramics and artifacts (despite strong local characteristics) in an area of strategic significance suggests that the city or its associated proto-state had been abruptly integrated into the Shang hierarchy. The ruler might have been a submissive local lord, but given the site's pronounced military character, localized weapons production, presence of stone weapons in some graves, and the repeated appearance of the character for dagger-axe, it seems more likely that a Shang clan force must have temporarily occupied the area. In addition to immediately erecting fortifications, just as at P'an-lung-ch'eng they would have brought ceramic and bronze designs, accounting for their similarity to Cheng-chou even though necessarily overlaying indigenous styles.

The changes visible at **Tung-hsia-feng** in the west and P'an-lung-ch'eng down in the Chiang-han plains provide clear evidence of the Shang's formidable power and their determination to secure peripheral areas previously controlled by the Hsia, including by utilizing Hsia fortresses that dominated access to crucial mineral resources. At Tung-hsia-feng the Shang incorporated the Hsia defensive ditches within a much larger, irregular, but presumably rectangular-shaped, early Shang-fortified town that was protected by 8-meter-wide walls. Constructed from tamped earth layers some 10 centimeters thick, the 1.8-meter-high walls are said to have been comparable to P'an-lung-ch'eng and no doubt served the dedicated purpose of projecting power and controlling essential routes from the Shang heartland out to vital mineral deposits.[74]

8.

CHAOS, CONTRACTION, AND RESURGENCE

◆◆◆━━━━━━━━━━━━━━━━━━━━━━━◆◆◆

ALTHOUGH DISAGREEMENTS ABOUND over the causes and extent, without doubt the middle Shang—which may be defined as post-Chengchou but pre-Anyang and therefore the reigns of Chung Ting to Hsiao Yi or perhaps P'an Keng—was a period of contraction. As indicated by the absence of Shang artifacts, resurgence of indigenous styles, and ascension of new groups such as the Chou, numerous settlements in Shanxi and the western fortresses of Tung-hsia-feng and Yüan-ch'ü were abandoned to local populations.[1] In addition, a number of "foreign states" or *fang-kuo* that would prove troublesome during the latter part of Wu Ting's reign, including Ch'ien, Sui, and Ching, were evolving in Shaanxi about this time.[2]

Yüan-ch'ü's abandonment despite persistent threats from the western quarter suggests that exposed fortresses lacked the tactical power necessary to function as control points in relative isolation.[3] The Shang may have withdrawn its forces as part of a revised strategic approach or simply decided that the bastion had become an indefensible logistical burden because of its inability to rely on locally produced foodstuffs. Nearby relatively mobile steppe peoples such as the Kung and T'u-fang may have already been exerting enormous pressure, but the bastion's disuse could equally be evidence of imperial weakness or debauchery. However, even after vanquishing the local aggressors, King Wu Ting apparently chose not to reoccupy it or station a permanent garrison there, a decision that suggests the ad hoc nature of Shang military efforts, the difficulty of exerting control at a distance, an overall contraction of

Shang military power, and perhaps a general disinclination to maintain standing border forces despite having erected numerous strongpoints on the perimeter.

Clan disorder and conflict over kingship have been proposed as the impetus for the Shang's preoccupation with internal affairs as well as the decision to shift the capital, abandoning the well-developed, fortified city of Cheng-chou. Whatever the cause, the massive fiscal and man-power expenditures required to construct an entirely new city, including expansive palaces and substantial fortifications, must have diverted vital resources from power projection, if not military affairs in general. How-ever, the development of highly productive mineral resources in the east and southeast probably diminished Tung-hsia-feng's formerly vital role in controlling the nearby mountains.

The Shang also withdrew from P'an-lung-ch'eng, but Wu-ch'eng continued to flourish long after P'an-lung-ch'eng's decline. Wu-ch'eng's material culture continued to reflect early Anyang developments to some degree, but it clearly had shed central control and was no longer an integral part of the Shang realm. Many Shang enclaves in Hubei were also displaced by indigenous peoples, and the only outward-oriented activity, perhaps a collateral result of moving the capital eastward back toward the ancient heartland, was a weak expression of power into the nearby Shandong area that early on brought Lin-tzu within the shifting periphery and saw the establishment of a few eastern outposts.

It has traditionally been held that the Shang moved their ritual and administrative capital five times after King T'ang vanquished the Hsia and initially ruled from Po. Chung Ting reportedly shifted it to Ao, Ho Tan-chia from Ao to Hsiang, Tzu Yi from Hsiang to Hsing, Nan Keng from Hsing to Yen, and finally, in the most famous move of all, P'an Keng from Yen to the Anyang area.[4] Although the exact locations con-tinue to be much debated, they are generally confined to a broad corridor from Erh-li-t'ou to Anyang itself. Despite proposals of martial motiva-tion, these shifts remain enigmatic and arguments have been simulta-neously advanced that certain ones represent a movement away or a movement toward confronting such threats.

According to the generally accepted account, after Cheng-chou had flourished for about a century, in his first year of rule Chung Ting ordered the populace to embark for a new capital known as Ao.[5] The usual justifications have been advanced for his precipitous decision, including internal disorder, flooding, and a desire to confront the Tung Yi more directly. However, the last two are somewhat improbable: Cheng-chou doesn't show much evidence of flood damage, and it would have been foolhardy to endanger the state's administrative and ritual assets by making them more accessible to enemy forces (despite subsequent thinking about "fatal" terrain) by moving to confront the enemy. Nevertheless, ongoing conflict with the Lan Yi, a subgroup of the Tung Yi, is suggested by a *Bamboo Annals* entry that states, "in the reign of Chung Ting, the Lan Yi made incursions, some revolting and some submitting," and "when Chung Ting ascended the throne he conducted punitive expeditions against the Lan Yi." More likely, as reported in the *Shih Chi*, either unrest among the nobility or conflict over the royal succession probably prompted Chung Ting's policy of compulsory emigration. However, it seems to have initiated nine generations of turbulence that ultimately caused the Shang's power to wither so dramatically that the feudal lords no longer felt compelled to pay homage at court.

The remnants of a large Shang fortified city at **Hsiao-shuang-ch'iao** have been proposed as the site of Chung Ting's capital of Ao, though not without controversy.[6] Located 20 kilometers northwest of Cheng-chou and twenty kilometers from the Yellow River against Mt. Mang (anciently known as Mt. Ao), it dates to the Pai-chia-chuang period but employed construction techniques remarkably similar to Cheng-chou. An expansive city of just under 1.5 square kilometers, it was protected by a 3.5-meter-wide, 1.75-meter-deep, essentially rectangular moat that runs 1,800 meters from north to south and 800 meters from east to west.

Whether it superseded Cheng-chou (Po) as the capital or was a simultaneously occupied secondary capital where the ruler may have chosen to reside to project power against threats from the northwest or escape internal turmoil remains unknown.[7] Insofar as conflict with the Tung Yi seems to have constantly troubled the Shang during the middle period, the discovery of several three-sided stone spades in sacrificial pits at Hsiao-shuang-ch'iao that originated in the Yüeh-shih culture identified with the Lan Yi would seem to be evidence that Chung Ting conducted a successful campaign against them.[8] A hundred heads from captives who were subsequently sacrificed have also been uncovered that show heavy cutting marks.[9] However, the movement to Hsiao-shuang-ch'iao actually shifted the capital away from the friction zone just when Shang victories in these eastern clashes may have forced the Lan Yi to disperse into Shandong, initiating a substantial displacement of Tung Yi.[10]

After a mere twenty or thirty years King Tsu Yi reportedly moved the capital from Ao to Hsing. One likely location for Hsing is **Ke-chia-chuang**, a site that certainly was not occupied until P'an-lung-ch'eng and Yüan-ch'ü had been abandoned.[11] This retrenchment in the northeast represents a thrust back toward the early Yü-pei/Chi-nan heartland, while Ke-chia-chuang's location coheres well with traditional scholarly projections of Hsing having been positioned near Hsing-t'ai in Hebei. However, this shift would have brought the Shang core much closer to the Tung Yi, presumably spawning more frequent confrontations. However, the other proposed site, Tsu Yi's capital of Hsing (or Ying)—Tung-hsien-hsien—is also located at Hsing-t'ai. Although only preliminary reports have appeared, Tung-hsien-hsien seems to have flourished between Cheng-chou's decline and Anyang's initial development and apparently continued as a regional center to the end of the dynasty. A few oracle bones have been recovered, and it evidences a cultural stage similar to Huan-pei, but surprisingly lacks walls.[12]

Next in the traditionally accepted sequence of Shang capitals is Yen, tentatively identified with a site out in Shandong amid the Tung Yi, near what would eventually became Ch'ü-fu in the state of Lu. It apparently served as the Shang center for some thirty-three years and may have been Nan Keng's residence before he became ruler. Once enthroned, he presumably designated it as the official capital, thereby inescapably continuing the Shang's confrontation with sometimes hostile Tung Yi cultural manifestations.[13] Tsu Chia retained Yen before P'an Keng initiated the famous move to the Anyang area, somewhat away from the eastern threats that the Shang had been successfully repressing.

Acrimonious debate continues over the nature of P'an Keng's motivation for again shifting the capital, the actual location, and the reliability of the cryptic historical writings chronicling it.[14] Until recently it had been assumed that he was the first to occupy the area south of the Huan River known from Chou times onward as the "wastes of Yin," even though the lack of any oracle bones predating Wu Ting's reign has proven problematic.[15] However, the discovery of a large, well-engineered bastion whose radiocarbon dates just precede the facilities at Anyang has led to the well-reasoned proposal that P'an Keng ensconced the Shang at **Huan-pei**.

Located 19 kilometers east of the T'ai-hang mountains on the northern bank of the Huan River just above Anyang, Huan-pei's walls form a slightly distorted square that extends 2,200 meters from north to south and 2,150 meters from east to west and encompasses a massive 4.7 square kilometers. Though neither ditches nor moats reinforce the perimeter, the walls vary from 7 to 11 meters in width and were erected upon deep, carefully layered foundations that began with an interior side trench and were then expanded with a conjoined trench on the exterior.[16]

Huan-pei's immense size naturally prompted archaeologists to deem it one of the Shang's intermediate centers, possibly Hsiang, rather than simply a military bastion.[17] However, it is more likely that P'an Keng, Hsiao Hsin, and Hsiao Yi ruled from Huan-pei before Wu Ting ordered the erection of a new administrative capital when he assumed power, perhaps because a conflagration had heavily damaged the ritual complex.[18] (Whether the fire was accidental or the result of enemy action is unknown. However, if it were viewed as an expression of Heaven's will or interpreted as a fatal omen, psycho-religious factors could have compelled the move.) This would explain the lack of pre–Wu Ting oracle bones at Yin-hsü but not the slight chronological discontinuity evident between it and Huan-pei.

Surprisingly, after being abandoned the vast complex of highly functional structures at Huan-pei was never reoccupied during the Shang's last centuries, despite being perfectly positioned to block attacks from the north. Even if the rather unrealistic conjecture that P'an Keng moved the capital twice[19]—first to Huan-pei then back to the Cheng-chou area or even to Hsiao-shuang-ch'iao—were to be substantiated, Huan-pei's abandonment as a military bastion would remain tactically and strategically puzzling.

LATE SHANG: THE ANYANG PERIOD

The first ancient Shang site to be systematically explored, Anyang has yielded many of the artifacts and most of the oracle materials that underlie current depictions of Shang history and culture.[20] A period of renewed fluorescence clearly began shortly after P'an Keng or one of his immediate successors moved the capital to Anyang, an area that had once been occupied by predynastic Shang culture.[21] The forcible emigration of the populace seems to have prompted vehement opposition, compelling the king to issue proclamations acknowledging the burdens and inconvenience, yet mandating acquiescence on pain of death because of the situation's exigency. Unfortunately, despite its vividness and detail, "P'an Keng," the famous text purportedly preserving his pronouncement, is clearly a late fabrication and therefore useless except perhaps as a vestigial memory of their discontent.

P'an Keng's reasons for initiating this incomprehensibly massive effort remain unknown but likely stemmed from a combination of internal strife and external military pressure, because the immediately preceding era seems to have been marked by weakness, contraction, threats, incursions, and rebellion. However, other factors have been proposed, including exhaustion of the land, pollution of the environment, irrecoverable fire or flood damage, and widespread debauchery that could only be rectified by the imposition of draconian Virtue.[22] P'an Keng thus seems to have been trying to reassert royal authority in a pristine environment, unencumbered by political and personal entanglements, while Anyang was well situated to exploit numerous natural resources, including water, wood, and minerals.[23]

The lack of engineered fortifications at Anyang other than a lengthy moat has provoked questions about its character as a capital. As already noted, according to traditional texts and subsequent historical thought, capitals are defined by the presence of the ruling clan's ancestral temple, fortified palace, and administrative complexes within a district demarked by protective barriers and substantial exterior walls, whether freestanding or systematically conjoined with internal and external moats.[24] However, Anyang's opulence and undeniable role as the Shang's administrative and ritual center over the last part of the dynasty have prompted historians to seek an explanation for the incomprehensible absence of substantial fortifications, especially in the light of lessons that should have been learned from Erh-li-t'ou's vulnerability.

Even though the Shang had a lengthy tradition of ensuring site security before initiating other projects, Anyang's early kings may have lacked the necessary power to coerce the populace into undertaking the onerous burden of wall building, particularly if they were still recovering from a disaster that had consumed significant resources. Conversely, the ability to order the excavation of the extensive moat that connected the river on the north and the south and partially defined and protected the imperial quarters, whether under his immediate predecessors or at the start of his reign, indicates that Wu Ting's authority was more than sufficient to mandate the erection of walls or order the assignment of prisoners to the task.[25] However, insofar as his aggressive reassertion of Shang power ensured decades of relative tranquility, perhaps his successors had the luxury of simply immersing themselves in the pleasures of empire that ultimately enervated the state, causing the domain's contraction.

In addition to claiming that Anyang's rulers were simply oblivious to external threats or the city was protected by as yet undiscovered fortifications, the lack of perimeter defenses has been rationalized as justified by superior Shang strength,[26] an aggressive policy of projecting power and preemptively striking enemies at a distance, and confidence in the natural strategic advantages provided by the terrain. Early in the Warring States period the great general Wu Ch'i reprised the area's dominant characteristics while rebuking his ruler for believing a state could rely on natural strategic advantages to survive:[27]

> Marquis Wu voyaged by boat down the West River. In midstream he looked back and exclaimed to Wu Ch'i, "Isn't it magnificent! The substantiality of these mountains and rivers, this is the jewel of Wei."
>
> Wu Ch'i replied: "The real jewel lies in Virtue, not in precipitous defiles. Formerly the Three Miao had lakes Tung-t'ing on the left and P'eng-li on the right but they didn't cultivate Virtue and righteousness and Yü obliterated them.
>
> The place where Chieh of the Hsia Dynasty resided had the Yellow and Chi Rivers on the left, Mt. T'ai and Mt. Hua on the right, the

cliffs of Yi-ch'üeh in the south, and the slopes of Yang-ch'ang to the
north. But in his practice of government he didn't cultivate benevolence and T'ang displaced him.

The state of King Chou of the Yin Dynasty had Mt. Meng-men
on the left, the T'ai-hang mountains on the right, Mt. Ch'ang to the
north, and the great Yellow River flowing to the south, but in his practice of government he didn't cultivate Virtue and King Wu killed him.

From this perspective (the state's jewel) is Virtue, not the precipitousness of its defiles. If you do not cultivate Virtue all the men in
the boat will comprise an enemy state.

Another version of this event, recorded in the *Chan-kuo Ts'e* (*Intrigues
of the Warring States*), which is even more explicit in asserting the inad-
equacy of terrestrial strategic advantages under corrupt governments
and accurately describes the geographic features about Anyang, observes that Anyang had "Mount Meng-men to the left, the Chang and
Fu rivers to the right," and that "the Yellow River belted it to the front
while it was backed by mountains to the rear."[28] Wu Ch'i then asserts
that "it was marked by this precipitousness, yet because Yin didn't exert
himself in the practice of government, King Wu attacked him," and
boldly concludes that "the precipitousness and expansiveness of the
rivers and mountains are not sufficient for security."

No doubt a fabrication, the conversation of course reflects Warring
States military science rather than Shang dynasty sentiment. Moreover,
despite the enormous protective value of fortified walls (as attested by
Sun-tzu's admonition to avoid injudiciously assaulting them), Wu Ch'i
pointed out that their own state of Wei had already vanquished several
fortified cities. Nevertheless, Anyang was somewhat protected by the
T'ai-hang mountains to the west and northwest; the Huan and Chang
rivers respectively to the north, with the former flowing downward into
the southeast where it joins the Yellow River; and the Yellow River itself
running along the south.

In view of these purportedly strong defensive advantages, several
analysts have recently cited additional Warring States military theory
to argue for the city's essential invulnerability. Traditional Chinese military thought early on emphasized discerning and exploiting inimical
and advantageous features of terrain, eventually codifying them into

the proto-science of what might best be termed "strategic configurations of terrain." The *Art of War*, traditionally attributed to Sun-tzu and the acknowledged progenitor of Chinese military thought, devotes two of its chapters to classifying terrain based on discernible features and correlated operational possibilities. Thereafter, Wu Ch'i analyzed the plight of field forces on difficult and constricted ground, and the *Liu-t'ao* (*Six Secret Teachings*) enumerated the tactics appropriate to employing component forces on various types of broadly categorized terrain.

Sections in the *Kuan-tzu*, also compiled in the Warring States period, preserve some general principles for locating cities in general. However, long before the advantages and disadvantages of locations were explicitly assessed, early settlement builders had already been exploiting natural water barriers by simply choosing to inhabit river- and lakeside areas for their access to water. Just as the *Art of War* would discuss, armies that attempt amphibious assaults find themselves not only hampered and discomfited, but also easy prey in midstream for archers arrayed on the shore.

Shang strategists probably pondered the question of topographical advantages in somewhat simpler terms, and Chou forces had little difficulty fording the Yellow River (albeit unopposed) some distance from Anyang when they launched the expeditionary assault that ended the dynasty. However, in antiquity it was axiomatic that one should "value high terrain and disdain low ground."[29] The T'ai Kung said: "Occupying high ground is the means by which to be alert and assume a defensive posture" and Sun-tzu admonished commanders, "Do not approach high mountains, do not confront those who have hills behind them.[30] If the enemy holds the heights, do not climb up to engage them in battle."[31] The eclectic *Huai-nan Tzu* added, "heights constitute life, depths constitute death. Mounds and hillocks are male, gorges and valleys female."[32] All these statements reflect the decidedly disadvantageous nature of mounting tiring uphill assaults that must ward off falling missiles and require soldiers to strike upward against enemies in a superior position. Even more foolhardy would be crossing a deep valley and then attempting to storm city walls with exhausted troops and reduced numbers, which would make it difficult to achieve the historically attested minimum of about five or seven to one required for an assault to prevail.

Cities that encompassed any form of higher terrain, such as moderate-sized mounds, were therefore considered strong, not easily approached or overwhelmed. Even if the walls could be penetrated, the heights would provide natural vantage points for the compressed defenders, particularly if large buildings remained that could serve as strongpoints for a concerted defense. However, Anyang was neither situated on high ground nor closely backed by formidable mountains, and no mounds or interior defensive structures have yet been discovered, only the foundations for palatial structures.

According to Sun Pin's characterization of male and female cities, "female" or strategically weaker ones can, and by implication should, be attacked, but the stronger or male ones should be avoided rather than assaulted or besieged. However, Anyang's features do not cohere with Sun Pin's characterization of male cities, and the T'ai-hang mountain range is too far away to be of any defensive use even though it obstructed steppe raiders descending from the north and northwest, especially if the few passes were well blocked by fortified barriers. (Shang oracular inscriptions show the passes were repeatedly penetrated, virtual conduits for passage.) Conversely, Anyang most closely conforms to Sun Pin's description of a female city, essentially confirming its susceptibility to attack.[33]

Furthermore, despite Wu Ch'i's disquisition, none of the defensive advantages propounded under the rubric of "strategic configurations of terrain" are actually present at Anyang. Having been abandoned, even the formidable bastion of Huan-pei just north of the river afforded no protection. Instead, the site was exposed in almost every direction, bereft of protection apart from the minor rivers and a single artificial moat. Moreover, rather than difficult ground, the level plains about it constituted "accessible" and "tenable" terrain according to the *Art of War*'s classification: land highly suitable for military operations, devoid of natural features that might impede aggressors or obstacles that might be exploited as primary defenses. Even the expansive Yellow River that flows some distance from Yin-hsü itself, just beyond the intervening plains, was fordable at more than one location. Although well situated to project power and control the trade and transport routes into Shandong and the passes through the T'ai-hang mountains, Anyang's geostrategic advantages were thus clearly insufficient, particularly for a weakened state that had mounted a precipitous move.

A more likely explanation is that despite their internal problems, Shang leadership may have felt confident in the ability of peripheral fortified towns such as T'ai-hsi to blunt incursions in conjunction with external barriers and frontier posts of unknown size and strength. The forces deployed at dispersed bastions and concentrated in the secondary capitals that flourished at the end of the dynasty, especially Chao-ko, were presumably deemed sufficient to intercept enemy invaders and minimize the damage they might inflict, if not vanquish them. Since the Shang never suffered deep penetrations or damage of any consequence until the Chou invasion and effectively carried the battle to their enemies throughout the dynastic period, their confidence was not misplaced.

Multiple capitals would frequently be erected throughout subsequent Chinese history. Sometimes they were occupied sequentially, at others simultaneously, in which case one often functioned as the ritual center, the others as secondary administrative foci or seasonal residences.[34] Generally referred to as regional, intermediate, or secondary centers by modern writers, they were often highly militarized or offered rulers, especially more pleasure-oriented individuals, an escape from the capital's constraints. However, although never mentioned, another possible reason for administering the realm from a secondary capital would be the ruler's ability to force the nobles to cater to his whims, to travel there from their more comfortable, older quarters, thereby psychologically solidifying his power while destabilizing them and depleting their fiscal resources, though no doubt at the cost of serious antagonism.[35]

Late Shang ancillary capitals included Han-tan to the north, an area previously occupied by predynastic Shang culture and subsequently the capital of the state of Chao as well as a crucial geostrategic location;[36] Chao-ko to the south; and Shang-ch'iu, often said to have retained its importance throughout the dynasty as the original ritual center and the location of the oldest and thus most important ancestral temple. No doubt they were all envisioned as buttressing Anyang's natural strategic advantages rather than acting as alternatives to Anyang. Although archaeological confirmation is lacking, minor residual forces may well have been retained at Yen-shih, Cheng-chou, and perhaps Yüan-ch'ü, though P'an-lung-ch'eng and the bastion of Huan-pei just north of the Huan River had by then been abandoned.

Traditional historical accounts suggest that Chao-ko was located south of Yin-hsü, somewhere between the theocratic capital of An-yang and the extended boundary at Mu-yeh. Several sites have been proposed, with one in the district of Ch'i slightly to the southwest but still north of the Yellow River being the most likely. Local tales claim that Chao-ko was protected by three encircling walls that described a rectangle of about 1,000 by 600 meters, but no definitive remains have as yet been recovered.[37]

Chao-ko was probably conceived and functioned as a military bastion long before Emperor Hsin transformed it into an opulent pleasure center, thereby dissipating its essentially Spartan character despite the certain presence of significant military forces and his personal bodyguard. A single passage in the *Bamboo Annals* about his enlargement of the capital records his penchant for architectural grandeur: "During his reign King Chou considerably enlarged his city so that it reached Chao-ko to the south, occupied Han-tan in the north, and extended to Shang-ch'iu, erecting detached palaces and secondary structures throughout." Chou dynasty accusations of debauchery and perversity generally identify him with Chao-ko, which remained a somewhat distant, freestanding city despite the *Annals'* exaggeration, and some accounts claim he futilely sought refuge there after being vanquished at Mu-yeh.

MARTIAL ACTIVITIES IN THE ANYANG ERA

In addition to the time-honored but dubious traditional accounts, four "irrefutable" sources exist for studying the nature and evolution of warfare in the late Shang dynasty: fortifications, archaeologically recovered weapons, a few cauldrons with commemorative inscriptions, and the divinatory materials preserved on comparatively fragile turtle plastrons and animal bones already noted. Apart from a few characters preserved on early ceramics, these famous oracular inscriptions comprise China's earliest written historical materials and thus the key source for any reconstruction of Shang activities.[38]

Despite their large numbers and the extensiveness of their subject matter, the inscriptions suffer from inherent constraints, including serious questions about their representational validity. Insofar as they only record activities significant or troubling enough to compel the ruler to query the ancestors or Ti—whether to shift responsibility, gain their sanction, or invoke their assistance—they are necessarily limited in scope. Many subjects no doubt fell outside the purview of such invocations or were simply too mundane to report to the ancestors. In addition, the vast number of oracle bones so far recovered may constitute only a small percentage of the total created during the process of divinatory inquiry, and further discoveries may reveal radically different concerns.

Debate also continues about whether they are records of real inquiries or simply charges to the spirits beseeching their aid or seeking confirmation of the moment's appropriateness for some already predetermined course, an impression certainly conveyed by many explicitly martial inscriptions. Finally, although they were certainly regarded with respect, even viewed as numinous, and therefore collected for burial at designated locations rather than simply burned or discarded, there is no sense that they were intended to be preserved for posterity, to function as an archive of some sort, particularly as written records supposedly existed that have been lost or have disintegrated.[39]

The process of scrutinizing thousands of prognostications for over a hundred years, at first perfunctorily but now intensely and in detail, aided by fortuitous discoveries of significant groups and numerous inquiries on single scapula, has produced a number of valuable chronologies and useful topical studies. From the several thousand inquiries that focus on martial activities the outlines, though perhaps not yet the entire substance, of Shang warfare can be gleaned.[40] In particular, the names of allies and enemy states,[41] the major campaigns, cycles of incursion and plundering, commanders, troop strength, component forces, and various vestiges of the military organization can be glimpsed. However, ignorance still prevails with regard to any tactics that may have been employed apart from straightforward assaults or perhaps two-pronged attacks, as well as actual battle sites, frequency of casualties, and numerous other questions integral to any serious study of military history. Similarly, despite much speculation and many views having been offered, the nature and composition of the troops remain more a matter of conjecture than definitive knowledge.

Discernible motives for undertaking outwardly directed military activities range from a desire to react to incursions and thereby thwart or punish enemies that were encroaching upon the Shang or preying upon its allies; to consolidating the Shang's position; to imposing or enforcing Shang will; and, less frequently, to simply projecting power. Although the oracle bones vividly depict a state beset by threats and responding to challenges, because martial events are great affairs entailing the lives of the populace and prestige of the ruler, they tend to be disproportionately emphasized in divinatory inquiries. In contrast, peace and tranquility can often only be inferred.

Expansive bronze workshops,[42] great hoards of cauldrons and weapons, and extensive fortifications all provide evidence that the Shang employed tens of thousands of men in mining, manufacturing, farming, and state labor projects. Military campaigns that might be undertaken in response to perceived threats or as part of power projection efforts, although still major events, were no doubt easily sustained and therefore not as significant in the life of the state as might be imagined. Nevertheless, the numerous raids and incursions it suffered, some more damaging than others, must have troubled and enervated the powerful Shang.

Although the final resolution of many of these conflicts cannot definitively be known—the Shang apparently prevailed most of the time— oracular inscriptions and archaeological discoveries do show the era's warfare to have been both brutal and bloody. Heavy casualties and doom were inflicted upon the defeated, especially the Ch'iang, who were frequently enslaved or sacrificed on Shang altars. Even the briefest encounter was never the ritualized confrontation that supposedly characterizes "primitive warfare" or would be depicted in traditional writings of later centuries, but an intensely violent clash primarily wrought with shock weapons intended to destroy the enemy as quickly and thoroughly as possible.[43]

Finally, it should be noted that despite the study of oracular inscriptions having advanced to being somewhat more science than art, even the simplest statements not only often spawn divergent readings and opinions, but also frequently become embroiled in arguments about the probable transcription of characters and pronunciation of names. Apart from the inconvenience of these and numerous other technical and interpretive issues, the conclusions being reached are also premised upon fundamental assumptions about the nature of Shang divination. Except when the results have been verified or can be deduced from subsequent inquiries, they essentially ignore the question of whether a contemplated course of action was actually implemented. Thus, for example, the conclusion that two or three different enemies were simultaneously attacked may be not only unjustified, but completely erroneous.

9.

KING
WU TING, I

◆·◆·————————————————————————————◆·◆

APPROXIMATELY HALF OF the inscriptions recovered from Anyang date
to King Wu Ting's era, traditionally ascribed to 1324 to 1265 BCE
or some fifty-nine years, but now basically consigned to 1251 to 1192
or the more likely 1239 to 1181, though some would limit it to just 1198
to 1181 BCE.[1] In contrast with the vibrancy and variety of the queries
undertaken during his extraordinarily energetic rulership, the divinatory
materials identified with later reigns rapidly become more routine and
circumscribed. Moreover, his years were punctuated by extensive martial
efforts as he personally took the field and frequently summoned others
to act as commanders of their own troops, sometimes even royal clan
troops, either in coalition actions or separately, but those recorded for
subsequent eras are relatively sparse. A comparatively detailed examination of Shang military activities during his rule thus proves more illuminating than extensive speculation about the nature of the relatively
few military events discernible in the latter part of the dynasty.

Prior to his assumption of power, a series of reputedly weak and
debauched rulers had allowed the Shang to be ignored, even insulted,
by supposedly submissive proto-states that not only failed to display the
proper ritual respect but also neglected to submit their tribute payments.
Contrary to claims that the south had remained untroubled, nearby
peoples mounted raids and made incursions from every quarter, seizing
goods and people. Former Hsia clans who had been compelled to acknowledge Shang rule centuries earlier certainly numbered among those
defying royal authority. Equally troubling, marauders repeatedly plundered many of the contiguous proto-states that still acknowledged
Shang authority.

Reacting to this erosion of prestige, the charismatic Wu Ting strove to reimpose royal control over nearby clan groups, tribes, proto-states, and remnant Hsia entities and revive the Shang's awesomeness throughout the realm.[2] Intermittent but intense efforts were made to respond to a range of affronts, some more blatant and damaging than others, and reestablish Shang dominance over the landscape, even though many forward outposts had long been abandoned. Contiguous rulers were coerced into appropriately submissive behavior through a combination of theocratically based political measures augmented by occasional displays of imperial force intended to psychologically subdue the recalcitrant. However, the truly belligerent ensconced in more distant realms became the target of military forays as Wu Ting launched expeditionary campaigns against dozens of states, turning from one theater of operations to another as he vanquished both the powerful and the merely troublesome.

Extensive study of the so-called oracle bones pertaining to King Wu Ting's reign with a view to classifying and dating their contents on the basis of sequenced events, relative date notations, names of ministers and commanders, and the mutual exclusiveness of diviners has suggested a three-part periodization as being the most fruitful.[3] This segmentation no doubt artificially obscures the consistent character of Shang warfare to some degree, and its validity depends on whether the king's reign was fifty-nine years, twenty-nine years, or just twenty years, from 1200 to 1181 BCE. (A shorter reign bespeaks a continuum with gradations, whereas a longer one is more amenable to distinct trends.) Nevertheless, even if merely a convenient means for grouping conflicts and responses, this periodization remains useful for discussion purposes.

Shang military activity in Wu Ting's era may also be deduced from the presence of readily identified artifacts and burial characteristics, particularly in Hubei and portions of Hunan during his later years, where Shang vessels and implements are found intermingled with clearly defined indigenous products.[4] Weapons, graves, and newly established, semipermanent strongpoints also provide evidence that Wu Ting's extroverted reign frequently deployed troops to Shanxi and northern Shaanxi. For example, a walled town at Ch'ing-chien Li-chia-yai in Shaanxi where both late Shang- and northern-style weapons have been recovered deftly exploited the terrain's features. Apart from being bounded by water on the south, west, and north, it also derived virtually insurmountable defensive advantages from hundred-meter cliffs bordering it on the north and south. To the east a defensive wall augmented these natural obstacles, completing the encirclement.

WU TING'S EARLY PERIOD

As might be expected, apart from one or two distant thrusts, Shang consolidation efforts during the initial period of resurgence generally proceeded from the nearby to the distant. Inscriptions recovered to date indicate that King Wu Ting contemplated attacking thirty or more enemies during this first period, though not all these campaigns were necessarily initiated. Prominent enemies from the outset included the Chih, Lung, T'u,[5] Ching, Li, Yin, Hsüan, Ch'ien, Chih, Ch'üeh, Kuei, T'ung, Hsien, Yüeh, Wang, Ch'iang, Kung, Hu, Kung, and at least twenty others. Military activity consisted primarily of short expeditions mounted to quell clearly defined threats and reassert authority over the recalcitrant. In most cases a single army, either royal forces led by the ruler himself or, more commonly, a commander supervising his own locally raised troops, went forth and successfully vanquished the enemy.

Generally these frequent but limited expeditions were not repeated, suggesting that most enemies were easily quashed or found early surrender enticing because the vanquished, in becoming prisoners of war, might be enslaved or sacrificed. However, in a few cases forces had to be levied again or otherwise dispatched, extending the campaign a month or two. Occasionally two armies were simultaneously mobilized, sometimes in conjunction with royal contingents, but directed against different objectives rather than operating as a single coalition army targeting the original enemy. However, not all the Shang's enemies succumbed so easily. Despite significant victories that occasionally elicited pledges of fealty, tribal peoples such as the Ch'iang were simply too powerful, remote, and mobile to suppress. For others such as the Hsüan and T'u-fang lengthy campaigns of six months or more and the participation of several generals in coalition-type forces would be required before they were conquered or compelled to disperse, often not until late in Wu Ting's reign.

Many of the conquered states were simply reintegrated into the Shang world of relationships, often becoming the site for Shang hunts or areas where the Shang developed agricultural fields[6] or built fortified towns to settle the people and stimulate economic expansion. Others such as Ch'üeh, Chih, and Yüeh evolved into loyal tributary states that furnished troops and commanders for future campaigns against other rebellious enclaves, causing them to become the subject of prognosticatory inquiry about their status and harvest. States that quickly capitulated, whether forcefully or not, were sometimes entangled through marriage alliances, and many of Wu Ting's consorts, including Fu Ching who came from the Ching, originated among them.

The prompt conversion of several early enemies into highly subservient (albeit perhaps unwilling) allies who could be employed to enforce Shang mandates significantly explains the Shang's ability to aggressively reestablish its authority and sustain field efforts against multiple enemies with little adverse effect. Even though Shang core forces sometimes participated in conjoined actions with these newly created allies, the Shang generally managed to avoid incurring casualties, exhausting their warriors, or depleting their treasury because the satellites furnished their own troops and provisions. As the portion of the world forced to acknowledge Shang suzerainty continually increased, these submissive but potentially dangerous, nearby entities had to expend their forces and resources. No doubt they accrued some benefits other than escaping destructive Shang vengeance, whether tangible material rewards or intangible recognition and integration into the Shang hierarchy, as compensation for their debilitation. Two or three local rulers became so esteemed that they were awarded command of royal clan troops or, such as Hsüan, were appointed to serve as one of the king's diviners.

The Chien-fang and the Chung were numbered among the persistent troublemakers in Wu Ting's early era. The former were defeated only through the concerted efforts of commanders from Yüeh and at least one other state, while the latter, located southwest of the Shang, were formidable enough for the king to personally lead 5,000 troops against them. Four other rebellious states that had to be vanquished or at least threatened with military force were Ch'üeh, Chih, Yüeh, and Wang. Reputedly located in Yü-hsi,[7] the Ch'üeh (also termed the Ch'iao) seem to have been the first to suffer Shang chastisement and quickly acknowledged Shang authority before their ruler became King Wu Ting's most active and successful first period commander.

Among Ch'üeh's first missions on behalf of the Shang was subjugating the proto-state of Chih, said to have been located in the Ling-shou area in the upper northwest corner of modern Henan.[8] Oracular inscriptions show that Ch'üeh and another commander named Fu were dispatched and that some three months were required.[9] Although the extent and ferocity of their conflict are unknown, Chih not only submitted but also subsequently became closely integrated into the Shang hierarchy of subservient states and sent in tribute. Chih's rulers also furnished two prominent commanders who were frequently called upon to direct punitive expeditions against the Shang's most powerful enemies in the middle and late periods, Chih Kuo and Chih Chia.

Two other groups targeted by Ch'üeh's campaigns in the early period, Yüeh and Wang, were reputedly located in the southeast rather than the west or northwest, the most frequent source of Shang aggressors. Situated on the upper reaches of the Huai River, Yüeh was attacked by Ch'üeh as part of an effort against them and the nearby state of Wang in the northwest corner of Anhui.[10] Both were eventually defeated—inscriptions query whether Yüeh wouldn't be seriously damaged in a coming attack[11]—then integrated into the Shang realm of obedient, contiguous states. The state of Wang would subsequently be noted for Wang Ch'eng, a highly effective commander in Wu Ting's middle period whose longevity was exceeded only by Yüeh, who served throughout the remainder of the king's reign.

SOUTHERN CAMPAIGN

In contrast to the violence reported on the northern, eastern, and western perimeters, it has long been held that the Shang had been largely untroubled by the proto-states located to the south prior to Wu Ting's reign. However, although written materials may be lacking and forceful attacks may have been avoided, artifacts provide decisive evidence that the Shang suffered a major contraction in the south immediately following the period of fluorescence at Cheng-chou, with P'an-lung-ch'eng and Wu-ch'eng returning to local control. Given the crucial mineral resources located in the south and the Shang's escalating appetite for luxury and ritual items fabricated from bronze, it is unlikely to have been an entirely voluntary retrenchment despite Shang resources being allocated to other priorities, including relocating the capital. As the economy expanded and the army mounted increasingly ambitious forays under King Wu Ting's aegis, attention was naturally turned to the vital south, an area that was routinely included in queries about prospects for the harvest and productivity in the four quarters.[12]

"Yin Wu" ("Martial Yin"), a Western Chou ode included in the *Shih Ching*, waxes rhapsodic upon a southern expedition attributed to Wu Ting's era that probably unfolded near the end of the first period:

> *Explosive the Yin's martial power,*
> *Fervently attacking the Ching and Ch'u.*
> *Deeply penetrating their narrows,*
> *Rolling up the regiments of Ching,*
> *Imposing order on their territory,*
> *Continuing T'ang's heritage.*

Several oracular inscriptions confirm that Wu Ting initiated an extensive southern campaign that required more than six months, was multipronged, co-opted the assistance of subordinate states, and proceeded through three routes of approach, including the now famous state of Tseng in the border region.[13] Nevertheless, rather than responding to actual incursions, the king seems to have been troubled by the threat potentially posed by proto-states in the "southern lands" (*nan t'u*) and their efforts to form alliances.[14] Ostensibly intended to reassert Shang authority over recalcitrant states and subjugate active enemies, the expedition almost certainly sought to regain direct access to crucial mineral resources in the middle and lower Yangtze and perhaps even marginally penetrate the upper Yangtze.

Several clan forces were mobilized and armies from the minor states of Tseng and Chü (located in Hubei, en route) enlisted for the effort.

Subordinate command was entrusted to two increasingly experienced leaders, Ch'üeh and Kung, but the king himself personally exercised overall authority from a forward position, reflecting the campaign's importance.[15] Whether because of the difficulty of the terrain or to minimize awareness of their approach, the forces were divided into three contingents. The king and royal clan forces are identified as being located on the right in accord with Shang esteem for the right, while the contingents from Chü and Tseng were assigned to accompany Shang components in the middle and left, respectively.[16]

The campaign's initial target was the proto-state of Yü, located somewhere in the Chü river valley in Hubei in what subsequently became Ch'u territory. Although largely unmentioned in traditional sources, Yü's forces must have constituted a formidable enemy because the king's prognosticatory inquiries evince real concern over their ability to inflict serious harm upon Shang forces and imply little reticence on their part to mount aggressive attacks.[17] Nevertheless, Wu Ting subsequently sacrificed some one hundred Yi in a single ceremony, which suggests far more prisoners were taken. Apparently conquered in just two months, the Yü simply disappeared thereafter.

Attacks were subsequently launched against two other proto-states located in ancient Ch'u territory, Kuei[18] and T'ung (also transcribed as Yung). The first thrust, directed against Kuei, was initiated in the eighth month and the second, targeting T'ung, in the tenth. The king regarded the conflict with the Kuei seriously enough to offer sacrifice for victory, and his prayers seem to have been answered because the Shang were once again able to redirect their efforts in just two months.[19] However, the T'ung were not so easily subdued, and Shang expeditionary efforts continued into the second month of the following year, when a powerful strike under the king's personal direction was planned that probably achieved victory, because the inscriptions begin speaking about "pummeling" them and Shang attention was soon turned to the Hu-fang.[20]

Although the T'ung required just four months to subdue, in contrast to many first period conflicts that saw one or another allied commander simply deployed, the overall southern effort exceeded a half year and entailed a major commitment of energy and forces, as well as the king's personal participation. Throughout Chinese history the south's numerous rivers, lakes, and marshes; dense entangling vegetation; and virtually impenetrable mountain ranges would always challenge and could easily immobilize armies experienced only in plains warfare. Moreover, as nomadic invaders from the cold, arid steppe would discover during the Sung, the heat, humidity, and rampant diseases always present but even more intensely in the summer and in semitropical areas rapidly debilitated men and horses unaccustomed to such conditions. As well attested by one of their seasoned commanders succumbing to these miasmic conditions, Shang troops must have experienced significant misery during the campaign.[21]

The warriors inhabiting these and other peripheral southern states were not just aggressive fighters who skillfully exploited the many features of their often inhospitable terrain, but also excelled in archery. (Members of the Yü/Yi were integrated into Shang campaign units after being vanquished because of their archery skills.)[22] Conversely, the heat and humidity pervading the south could rapidly render bows fabricated from northern materials essentially useless and often compelled armies to discard their missile weapons and rely on close combat. Nevertheless, the Shang managed to so severely vanquish their opponents that they never reappear in Shang consciousness.

Another southern state, the Tiger Quarter (Hu-fang), was probably centered somewhere between Tung-t'ing and P'o-yang lakes, though their culture, particularly as manifest in ritual bronze vessels with stylized tiger motifs, extended over a wider area, encompassing the well-known sites of Wu-ch'eng and Hsin-kan.[23] Situated in an area anciently noted for its numerous copper mines, the Hu early on developed a largely indigenous metal tradition marked by localized designs, unique weapon types such as a hooked *chi* and highly lethal arrowhead, and a far lower alloy content than that of the Shang. Even after the Shang method of employing ceramic molds for ritual pieces was adopted, the Hu-fang continued to use stone molds for utensils and weapons. Following the post Cheng-chou retrenchment, Shang decorative motifs were increasingly modified.

Because they were ethnically and culturally distinct, advanced enough to have their own basic writing system (composed from symbols apparently inherited from both the Hsia and Shang), and situated in an area of crucial mineral resources, the Hu were undoubtedly seen as a threat. However, the extent and course of their conflict remain unknown because only one set of inscriptions actually speaks about attacking them. An entry for an eleventh month states that the king will order Wang Ch'eng and (the ruler of) Chü, a peripheral state in northeast Hubei on the river Chü who previously participated in the more massive southern campaign mounted under Wu Ting's personal direction, to strike them,[24] but three others record the king as reporting to the ancestors (no doubt in a quest for their blessing) his deputation of Chü alone.[25] Whether Wang was victorious and the Hu submitted or the Shang beaten off and deterred from further action, as suggested, is uncertain.[26]

WU TING'S MIDDLE PERIOD

Wu Ting's middle period saw the Shang grappling with and generally subduing more serious enemies, including the Lung and T'u, generally referred to as the Lung-fang and T'u-fang in the inscriptions. This was also the period in which the king's famous consort, Fu Hao, became visible as the Shang's most active and successful commander. Several states were still vanquished with armies as small as 3,000 men, after which their lands were turned into hunting areas or agricultural terrain, and at least one of them thereafter furnished horses and provisions in tribute. But to conquer the Lung-fang, T'u-fang, and other strong enemies, several forces and field efforts of up to six months rather than single, decisive clashes were required. Whether this was because the enemy avoided decisive battles, numerous engagements were required to defeat them in detail, or a combination of both remains unclear.[27]

REBELLION OF THE HSÜAN, JUNG, AND WO

These formerly, if nominally, submissive states rebelled one after another while the king was preoccupied with attacking the T'an in the seventh month.[28] Provoked by an incursion in the sixth month, the Shang at first mounted a defensive effort against the T'an.[29] The king initially contemplated leading the attack (or going westward),[30] but seems to have deputed royal forces and then Ch'üeh that month, followed by others, including possibly Lin and Chih together.[31] Suppressing the T'an seems to have required a fairly short but intense period of activity[32] that produced a victory or two (as indicated by queries about capturing prisoners and the king being victorious) before Lin achieved the final conquest late in the eighth month. After this the T'an essentially disappear except for a final reference to Ch'üeh and Ko mounting an attack in the twelfth month, perhaps a residual effort to confirm the finality of Shang suppression.[33]

Various locations have been proposed for the T'an, ranging from a nebulous spot in the west to a rather specific site in the opposite direction in Shandong. Although proponents of the western site are in a majority, geostrategic concerns and historical antecedents argue strongly for the east.[34] One inscription shows Wu Ting expressing concern that the T'an were about to inflict severe damage upon the pivotal state of Ts'ao, a Shang outpost reputedly established by King T'ang himself when the Shang vanquished the San-tsung.[35] Archaeological evidence confirms an overwhelming Shang presence at a citadel near Ting-t'ao that was continuously occupied throughout the Cheng-chou and Anyang phases.[36] Since Ts'ao served as a crucial control point in the east, any detrimental action undertaken by the T'an would naturally evoke Shang concern. It also well coheres with Hsüan aggression in the seventh month, because they would have been more likely to exploit a power vacuum resulting from Shang forces being deployed eastward than if the latter were operating in a contiguous western area and could then easily be diverted to strike Hsüan population centers while their fighters were externally deployed.

The conflict with the Hsüan, located about twenty kilometers west of Yüan-ch'ü,[37] began when they and the Jung both rebelled late in the sixth month just after the campaign against the T'an had gotten under way. Whether their aggressions were merely opportunistic and independently conceived or part of a deliberate coalition action cannot be known.[38] However, if only out of self-interest and survival, the peripheral proto-states must have been aware of local military developments and consciously undertaken efforts to acquire relevant intelligence as well as cooperate with contiguous peoples not otherwise embroiled in relationships of mutual antagonism.

The damage inflicted by their incursions prompted the king to ponder deputing various peripheral commanders and officers to subjugate them. Suppressive efforts were initiated early in the seventh month while Lin coped with the T'an, and the Shang deputed the Wo and Ma to counter the Jung.[39] However, their efforts must have proven insufficient because the rebellion continued into the eighth month and even inflicted damage upon the client state of Ku. Ultimately numerous commanders were considered for dispatch and several, alone or in combination, no doubt ordered forth beginning in the eighth month, including Ko, Yüeh, Ch'üeh, Chih Kuo, Lin, Ku, and others, as well as the Dog Officer. Ch'üeh, who was most often appointed to sole authority,[40] seems to have requested the assistance of royal clan forces.[41]

In the twelfth month the king dispatched Ko and the indefatigable Ch'üeh to attack the Hsüan, but about this time the Wo discerned an opportunity in the chaos to escape the Shang yoke and similarly rebelled. Nevertheless, the Shang managed to prevail because inscriptions suddenly begin to ask whether certain generals will "get" or "obtain" the Hsüan and Wo,[42] including one pair that suggests Ch'üeh and Yüeh may have both been in action against them[43] and others that show Ch'üeh pursuing and capturing the enemy.[44]

Even though they were still capable of inflicting damage upon Shang forces at the end,[45] the Hsüan were eventually forced to capitulate, and their leader, who had previously served as a diviner for King Wu Ting, surprisingly was able to resume his role at the Shang court despite the severity of his defiance.[46] Their continued loyalty thereafter is also evident from the Chou deliberately targeting them during the drive to conquest, though the strength of the fortifications discovered in this area no doubt enhanced their importance as a Shang strongpoint.

THE CHI-FANG CAMPAIGN

A campaign tentatively dated to 1211 to 1210 BCE targeting the Chi-fang, insightfully reconstructed from chronological information embedded in oracular inscriptions that show several, often simultaneous expeditionary strikes being mounted against discrete enemies, provides an example of Shang efforts late in the first or early in the middle period.[47] In aggregate, even allowing for the likelihood that many queries merely reflect action contemplated rather than undertaken, the impression conveyed is one of unremitting (though sporadic) efforts involving multiple armies immersed in virtually continuous conflict.

Stretching over some six months from the twelfth month of one year through an intercalary thirteenth month and into the fifth month of the following year, Shang suppressive efforts against the Chi-fang, whose leader was known as Fou, unfolded amid multiple ongoing efforts against the states of Hsi (also transcribed as Chou and Tzu) and P'ei (similarly identified as P'i and Pu).[48] An initial though apparently futile inquiry sought to determine whether a strike mounted under the king's personal direction would inflict damage upon Hsi over the coming four-day period.[49] This was then followed by a more fruitful prognostication ten days later questioning if the Shang would be successful the following day, the militarily significant *chia-tzu*. A subordinate commander named Chu is then recorded as having prevailed through an attack that commenced late in the day and continued into the early hours of the next, *chia-tzu*. This clash is noteworthy for having been undertaken so late, perhaps close to dusk, and turning into a night encounter, showing that even though night battles were generally eschewed in antiquity, they sometimes occurred through happenstance or tactical acumen.

An inquiry early the following year after Hsi's subjugation indicates the king had been assessing the prospects for attacking P'ei and Chi-fang. He contemplated exercising command in both attacks, and a sequence of prognostications suggests that an unusually intense and prolonged clash with Fou unfolded early on at a place called Shu (or perhaps Hsün) under the king's direction.[50] However, he subsequently deputed responsibility for the Chi-fang to two well-known military figures, Ch'üeh and Prince Shang, who apparently mounted successive assaults against Fou in the second and third months.[51]

Particularly interesting is an inscription from the fourth month that clearly shows Wu Ting perceived a significant threat in the Chi-fang having begun work on external fortifications. (The query states: "We ought not [to allow] Chi-fang Fou to erect a wall. Prince Shang should destroy it.")[52] The immediate result was an aggressive, essentially peremptory strike, perhaps the first known historical example of defensive augmentation having prompted aggressive military action. Some ten days later, in the fifth month, the prince apparently succeeded in damaging the wall, and there are indications that Fou was captured and sacrificed to the ancestors, concluding a campaign that stretched over six months and entailed multiple objectives under the direction of several commanders, including Ch'üeh, Lin, Tien, and Prince Shang.[53] Whatever forces were employed, it probably typifies the sort of ongoing effort intermittently but persistently mounted against resilient enemies located in more distant quarters during Wu Ting's middle period.

THE CH'IANG

The Ch'iang, also known as the Hsi (West) Ch'iang or sometimes even the Hsi Jung, were a strong, energetic people who supposedly numbered among those who had originally acknowledged King T'ang's authority immediately after he conquered the Hsia.[54] Encompassing several large clans or tribes, the Ch'iang were dispersed throughout an extensive arc of territory west and northwest of the Shang, ranging from the eastern part of Qinghai through Gansu, Ningxia, southern Inner Mongolia, northern Shaanxi, and eastern and northern Shanxi.[55] Although southern Gansu was agriculturally productive,[56] the semiarid conditions prevailing in many northwestern locations compelled them to intermix pastoral, hunting, and agricultural practices, the more distant generally being more nomadic. However, this integration no doubt contributed to their resiliency and adaptability over the centuries.

If, as traditionally claimed, their ancestors included Yü the Great and elements of the Hsia populace that had been compelled to disperse onto comparatively inhospitable terrain after being defeated by the Shang centuries earlier, the Ch'iang would have certainly been predisposed to view the Shang with enmity.[57] Archaeological finds and inscriptional records show that, unlike the vanquished from almost all other groups, Ch'iang prisoners were enslaved or sacrificed as obliviously as cattle and pigs, in large numbers ranging from one through several tens to even three or four hundred.[58] Moreover, in addition to being the most frequently named group in the oracular inscriptions, the Ch'iang were the most often sacrificed, giving the appearance of a virtual genocidal campaign, particularly as members of the Shang ruling clan and various officials deputed on military campaigns were specifically tasked with capturing Ch'iang prisoners.[59] Those not eventually slaughtered in sacrificial rites were employed in farming, as household servants, and perhaps even in military posts, since one commander apparently came from the Ch'iang, though it is not known whether he had voluntarily emigrated to the Shang or was a slave or former prisoner.[60]

Prompted by Shang ferocity and perhaps inherent ethnic animosity, the Ch'iang (and particularly the Chiang clan with whom the Chou royal family frequently intermarried) eventually furnished important allies for the Chou, then similarly ensconced in the west.[61] Later traditions further assert that the Chou were anciently related to the Hsia and had deliberately imitated Hsia administrative and agricultural practices, possibly another factor contributing to their mutual affiliation.[62]

Fiercely independent even as prisoners,[63] like most steppe peoples they tended to be quiescent when the Shang was strong but readily exploited weakness and martial preoccupation to mount incursions that targeted both the core enclave and subordinated Shang peoples. Thus, the incipient dynamics of the steppe/sedentary or, as subsequently labeled by central government authorities, "civilized"/"barbarian" conflict that would plague Imperial China were already present in the Shang, the relationship between the two parties at any moment being defined by their relative imbalance of power. (Steppe aggressiveness thus becomes synonymous with imperial weakness rather than simply being a blind manifestation of inherent, anticivilized tendencies.)

Highly resourceful and tenacious, the Ch'iang proved troublesome throughout King Wu Ting's era and continued to be aggressive throughout the Shang, being identified under the last rulers as one of the *ssu pang* or "four allied states." Because of their remoteness, only minor clashes seem to have arisen during Wu Ting's initial period, when the king was just beginning to reassert Shang authority. However, they apparently mounted noticeable raids, as they were frequently the subject of inquiry, and generals such as Ch'üeh and Chiang, who were dispatched both individually and jointly to quell them, took some captives.[64] In contrast, Wu Ting's middle period must have witnessed violent, extensive incursions, as the king mobilized the most massive response of his reign to quash them. For example, in one sequence Chih Kuo commanded the initial response in the tenth month; a second force under

Shih Pan, dispatched in the twelfth month, undertook a pursuit that may have resulted in a significant victory; and yet a third effort under another commander captured prisoners in the first month of the following year.[65]

Nevertheless, perhaps because they were more dispersed than most of the Shang's other enemies, subduing the Ch'iang seems to have been an almost insurmountable task that intermittently required major campaigns throughout Wu Ting's era.[66] Troops were conscripted in quantities of 3,000, and the largest aggregate force recorded in the oracle bones, some 3,000 troops under Fu Hao together with the royal army, which just this once numbered 10,000 (and presumably went forth under the king's personal command), for a total of 13,000 men, was dispatched.[67] Thus, it is hardly surprising that some thirty generals, nearly all those known to have been active in this period, would eventually participate in campaigns against the Ch'iang, including Fu Hao, Chiang, Kuang, Lung, Wu, Yüeh, Chih Kuo, Ch'üeh, Ko, T'u, Wang Ch'eng, Shih Pan, and a number of officials, though generally not the king himself.[68]

As indicated by frequent queries about generals such as Yüeh seizing prisoners, these concentrated efforts apparently deterred the Ch'iang from making incursions and may have compelled them to temporarily withdraw further into the steppe to avoid decisive engagements. Nevertheless, neither increased force levels nor significant victories barred them from launching further attacks in Wu Ting's final period, particularly when the king was engaged in battling the Kung.[69] However, these later incursions seem to have been mounted on a reduced scale, and key commanders such as Ch'in readily achieved local victories in which numerous Ch'iang prisoners were seized.[70]

THE HSIA-WEI

The struggle with the Wei or Hsia-wei located in the southeast required a major effort during King Wu Ting's middle period,[71] just when the Shang was preoccupied with multiple challenges from the Meng, Ch'ing, Lung, Pa, Yi, and T'u-fang.[72] Wu Ting initiated countermeasures in the second month when he divined about ordering the Chief of Prefects to lead an attack against them in association with the well-known commander Wang Ch'eng.[73] Whether this campaign materialized or not, the king levied 3,000 men shortly thereafter[74] and then personally directed a field effort roughly a month later in which he was accompanied by Wang Ch'eng,[75] who seems to have become somewhat of a specialist in Hsia-wei warfare.[76]

After unknown events over the intervening eight months, which included requisitioning support from the allied state of Hsing,[77] Wu Ting ended the Hsia-wei threat in the eleventh month when, accompanied by Wang Ch'eng, he launched an expeditionary assault.[78] Thereafter the king is noted as hunting in their lands[79] and inquiring about their welfare.[80] Apart from an incident in Ping Hsin's reign, they seem to have remained quiescent, and Emperor Yi encamped there en route to attacking the Jen-fang. Nevertheless, a major effort had been required, whose importance is shown by the large number of pigs sacrificed for its success.[81]

Immediately after the Hsia-wei campaigns, one group of the Yi located in the west (as distinguished from the well-known eastern groups) was also targeted for punitive efforts under Fu Hao and other prominent commanders.[82] Apparently they were quickly defeated or chose to acknowledge Shang suzerainty, because they are subsequently noted as participating in the coalition actions against the Pa-fang, described below.[83]

THE LUNG-FANG

Located between the Shang and the Ch'iang but probably considerably closer to the latter,[84] the Lung or Lung-fang could not avoid becoming entangled in complex relationships with both powers. At times they were so subservient to the Shang that they undertook or participated in joint military actions against the Ch'iang, but at others they asserted their independence and mounted troublesome border incursions either by themselves or in alliance with the Ch'iang.[85] In order to confront and defeat the Ch'iang, Wu Ting deemed it necessary to first subjugate the Lung, thereby securing the nearby perimeter and neutralizing a potentially lethal enemy before venturing past them on a distant campaign.

The king personally directed attacks that apparently deterred them somewhat in the early period.[86] However, they retained the potential to be troublesome and continued to plague the Shang whenever Wu Ting became preoccupied with other peoples and areas, particularly late in the second period when the king's attention was diverted to the Ch'iang, Yi, Pa, and others. Several inscriptions show that the Shang would have to mobilize its best commanders, including Shih Pan,[87] Chih Kuo,[88] Fu Ching,[89] and others,[90] and dispatch them in multiple campaigns that targeted the Lung alone and in conjoined assaults on the Ch'iang[91] until they were defeated and apparently acknowledged Shang authority.[92] Thereafter, Wu Ting ordered them to conduct a hunt[93] and inquired about their well-being during a Shang campaign against the Ma-fang,[94] presumably because they were furnishing troops or had previously participated in actions against the Ch'iang.[95] Nevertheless, they apparently mounted at least minimal incursions late in Wu Ting's reign while the king was embroiled in the lengthy campaign against the Kung-fang, requiring the dispatch of suppressive forces.[96]

T'U-FANG

Clashes with the T'u-fang, which probably commenced early in Wu Ting's reign, continued sporadically until being resolved toward the end of the middle period.[97] More than a hundred inscriptions record numerous levies and the deputation of several commanders to quell their opposition, attesting to the ferocity and swirling nature of their conflict.[98] Believed to have been centered in the north, especially northern Shanxi and Hebei or about two weeks' march from Anyang, the T'u-fang were predisposed to raid the northern and western parts of the Shang despite the immense defensive power concentrated nearby.[99] No doubt the ready seizure of material goods and captives made plundering appealing, but an additional motivational factor may again have been the heritage of animosity engendered when their ancestors, the Hsia, were forced out into the semiarid steppe after being vanquished by the Shang.[100]

At least some of their incursions seem to have been mounted by surprisingly small forces, yet they proved to be extremely troublesome,[101] compelling Wu Ting to mount a strong response.[102] Apart from the king himself, who seems to have frequently exercised overall coalition command,[103] Fu Hao,[104] Yüeh,[105] Wang Ch'eng,[106] and especially Chih Kuo,[107] leading between 3,000 and 5,000 troops each,[108] were dispatched to attack T'u-fang forces over roughly a year and a half. Eventually the T'u-fang were defeated, their leaders slain, and numerous soldiers taken prisoner, prompting the remnants to either submit or move away to avoid decimation.[109] Their lands were then integrated into the Shang domain and opened for farming or maintained as hunting areas, resulting in a significant northern expansion.

Which inscriptions are deemed relevant and how they are arrayed can result in significant variations in the probable chronology of the T'u-fang campaign.[110] However, a likely sequence that easily encompasses the many Shang efforts mounted to combat them finds the first significant measures being initiated just when the Hsia-wei and Kung-fang were also proving troublesome and the king levied troops in the eleventh month for a punitive campaign against the T'u-fang.[111] In the twelfth month the king even deputed Chih Kuo in command of the San-tsu (three royal clan troops) to attack them,[112] and the next month, the first of the new year, the noted commander Yüeh apparently managed to take some captives, a likely indication of having achieved a limited battlefield victory.[113]

Nevertheless, the T'u-fang must have still posed a serious threat because suppressive efforts continued into the third month, when another 5,000 men were summoned under the king's direction, greatly augmenting the forces already deployed.[114] In the fourth month the king himself, accompanied by Chih Kuo, launched an expeditionary attack.[115] (Oracle records confirm that the king was sufficiently troubled by the T'u-fang challenge to offer sacrifice in the temple for the campaign's success.)[116] In the fifth month men were again levied, and the king's illustrious consort, Fu Hao, was ordered into the field for an assault.[117] However, even her efforts must have proved inadequate, because additional troops had to be called up yet again in the seventh month prior to the king initiating another attack.[118] Thereafter, in the ninth month the king once again directed an attack.[119]

After nearly a year of multiple conscriptions and numerous planned assaults, some measure of success must have been realized because the oracle strips dating to the tenth month speak about "damaging" the T'u-fang.[120] Little is known about the subsequent months, but the enemy could not yet have been completely quashed, because on the fourteenth day in the third month of the succeeding year, reports of multiple, apparently coordinated, Kung-fang and T'u-fang attacks were received, the most famous being from the esteemed commander Chih Kuo, which stated, "the T'u-fang have mounted a punitive attack[121] on our eastern suburbs, seriously damaging two towns. The Kung-fang have also invaded the fields in our western suburbs."[122] The king apparently levied another 5,000 men for the campaign in this same month,[123] but inimical reports were once again received in the fifth month.[124] Accompanied by Chih Kuo, Wu Ting then mounted another assault against the T'u-fang.[125] Perhaps it was this last effort that vanquished them, because they are rarely mentioned thereafter, even though some undated fragments speak about pursuing and capturing them,[126] and the Shang reportedly built a citadel out in their territory.[127]

OTHER CONFLICTS

Several states active in Wu Ting's middle period vacillated between being submissive (and accordingly entrusted with defensive functions) and rebellious, particularly in times of turmoil, and therefore aggressively targeted. One particularly visible group, previously mentioned, were the Ma-fang, located in the northwest,[128] reputedly slightly south of the Ch'iang, as evidenced by attacks that simultaneously targeted them both.[129] As attested by reports of Ma-fang activities and royal inquiries about the prospects for attacking them, a major clash apparently unfolded after the Lung-fang campaign during the winter.[130] Apart from being entangled in Ch'iang activities, they also acted as a Shang perimeter force and are recorded as acting in conjunction with the Wo to capture the Jung and being ordered to undertake archery in the north.[131]

Finally, a concerted campaign was mounted against the Pa-fang at the end of Wu Ting's middle years or early in the last period of his reign amid challenges simultaneously posed by the Yi, Lung, Meng, and Hsia-wei.[132] Pa-fang's location remains a matter of controversy, for rather than Sichuan, site of famed Pa and Shu cultures, they apparently inhabited the near southwest, and numerous Shang artifacts have been recovered in the Ch'eng-ku region of Shaanxi, presumably evidence of

campaigns into the general region of the Pa, Shu, P'u, and Ch'iang during the Shang.[133]

The campaign apparently commenced by deputing Chih Chia, presumably leading his own clan forces, to mount the initial response,[134]
but further measures proved necessary when they encountered
difficulty.[135] Queries indicate that rather than relying on the standing
forces, men had to be levied and more than one general dispatched. The
king asked whether he should attack the Pa in association with Chih
Chia[136] and, separately, a commander named Hsi, or whether Fu Hao
should do so in conjunction with Chih Chia.[137] (Two inquiries recorded
on the same plastron close together indicate that the king was trying
to decide between attacking the Pa with Hsi and the Hsia-wei with
Wang Ch'eng.)[138] This flurry of activity apparently vanquished the Pa,
as inscriptions ask if the Pa are about to be defeated[139] and suggest the
final crushing blow was about to be delivered by the king in association
with Chih Kuo.[140]

10.

King
Wu Ting, II

❖◆———————————————◆❖

WU TING'S LATE PERIOD

Rather than simply reacting to threats, thwarting future incursions, or deterring peripheral states with limited campaigns, efforts in the third period focused on projecting power and thoroughly eliminating enemies. Perhaps because of the considerable success achieved against the T'u-fang at the end of the previous period, the final years of Wu Ting's reign were marked by campaigns against major enemies highly capable of withstanding Shang pressure, particularly the Kung-fang. A few experienced second period commanders remained active, but several new figures assumed martial prominence. Prefiguring subsequent conflict, the Chou became troublesome when they may have sought to exploit the power vacuum resulting from Shang preoccupation with the T'u-fang and Kung-fang. However, their rebelliousness merely elicited a strong response that coerced them into renewed submissiveness.

THE FANG

Another of the major peripheral powers that proved troublesome to the Shang both early and late in Wu Ting's reign, the Fang were located somewhere in the general area of middle and southern Shanxi, perhaps centered in modern Hsia-hsien.[1] Variously identified as members of the P'eng or Yi, though not necessarily either, they would remain formidable enemies in every reign from Wu Ting to Emperor Hsin.[2] The Fang were easily capable of mounting damaging invasions that penetrated right to the area of the capital Ta-yi and apparently did so on several occasions.[3] Uncertainty about the outcome of engagements—who would prevail, who perish[4]—prompted a quest for intelligence about them[5] and repeated queries about whether they were coming forth in large numbers.[6] During Wu Ting's reign they plundered allied Shang states[7] and invaded the core domain, taking small numbers of prisoners in some raids[8] and destroying towns in others.[9] King Wu Ting was compelled to levy troops and initiate resistance efforts, and on at least one occasion the Chief Horse Commander led the chariots in a rapid response.[10] Several generals and a few officials such as the *li* were eventually dispatched to quell them.[11]

Relatively quiescent during Wu Ting's middle period, the Fang came forth aggressively in great numbers from the seventh through thirteenth months one year late in Wu Ting's reign.[12] (Incursions had already occurred in the fourth, sixth, eleventh, and twelfth months of previous years, showing an absence of seasonal preference.) During this uprising they assaulted and damaged the incipient Chou, a subgroup of the Yi, and finally the Shang itself.[13] Wu Ting took the field to vanquish them[14] and also deputed others in command of separate forces, including his *shih*,[15] who apparently suffered a severe defeat,[16] prompting another commander's dispatch in the eighth month,[17] and then Yüeh, who apparently succeeded in capturing the leader and temporarily ending the threat.[18] (Their submission is confirmed by another inscription showing them once again acceding to Shang dictates.)[19]

THE KUNG-FANG

Although it has been suggested that the Kung or Kung-fang were centered in the Chung-t'iao-shan region or northwest of the T'ai-hang Mountains, they seem to have inhabited an area northwest of the Shang in southern Inner Mongolia, Shaanxi, and northern Shanxi.[20] Various Kung groups frequently raided Shang allies and subordinated peoples including the Yu, Ch'üeh, T'ang, Hua, Chih, Ko, Fang, Ching, Lü, and Fu, proving that strong peripheral proto-states and tribal peoples could survive even in the face of the Shang's great, but obviously still circumscribed, power.[21] They not only plundered towns, sometimes several at once, and confiscated provisions, but also seized prisoners and cattle,[22] the latter having become readily obtainable, conveniently self-powered objectives as a result of Wu Ting's encouragement of agriculture and animal husbandry.

Kung raiding parties also rampaged eastward into the Shang heartland. Even though the contingents were small, the frequency and geographical scope of their aggressiveness forced Wu Ting to respond dramatically or suffer a debilitating loss of prestige. As chronicled by some 400 oracular inscriptions, these intertwined incursions and responses constituted the longest-running conflict of his era. Early incursions must not have been considered particularly damaging, because Wu Ting had essentially ignored them in favor of assaulting the Pa, T'u, and Hsia-wei.[23] However, they not only continued throughout his reign but also grew more intense after Fu Hao's death, finally concluding after a three-year concerted effort largely under the king's personal direction.[24] Prior to initiating suppressive measures, Wu Ting neutralized the T'u-fang so as to ensure that they would be unable to exploit the army's absence and make further inroads.[25] Before the Kung-fang would finally be vanquished, numerous generals had to be summoned, sometimes participating in campaigns with the king or individually deputed to go forth and crush the enemy.[26] These include Fu Ching;[27] Ch'in;[28] Wang Ch'eng;[29] Chih Kuo,[30] who seems to have led the king's advance forces[31] and whose fate was a subject of inquiry;[32] Yüeh, who similarly attacked the Kung[33] but was also severely imperiled by them;[34] Fu;[35] several princes, including Hua[36] and Shu;[37] Shih Pan;[38] a number of high ministers; and at least two members of the proto-bureaucracy, the Tuo Ch'en[39] and Tuo P'u.[40] Troops were specifically levied at least three times and the various commanders sent out in succession, whether to continually harass the enemy, reinforce field troops, or act as separate forces to confine the obviously mobile enemy.

The final, concerted campaign against the Kung-fang, which required more than two full years to conclude, has been reconstructed from a number of oracular inscriptions.[41] Having previously encountered little opposition because of Wu Ting's preoccupation with the T'u-fang, the Kung-fang were particularly vigorous and troublesome. In the thirteenth month of a year late in his reign, the king again received a report that the Kung had come forth and plundered four towns, including Hsien,[42] a site that would eventually serve as the springboard for their counterattack in the first month of the next year.[43] Whether the forces dispatched in response failed to intercept the enemy or the engagement proved in-decisive, the Kung reappeared eighteen days later.[44] The third month then witnessed simultaneous, apparently coordinated attacks by the Kung-fang and T'u-fang on Chih's eastern and western borders, as al-ready noted in the T'u-fang account.[45]

The Kung mounted another incursion in the fourth month, seizing ten prisoners.[46] This affront must have particularly rankled because the king quickly initiated a major suppressive campaign, levying troop num-bers of 3,000 and 5,000 men and personally going forth on what might be interpreted as a heavy reconnaissance effort (since the inscriptions specify he was going to "look into" the situation), no doubt intended to result in a major engagement.[47] Even though this was followed by a so-called punitive campaign of uncertain success, in the fifth month the Shang again suffered a Kung attack. However, the king somewhat surprisingly ordered Chih Kuo to accompany him in an assault on the T'u-fang.

In the seventh month at least three more sites were attacked, in-cluding the Wo, who were probably situated near Cheng-chou, well within the Shang heartland.[48] Since several inscriptions indicate that in the eighth and ninth months the king entreated the spirits for aid, the challenge posed by the Kung must have dominated martial con-cerns.[49] However, the spirits apparently remained unresponsive, be-cause a highly visible enemy offensive in the tenth month compelled the king to dispatch Yüeh and other experienced commanders to counter their unremitting aggression.[50] This suppressive thrust con-tinued into the thirteenth month, when the king and Yüeh again led forth campaign armies.[51]

The first month of the new year witnessed unabated Kung incur-sions.[52] Although datable records are lacking for the intervening period, even in the tenth month Yüeh, the campaign's commander, was still in the field battling them.[53] King Wu Ting further reinforced his efforts by dispatching coalition infantry forces to mount an attack in the twelfth month.[54] Finally, following the mobilization of further reinforcements in the fourth month of the succeeding year, inscriptions show that Shang forces were reaching and presumably pummeling the Kung, evidence that the conflict must have been nearing a conclusion. From the inquiry's negative formulation it appears the Kung were no longer capable of mounting an attack, and the main question had become whether they wouldn't be captured.[55] Thereafter, although their conflict probably did not fully conclude until the king's successor's reign (after which they are only sporadically mentioned), the Kung apparently migrated further out into the steppe, abandoning the inner lands to Shang occupation.

KUEI-FANG

Prior to the discovery of inscriptional hoards it had always been believed that Wu Ting's most illustrious military accomplishment was his victory over the Kuei-fang (Ghost Quarter), thought to have inhabited an extensive region across Inner Mongolia, northern Shaanxi, and northwest Shanxi.[56] According to line judgments found in the last two hexagrams of the *Yi Ching*, three years were required to conquer them: "Kao-tsung [Wu Ting] attacked the Kuei-fang and after three years conquered them" and "Fervently employ [the military] to attack the Kuei-fang and in three years there will be rewards in the great state."[57]

Perhaps because of the images invoked by their name and its connotations, the conquest of the Kuei-fang has always been far more symbolically charged than victory over any other peripheral predator. Given the apparent magnitude of the conflict, the absence of oracle records comparable to those found for the Ch'iang or Kung-fang is puzzling. Furthermore, the few inscriptions that refer to the Kuei-fang depict them as a reconciled people who submissively undertook the duties incumbent upon an allied lord, including fighting the Ch'iang and capturing prisoners, and were therefore a subject of the king's solicitous inquiry.[58] At least one Kuei leader played an active sacrificial role in Wu Ting's time and thereafter, seemingly precluding the possibility of enmity between them.[59]

Two theories have been offered to explain this apparent anomaly. The simplest yet most encompassing version concludes that since the Kuei-fang's distribution in the northwest coincides with that projected for the Kung-fang, T'u-fang, and Ch'iang, and the Shang required more than three years to extirpate them all, the term "Kuei-fang" must be a broad form of reference, one that encompassed all these tribes. Accordingly, the *Yi Ching* should then be understood as referring to the monumental totality of these conflicts.[60] Variants of this explanation limit the scope of the term, making "Kuei-fang" synonymous with "Kung-fang," the Shang's most prominent enemy,[61] or conversely, consider it to be similar to "Ch'iang," a sort of generic name for the tribal groups out on the steppe.[62] Because the *Yi Ching* was composed in the Western Chou several centuries after Wu Ting's era, the term "X-fang" is thus thought to have already become a rubric for any of the numerous peoples in the west or northwest with whom the Chou itself had come into conflict in its predynastic days rather than a referent to any specific tribe or enemy people.[63]

The second approach, which focuses on the nature and intent of the *Yi Ching* itself, emphasizes that previous events are cited for prognosticatory purposes rather than the creation of a historical record.[64] Moreover, contrary to traditional belief, it is assumed that the two lines refer to two different events rather than a single lengthy engagement with the Kuei-fang.[65] The most famous conflict of Wu Ting's era, his war with the Kung-fang, therefore somehow became the basis for the first entry and the name was simply transformed.

However, the second *Yi Ching* entry, which speaks about fervently rousing oneself to mount an attack that will require three years to succeed, focuses on trepidation: "With trembling they attacked the Kuei-fang and after three years received rewards from the great state." This is thought to refer to a conflict recorded in the *Ku-pen Bamboo Annals* for Wu Yi's thirty-fifth year: "Chi Li of the Chou attacked the Hsi-luo Kuei-fang and captured twenty barbarian kings." This interpretation turns on the comparative sizes of the Chou, Kuei-fang, and Shang, the latter being the "great state" that will eventually grant the rewards, while the Kuei-fang, also being larger than the Chou, will compel the Chou to evince an attitude of serious commitment. Moreover, the campaign was probably a crucial step in an expansionist thrust that initially targeted the very tribes that had compelled the Shang to move to Mt. Ch'i in order to gain control over the greater area and stabilize it.[66]

A subsequent expansion of this approach stresses the contextual implications of the two hexagram names, "Completed" and "Not Yet Completed," in interpreting the historicity of the lines.[67] In the first case, King Wu Ting, an appropriately powerful figure for undertaking a martial campaign, defeats an enemy known as the Kuei-fang, synonymous with the Kung-fang. However, the second hexagram describes a situation long prior to completion in which trepidation is required because the conquest of the "barbarians," a people marked by their "otherness," cannot but be difficult.[68] The psychological and sequential implications being manifold, the line's subject must prepare and strive to achieve the goal with its promised rewards. The hexagram is thus fraught with troublesome implications, the historical actuality of Chi Li's campaign less so.[69]

Finally, the *Chin-pen Bamboo Annals* contain an entry for Wu Ting's thirty-second year in which an attack on the Kuei-fang is noted,[70] and another for his thirty-fourth year that simply states his armies conquered them.[71] Although these entries would seem to confirm the *Yi Ching* account and provide a fixed date for these events, they may well have been derived from the latter rather than preserving an independent record. Nevertheless, despite the absence of oracular inscriptions, these classic entries asserting that an epic battle with the Kuei-fang occurred can perhaps be understood in the combined light of these theories. Obviously a significant clash occurred, no doubt the one between the Shang and Kung-fang, but the enemy's identity somehow became transmuted into the Kuei-fang. The *Yi Ching* authors must have been sufficiently struck by the intensity and scope of the conflict to envision it as the core referent for hexagram sixty-three, "Completed," in contrast with the idea of "Not Yet Completed" in hexagram sixty-four, itself illustrated by a later Chou campaign undertaken in radically different circumstances.

THE CHOU

Questions of inscriptional dating underlie any characterization of Shang and Chou relations, but they seem to have fluctuated among cordiality, concealed antagonism, and open belligerency.[72] It is well-known that two of the last three Chou kings, Chi Li and Wen, married women from the Shang, and that King Wen's wife was Emperor Yi's daughter. However, as attested by references to a "Fu Chou," one or more of Wu Ting's consorts must have originated among the Chou, and another Fu Chou appears in the inscriptions dated to the conjoined reigns of Wu Yi and Wen Ting.[73] Conversely, recently discovered inscriptions from Wu Ting's era inquiring about the possible death of a woman from the Shang royal house who had become the wife of Chung Chou's ruler indicate that marriages were made in both directions, no doubt to solidify the relationship.[74]

The Chou are sometimes noted as having sent items in tribute, including shells, shamans, and beautiful women, the latter perhaps a precursor of their subsequent efforts to exploit Shang licentiousness.[75] Shang rulers occasionally employed prognostication to inquire about their general well-being, including whether they would not be harmed by enemies such as the Fang or Ch'üan.[76] At times the Chou were referred to as the Chou-fang, indicating recognition but nonmembership in the Shang sphere; at others the ruler was termed a Hou (Lord) and even ordered to undertake some action[77]—both evidence of subservient status and having been at least somewhat integrated into the Shang hierarchy. A commander Chou, presumably the Chou leader at the head of his own troops, is known to have acted in a battlefield capacity similar to commanders from other subject states in Wu Ting's early period.[78] Ironically, King Chi Li was designated as a Mu-shih ("Shepherd"), sort of the regional chief of the nearby states, in recognition of his aggressive martial activities, but was eventually executed. His son King Wen was appointed as the Hsi Po, a term that means "Duke of the West" but is generally understood as connoting Western Protector, though both of them also submitted to being imprisoned by the Shang, undeniable proof of their subservient status.

A fairly high-intensity campaign that targeted the proto-state of Chou and succeeded in ending their overt rebelliousness for the remainder of the dynasty seems to have been undertaken late in Wu Ting's reign.[79] Whether it stemmed from a collision of Shang and Chou interests as a direct result of the Chou's growing presence at Pin or was triggered by a particular event is unknown. However, contrary to traditional accounts the Chou may have been trying to exploit the power vacuum created by Wu Ting's preoccupation with the nearby T'u-fang and Kung-fang to aggressively expand.

One possible sequence concludes that the Shang's efforts were initiated in the twelfth month when the leader of the Ch'üan (Yi), accompanied by his troops, mounted a destructive attack on the Chou.[80] Thereafter, from the third to fifth months the king dispatched him, several of the royal clan forces,[81] and others to continuously attack,[82] and had yet another minor lord assault the Chou in the eighth month.[83] Although Wu Ting remained aloof, additional Shang and allied commanders such as the ruler of Meng and Lord of Ts'ang were also dispatched at various times, indicative of the conflict's intensity.[84] Whether the Chou finally submitted to these onslaughts turns upon the interpretation of one fragmentary strip inquiring whether the Fang will severely harm the Chou.[85] Some scholars interpret the king's interest as indicating that the Chou had again become submissive, but others contend that they disappear from the oracle records because they were no longer a matter of concern.[86]

This sudden inscriptional absence has also been tentatively explained with reference to late literary records that assert the Chou ancestral leader known as Tan Fu moved the populace from their original site at Pin (probably in the Fen river valley) out to Mt. Ch'i in order to avoid "barbarian" ("Ti") pressures.[87] Insofar as this transfer presumably shifted them beyond the Shang's immediate sphere of concern, it is thought to account for their apparent quiescence.[88] Although the oracle bones indicate that numerous groups apparently assaulted the Chou, since the Ch'üan have been identified as being the K'un-yi (who numbered among the Shang allies conquered by the Chou) and as members of the Ti, they may well have been the aggressors responsible for the Chou shift.[89] Ironically, it would be ongoing pressure and finally an invasion by the Ch'üan Jung that would eventually compel the reputedly dissolute Chou ruler to abandon their dual capitals of Feng and Hao in 771 BCE and move eastward, precipitating the historical schism between the Western and Eastern Chou.

Traditional materials that record this shift, particularly the *Bamboo Annals*, date the Tan Fu's migration to Wu Yi's initial year, roughly fifty years after Wu Ting's demise. No doubt on the assumption that this sort of precursor to the ongoing steppe–sedentary conflict witnessed throughout imperial history had already become a constant factor, it has been suggested that the migration was the result of another attack mounted by the Ch'üan, prompted by a breakdown in Shang authority rather than by Shang enmity.[90]

Even though both explanations—defeat and displacement—are plausible, given the animosity marking Shang–Chou relations as chronicled by the *Bamboo Annals*, the martial nature of the Chou's eventual multidirectional expansion, and their formalized role as a frontier bulwark prior to the dynasty's overthrow, the Chou's almost complete absence from the late oracle records remains puzzling. A significant number of oracle bones have been found at Chou-yüan, though fewer than 200 preserve inscriptions and considerable controversy marks their interpretation.[91] Disagreement about their provenance has also resulted in confident assertions (based on their distinctive terms and a style of language similar to the Chou bronze inscriptions) that they must have originated in the Chou as well as equally vociferous denials that the content and perspective could not possibly be Chou, that they must have been left behind by visiting Shang rulers. (The second claim essentially focuses on these few in isolation, thereby ignoring the roughly 17,000 other bones that lack inscriptions, a number far too large to have been Shang remnants.)

The contents of three key inscriptions remain puzzling. The first records that the querent was planning to sacrifice two women, three rams, and three pigs to Kings T'ang and Yi of the Shang. One explanation holds that the sacrifices, presumably being offered by the Chou ruler, although unusual, were not impossible because the Chou were seeking protection from the ancestral spirits of their overlords. (This contravenes the idea prevailing from Confucius onward that sacrifices could only be properly offered within the clan and to one's own ancestors, though this proscription may simply have formalized desirable practice.) Moreover, because the Chou were also closely tied to the Shang through marriage relations, King Wen's consort having been the youngest daughter of Emperor Yi and his mother also having come from the Shang, he would have been doubly justified in seeking the blessings of the high Shang ancestors, especially his recently deceased father-in-law. Yet it must have caused him considerable consternation because his father had been slain by the Shang.

A second inscription beseeching the early Shang ancestor known as T'ai Chia for protection is problematic in the way it refers to Chou-fang

Po or the Duke of the Chou-fang. Interpretations vary dramatically, one being that it records a request to protect the Chou-fang Po either by King Wen himself or the Shang emperor upon appointing him as Western Protector, another that it preserves a query by the emperor about sacrificing him to the great Shang ancestor T'ai Chia.

Finally, the third, asking about prospects for a forthcoming hunt, is assumed to record a Chou divinatory query undertaken prior to the Shang emperor's arrival and is therefore cited as evidence that the Shang still retained confidence in the Chou's submissiveness.[92] However, the frequently voiced hypothesis that late Shang rulers avoided problematic areas and confined their hunts to their ever-dwindling, secure domain is extremely dubious because these hunts, though certainly motivated by a quest for pleasure and aggrandizement, had a marked military character. Apart from offering a chance to personally gather intelligence and overawe people, the hunt could even serve as a precursor for sudden military action and thus can be seen as confirming Shang suspicions.

In aggregate, accepting the majority view that these inscriptions originated in the Chou, it appears that its rulers were maintaining the façade of loyal allies by participating in Shang ritual activities even as they considered themselves kings and sought to augment their power. Having diverted their attention eastward, the Shang's last despots were probably content to avoid battlefield confrontations with the Chou as long as a nominal or at least fictive submissiveness existed.

WU TING'S COMMANDERS

Although various individuals, clan princes, and officials from the incipient bureaucracy were frequently entrusted with field responsibility, Shang kings such as Wu Ting personally participated in numerous battles, sometimes in sole command, at other times accompanied by well-known figures, whether as allies or in a subordinate role. Overall, it appears that Wu Ting commanded perhaps a third of their expeditionary campaigns, primarily those directed against formidable enemies such as the Kung-fang and T'u-fang, particularly whenever initial Shang efforts proved inadequate.[93] However, he also relied on a small group of commanders who received frequent assignments and were repeatedly assigned to battle certain enemies, undeniable evidence that martial knowledge and expertise were being increasingly recognized.

As might be expected, a number of commanders who rose to prominence early on in Wu Ting's lengthy reign perished or faded away before its conclusion. Martial responsibility was often assumed by leaders from allied states that had initially opposed the Shang, including four already noted: Ch'üeh, Chih Kuo, Yüeh, and Wang Ch'eng.[94] However, without question, the outstanding figure of the first period was the leader known as Ch'üeh (or Ch'iao), chief of a tribe or proto-state located to the west of the Shang, perhaps in either Chin-nan or Yü-hsi.[95] Vigorous throughout Wu Ting's middle period as well, he was repeatedly deputed to conquer such rebellious proto-states as Hsien,[96] Hsüan,[97] Mu,[98] Fu,[99] Sang,[100] Ko, Ch'ing,[101] Chi-fang,[102] T'an,[103] Wo,[104] Ch'iang (in the early period), Ts'ai,[105] Chi,[106] Wang[107] and Yüeh[108] (whose leaders both became famous Shang generals after submitting), Yu,[109] and others,[110] including a few whose names (characters) lack modern equivalents and thus remain unpronounceable, as well as clan states with the Shang surname. In total, he undertook more than twenty efforts on the king's behalf, sometimes even in conjunction with Fu Hao, as well as important thrusts with the king to the south in the initial period, far exceeding the four or five each directed by all others.[111] Even though he was not a royal clan member, he was still entrusted with authority over the royal army,[112] and his fate was frequently the subject of oracular inquiry.[113]

Although the tribal leader Chih Kuo became one of Wu Ting's four major commanders in his middle and late periods, he was initially counted among the Shang's enemies and similarly had to be forcefully coerced into submitting.[114] His participation in several important campaigns, including those mounted against the T'u-fang,[115] Pa-fang,[116] Ch'iang,[117] Hsia-wei,[118] and Kung-fang,[119] has already been recounted, but he was also instrumental in efforts against other proto-states, including the Jen-fang[120] and T'an.[121] Furthermore, he seems to have been chosen far more often than other prominent commanders to accompany the king and Fu Hao, suggesting that he specialized as an executive officer for the main commander or as the subcommander for a second, discrete force charged with independent but coordinated battlefield action.[122]

Wang Ch'eng became a staunch Shang ally after having been the objective of suppressive campaigns led by Ch'üeh and others. He eventually proved the key commander in the victorious effort mounted against the Hsia-wei, nearly always taking the field with the king.[123] Despite clearly specializing in Hsia-wei warfare, he participated in critical actions against the T'u-fang, Hu-fang,[124] and Kung-fang,[125] and in a few other conflicts, generally in alliance with the king.[126]

A number of women clearly played a vital role in Shang administrative and military activities during Wu Ting's reign. They were designated by the term "*fu*," an important character with the fundamental meaning of "broom," which, with the evolution of the written language, came to function as a signifier in the composition of Chinese characters, connoting "consort" or "wife." (In the Shang the concise form is already functionally identical to this later character.)[127] One theory holds that individuals designated as *fu* originated among the king's attendants, with the role perhaps having been hereditary among clans, accounting for the occasional reappearance of a name across generations.[128] However, more likely the term formally signified the king's highest-ranking consorts, with the second character in "Fu X" indicating her clan or state of origin.

About one hundred or so *fu* appear in the oracle inscriptions, including Fu Ching, Fu Hsi, Fu Lung, and Fu Chi. A few clearly attained significant status and exercised surpassing, though derived, authority. Perhaps after initially being responsible for managing ritual and internal palace functions,[129] they came to be entrusted with directing a wide range of external activities, including marshaling troops for military campaigns,[130] exercising active command in the field,[131] administering external areas,[132] providing for border defense,[133] and overseeing agricultural activities.[134] Most of the "Fu" who appear in the oracular inscriptions were the formal consorts of Shang kings, but a very few wives of powerful local lords, clan chiefs, and even high officials managed to acquire similar status. Collectively referred to in the inscriptions as the "*tuo fu*" or *chu fu*—literally, the many *fu* or more simply "the *fu*"—as a group they were sometimes entrusted with specific military responsibilities such as providing for regional defense.[135]

Apart from Fu Hao, one of King Wu Ting's three consorts, even the most famous Shang dynasty commanders are known solely through oracular inscriptions. However, the remarkable artifacts preserved in her tomb, only recently discovered, flesh out her portrait and dramatically verify the extensiveness of her political and martial activities. Fortuitously still undisturbed by robbers, Fu Hao's tomb included unusually massive cauldrons of superlative quality and entire sets of bronze vessels otherwise unseen. Totaling some 3,500 pounds, they provide evidence not just of her status and wealth, but also of Shang opulence and the vast scale of its bronze industry.[136]

Hundreds of jades and numerous weapons, including eighty-nine *ko* or dagger-axe blades, many with the identifying mark of Fu Hao or her posthumous title Mu Hsin, and a number of sacrificial victims have also been recovered. Two large axe blades, similar to those employed during ceremonies in subsequent ages to formally bestow martial power, inscribed with her name have also been discovered. These astonishing findings have prompted a closer examination of her life and her unique career as a martial hero as chronicled by some 250 oracle inscriptions.[137]

Most analysts believe she was active only during Wu Ting's era and died before the Kung-fang campaign, though whether in battle or from natural causes was not recorded.[138] Based on the composition of *"hao"*— a character composed of the character *"tzu,"* referring to Shang kinsmen, and *"nü,"* the basic character or signifier for *"woman"*—she may have originally been from the king's clan and probably entered Wu Ting's household quite young, in accord with regional custom in the period.[139] Prognostications indicate she gave birth to several children, both male and female;[140] was otherwise the subject of the king's inquiries about her health and security;[141] and became the recipient of sacrifices after her death[142] and even perturbed the king by appearing in his dreams.[143]

Fu Hao conducted numerous sacrificial ceremonies,[144] authorized divinations, and also sent in turtleshells to be used for divination while undertaking military duties out in the field.[145] Moreover, she apparently enforced clan discipline, even acting in a judicial capacity to apprehend a member who had committed a capital offense,[146] and handled inner court functions and foreign relations.

From the numerous inscriptions focusing on the auspiciousness of her military expeditions, she not only directed her own troops, but also served as overall commander in campaigns that included one or more experienced, well-known generals such as Chih Kuo, apparently acting as the king's alter ego in these efforts.[147] As already noted, she also had charge of (possibly) the largest recorded Shang force, some 13,000 troops, or more than four times the usual levy of 3,000 men. However, the actual nature of her role remains unknown, prompting some historians to claim it was merely symbolic. Nevertheless, since it is spoken of in the same terms as the king or any other general exercising command, such doubts seem unfounded. Despite the many weapons found with her name inscribed in tomb number five, including symbolic axes of power, whether she actually wielded any weapon, perhaps a bow, is another question.

Although Fu Hao participated in virtually every important campaign from Wu Ting's middle and early late periods, the great confrontation with the Kung-fang apparently came after her death.[148] She successfully attacked the T'u-fang,[149] Pa-fang (accompanied by Chih Kuo),[150] Yi (accompanied by Hou Kao),[151] Ch'iang (the famous campaign in which she was ordered to conscript 3,000 men, 10,000 more men were levied, and a number of prisoners were taken),[152] Lung,[153] and others.[154] In addition, she assembled and led an advance force that prepared the way for the main army in several instances[155] and also executed a coordinated, two-pronged attack in conjunction with the king in another.[156] Her martial career was thus extraordinary in terms of the number of campaigns directed, victories achieved, and realm of activity, because she commanded efforts in every quarter but the south. During her life she clearly enjoyed higher status than the other *fu*,[157] and around the time of her death she was already being emulated to some extent by Fu Ching, evidence that her role as a commander, though surpassing ordinary mortals, had not been an aberration.[158]

11.

THE
LAST
REIGNS

❖·❖————————————————————❖·❖

W U TING'S REIGN can be summarized as a progressive reassertion
of authority, expansion of the domain of control, and the projection of power beyond the core states in order to stifle independent
peoples who had sought to exploit increasing weakness by mounting
incursions and plundering the state's growing wealth. Benign trading
relations no doubt predominated, but the idyllic portrait of a virtuous
dynasty whose charismatic power enthralled contiguous states is certainly a highly romanticized construct of later centuries.

Many enemy actions were simply the sort of minor predatory foray
that would interminably plague Imperial China throughout its history.
As ephemeral as the bite of a mosquito, they were easily absorbed
and had no discernible impact on the central court apart from the
king perfunctorily ordering some minimal action in response to incoming reports. Others, particularly the challenges mounted by the Ch'iang,
T'u-fang, and Kung-fang during the latter half of Wu Ting's reign, constituted a matter of focal concern. Nevertheless, unlike the subsequent
Spring and Autumn and Warring States periods, when staggering fiscal
amounts and massive manpower would be allotted just to ensure a
state's survival, martial activities didn't extensively draw upon or deplete
the Shang's resources.

Comparatively fewer inscriptions have been recovered for the dynasty's remaining century and a half, but they are sufficient to discern
a general contraction of Shang power and a shift in military focus eastward. After Wu Ting's vibrant era, nine kings reigned from Anyang, in-

cluding the much-demonized Emperor Hsin, many of whom received the epithet of "martial," indicative of ongoing military activity. Based on the PRC chronology project, they have been assigned the following dates:

Tsu Keng, Tsu Chia, Ping Hsin, K'ang Ting, forty-four years, 1191–1148
Wu Yi, thirty-five years, 1147–1113
Wen Ting, eleven years, 1112–1102
Emperor Yi, twenty-six years, 1101–1076
Emperor Hsin, thirty years, 1075–1046

Under their aegis the Shang has traditionally but simplistically and incorrectly been depicted as descending into inebriation and irreversible weakness before finally collapsing.[1] Without doubt Shang authority receded in the north, northwest, west, and south, any strongpoints that had been established in the initial period of fluorescence at Cheng-chou or by campaign forces during Wu Ting's resurgence generally being abandoned. Formidable Shang towns like Chin-hsi and minority cultural enclaves such as Hsia-chia-tien (which derived its bronze and stamped earth techniques from the Shang) enjoyed newfound independence and a period of indigenous cultural resurgence. Despite military outposts and strong bastions that anchored what might be viewed as fingers of power, Shang awesomeness diminished somewhat even in the northeast, and the territory in which the king could freely hunt visibly contracted.

However, contrary to impressions of weakness and ineptitude, the Shang not only remained militarily active in the east and southeast, where the states and peoples would never be more than nominally submissive, but also increased their efforts out of various motives, including a quest for natural resources such as salt.[2] Their continued ability to undertake sustained expeditions eastward and down into the Huai River area suggests that any failure to maintain their earlier dominance over the realm stemmed from factors other than incompetence, corruption, or formidable external challenges.

Although the Shang probably originated in eastern China, and numerous artifacts and practices that show significant interaction and cultural intermixing indicate that the Yi were close, predynastic allies of

the Shang, both Shandong and southeast China became zones of contention during the dynastic era.[3] Relations with groups in these areas
varied from voluntary submission through indifference and outright rebellion. The submissive continued to accept Shang values, customs, and
aspects of material culture and were accordingly rewarded with nominal
participation in the Shang hierarchy. However, depending upon such
strategic factors as location, mineral resources, and degree of threat
posed, Shang clan forces were apparently deputed to establish control
over more hostile regions by exploiting the enticements of material
culture, awesome displays of power, or brutal force. The presence of
these Shang military colonies is well attested by ritual items, particularly
cauldrons marked with clan names such as Chü otherwise unknown
from oracular inscriptions, and disproportionate numbers of bronze
weapons, including oversized axes of authority and other dramatic symbols of power.[4]

Shang expansion into the east and southeast occurred in two waves,
the first while ensconced in Yen-shih and Cheng-chou and the second
during the Anyang phase, when aggressive actions and the development
of dozens of sites accompanied retrenchment.[5] Oracular inscriptions
inquiring whether certain commanders (including Fu Hao) should be
dispatched against the Yi or whether the king himself should assume
command show the first campaigns were initiated in Wu Ting's era.[6]
According to traditional records, during Wu Yi's reign the Tung Yi
moved back into the Huai and Hsi, areas previously dominated by the
Shang, and began to prove generally troublesome.[7] Some inscriptional
and archaeological evidence indicates that in addition to two well-known
expeditions that were undertaken late in the era, campaigns were
mounted against the Yi in all the other reigns.[8]

Late Shang relations with the Yi who inhabited the upper Huai were
equally characterized by friction and turbulence. Shang determination
to control the area is attested by the shift of the Ch'ang clan, a prominent
martial family, to the Lu-yi district of Henan sometime during the last
few reigns at Yin-hsü.[9] (Evidence for the family's military importance
is visible in a munificent Anyang tomb dating to late in Wu Ting's reign
that ranks second only to Fu Hao's in the number of ritual bronzes,
weapons, and jades recovered. Among the more than 310 bronze objects,

including square cauldrons emblematic of power, are seven axes, three large knives, seventy-one dagger-axes, and seventy-six spearheads, many of which are discussed below. Seven jade *chi*, two jade axes, and seven jade dagger-axes were also found.) Sacrifices were also being made to a commander Ch'ang immediately following Wu Ting's reign, and the clan may have been related to the king.

The discovery of a Lu-yi tomb containing numerous Shang ritual bronzes quite similar to those recovered at Anyang is interpreted as evidence that the clan controlled the minor state of Ch'ang located in this region. Although archaeologists have avoided speculation, it seems likely that the clan had been dispatched to a problematic area for security purposes rather than having originated there and early on provided the screening actions for which they were subsequently honored. Somewhat surprisingly, a number of early Chou dynasty bronzes were also discovered in the tomb and certain Chou burial characteristics noted, suggesting that the clan managed to remain powerful after the Chou conquest by acknowledging Chou authority.

Late Shang artifacts have similarly been recovered in Shandong at Chi-ning, Ho-che, Lin-hsi, and other locations in a pattern of diminishing eastward prevalence. The highest density has been found around two slightly separated Shandong locations, T'eng-hsien Ch'ien-chang-ta, which may have been the location of the former Shang capital of Yen or Pi and where the Shih clan seems to have exercised administrative and military control, and T'eng-chou Ching-hsiang, where a royal clan member, possibly even one of Wu Ting's brothers or sons, was ensconced.[10] Whether the indigenous peoples were nominally submissive or actively repressed remains unclear. However, wherever found, Shang culture primarily affected the upper classes, particularly those controlling the Yi clan states.[11]

In the Hai-tai region the struggle for dominance between the late Shang and the Yi in their incarnations as the Jen-fang, Yü-fang, Huai Yi, and various minor states such as Ku, Feng, Hsü, Ts'ai, and Yen saw conquest and varying degrees of displacement and amalgamation.[12] Extensive Shang artifacts have been found in the Chiao-tung peninsula, in the eastern Chiao-lai plains area (extending to Lin-tzu), in Anhui and northern Jiangsu down as far as the Yellow Sea coast, and even around the

Huai River. They are especially prominent in the area of Lu and around Chi-nan City, as well as Su-fu-t'un.[13] Whether Su-fu-t'un was a satellite state with foreign kingship or a purely Shang military site remains unknown, but the governing clan clearly must have been entrusted with military responsibilities because great symbolic axes (*yüeh* inscribed with "commander Ch'ou," a name that also appears in the oracular inscriptions) have been recovered.

CONFLICTS AND CAMPAIGNS

Inscriptions attributed to this era convey the general impression that the level of military activity, though intermittently intense, was only a fraction of that during Wu Ting's reign. Although this may be because insufficient numbers have been recovered to accurately characterize martial developments, it more likely stems from warfare (understood as an ongoing effort to defend the integrity of the borders and keep contiguous peoples reasonably submissive) having become more routine and thus normally assigned to standing units rather than undertaken on an extemporaneous basis. Even though the Shang had the power to undertake extensive campaigns, few appear to have been mounted, with those discussed below being exceptional. Furthermore, Shang kingship did not consist of static governance exercised from a single imperial site as in later history, but was highly peripatetic, essentially a form of "rulership in movement" intended to display royal power and facilitate, if not ensure, personal participation in local issues.

Even though comparatively fewer inscriptions can be attributed to the post–Wu Ting era and difficulties remain in assigning them to arbitrarily schematized reigns, Anyang military activity tends to be understood in terms of five distinct periods, the first devoted to Wu Ting, the remainder encompassing four pairs of rulers each. These late monarchs varied in their aggressiveness, some of them being particularly noted for provoking enmity among otherwise quiescent peripheral peoples. However, analyzing Shang actions purely in terms of these segmented reign periods obscures the essential continuity of regional dynamics and implies a causative relationship where none may have existed. Steppe/sedentary interactions over the centuries would always be highly complex, never a simple reflection of Imperial Chinese actions.[14] Sudden changes in peripheral aggressiveness may have resulted simply from internal issues (such as leadership clashes) or from external factors totally irrelevant to Shang attitude and policies, especially ongoing weather changes that may have caused food shortages, compelling predatory actions.[15]

Given the decline in temperature and moisture that resulted in some drying out and harsher conditions in the semiarid regions following Wu Ting's reign, it is hardly surprising that old enemies such as the Ch'iang and Yi were active following his demise. Others reappeared after apparently having been vanquished, and several new names emerged to command Shang attention, some briefly, a few across the era. Of particular interest is the apparently coordinated action undertaken by three or four states in temporary alliance, leading to them being termed the *san pang* and *ssu pang*, the "three allies" and "four allies."

Little is known about the relatively brief second period that basically refers to Tsu Keng and Tsu Chia. Although several clashes with the Kung-fang seem to have arisen, the major nemesis continued to be the Ch'iang, who mounted numerous minor incursions. Thereafter, despite indications that Tsu Keng and Tsu Chia had hunted in Hsia-wei territory, the Hsia-wei apparently rebelled and had to be suppressed during the third period identified with Ping Hsin and K'ang Ting. The king and the Hsiao-ch'en Ch'iang led a coalition campaign that captured weapons, chariots, and four of their leaders, who were subsequently sacrificed to the ancestors.[16] Cowed by their defeat, they apparently remained submissive right through the Chou conquest, because the last two Shang emperors again conducted hunts in their territory.

Infantry campaigns were mounted against the Hsiu-fang and the Hsiang-fang, both of whom would later reappear as members of the *ssu pang*. Hsiu-fang prisoners were captured,[17] and warriors seized from the Hsiang-fang were sacrificed by the Shang.[18] However, K'ang Ting apparently directed the era's major efforts toward the various Ch'iang,[19] who must have again become powerful as well as aggressive because repeated inquiries show considerable concern about the fate of expeditions mounted against them.[20] Multiple queries on the same plastron suggest the king carefully pondered what units to employ, which commander to appoint, what the response might be, and whether to further augment his field forces.[21]

Rather than ordering Shang allies forth as in Wu Ting's era, various clan forces, including the five clans, were deputed in response.[22] In addition, the *shu* (border specialists) were dispatched, indicative of a long-term defensive commitment over and above any effort to extirpate the enemy as well as further evidence of an ongoing shift from extemporaneous response to routine engagement.[23] Although their strength and mobility made the task difficult, allowing the Ch'iang to shift, reorganize, and resurge, they were eventually engaged in their heartland and vanquished. Apart from those slain, a number of prisoners were taken and two Ch'iang leaders sacrificed.

Despite the usual difficulties involved in apportioning the extant inscriptions to Wu Yi and Wen Ting, both fourth period rulers were noted for their aggressiveness, expansion of the standing forces, and love of hunting.[24] Later (and therefore merely speculative) writings also claim that Wu Yi was brutal and repressive and thus provoked revolts among otherwise subjugated peoples.[25] Wen Ting is generally known to posterity as Wen Wu Ting, the *wu* (meaning "martial," just as in Wu Ting's and Wu Yi's names) indicating his penchant for military activity. As usual, clashes that resulted in the capture of Ch'iang prisoners are intermittently noted, indicating that Ch'iang groups continued to be troublesome until Wu Yi's successor.[26]

Wu Yi's lengthy reign of thirty-five years (1147–1113 BCE) was punctuated by a few campaigns of note, including one against the Chih-fang who, despite having been quiescent since Wu Ting's era, now had to be vanquished by troops from the five clans.[27] Similarly, the Fang-fang, previously suppressed by Wu Ting, began to make incursions from out of the north,[28] prompting an active Shang response.[29] Even members of the Yi dwelling in the northwest had to be targeted anew for repressive measures that carried Shang forces out into Sui territory.[30] (The Yi presence in Sui is generally cited as proof of the proximity of the two groups, but their mobility and apparent mutual cooperation are more noteworthy, evidence that loose coalitions of contiguous states had begun mounting coordinated actions against the Shang.)

The Ch'a-fang, Hsiang-fang, and Sui-fang are similarly noted as having acted in concert, and the Ch'a also joined with the Man in making incursions.[31] (The Man had previously appeared in Wu Ting's period down in the south, especially around the area that would become Ch'u, but they also populated more northern areas. Just like "Ch'iang," the term came to encompass numerous peoples.)[32] Suppressive efforts mounted by augmented border defense units (*shu*) succeeded in neutralizing the threat from the Ch'a, who temporarily disappeared only to be counted among the *ssu pang* (along with the Sui) in the fifth period.

Little is known about Hsiang-fang (or Hsing-fang) activity in this period, but they are recorded as being jointly targeted with the Li[33] and were apparently vanquished since the Shang eventually opened fields in their territory.[34] However, without doubt Wu Yi's most important military effort was directed against the Li, a formidable enemy reputedly located in the Hu-kuan (Hu Pass) area of Shanxi that occasioned much Shang consternation, compelling the king to take the field.[35] He not only exercised sole command[36] but also directed Shang measures in conjunction with well-known commanders such as Chih Kuo[37] and Ch'in.[38]

A number of inscriptions inquire about pummeling the Li,[39] having the king's clan forces and the three clans pursue them,[40] seize them,[41] clear them out,[42] and even "kill" or "slay" them, a very uncommon term,[43] all of which indicate they must have been convincingly defeated. Based on an entry in the *Bamboo Annals* that indicates severe clashes between the Li and the Chou while the latter served as the so-called Hsi Po or Western Duke, at least one analyst not unreasonably believes that the conflict's intensity reflects a battle to control the intervening area.[44]

Few enemies or concerted martial efforts have yet been assigned to Wen Ting's brief era apart from routine repressive activities mounted against the Ch'iang.[45] The last two reigns of Emperor Yi (1101–1076 BCE) and the notorious Hsin (1075–1046) in the fifth period were each marked by lengthy campaigns intended to eliminate Eastern Yi threats.[46] Some one hundred inscriptions dating to Yi's era allow reconstruction of a punitive expedition personally commanded by the emperor against various members of the Yi, including the Jen-fang, in the east and southeast just across the Huai River.[47] The forty-one queries that can be attributed to the outward journey and thirty-seven to the return encompass twenty-five 10-day periods, commencing with the ninth month of his tenth year and extending into the fifth month of the eleventh year, or slightly more than eight months, variously calculated as 250 or 260 days, the former

expended in actual campaigning plus an additional 10 days for final sacrifices and presumably ceremonial demobilization.

After an initial detour northwest to conduct preliminary sacrifices
and announce the campaign to the ancestors, the army was marshaled
and ordered in the traditional hunting area of Ch'in-yang slightly southeast of Anyang. Rather than cutting overland and thereby being compelled to contend with numerous mountains and bodies of water, they
marched along relatively flat riverside terrain after emanating from
Ta-yi Shang.[48] Proceeding past Wei,the army reached You, a clan state
whose forces under the Earl of Hsi joined them and (in one possible sequence) then subjugated important Yi forces (discussed below). Continuing their progress, the conjoined forces turned down and crossed
the Huai River to engage their final enemy, the Lin-fang. From inception
to final victory consumed 106 days, but the army moved in spurts interrupted by encampments of a day or more for rest, training, hunting,
and ritual sacrifice. Furthermore, the inscriptions specify that they were
marching, implying only a very small chariot component was present for
command and control purposes, not unexpected because chariots were
ill suited for operations in the comparatively moist Huai river valley.

In contrast to the lengthy period required to reach their objective,
time on target was apparently minimal, because they are known to have
been at Ling-fang on *keng-yin* and returned on *kuei-mao*, the twenty-
seventh and fortieth days of the sixty-day cycle, respectively. The slightly
less than two weeks available at the southernmost point implies that
the conflict consisted of a single clash or at most a short sequence of
brief but decisive battles. After being joined by the Earl of Hsi, Shang
forces scored a preliminary victory over the Lin-fang, another member
of the Eastern Yi, and then vanquished the Jen, who must have been
startled at being attacked by a Shang army so far afield.

Unfortunately, the inscriptions are silent on the nature of the actual
encounters, only a couple of fragments suggesting that Shang forces
managed to capture a local chief and burn an enemy encampment,
certainly the first recorded use of incendiary measures. However, additional inscriptions indicate that the Huai River clashes had been preceded by what might be termed a "clearing-out" operation around the
Wei and Hua rivers in Shandong that focused on the area around Lin-

tzu and targeted the enemy's city of Ch'iu (or Chü) for attack. The assault had to employ a variety of forces, including left, right, and center border protective units (*shu*), masses (*chung*) from the left and right *lü*, and the king's own clan troops, to severely pummel and suppress the enemy.[49]

The return march paralleled the outward journey but initially progressed several kilometers south of the original route, often on the opposite side of local rivers, presumably to facilitate foraging and resolve other logistical problems, before revisiting most of the major fortified towns initially traversed. Despite making frequent, often lengthy stops to feast, hunt, sacrifice, and no doubt manifest the monarch's enhanced awesomeness, only ninety-nine days were required. Based on a limited number of inscriptions (including one on a skull) and a bronze commemorative vessel, it appears that a second campaign targeting the Eastern Yi was undertaken in Yi's fifteenth year. However, little is known beyond the fact that two tribal leaders were captured and sacrificed.[50]

Whether the Jen-fang campaign is attributed to Emperor Yi's reign or Emperor Hsin's, it is certain that Emperor Hsin commanded a Shang punitive expedition eastward into Shandong to repress the Yü-fang, another troublesome member of the Eastern Yi that had been growing in strength and attacked Shang interests at Mao-shih and Kao not far from Shang-ch'iu.[51] After starting out once again from Ta-yi Shang, imperial forces were augmented by clan troops from several nearby nobles. Despite the target being considerably closer, the effort consumed an entire year, during which the emperor exploited the intermediate terrain for logistical support, implying that displaying his awesomeness among the faithful might have been an equally important objective.

Scholars who ascribe three campaigns to Emperor Hsin's reign have frequently claimed that these lengthy campaigns exhausted the state,[52] yet in contradictory fashion contemporary PRC historians have blamed their success for fostering greater arrogance in an already dissolute ruler. Impetus for the former derives from two statements in the *Tso Chuan*, "the Tung Yi rebelled when Chou of the Shang held a martial convocation at Li," and "King Chou conquered the Tung Yi and lost his life."[53] The first instance is cited in the context of enumerating arrogant or presumptuous behavior in kings who coerced other states into gathering for a military assembly. (In this case the state of Ch'u not only held one, but then foolishly proceeded to invade the state of Wu.)

However, having unfolded in the eighth and ninth years of his lengthy reign,[54] Emperor Hsin's expeditionary effort would have occurred far too early to constitute a debilitating factor in the Shang's demise. In addition, once in the field the army certainly lived off the land and provisions furnished by submissive allies, reducing the direct burden on the state. Victories should have resulted in the acquisition of valuable spoils and perhaps large numbers of prisoners who could be forcibly employed in productive labor (if not sacrificed), increasing the Shang's wealth rather than depleting it. Without doubt these lengthy efforts attest to the power and willingness of the last two "debauched" emperors to mount extensive campaigns, whether simply motivated by the insult of obstreperous behavior or intended to be preemptive.

The Tung Yi clearly played a pivotal role in the Shang's ascension to power even though the Shang had acted on behalf of the Hsia in an attack on the Yu Yi as they were developing their own military power.[55] Perhaps from their inception, predynastic Shang and Yi (Yüeh-shih) cultures had been closely intertwined, many Yüeh-shih elements including core divination practices being incorporated by, or at least common to, the Shang.[56] Yüeh-shih participation in an uncertain but probably active conquest role is indicated by the presence of their artifacts among Shang deposits in the postconquest fourth period at Erh-li-t'ou[57] as well as around Yen-shih.[58] Prior to the final campaign the Yi had been known as allies of the Hsia,[59] their submissiveness being attested by Chieh's expectation that they would attend military conclaves when summoned. According to Mo-tzu's account, when King T'ang had first acted in a rebellious manner, the Nine Yi's response to Chieh's summons was all that deterred him from initiating an attack. Only after further developments, including a Hsia attack on a Yi proto-state, saw a similar summons fail did Yi Yin, reputedly a member of the Yi, conclude the moment to strike had finally arrived.

King T'ang thus not only neutralized the potential Yi threat but continued the Shang's close relationship with them and somehow persuaded or cajoled them into participating in the final campaign. Nevertheless, before the end of Hsin's reign the Shang would subjugate the area to the east, eliminate Yüeh-shih culture, and penetrate the southeast. However, the rapidity of the process and the exact nature of the displacement or conquest of Yüeh-shih groups, resulting in the amalgamation of Shang and Tung Yi cultures, remain open to question. Without doubt the process was clearly affected by topographical factors, being easier on the plains of Shangdong than in the mountainous region of the upper Huai, with its many tributaries and streams.[60]

Serious conflict apparently began during Chung Ting's reign when, despite reputed internal dissension and a general withdrawal from the P'an-lung-ch'eng region, artifacts recovered from Ao show a major victory must have been achieved and the immediately contiguous area in nearby Shandong subjugated. Sporadic military efforts over the Shang's two centuries of reign from Yin-hsü culminated in lengthy expeditionary campaigns under Emperors Yi and Hsin that concluded the archaeologically attested process of bringing the greater Hai-tai region and much of the upper Huai river basin under Shang control. The latter were never integrated into the Shang hierarchy of proto-states and military bastions;[61] nevertheless, these Shang thrusts initiated the separation and displacement of the Yi subsequently known as the Huai Yi, a process that would be completed in the early Chou.[62] Although it may be fashionable to attribute the Tung Yi's extirpation to simple cultural superiority, Shang violence, force, and outright military aggression thus played the crucial role in expanding its dominance to Shandong, northern Jiangsu, and upper Anhui.

Enmity with the *ssu pang* or four allied states is also frequently mentioned with reference to the final period. The four are normally identified as the Ch'iang, Ch'a, Hsiu, and Sui. The Ch'iang are of course well-known but apparently suffered significant defeats just prior to or during the early part of Hsin's reign, because the emperor is noted as hunting in their former domain[63] and performing divination there. The Sui had previously been active in K'ang Ting, Yi, and Hsin's eras, and even Wu Ting had deputed Ch'üeh and his own infantry to strike them.[64] The Ch'a, who were located near the Ch'iang and thus a western force,[65] had previously cooperated with the Sui during Wu Yi's reign and been temporarily dispersed. Threats from the Hsiu had also been deflected in the third period.

The era ultimately ended with the Chou's precipitous victory at the Battle of Mu-yeh following a rapid but direct march from their homeland in the Wei river valley. For whatever reason, Emperor Hsin's defenses were easily shattered and his fate quickly sealed. Extensive mopping-up activities and forays onto contiguous terrain concluded the effort, leaving many inexplicable questions, including the reason for the execrable performance of the Shang forces; Hsin's failure to anticipate the campaign in view of highly transparent Chou intentions and an aborted, earlier thrust; and the unfolding of the battle itself. However, these and other aspects, such as the nature of the forces and tactics employed, are best discussed in the context of the Chou's rapid rise and astounding conquest in our succeeding volume, *Western Chou Warfare*.

12.

THE SHANG
MARTIAL
EDIFICE

◆◆◆────────────────────────────◆◆◆

M ILITARY POWER DERIVES FROM NUMEROUS FACTORS, including administrative organization, the ruler's talent and charisma, the polity's material prosperity, the culture's martial spirit, and any propensity to control others and resolve disquieting situations with violence. Whether the monarch can act despotically in initiating aggressive activities, must gain the support of key clan members, persuade an extended circle of influential citizens, or even cajole the general populace into participating strongly influences a state's military character and its bellicose tendencies.

Although the early Shang kings probably monopolized power through the usual methods of successful chieftains, by Wu Ting's era the lines of succession seem to have been reasonably well established and hereditary authority fundamentally institutionalized despite the turbulence marking the middle period. Personal charisma, physical prowess, and martial skills no doubt remained vital to dissuading challengers and thwarting assassination. However, in the interstice between regicide and absolute obedience, major actions probably still required the acquiescence of clan members, as well as the immediate obedience of subjects and subordinates. Shang rulers ensured that these were forthcoming by monopolizing theocratic power and wielding authority over life and death.[1]

The right to communicate with the ancestors or spirits on high, entities who were believed capable of affecting every aspect of life ranging from personal illness through weather, plague, drought, and military incursions, was reserved to the king. (A few high-ranking clan members, including Fu Hao, the king's consort, sometimes also conducted divinatory inquiries, but the privilege was clearly derivative.) Even though divination became more perfunctory in later periods, insofar as the Shang populace acknowledged the preeminence of his transcendent authority, the king was empowered and his actions sanctified, none daring to violate the supreme will of the spirits.

Shang kings arbitrarily decided the fate of individuals and groups, selecting people ranging from relatives to prisoners for sacrifice and ordering punishments that they sometimes directed, including castration and decapitation. They could compel clan members and subordinates to undertake projects such as land reclamation, external missions, and military affairs, and their power even extended to the submissive protostates. Authority over external areas was maintained through a variety of means, including gifts, acknowledgments, emissaries, reports, inquiries, and military support for the endangered. Every successful activity, particularly military expeditions because they entailed the imposition of command and control measures, reinforced the king's power.

Insofar as the Shang conquered the Hsia and then claimed dominion over the nearby realm primarily through alliance building and intimidation rather than the dispersion of clan members who established incipient states, unremitting political and military efforts were necessary to preserve and enhance its position. Marriage relations, primarily achieved by the king taking a female member of an important clan as one of his consorts, were among the means systematically employed to strengthen ties with contiguous states. (King Wu Ting was especially prolific in this regard, at least fifty-two of his consorts being known from the oracular inscriptions, whereas only twenty-two can be ascribed to Kings Wu Yi and Wen Wu Ting in the fourth period.)[2] In addition to suffering a high degree of psychological intimidation, states were thereby entangled in marriage alliances and expected to furnish troops and logistical support upon demand.

The king undertook numerous peregrinations to assert his power and continuously interact with the disparate entities and peoples within the Shang's perceived realm. His rambles were undoubtedly multipurpose; armed forces always accompanied him, whether to engage in hunting or simply to provide protection and companionship. No doubt a primary objective was manifesting the Shang's "awesomeness," an objective that was accomplished through military display and the conspicuous show of ritual bronzes and other trappings of prestige, wealth, and power. Through ostentation the susceptible were psychologically subjugated, the hierarchical order consolidated, and the Shang's destructive potential impressed upon the world.[3] In fact, the term *"te fang"* seen in the oracle bones—literally meaning "virtue/quarter"—apparently entails the idea of conducting an imperial procession or tour of inspection intended to put the *fang* (external states) in order.[4] Gathering military intelligence and providing an opportunity for the king to assess subordinate states and semi-independent rulers in person were almost certainly collateral objectives.[5]

As attested by numerous divinations inquiring about the auspiciousness of going forth to hunt in various areas, hunting was certainly a major preoccupation, so common that it was denounced by the Chou when it charged the Shang with perversity.[6] More than two hundred place-names can be identified, many mentioned only once, though others recur dozens of times and are also listed as areas for pasturage.[7] Conducted on a large scale by the equivalent of a regiment on maneuver over extended periods of up to thirty or forty days, hunting similarly had multiple functions. Personal enjoyment and the training of core military forces were certainly foremost, but secondary objectives probably included evaluating capabilities and performance, especially archery, which was the main form of attack apart from nets and pits;[8] imbuing authority and accustoming men to a chain of command; developing coordination and cohesiveness; becoming familiar with the terrain prior to seizing territory;[9] and eliminating dangerous animals. The hunt's essential military nature is perhaps made most evident by instances of the king diverting the limited forces already in the field to attack a foreign proto-state.[10]

Hunting still constituted a viable method for rapidly acquiring significant amounts of protein because the population density was low and large forested and marshy tracts remained.[11] The large numbers of animals slain and captured furnished an essential part of the requirements for the Shang's frequent sacrifices and extensive feasting, even though domesticated herds had been developed. The victims included tigers, some sort of wild ox, at least two species of deer, and wild boar. For example, one report lists 1 tiger, 40 and 159 respectively of two different types of deer, and 164 pigs;[12] another 11 wild oxen and 15 boar;[13] and a third 6 wild oxen, 16 boar, and 199 deer.[14] Based on the numerous inscriptions from Wu Ting's reign, it appears that rather than observing the sort of seasonal proscriptions described in late Warring States writings, hunting was conducted throughout the year.

The ruling elite confronted both internal and external challenges as the Shang evolved from a chiefdom into a state amid a generally hostile environment. More warriors than administrators, contrary to traditional depictions the Shang well understood the importance of battlefield achievement, valued physical prowess, and enthusiastically embraced military talent. Their strong martial orientation is reflected in elaborately decorated bronze and highly polished jade weapons, stylized metallic human and animal face masks, and other symbols of authority and achievement, including great axes.[15] The weapons and massive bronze vessels found in Fu Hao's tomb and those of other important military commanders throughout the Yin-hsü years show that they were not just employed by the living, but also interred with the dead to honor martial prowess and authority, a practice that continued to the end of the dynasty.[16]

The ruling clan also monopolized the acquisition and exploitation of the mineral resources necessary to fabricate metal weapons and the bronze vessels and other valuable objects that were employed to reward the faithful. Massive cauldrons adorned with intricate designs of religious significance, cast in alloys gleaming with a burnished golden color, contributed to an impression of overawing opulence. Weapons, whether of stone or metal, were apparently produced solely in government workshops and even handicrafts remained under central control, ensuring that only the compliant would have access to these products.

The Shang's ascension was marked by even more extensive and aggressive martial efforts than the Hsia in the area of resource acquisition. The sudden expansion seen at P'an-lung-ch'eng reflects the generally increased output evident at the extensive copper deposits found in this general area, particularly Jiangxi Tuan-ch'ang T'ung-ling-kuang and

Hubei Ta-yeh T'ung-lu-shan-k'uang.[17] Further evidence of this quest is visible in the numerous small Shang enclaves that appeared near limited but rich deposits of copper and lead in western Henan on the upper reaches of the Luo River. Eighteen Erh-li-kang (Shang) settlements that suddenly replaced three Erh-li-t'ou sites could, although averaging only 40,000 square meters each, easily ship locally smelted metals via the region's numerous waterways to Yen-shih.[18] When Shang power contracted in the west, widespread but highly concentrated ore deposits in the southeast came to replace western sites.

SHANG MARTIAL COMMAND

Historians analyzing the Shang generally claim that the rudimentary staff positions then appearing were never differentiated into civil and martial and that all military functions were performed on an ad hoc basis by civil administrators.[19] These assertions, written from a perspective prejudiced toward the "enlightened" nature of later ages when the civil reputedly dominated, inherently assume that the first positions to evolve were civil (and therefore more progressive) rather than martial.[20] It is then disparagingly concluded that Shang civil officials were compelled to reluctantly accept military burdens. However, not only did specialized military positions clearly exist in the Shang, but it was an era pervaded by concerns with domination and defense, more conducive to a purely military hierarchy than any gestation of civil functionaries.

The Shang was a warrior elite culture that required participants to embrace a vigorous lifestyle and the martial values of a large, evolving, but still clan-based chiefdom. Inscriptions on bronze vessels such as the *Hsiao-ch'en Yü Ts'un* show that the king granted generous rewards, including substantial plots of land, for military merit, and also rescinded "fiefs" for failure.[21] Rather than the sort of glittering cultural manifestation subsequently portrayed, it was a brutal, bloodthirsty age of frequent, aggressive warfare in which people were slain, enslaved, and sacrificed without compunction. Furthermore, contrary to later depictions of a purely moral effort marked by an overriding civilian orientation, of virtue and civility having been interrupted by the unruly and baleful face of war, the Shang dynasty was founded through decades of combat and a brief period of sudden conquest. The Shang didn't just displace the "one man," their ostensible target, through a simple punitive attack, but systematically extirpated the Hsia throughout the realm.

Heritage has immense impact in shaping values and determining mindset. Martial spirit, once unleashed, doesn't necessarily diminish, accounting for the *Ssu-ma Fa's* emphasis on performing ceremonies designed to reintegrate combat weary soldiers into civilian life.[22] The exhilaration of victory and admiration of military prowess clearly pervaded the Shang's early years, deeply influencing the establishment and monopolization of positions of power. (It should not be forgotten that even the last, reputedly debauched emperor, Hsin, had a reputation for great strength and martial ability.) Survival ranked paramount, and "civil" functions, although necessary to ensure the state's fiscal and material prosperity, were certainly derivative. Many individuals within this highly charged martial context no doubt deemed administrative tasks a distraction and an annoyance.

The absence of pre-Anyang writings makes it impossible to characterize the exact nature of early Shang rulership, but the dynamics seem to indicate a gradual transition from chieftain to monarch subsequent to the Hsia's defeat. Despite internal problems and intrigue, Shang rulers acquired the despotic powers already described and served as the final arbiters of military affairs. Anyang oracular inscriptions already show the king performing all the duties generally apportioned between twenty-first-century commanders-in-chief and ministers of defense, as well as frequently serving as battlefield commanders.

Regents from Wu Ting onward decided whether military actions should be undertaken, which enemies should be struck, the campaign's objectives, the number of men to be levied or deputed, the allies to be summoned, the commanders to be appointed, the manner or tactics of attack, resolution of the conflict, treatment of the vanquished, and the disposition of their land. Allies might also be ordered to undertake offensive or defensive actions alone, in coalition with others, or in conjunction with the Shang itself. Combat appointments were solely at the monarch's discretion, all field authority being derivative.

With a few exceptions such as Fu Hao, the king initiated divination procedures to inquire about the appropriateness of these military actions and seek the sanction of the ancestors. Whether Shang warriors were as reluctant as the early Greeks to undertake military action without auspicious indications remains unknown, but in the context of Shang religious and ritual emphasis it seems likely.[23] Prognostication being a powerful psychological tool, the king's queries were probably intended to coerce or persuade others as much as to appeal to the spirits and perhaps reduce the responsibilities of decision making.

Although the king could have always led in person, in about half the recorded conflicts he opted to appoint others to command coalition forces and their segmented contingents. Royal clan members, important members of other esteemed clans,[24] and close officials such as the *tuo ch'en* ("chief subordinate") were often entrusted with this responsibility, with those who proved successful being repeatedly employed. Despite shouldering other responsibilities such as overseeing hunts and directing economic projects such as land reclamation, many evolved into de facto military specialists who could plan, organize, and lead campaigns. Rulers of allied and subservient states were also dispatched on expeditionary missions, normally at the head of their own military forces, with the successful tending to be reappointed if precampaign prognostications boded well.

As early as Wu Ting's era the army (*shih*) was already an identifiable operational force. Although the title *chiang*, usually translated as "general" in accord with Western convention, does not appear in the Shang, the character *shih* is used in naming certain commanders. (The usual format is *shih* plus clan name such as Pan, so essentially "General Pan.") In addition, *shih chang* or "leader of the *shih*," which is mentioned in the *Shang Shu* but not the oracular inscriptions, may have been a functional title, particularly late in the dynasty in accord with the army's increasing prominence and formalized organization.[25]

Shang oracle writings indicate formalized ritual procedures were observed whenever the king appointed someone to direct a campaign.[26] Although dating a thousand years later, something like the ceremony described by the T'ai Kung in advising King Wu how to properly empower his field commander and transfer the necessary authority apparently occurred in the ancestral temple:[27]

> When the state encounters danger, the ruler should vacate the Main
> Hall, summon the general, and charge him as follows. "The security

or endangerment of the Altars of State all lies with the army's commanding general. At present a certain state is not acting properly submissive. I would like you to lead the army forth in response."

After the general has received his mandate, command the Grand Scribe to bore the sacred tortoise shell to divine an auspicious day. Thereafter, to prepare for the chosen day, observe a vegetarian regime for three days, and then go to the ancestral temple to hand over the *fu* and *yüeh* axes.

After you have entered the gate to the temple, stand facing west. The general enters the temple gate and stands facing north. You personally take the *yüeh* axe and holding it by the head, pass the handle to the general, saying "From this to Heaven above will be controlled by the General of the Army." Then taking the *fu* axe by the handle you should give the blade to the general, saying "From this to the depths below will be controlled by the General of the Army. When you see a vacuity in the enemy you should advance; when you see substance you should halt. Do not assume that the Three Armies are large and treat the enemy lightly. Do not commit yourself to die just because you have received a heavy responsibility. Do not regard other men as lowly because you are honored. Do not rely upon yourself alone and contravene the masses. Do not take verbal facility to be a sign of certainty. When the officers have not yet been seated do not sit. When the officers have not yet eaten do not eat. You should share heat and cold with them. If you behave in this way the officers and masses will certainly exhaust their strength in fighting to the death."

After the general has received his mandate, he should bow and respond to the ruler: "I have heard that a country cannot follow the commands of another state's government, while an army [in the field] cannot follow central government control. Someone of two minds cannot properly serve his ruler, someone in doubt cannot respond to the enemy. I have already received my mandate, and taken sole control of the awesome power of the *fu* and *yüeh* axes. I do not dare return alive. I would like to request that you condescend to grant complete and sole command to me. If you do not permit it, I dare not accept the post of general." You should then grant it and the general should formally take his leave and depart.

Composed in the late Warring State period, this excerpt reflects late Spring and Autumn and early Warring States thought first expressed in the *Art of War* about the commander's necessary independence in the field. Nevertheless, generals were to be commissioned before the ancestors and properly sanctioned by appropriate prognostications, and their appointments were recorded on wooden tablets, just as oracular inscriptions indicate occurred in the Shang. Even King Wen of the Chou was reputedly appointed as the Western *Po* (duke) with the awarding of a bow, arrows, axe, and *yüeh* axe, all symbolic of conferring authority.

Although many military needs were fulfilled by clan members and others who temporarily but repeatedly assumed prominent combat responsibilities, officials also staffed a number of recurring, obviously martial positions.[28] The incipient administrative structure discernible even in the earliest oracular inscriptions clearly evolved over the centuries at Anyang, becoming more formalized and specialized. Apart from being consistently entrusted with functionally differentiated tasks such as reclaiming lands on a repetitive basis, members of the Shang warrior elite primarily exercised military authority.[29] Without their modifiers (such as *tuo* for many, generally indicating a higher or supervisory position and thus meriting the appellation "chief," and *mou* for "directing," later "planning"), these officials included the *ma* (horse), *ya* (commander), *fu* (which means "quiver" but whose role is unknown),[30] *she* (archer), *wei* (protector), *ch'üan* (dog), and *shu*. Although their exact responsibilities remain nebulous, their titles imply that they had charge of well-defined battlefield aspects ranging from animals through weapons and, in addition to logistical responsibilities, probably directed training and led units as subcommanders on the field.

The position of *ya* or "commander," seen alone and as *tuo-ya* and *ya-mou*, without doubt was a purely military position with encompassing martial responsibilities and authority. It not only appears numerous times in the oracular inscriptions, but also on late Shang bronze vessels and great symbolic axes of power that have been recovered from military enclaves on the Shang periphery.[31] Individuals with the title *ya* were deputed to border locations, where they performed defense-related functions and could appoint subordinate officials and assign command responsibilities to local nobles, evidence of their considerable authority;

undertook command of field armies entrusted with attacking and damaging enemies such as the Ch'üan; and shouldered responsibility for the king's protection.

The discovery of not just individual chariots but also multiple vehicles buried as marks of martial prestige with high-ranking nobles as well as in burial mounds solely of horses and chariots clearly proves that they existed in sufficient numbers to be of more than occasional use for prestige transport. However, they were expensive, fragile, complicated to manufacture, and probably functioned mainly as dispersed platforms for archery and exercising command over units of closely integrated infantry. There are few references in the oracle inscriptions to employing them on the battlefield, and the maximum number ever specified is only 300. (In the latter case they may have comprised discrete chariot units that could have acted as penetration or flanking forces if used en masse, or simply highly mobile archery platforms.)

Because horses provided crucial motive power, the *ma* or horse officer seems to have been entrusted with proportionately greater authority. (The term *ma* is interchangeably used to indicate the physical presence of "horses" and the warriors or officers who employed them, whether charioteers or perhaps even cavalry riders.) Highly knowledgeable and presumably experienced in equine-centered military affairs, in normal times the *ma* was entrusted with raising, training, and evaluating the state's horses, including those sent in as tribute.

Tuo ma, literally "many horse" but clearly a functional equivalent of "horse commander" or "chief *ma*," also appears. To the extent that chariots formed a core component of the army, the *tuo ma* appears to have controlled contingents of the standing army; assumed a command role on the battlefield, being deputed on campaigns to attack enemy states; been entrusted with defensive responsibilities that extended to the ruler; and also directed the hunt on occasion. Sometimes even *tuo ma ya*, or "supervisor or commander of the *tuo ma*," is recorded in the inscriptions. However, the esteemed title *ssu-ma*—"supervisor of horse," but later, in the Chou feudal hierarchy, functionally the "minister of war"—though probably equivalent to *"tuo ma ya,"* had not yet evolved.[32]

The "canine officers" (*ch'üan*) probably had their origin as kennel masters for the king's dogs, but their number multiplied and their authority expanded as the role of dogs increased in protection, the hunt, and perhaps the battlefield,[33] and they apparently performed intelligence-gathering functions.[34] The "chief canine officer" (*tuo-ch'üan*) also appears in the inscriptions, but most of the entries refer to the *ch'üan* supervising the hunt and assembling the "new" archers, commanding attacks on enemy states, and offering Ch'iang prisoners in sacrifice to the early Shang ancestor T'ai Chia, a remarkably rare privilege. The term *ch'üan-mou* also appears in a few instances, apparently designating either a canine officer named Mou or one entrusted with planning responsibilities.

Archery was highly esteemed in the Shang, and two official titles appear, *she* and *tuo-she*.[35] Just as the term *ma* (horse) has an extended meaning as the horse commander or official in charge of horses, the character for *she* has been generally interpreted as a title that clearly derives from the archer's basic role. Apart from whatever responsibilities they must have had for archery contingents in combat, the *tuo she* undertook broader responsibilities for the realm's protection, often in conjunction with the *wei* or protector.[36] However, their role seems to have been more circumscribed than other military officials.

Unexpectedly, the position of "*shih*," functionally "historian" or "astrologer" in later ages (especially with the honorific "*t'ai*" or "grand" preceding it as an official title), in this early form meaning "emissary," also seems to have been responsible solely for military activities in the Shang.[37] Inscriptions indicate that *shih* were being dispatched to the various quarters, especially west and south, and to designated locations for both offensive and defensive purposes. In some cases they commanded expeditionary campaigns such as against the Kung-fang during Wu Ting's reign; in others they were entrusted with mounting some sort of standing defense against external threats, particularly from the steppe, thereby becoming the first known border commanders.

Queries about whether a particular *shih* would return from a long-standing assignment on a certain day or be successful in capturing prisoners indicate their importance in the king's consciousness. Granting of martial authority to them similarly was formalized with ceremonies that had to be held on auspicious days, no doubt much as described above. Regional designations also seem to have appeared, such as "*nan shih*" or the "southern shih" and at least two ranks, *ta* (great) and *hsiao* ("little," "minor," or perhaps "junior") *shih*. For one southern campaign they were further differentiated as center, left, and right, clearly in accord with Wu Ting's initiation of tripart field forces, implying that the south had become a troublesome area.

Although it has been suggested that the term *shu*—which means something like "guarding" or "protecting"—designated a border contingent, in many usages it clearly refers to an officer entrusted with command of units responsible for defending the periphery. The character itself is seen as being composed by a man and a dagger-axe (*ko*), the same elements as *fa*, "to attack." However, in the latter the man is holding the *ko*, whereas in *shu* he is standing underneath the *ko*.[38] In various inscriptions the *shu* are ordered to exercise command functions in the field, sometimes in association with other normally subordinate officers such as the *ma* or *she*; dispatched to attack and damage enemy states; and assigned responsibility for ordering and commanding the *chung*, especially the king's *chung* (for whom they may have been the only commanders).[39] *Shu-mou* and *wu-tsu-shu* are also seen, with the latter probably referring to the commander of a group of units from the five major clans that had been deputed to shoulder perimeter responsibilities. Whether referring to actual contingents, as in Wu Ting's era, or an official title, the common tripart designation of left, right, and center also appears.[40]

Finally, depending on how titles such as *tuo ma* are interpreted, some analysts claim to be able to discern the existence of a fairly structured military hierarchy even in the Shang, though certainly not the one depicted in systematic Warring States idealizations. Although not necessarily unexpected because minimal lines of battlefield authority would have been required for the army to perform effectively, the more important question would seem to be how rigidly defined they may have been. Although the term *tuo* can simply refer to the many officials of a certain type, it usually indicates a superior position or commander for such officials, with even the *ma hsiao-ch'en* (junior official for horses), for example, serving under the *tuo ma*. The appellative *ya* should then designate an even higher-ranking position within a hierarchy of defined functions. Presumably everyone on the field of battle would have been subservient to the overall army commander, whether the king, minor ruler, or specialist such as the *shu* (when out at the border), but subordinate authority may have been more fragmented, with less clarity in the relationships prevailing among the commander of the dog officers, archery commanders, and others.

The intermixture of the *lü* and *hang*, regiments and companies respectively, increased the complexity. Insofar as the clan units seem to have existed as independent units for another half millennium, their commanders—especially those commanding the king's *tsu*—may have escaped the nominal hierarchy or, relying on their personal charisma and power, simply refused to accede to delegated authority. Although such chaos should not have been tolerated, unless their forces were somehow integrated into the *lü* that were then emerging as an operational contingent, it is likely to have persisted, considerably muddying the overall, somewhat ill-defined authority structure.

Another title known to exist in the Shang was *yin*, as in the famous Yi Yin.[41] Though generally assumed to be a civil position such as chancellor, minister, or even the world's first national security advisor, it also appears in conjunction with military contingents, such as *yu tsu yin* or the *yin* for "the clan force on the right"[42] and *tuo shih* (arrow) *yin*.[43] Based on the *Shang Shu*, the various subunits from 100 up through the *hang* and *lü* also had *chang*, leaders or subofficers. In contrast to the Chou, who esteemed the left in government offices and divination, the Shang emphasized the right.[44] Thus the famous Yi Yin was reputedly the *yu hsiang* or "minister on the right," and the three armies were generally enumerated as right, middle, and left.[45] However, the king still commanded from the center.

SHANG MILITARY CONTINGENTS

Military units and structural organization slowly but continuously evolved from the inherited, extemporaneous methods that characterized Wu Ting's early era to more permanent forces and organizations. Nevertheless, as Western military history shows, a certain amount of "organizational fluidity" (more aptly termed chaos) invariably arises during periods of change and retrenchment. Carefully crafted units designed to replace presumably outmoded variants often fail to fully displace the latter or to be tightly integrated into the revised hierarchy, unexpectedly resulting in the simultaneous existence of vigorous new contingents and antique remnants. Factors other than simple military conservatism can also intervene, including a need to preserve clan-based forces to maintain internal control and manipulate power. Assumptions of structural rigor and homogeneity are therefore often fallacious and counterproductive.

Several new organizational units were created in the late Shang, but they essentially drew upon the same personnel, just differently grouped and segmented. Whether these innovations in affiliation and command resulted in an increasingly formalized hierarchy of company, brigade (or regiment), and army (*hsing*, *lü*, and *shih*), similar in structure to that found in the Spring and Autumn period, remains a much-debated question. Unfortunately, even when supplemented by later textual tradition the evidence that can be derived from the oracular inscriptions remains insufficient to determine unit strength or discern unit composition. Nevertheless, despite the appearance and predominance of the "army" (*shih*) and "brigade" or "regiment" (*lü*) as fundamental operational units, rather than a thorough organizational revolution, an admixture of units evidently persisted.[46]

Insofar as historians have expended enormous energy in subsuming these presumably distinctive contingents into encompassing schemes, their probable nature merits brief consideration. Preliminary to reviewing the possibilities, it should be noted that a unit's full complement may not have been mobilized. Because even a few absences would dramatically impact small contingents, serious deficiencies in units such as the 100 archers would have been highly visible and therefore subject to severe accountability. However, the larger units of 1,000 and especially 3,000 might have only achieved the staffing levels of 700 or 800 per thousand often seen in later ages. The accuracy of figures cited for this period therefore depends on how strictly the officials entrusted with responsibility for the levies were held to consistent standards. In contrast, despite a natural inclination to inflate the count and thereby garner greater rewards, reports of the numbers slain and taken captive in actual engagements, being fairly low and highly specific, were probably more accurate.

Based on inscriptions that indicate the existence of three *shih*, the first use of the character *shih*, the traditional term for "army," has been attributed to Wu Ting's era.[47] The character has been interpreted as originally depicting an accumulation of men, closely related to a mound or hillock, and derived from the character for town (*yi*). Without doubt, armies in the late Shang were primarily associated with towns and functioned in a protective capacity when not deputed for field duty, including guarding cities at night.[48] However, the range of activities they might undertake would have been constrained by the contingent's composition and size.

On the assumption of institutional continuity, another pronouncement embedded in a prognostication generally ascribed to the consecutive reigns of Wu Yi and Wen Ting, stating that the king "created three armies," has been interpreted as signifying the addition of three more armies, bringing the total to six, rather than the formalization or reinstitution of a three-army system that had perhaps fallen into disuse in an interim of reduced military activity.[49] Primary justification for attributing six operational *shih* to this era is derived from the conspicuous existence of six armies in the early Chou, the latter's institutions being deemed reflective because traditional literature and relatively early bronze inscriptions indicate that the Chou imitated many Shang organizational practices.

It has also been argued that Wu Yi, the *"wu"* emphasizing military prowess, not only conducted numerous aggressive campaigns against external enemies but also was greatly enamored of martial values and practices, accounting for his apparent addiction to hunting, an activity that reportedly claimed his life. His spirit and commitment would have nurtured a highly charged martial ethos conducive to army building, one that should have persisted even if the actual edict was issued by his successor, Wen Ting (who also merited a posthumous *"wu"* in his designation as Wen Wu Ting), as some analysts claim.[50]

Whether the army existed before Wu Ting's ascension, he deliberately created it through a conscious act, or it simply evolved during his reign, the *shih* first becomes visible in inscriptions from his era.[51] His frequent summoning of men in units of 3,000 from the very beginning of his monarchy suggests that the contingent of 3,000 was already a fairly well-defined, functional unit and a likely candidate for *shih* despite other numbers such as 1,000, 5,000, and 10,000 also having been levied. However, a certain degree of flexibility probably characterized these extemporaneously constituted field forces, and units of almost any size may have conventionally been designated as *shih*. Perhaps, as suggested, the *shih* originally numbered just a thousand before expanding in accord with the historically attested tendency of units to strengthen over time. This would partially account for the persistence of 1,000-man operational field units.

In addition to multiplying the army's logistical burdens, Wu Ting's increasingly lengthy peripheral engagements must have imposed escalating manpower requirements and stimulated military specialization. These voracious needs seem to have compelled a gradual shift from temporary levies to the *shih*'s more consistently organized and stable forces, explaining the disappearance of the term *teng* ("levy") after his reign. However, though certainly expressive of an intent to reorganize and probably a step toward greatly expanded standing forces, there is no evidence that these armies were permanent or that the designated members were extemporaneously mobilized.

Although no formal pronouncements have been recovered, even in Wu Ting's era Shang field armies already consisted of a central force bolstered by two flanking commands. Segmentation into left, right, and center characterized not only the army but also the *lü* (regiment), archers, chariots, *hang* (companies), and *tsu*. As early as his famous southern campaign three armies denoted as left, right, and "my" (or the king's) forces were dispatched, and references to the "army of the left" and "army of the right" are seen throughout his reign, attesting to both the concept and actuality of their existence. However, three armies need not always have been fielded, and any army could be segmented for operational purposes into its constituent components.

Late in Wu Ting's reign, when an astonishing 23,000 troops may have been summoned in a short period for an expeditionary campaign against the T'u-fang, a basic operational field force of 9,000 comprising three 3,000-man armies would have been both reasonable and readily sustained by expanding Shang economic resources. However, in accord with Shang decade-based practices, it has also been suggested that the *shih* may have numbered as few as 100 men[52] (which seems extremely unlikely) to as many as 10,000 men, though the latter would have required a massive, highly unlikely increase in manpower and constituted a dramatic shift away from the basic 3,000-man complement.[53]

Nevertheless, in the face of eastern challenges and given the martial character of the Shang's despotic rulers, it would be surprising as well as contrary to the natural tendency of states to constantly augment their military power if the army had remained petrified at 3,000 men. Although the numbers reputedly fielded by the Shang at the Battle of Mu-yeh must be severely discounted because they transcend the realm of possibility, they should still be understood as indicating the presence of a massive force rather than be completely dismissed. Ten-thousand-man armies that were probably unattainable in Wu Ting's reign may well have been realizable by the dynasty's ignominious end, when a professional military force that proved capable of sustaining campaigns down into the southeast had clearly evolved.

Near the end of its existence the Shang deployed a number of armies at its secondary capitals. Although force levels of 3,000 men would have been more easily encamped, armies of 10,000 would certainly have cohered with the ruler's personality. Six such armies would have provided the Shang with a core force of roughly 60,000 at the final clash, to which would be added whatever strength their allies might have provided. Given the urgency of the developments, the latter may not have exceeded 10,000 men, accounting for the 70,000 traditionally (and more realistically) said to have confronted the Chou.

Conversely, being an emerging power, the Chou had to rely on allied contributions to bulk out their own highly motivated forces. At this time their *shih* are more likely to have still numbered a traditional 3,000, or perhaps only 2,500 if based on pyramids of five, as is sometimes claimed.[54] Six armies with a nominal strength of 3,000 men augmented by 3,000 elite tiger warriors and the contributions of their allies, presumably at significantly diminished force levels because of their comparatively smaller size, would have resulted in a force somewhat less than half the Shang's vaunted might.

Next in the hierarchical order stood the *lü*, whose character has traditionally been understood as depicting two men under a pennant (clearly referring to a military unit from inception) and entailing a sense of "multitude." Discounting the validity of the one inscription that indicates 10,000 men had been levied, it was a unit whose numbers seem to have varied greatly but probably averaged 500 or 1,000 men.[55] In subsequent eras it would be an intermediate organizational unit that would integrate several companies into a *shih*, the latter then being best understood as a regiment within the context of the new term for army, *chün*. *Lü* then, separately and in combination with *chün*, would come to designate the army in general, appearing in such terms as *chen-lü*, "review and order the troops," a ritual that apparently was already being implemented in the Shang before dispatching the troops on campaign. Although numerous questions remain about its composition and function, in the Shang, *lü* can be envisioned as an independent brigade or regiment that was fielded in conjunction with levied armies, but was not subsumed under the latter's organizational umbrella.

References to the unit remain sparse, but from the famous (if perhaps dubious) inscription "levy Fu Hao with 3,000 and levy *lü* 10,000," it has also been concluded that the *lü*'s composition somehow differed from that of a normal levy and was still an ad hoc unit in Wu Ting's era.[56] (This ad hoc nature and its early appearance would argue against claims that the *lü* represents a shift toward military formalization and specialization.) Moreover, based on a notation that "the *chung* should be conjoined with the right *lü*,"[57] it is evident that their members were distinct. Even though the term had already appeared in King Wu Ting's era, *lü*'s operational inception seems to date to the reigns of Ping Hsin and K'ang Ting, reflecting a shift toward expanded operations.[58] If the king's *lü* is understood as the "middle" force, inscriptions from Ping Hsin and K'ang Ting's era suggest that the three standard components of left, middle, and right were all being fielded, though not necessarily simultaneously.[59]

The few inscriptions referring to *lü* mention them being called up for both training and field action.[60] Suggestions have been made that in the later reigns they continued to provide an operational umbrella for forces temporarily called to duty and thus represent a major step toward the concept of "people's soldiers," in comparison with the essentially professional warriors populating the government and forming the core of the semipermanent military forces. It has also been asserted (without substantiation) that they subsumed the clan armies within their structure.[61] However, this would not only constitute a shift away from discrete, individual martial entities to a true state force, but also contradict the tendency toward these people's forces. In addition, inscriptions referring to both the *lü* and *tsu* (clan forces) are not uncommon, evidence that they continued to coexist as operational contingents.[62]

The other large entity frequently ordered onto the battlefield was the *tsu* or clan regiment, whose character has traditionally been interpreted as depicting an arrow under a pennant. The oldest of all units, it must have originally drawn its fighters from among the physically qualified clan warriors who had the privilege and responsibility of serving. Insofar as the Shang loosely encompassed other prominent clans apart from the ruling Tzu house and its collateral lines, including some that had originated among the Yi or other early allies with whom they intermarried and enjoyed cultural exchanges, there were a number of such entities.[63]

Most prominent among them was the *wang tsu* or king's clan, also referred to as "my clan" in the king's prognostications, but the *tzu tsu* and *tuo tzu tsu*, referring to the royal house and many princely clans respectively, also played major battlefield roles.[64] Naturally the king's clan served as the central force when they fielded the full complement of left, right, and middle, but prognostications referring to just the left or right *tzu tsu* suggest they were also deployed singularly and together. The *tsu*'s basic size has been suggested as 500 men or roughly a battalion, in comparison with the *shih* and *lü* when, in comparison with *chün*, *lü* is understood as brigade or perhaps regiment.

Because these clan forces must have shouldered core responsibility for the campaign that overthrew the Hsia, postconquest their members certainly would have been reluctant to give up their privileges and honors for routine administrative duties. The other important clans contending for influence (or even survival) amid the evolving Shang state must have held similar views. Some sort of clan-based standing force had to be deployed from the outset to dominate the Hsia enclaves at Yen-shih and the vital crossroads to mineral-rich areas such as Tung-hsia-feng and P'an-lung-ch'eng. Clan forces would also have been required to protect the ruler and the interests of the ruling house. The Shang also took thousands of prisoners through combat, apparently retaining and employing at least some of them for domestic service and productive labor. Even if the Shang was not fundamentally a slave-based society, clan forces would still have been required to maintain control over enslaved elements of the populace and ensure internal security.

As military needs escalated, larger units evolved that necessarily drew on an expanded population base. No doubt the Shang began to include non-clan members who served the major clans or were otherwise associated with them through marriage relationships, before eventually reaching down to the ordinary inhabitants, peasants, and perhaps even slaves, some of whom may have already been accompanying their masters into battle.[65] However, the multiplication and formalization of new field units created an additional problem: whether the older, high-prestige heritage units would be subsumed into the regular forces, perhaps serving as an active core, or continue to operate independently as well as act as the king's guard. As already mentioned, oracular inscriptions and subsequent historical materials indicate that imperial clan forces played a persistent field role well into the Chou, and the five *tsu* were even charged with defensive responsibilities along the frontier in the last decades of the Shang.

One other unit, the *hang*, seems to have emerged and played a somewhat nebulous battlefield role late in the Shang. (The Chinese character for *hang*, best known under its more common pronunciation as *hsing* except in a military context, fundamentally entails the idea of movement but also came to mean a row or line in later ages.) A paucity of relevant inscriptions has prompted considerable speculation about its exact nature, many commentators noting that it would be deviously used during the Spring and Autumn to designate an army-sized force without actually employing the term *shih/chün*, to avoid infringing upon still nominally acknowledged royal prerogatives.[66] In subsequent ages it would become a subunit within the army, something like a company, but it seems to have operated as a separate battlefield entity in the Shang.

Inscriptions indicate that the usual tripartite deployment of left, right, and middle applied to the *hang*, and another pair designated as east and west seems to have existed. In addition, the term *ta hang* (large or great *hang*) appears, presumably referring to a battlefield entity that integrated the three component forces of left, right, and middle *hang*.

Although there are scattered references to the *hang* in Wu Ting's period, it only became more common during the subsequent era of increasing military specialization and formalization. Definitive numbers are lacking, resulting in assessments ranging from 100 to a very unlikely 1,000 and even claims that it exceeded the *shih*, though the latter would have to be conceived as a mere 100 men.[67]

These larger field units were frequently supplemented by at least two highly specialized contingents, the archers and chariots, both generally ordered forth in units of 100 or 300.[68] Their mode of reference implies that the chariots served intact rather than being dispersed, contrary to claims that they represent aggregate figures for apportionment among the 1,000 or 3,000 serving in the army or that each chariot was assigned some fixed number of fighters ranging from five to twenty-five. Although the Chou would see the evolution of the chariot-centered squad, chariots were at a premium in the Shang and therefore reserved for command purposes. A regiment of 100 chariots, unhampered by attached fighters, could have proved a decisive force for penetration and flanking on the dispersed battlefields of Chinese antiquity.

There are several references to "300 *she*," suggesting that in addition to archers exercising a command function atop the chariots, dedicated regiments of archers were deployed.[69] Assuming that the tripartite segmentation witnessed in the era's armies also applied to the archers, the 300 would encompass three companies of 100. In an effort to envision a cohesive military hierarchy, it has been further asserted that their being called up in numbers identical to the chariots—100 or 300 at a time—indicates that these are in fact the archers known to have manned the chariots.

However, not only is this an unsubstantiated assumption, but archery was normally the prerogative of the chariot commander. Furthermore, as will be discussed in a subsequent section, to function effectively the chariot crew—whether consisting of just an archer and driver or accompanied by a weapons man on the right—would have had to train together as a team before they could achieve the minimal coordination necessary to function on the battlefield. Archers could not simply be assigned at the last minute to chariots that were manned solely by a driver and would probably prove useless for military purposes.

Instead, whatever their operational size—10, 25, or 100—archery companies were almost certainly employed as discrete units on the battlefield to provide the mass volley fire needed to decimate the enemy and shape the battle space. Unfortunately, historical materials prior to the late Warring States have failed to preserve any passages on the early employment of archery contingents, with only Sun Pin advising that roving crossbow companies be used "to provide support in exigencies."[70]

13.

TROOPS,
INTELLIGENCE,
AND TACTICS

◆‑◆————————————————————◆‑◆

E VEN MORE QUESTIONS PLAGUE attempts to characterize the men who served in the various military units than the task of outlining the Shang's command structure. It has always been axiomatic that only men engaged in combat, but the dramatic command role exercised by Fu Hao and Fu Ching, coupled with legends about the T'ai Kung's daughter having led forces in the early Chou, perhaps a Chiang clan characteristic, has even prompted (totally unsubstantiated) claims that Fu Hao's contingent was composed solely of women.[1]

In the immediate postconquest days, when hundreds of allies reputedly acknowledged their authority, Shang martial requirements no doubt consisted simply of deploying holding forces to strongpoints and maintaining order in restive areas. Even though some personnel must have been engaged in agricultural and administrative duties, members of the core and extended clans no doubt proved capable of providing the few thousand men necessary for these small field contingents and the royal protective forces that enforced the king's will, including dragging people off to be sacrificed.[2] However, in response to an unremitting escalation in military needs, the army's composition would gradually shift from relying on clan warriors to relying on "soldiers" drawn from the ordinary inhabitants of the growing towns, farmers in the surrounding area, and even slaves.

In accord with theoretically mandated interpretations, Marxist-oriented PRC scholars generally view the Shang as having been a slave-based society in which massive numbers of slaves were employed in domestic tasks, productive work, agriculture, and even the hunt.[3] However, whether they or the lesser nobility and common people constituted the core workforce or even provided any noticeable labor remains problematic.[4] Certainly the Shang was a tightly controlled, essentially theocratic society in which rank largely dictated a person's power and influence and an individual's freedom diminished in direct proportion to lack of hereditary position or close relationship to the increasingly autocratic kings. Consequently, the populace was composed of royal and other clan members of varying degrees of distinction, common people, a variety of subservient classes, and certainly some slaves, all of whom seem to have been liable for military service.

In this context questions about the nature and role of the *chung*, designated by a character that came to refer to the "masses" or common people and the "troops" in the increasingly vast armies of later periods, have stimulated acrimonious debate.[5] Even the character's original meaning, commonly believed to have been a depiction of three people laboring under the sun, is contentious.[6] Based on inscriptions discussing the possibility of extinguishing a state and turning the people into *chung*, it has been suggested that they largely originated as war captives.[7] Moreover, it is clear from the inscriptions that the term *chung* refers to a particular status (such as persons serving in a dependent role) rather than some indeterminate military grouping, with further confirmation of their menial status being seen in their having been sacrificed and slain without compunction.[8]

However, such treatment was hardly unique, because everyone seems to have been subject to peremptory execution or sacrifice in the Shang, even the nobility and a few feudal lords falling under the axe. Conversely, some positive measures regarding the *chung*'s welfare seem to have been enacted: some were allotted the use of land, a few gained a degree of derived authority, and the auspiciousness of mobilizing them was the subject of prognostication. Inquiries about the possibility that they had perished or suffered harm would certainly seem to attest to the king's concern over their welfare, whether out of compassion or simple military efficacy.[9]

Even if they ultimately constituted a significant portion of a large segment of the Shang's inhabitants, being mobilized only in small numbers, they could not have played more than a minor role in the hunt and military activities.[10] Instead of combatants, the *chung* seem to have acted more as support personnel,[11] perhaps something like the servants who accompanied their masters into battle in other cultures and served in ancillary roles. No doubt the king's servants would have accompanied him whenever he exercised command, and repeated mobilizations may have solidified their presence on the battlefield. In addition, their apparent formation into defined contingents seems to have been a temporary measure, implying that they were not integrated into the hierarchy of standing units that evolved in Wu Ting's era and thereafter, though it has also been suggested that they were a sort of semipermanent military group with somewhat elevated status.[12] The most ever mobilized were a mere hundred, basically the same as the other known specialized units of archers and charioteers, and there are battlefield references to left and right *chung*, confirming that they comprised distinct functional units for operational purposes.

The rapid escalation of external martial activity witnessed during Wu Ting's reign, in requiring the frequent summoning of the realm's warriors, must have seriously stressed the manpower system. After his reign the absence of such levies implies that larger, more permanent numbers of men were maintained under arms, and it is during the late Shang that the *chung* assumed an expanded role. The term *chung jen*, almost universally interpreted as synonymous with *chung*, also became more common under the last Shang kings. However, in terms of function and military liability, the *chung* and *jen* were originally distinct, and the *jen* were mobilized far more frequently and in larger numbers than the *chung*. Although the exact sense and scope of *jen* similarly remain uncertain, the term apparently designated what might be considered the "free" people—to the extent that anyone in the Shang might be free—and therefore encompassed low-ranking clan members, various dependents, farmers, and others subsumed within the Shang apart from slaves.[13]

Through an uncertain process the distinction between the *chung* and *jen* began to erode after Wu Ting's reign, the *chung* expanding in scope and numbers and acquiring a major role in court-based military activities.[14] By the end of the dynasty they are thought to have been

furnishing the soldiers (rather than warriors) needed for external campaigns, the terms generally employed being *chung* and *chung jen* rather
than *jen* alone.[15] Though not unreasonable, this sort of explanation is
not fully encompassing unless they comprised the manpower for the
other types of units such as the *shu* and *lü* that had become prominent.
Nevertheless, even if the elite warrior nobility thereby lost many of
their military privileges under the Shang's highly despotic rulers (and
their redirection to administrative duties), clan units—especially the *wu
tsu* or five clans—continued to be fielded throughout the Shang and
would even form the core of Ch'u's forces at the Battle of Ch'eng-p'u
in the Spring and Autumn period.

SHANG MILITARY INTELLIGENCE

The development of an extensive intelligence system capable of effi-
ciently transmitting crucial economic and military information from
every quarter was another vital achievement that marked the emergence
of the centralized Shang state. Written administrative reports discussed
unusual weather conditions, eclipses, prospects for the harvest, and tribute items being forwarded to the capital, including horses and prognosticatory media.[16] Military dispatches tended to emphasize disruptions
and other urgent issues, especially the activity of raiders plundering the
border or more serious incursions being mounted by contiguous peoples, and thus frequently prompted regal action. Even in their absence,
the possibility that inimical events might be occurring clearly troubled
the king, because he frequently queried the ancestors whether he would
not soon receive dire news from the periphery.[17]

Reports of enemy action were rapidly transmitted over an incipient
network of roads and rivers by utilizing widely scattered state guest-
houses and hostelries where horses, provisions, and lodging were maintained. In addition to boats, chariots, and runners, it is claimed that
some sort of "pony express" may have existed, horses in the Shang
being ridden primarily for this sort of mission and possibly battlefield
command rather than employed for cavalry.[18] The system's efficiency
is well attested by reports of T'u-fang aggressiveness being routinely
received within twelve days from 1,000 *li*, or about 350 miles away.[19] In

a crisis the Shang also had a system of drums (and possibly signal fires) that could quickly warn of approaching enemy forces, though the amount of information that might be conveyed would have been minimal. Nevertheless, one or two early written characters associated with drum warnings were also employed in an extended sense to indicate the transmission of urgent information.[20]

This emphasis on intelligence gathering may have derived its initial impetus from King T'ang's exploitation of various reports on the Hsia prior to the Shang's uprising. Numerous inscriptions reveal that the gathering and transmission of intelligence had become well established and highly organized by the Anyang period. Despite being enigmatic, the inquiries often contain condensed reports prefaced by the term *yüeh*, a character that basically means "to say," as in the king "saying" or putting a proposition to the ancestors, but also designating something that had been reported.[21] The querulous king then ascertained the report's veracity through prognostication, such as when it was reported that the Kung-fang had made an incursion and he asked whether any damage had been suffered, whether they had really acted aggressively. (Although divination would often be condemned as a means for acquiring knowledge from Sun-tzu onward, in the Shang it was not only used to pose general questions of military intelligence—whether a certain enemy would attack or a certain quarter suffer disorder—but also to evaluate reports gathered through human agency.)[22]

Other terms used for reporting and transmitting important information (orders) outward via the same system include *wen*, "to hear," synonymous with "to learn" and "cause to be heard" or "inform"; *kao*, "to report (from below)" and "announce" or "proclaim" (especially for statements originating with the ruler); *t'eng*, apparently a report transmitted by horse, whether ridden or yoked to a chariot; and *hsin*, a character that now means "letter" or "information" but can simply be understood as knowledge transmitted by envoy.[23]

OPERATIONAL TACTICS

Plastrons that preserve several queries undertaken in succession, sometimes on a single day, others across a very few days, indicate that the king pondered various alternatives prior to commencing a campaign. Rather than deputing forces in a simplistic "up and at 'em" mode, threats were assessed, options evaluated, forces chosen, commanders appointed, and subject states co-opted, all based on incoming reports and previous experience.[24] In addition, routes of march and means of advancing (especially if rivers were to be forded or boats employed) had to be determined and logistical support arranged. Dealing with major enemies such as the T'u-fang and Kung-fang almost always required the formulation of more ingenious and extensive measures.

When confronting multiple enemies that posed a threat, queries were closely initiated about each of them in an attempt to determine the most likely possibility for success. Whether an attack would be auspicious; which enemy to attack;[25] whether the king should command in person; which ally or subservient state should be deputed or, later, which forces from among the army, *lü*, clans, or border defense units; who should accompany the king;[26] who should be appointed as commander;[27] and how many forces should be employed were all matters for contemplation.

In this regard Wu Ting seems to have enjoyed greater flexibility, because he frequently employed armies from the subservient states to undertake punitive efforts rather than expending Shang core resources. Perhaps the most complex example dates to his era, when actions against four different enemies—the Pa-fang, Yi, Lung-fang, and Hsia-wei—were simultaneously pondered and at least two commanders, Wang Ch'eng and Chih Kuo, considered for overall leadership.[28] The inscriptions closely recorded on another plastron similarly show four possibilities being weighed: Fu Hao attacking the Pa with Chih Kuo, Fu Hao attacking the Yi with Hou Kao, the king attacking the Chung, and Wang Ch'eng and others attacking the Hsia-wei.[29]

Even though the larger field contingents were cobbled together from several discrete components presumably capable of independent maneuver and battlefield redirection, the nature of the mission affected the forces to be mobilized and the tactics to be employed. Moreover, despite traditional claims that Shang martial efforts were focused on exterminating enemies and annexing land, the multilayered structure of Shang political relations resulted in objectives that ranged from chastisement to extermination. A few Shang forays appear to have been simple demonstrations of power mounted to cow nearby lords into remaining obedient and to overawe foreign peoples, but the majority were aggressive and specifically targeted. Even then, some attacks were simply coercive, being intended to persuade contiguous peoples or proto-states into submitting, then employing them as defensive screens or calling upon them to provide active fighting forces, roles they could effectively perform only if they had not been decimated.

Following the initial period of postconquest expansion and consolidation, a number of field efforts were undertaken to dominate resource-rich areas, but most were prompted by perceived threats. Numerous groups were targeted to reduce their military might or drive them further outward, and some even became the focus of punitive onslaughts designed to annihilate them. In addition, a few campaigns clearly had the sole objective of acquiring prisoners for long-term enslavement or sacrifice.

Insofar as Shang military planning suffered the constraints of uncertainty, its external campaigns must have been characterized by a high degree of operational flexibility. Shang expeditionary armies often consisted of multiple contingents that could be employed in segmented operations and limited maneuver, and field commanders probably had considerable freedom in selecting the actual tactics, however rudimentary. Furthermore, after being initiated, campaigns seem to have been largely opportunistic. Unplanned attacks pummeled the enemy and in contrast to immediately targeting two objectives, additional states were sometimes assaulted after the original objective had been vanquished.

The extended range of their expeditionary efforts and the vastness of the contiguous territory further complicated Shang military operations, inescapably affecting the choice of tactics. Even though the semi-mobile peripheral peoples might readily mount incursions against well-known, highly visible Shang targets, these aggressors had to be located before they could be engaged. The Shang probably had reasonably accurate information about their more permanent settlements, but their warriors might be deployed anywhere in the often hilly terrain, prompting uncertainty about where and how to engage them. Enemies who deployed minimal forward reconnaissance could easily avoid outward-bound Shang armies long before their own location could be discovered.

Despite archaeological excavations having recovered numerous weapons from the era, the actual nature of Shang dynasty combat remains a matter of speculation. Whether any sort of ritual preceded or governed the era's battles, limiting the types of weapons and character of the engagement, is unknown, but from the proliferation of arrowheads and dagger-axes it can be concluded that they both played fundamental roles. Moreover, contrary to the Greek notion that employing missile weapons was cowardly, archery seems to have always been greatly esteemed in China. Embedded arrowheads and skeletal wounds clearly show that the era's reflex bows possessed adequate power to impale bone and easily slay enemies who came within range.

Close combat was no doubt preceded by archers firing at enemies fifty to one hundred yards away, but their arrows would have been even more deadly at close range, right up to the point of engaging with long-handled dagger-axes and spears. The archers ensconced on the chariots would have enjoyed a sufficient height advantage to shoot unimpeded over any accompanying infantry and deployed forces, while the roving companies of a hundred, firing en masse, could also inflict great damage because the troops were only minimally protected by armor and carried comparatively small shields.

Piercing and shock weapons must have predominated once the forces moved forward and engaged individually. Spears and dagger-axes were the main piercing weapons, battle axes were the main crushing implements, and short daggers served as an extreme last resort. (Swords had not yet come into existence and thus were never employed, contrary to numerous claims.) To the extent that the two sides didn't simply deploy before rushing headlong at each other, battlefield tactics at the outset were oriented to obtaining an advantageous position, achieving surprise, and concentrating forces for greatest effectiveness.

When formidable enemies thwarted initial efforts undertaken by single contingents of perhaps 3,000 troops, the Shang was compelled to resort to repeated or sequential attacks mounted by individual forces or to employing an augmented field force created by temporarily conjoining two or more individual units to inflict a crushing blow. These larger forces generally attacked in straightforward fashion and were thus operationally indistinguishable from a single army.

The different targeting and operational possibilities preserved on single plastrons further attest to attempts to formulate and exploit simple operational tactics. The two main tactical variants appear to have been having two or more forces—whether different types or simply components of the same type such as the left and right *lü*—meet at a designated, somewhat distant location before combining for a unified attack, and having two or more armies proceed separately before attacking from different directions, either simultaneously or sequentially. Conjoined force assaults and simultaneous, multidirectional strikes were probably structured to emphasize the element of surprise, whereas staggered but scripted sequential attacks were presumably designed to provoke the enemy into concentrating their forces against the immediate threat before penetrating the undefended areas with a second contingent, just as famously formulated in the *Art of War* and other Warring States military writings.[30]

A few instances appear to provide evidence that, in addition to surprise, the Shang had already begun to think about concealment and ambush, both integral aspects of the much practiced hunt. In one of several efforts that would be mounted against the Pa-fang, the king personally led a contingent from the east intended to provoke the enemy into responding in a predetermined manner so that they could be ambushed by Fu Hao and Chih Chia's forces, already deployed in advantageous positions.[31]

Several terms denoting military action indicate that commanders were expected to employ various levels of aggressiveness from the outset. Attack methods ranged from ordinary strikes through harassing pursuits and strong punitive measures, though the simplest and most commonly seen was an assault or attack, *fa*. (The term also means "behead" in the context of ritual sacrifices and clearly entailed the idea that the enemy would be slain.) Two modes were possible, a direct strike mounted by a single force against a town or temporarily encamped enemy and an open field engagement, the latter primarily resulting when the Shang attacked forces in movement or enemies that came forth to resist.

A frequently noted variant was the punitive campaign or *cheng*, a term that came to describe "campaigns of rectification" that were mounted against the rebellious, especially external peoples who were viewed as "barbarians" in the Spring and Autumn and Warring States periods. However, given that the oracular inscriptions sometimes employ the same term to record attacks by peripheral peoples on the Shang, *cheng* had not yet acquired a strongly punitive sense.

Terms that connote more destructive intent include *p'u, tsai, shun,* and *t'u.* The first of these can probably be translated as to "pound" or "pummel," while *tsai,* to "damage" or "harm," is normally understood as meaning to "hunt," "damage," or "wound with weapons."[32] Though sometimes synonymous with "attack," *tsai* usually entailed an intent to inflict serious damage rather than just slay or capture and was even used when referring to the damage visited on two Shang cities by the T'u-fang.[33] The term *t'u,* to "massacre" or "slaughter," also means something like "punitively attack" but may have merely signified going out to resist the enemy or manifesting awesome power.[34]

Tun, which came to mean "substantial" or "solid" and even "to break" in *Chuang-tzu,* is perhaps best understood as to "pound" or "pummel." *Yü* would later mean "defend against," but in the Shang it seems to have referred to highly aggressive actions mounted on the periphery to extirpate the enemy. Armies were frequently dispatched to *chui* or pursue enemy contingents, implying that the latter had already been either defeated or scared off, whereas *ch'ü,* which simply means "to take" or "to seize," is occasionally used in a field context to indicate the taking of a town or prisoners. Other terms less frequently seen include characters that seem to mean "to ram," "capture," "search out and destroy," and a couple of uncertain but clearly aggressive meanings.[35]

The idea of reconnaissance, fundamental to any intelligence reporting system but also a separate function on the battlefield, had already appeared. Surprisingly, whether or not small numbers of scouts were forward deployed, it is large contingents that were dispatched to "look at" the enemy, though the term *shih* almost certainly includes the concept of assessment rather than simple observation. Late in the dynasty the *shih* (army) occasionally seems to have acted as a reconnaissance in force, either to probe (as later articulated in the *Wu-tzu*) or to provoke a meeting engagement.[36]

Apart from defending towns subject to border incursions, the *shih* (army) and other contingents that primarily engaged in combat missions could also be deployed to act as forward or defensive screens.[37] However, as the dynasty progressed the concepts of segmentation and functionality evolved, and the various contingents were sometimes denominated as forward, center, and rear. For example, in one late case actions by an advance unit were immediately followed by an enveloping attack (*chou fa*) mounted by border defensive units (*shu*).[38] However, even in Wu Ting's era the dispatch of a single army or levying of a subject state's forces was often a preliminary move, a strike that might achieve victory but would probably need reinforcement and therefore was only intended to open the way (*ch'i*).

Movements to surround enemies out in the field were also undertaken, sometimes by more than one army, and cases of enemy armies surrounding Shang units are similarly known, showing that steppe forces were not incapable of coordinated action.[39] In addition to assaults on settlements, Shang attacks were attempted against fortified towns, but little is known about them and they are unlikely to have been frequent because the defenders would have enjoyed an overwhelming advantage as long as they remained ensconced behind the massive walls, siege equipment not yet having begun to evolve.

Evidence for their relative impregnability is clearly visible in Wu Ting preemptively attacking an enemy proto-state because it had commenced fortifying its town with walls. Even the *Art of War* subsequently compiled in the formative period of siege technology admonished commanders not to waste their resources in foolishly assaulting fortified enclaves because a third of the troops would perish.[40] Fortunately for the Shang, pastoral practices predominating in the steppe precluded the seminomadic peoples from undertaking substantial, fixed defenses even though stone fortifications were employed for some villages in Inner Mongolia.

TRAINING

Based on documented practices of subsequent ages, it has often been asserted that weapons were a royal monopoly in antiquity, being furnished to the combatants only when under attack or being mobilized for an external expedition. Many later dynasties even prohibited weapons possession among the populace because they furnished the sole means of challenging the ruling family. However, particularly with the shift to bronze, even though their fabrication was carried out under governmental auspices, there is no evidence that this sort of proscription was ever implemented in the Shang. Conversely, being a rather brutal era in which the king gradually evolved into an unchallenged despot and martial values had long been esteemed, it seems more likely that members of the regal clans at least possessed weapons, if they did not routinely carry them. Moreover, the enormous number of weapons interred with the deceased throughout the Shang implies sufficient availability for them to be wasted in this fashion. (The late Shang gradually shifted toward the use of replica weapons and bronze ritual items that employed a larger component of more easily formed lead and less copper, thereby conserving the latter while minimizing the labor involved in sharpening and finishing.)

The inscriptions provide vestigial glimpses of formalized training measures in these weapons that apparently carried over into the early Western Chou. Only by becoming familiar with the features of their weapons and practiced in their employment could combatants survive on the battlefield and be contributors rather than liabilities. In addition to acquiring experience in the coordinated employment of the shield and spear or shield and dagger-axe, every warrior had to develop the strength necessary to adroitly wield his piercing or crushing weapon and sustain the effort under combat conditions.[41] The Shang also had martial dances that were performed with weapons that no doubt contributed to the overall development of martial spirit, but whether they had any training function (such as practicing coordinated movement) is unknown.[42]

The degree to which the privileged warrior class may also have been educated in writing, the techniques of command and control, or the rudimentary administrative skills increasingly needed to direct Shang farms or other enterprises remains unknown. Some historians claim that chariot driving, which would become one of the "six arts" or essential accomplishments of every gentleman in the Chou, had already begun to be important. Formal training would also distinguish warriors who had acquired additional capabilities, making them more qualified in some general sense for broader responsibilities.[43]

Because archery was highly esteemed and extensive practice is required to develop the skills necessary for firing quickly and accurately in the heat of battle, sons of the nobility certainly underwent formal training. In the middle Warring States period Sun Pin would comment that "those who excel at archery act as the left (of the chariot), those who excel at driving act as drivers, and those who lack both skills act as the right."[44] A good archer could easily fire several arrows per minute from his reflex bow, quickly expending the quiver of ten normally carried. A few inscriptions refer to archery schools, officials being entrusted with training people in archery, archery officers who exercised command functions in combat, and new archers being deployed on the battlefield.[45] Although there may have been archery contests under royal auspices such as convened in the early Chou or localized competitions that became the basis for the highly esteemed communal and ritual archery ceremonies that subsequently developed, evidence is lacking.

Nevertheless, oracular inscriptions fail to support claims which are based on late writings that Shang military training was already highly structured and carried out under government supervision. The use of common weapons such as the spear and dagger-axe was probably taught in the time-honored way, by older warriors and low-ranking officers skilled in their use, but again the inscriptions offer no further information. Conflict having been virtually a normal part of warrior life in the Shang, particularly under Wu Ting, a certain amount of "training" no doubt occurred in the family, from early age, to equip men with the necessary skills to fully participate in the society. On the battlefield itself, more skilled and experienced fighters invariably played the leading role, allowing novices to learn under life-threatening conditions and become effective warriors or soldiers, presuming they survived.

Furthermore, whether ensconced in a chariot or fighting on the ground, battlefield clashes required more than warriors simply wielding individual weapons. Early engagements may have rapidly disintegrated into hundreds or thousands of individual clashes and become nothing more than a chaotic melee, yet a tendency toward some sort of cohesive approach and the formulation of basic tactics that might be executed upon command is thus already visible in the Shang. But deploying and maneuvering to create tactical advantages required discipline and the creation of fundamental organizational units. Whether the clan forces, *lü*, and contingents of 3,000 consisted of numerous squads, platoons, or companies as in later eras and the soldiers trained together in small units for greatest effectiveness remains unknown. The only group training visible in the inscriptions remains the hunt, though there are indications of night exercises as well.[46] Claims that complex battle formations (which would have required far more extensive training) were employed in the Shang lack all substantiation, as does their projection back into the legendary era of the Shang's supposed progenitor, Fu Hsi.[47]

14.

METALLURGICAL EVOLUTION IN CHINA

✦✦✦━━━━━━━━━━━━━━━━━━━━━✦✦✦

FOR REASONS BOTH obvious and subtle, the discovery of metals has always been viewed as a turning point in the history of warfare. Whereas naturally occurring materials such as stone must be laboriously worked and impose numerous constraints because of their weight and inherent characteristics, the ductile properties of metallic alloys convey considerable freedom in designing and fabricating weapons.[1] Hammering and forging, the first measures, increased productivity, but casting even in simple individual cavities immediately multiplied the quantity and ensured the uniformity critical to combat. (The slightest change in weight or balance can cause fatal awkwardness when a new weapon is first employed, and arrowheads can stray far off target.) Thereafter, multiple casting dramatically increased productive efficiency, particularly for small, expendable arrowheads.

Coincident with increased population, economic prosperity, and centralized administration, China witnessed a sudden surge in mining, smelting, refining, and casting after a lengthy period of incipient development that resulted in industrial-scale production both in dedicated urban workshops and at a few distant fabrication points. This newly organized, "mass scale manufacturing" did not simply augment older methods, but replaced them, enabling warfare to escape the limitations imposed by craft methods that had relied on laborious hammering, chipping, and shaving. Sustained and perhaps stimulated by the increased availability of effective weapons, the scope and intensity of warfare had already begun to escalate in the Hsia, no doubt prompting increased demand for bronze weapons in a self-reinforcing loop.[2]

Although the emergence of metal weapons constituted a monumental step in the evolution of warfare, the impact of copper-based versions in the Hsia and Shang should not be overestimated. Effectively sharp edges can be produced in stone, bone, and surprisingly even bamboo (which can easily be lethal), and deadly spears with stone tips continued to be employed for centuries even after adequate metallic resources had become available.[3] Arrowheads were still being fabricated primarily from bone in the early Chou, and bone and stone continued to predominate in economically impoverished cultures and peripheral areas bereft of metallic resources for centuries more, especially for agricultural implements.

The relatively late appearance of fabricated metal objects in China, in comparison with Russia and the West, has prompted irresolvable arguments about indigenous origination versus diffusion or the hybrid known as "stimulus diffusion." Apart from issues of national pride, the idea that metallurgical insights are so complex as to be discoverable only once rather than being a common experience of mankind, one not just replicable but repeated in different environments and disparate times, has fanned the dispute. Nevertheless, many Chinese scholars believe that the unique, piece-mold bronze-casting techniques extensively employed to fabricate complex ritual cauldrons in the Shang must have evolved out of advanced ceramic methods and pyro-technology and therefore conclude that Chinese metallurgy is the result of independent discoveries.[4]

Fortunately, in comparison with questions about technological and productive capabilities and within the greater context of Chinese warfare, this intriguing issue may be deemed somewhat irrelevant. Nevertheless, it should still be noted that Xinjiang in the northwest shows considerable external influence in both alloy composition and object style. Conversely, the metallurgical tradition discernible in more central areas seems to have a strongly indigenous character and may have evolved separately despite the inevitable, sometimes extensive cultural interaction known to have occurred. Furthermore, the advanced bronze technology found in the K'a-yao culture that developed in the Huang-shui river valley south of the Yellow River is itself marked by readily identifiable local elements intermixed with many common to the central plains, north China, and even northern Eurasia.[5]

Local variation stems from several factors, including environment, lifestyle (agriculturally based or seminomadic), and accessible metal resources, though productive specialization is not necessarily limited by the latter's availability.[6] Copper, the crucial metal in the so-called Bronze Age revolution, is generally found intermixed with other metals, including tin, lead, arsenic, and antimony, resulting in what might be termed naturally occurring alloys when smelted prior to the evolution of more thorough understanding and craft practices.[7] Because human resources and fully processed metallic ores were never unlimited, warfare's importance within society clearly affected the purpose toward which technological capabilities were directed, including the production of daily utensils, decorative items, ritual bronzes, agricultural implements, and weapons.[8]

Thus, even though Shang production levels quickly escalated, the supply was neither inexpensive nor unlimited, no doubt key factors in the development and use of "semblance" artifacts late in the dynasty. Intended purely for display and for accompanying the dead, these implements never received the usual detailing, sharpening, and polishing associated with Shang dynasty weapons and ritual bronzes. Moreover, being cast from an inferior alloy with a much higher lead content and correspondingly reduced tin portion, they were fundamentally incapable of being perfected to the same degree. (Both rare and expensive, tin provided the essential characteristics of hardness and brittleness, whereas lead facilitated the flow during casting but resulted in a softer product less capable of being sharpened.)[9] However, other factors may have contributed to this tendency to inter inferior-quality bronzes, ranging from diminished reverence for the deceased, to growing disdain for spirits of the departed, to an increasingly insatiable demand for copper (especially for weapons), to just outright greed, since semblance bronzes were increasingly seen even in opulent graves.[10]

Even though other prestige materials such as jade (which played a critical prestige role in Liang-chu culture) might have served equally well, bronze metallurgy soon fulfilled a crucial role in producing ritual objects that could be manipulated by the ruling elite for political purposes. Bronze also became indispensable because it allowed rapid casting of the weapons needed to dominate an increasingly hostile world and critical chariot components. However, although copper smelted with zinc produces brass, a material with especially conducive characteristics for moving components, the fabrication process is far more complex. Despite the recovery of a few (presumably accidental) specimens, it remained far beyond Shang technical capabilities.

Mining activities rapidly expanded, and bronze production soared during the initial reign period at Yen-shih and Cheng-chou. The government established far-flung outposts at Tung-hsia-feng, P'an-lung-ch'eng, and other locales to ensure the security of the raw materials; embarked on predatory campaigns against the Yi to acquire them;[11] and apparently refrained from hostile gestures toward Shu so as to ensure an uninterrupted supply of copper from the Sichuan plains and lead from farther afield in Yünnan through Shu's intermediation.[12]

The scope of the bronze production facilities at the last capital at Anyang is equally astonishing. Two major, segregated workshops have been excavated, one north of Miao-p'u and the other southeast of Hsiao-min-t'un, which seems to have specialized in casting ritual vessels. Copper-smelting furnaces, molds for bronze casting, and various implements for preparing clay molds and for finishing and polishing the final products have all been found in the 10,000-square-meter work area. A staggering 30,000 molds have also been recovered, many of them composite, as well as numerous cores, including some that produced unusual vessels previously attributed to the early Western Chou.[13]

The incipient period of Chinese metallurgy ending with the Shang witnessed a progression from small copper decorative items and simple tools such as knives and awls to weapons and large ritual vessels. However, there seems to have been little inclination to divert highly valued metals to agricultural implements despite increasing reliance on agriculture. On the assumption that the Hsia and Shang were slave-based societies, it has been claimed that agricultural implements were never produced because the ruling class feared providing the downtrodden with metal weapons. Although recognizing the essentially convertible nature of agricultural equipment, this explanation amounts to nothing more than an idle projection of envisioned fears and is completely unfounded, because there would be no significant difference in the general effectiveness of stone and metallic variants.

Somewhat akin to the controversy over the nonexistence of iron swords, it has also been asserted that bronze's immense value mandated that broken and worn-out tools be melted down, thereby presumably explaining the absence of bronze agricultural implements at Shang archaeological sites. In contradiction, the recovery of highly decorated, symbolic farm implements presumably employed in ritual performances from a few Shang graves indicates that at least a few molds existed, implying some degree of production. Functional specimens such as plows, adzes, simple spades, shovels, and mattocks have also been found, particularly in peripheral areas where warfare played a lesser role, as well as highly specialized mining tools lying about ancient shafts.[14] Accordingly, it would appear that the Shang emphasized weapons of war and the ritual vessels essential to power, resulting in pedestrian agricultural implements continuing to be fabricated from wood, stone, and bone despite more effective shapes, greater sharpness, and greater resilience being possible with metal plows or hoes, but not to the complete exclusion of agricultural needs.

Although disagreement over the origins of Chinese metallurgy and the date of the first identifiable artifacts continues, the Shang clearly benefited from a lengthy heritage of technological development stretching back to the Yangshao (4400 to 2500 BCE) or possibly earlier. A general trend toward achieving a working knowledge of the properties of different metals and mastering the requisite working techniques is apparent from 3000 to 2000 BCE, coincident with the Lungshan period, when the stage of minimal productivity was realized. Nevertheless, arguments about when one or another culture crossed the horizon from the Stone to the Bronze Age and whether to characterize certain centuries as dual use have similarly not abated. However, the question of when the number of bronze implements in circulation became significant enough to label the era "chalcolithic" is largely irrelevant for Chinese military history, because the earliest weapons imitated lethal stone versions, and metal's ritual role dwarfed its military application, copper and bronze being allocated or diverted to weapons only with warfare's rising intensity.[15]

Rather than substantial copper or bronze objects, core evidence for the initial stage of metallurgical development is provided by metal fragments, ore residue, and smelted globules primarily of copper and crudely processed ores. Small items such as small decorative pieces, jewelry, pins, awls, and knives rank next in importance. Again, for the purposes of military history singular appearances are anomalous and irrelevant; only the widespread adoption of new materials in producing weapons has discernible impact. However, in conjunction with the erection of defensive fortifications, early attempts at manufacturing bronze weapons certainly imply a growing concern with external threats and a probable escalation in conflict.

Ancient China has long been recognized for the superlative quality of its massive vessels, precisely cast weapons, and other objects fabricated from various bronze alloys. Although silver did not appear until much later, gold was employed for small decorative items as early as the Shang,[16] yet it was glistening, highly burnished bronze that formed the very basis of power. The recovery of a broad axe with a meteoric iron blade affixed in a copper mounting clearly shows that Shang metallurgists recognized iron and were cognizant of its superior hardness. (An early Chou dagger-axe of bronze with a meteoric iron point has also been found.) Nevertheless, despite occasional claims based solely on traditional literary sources that the Hsia and Shang had already commenced smelting and employing it to produce weapons, iron would not be produced until well into the Chou.[17]

Discerning the existence and effects of the various components in China's bronze alloys is complicated by the impure nature of the minerals in situ, elements such as tin, arsenic, sulfur, antimony, zinc, and even gold and silver often being found intermixed in copper deposits.[18] Alloys combining two or more of these elements in apparently functional proportions may have inadvertently resulted from their presence in the ore. These occurrences tend to obscure the "normal" developmental sequence from copper through copper/tin and copper/lead and then ternary variants; deliberate but collateral intermixing of copper with arsenic and accidental brass formulations further add to the complexity. Only with the passage of centuries did a working knowledge of alloys emerge, enabling the Shang to consistently cast large ritual bronzes and weapons with deliberately chosen, varying degrees of hardness and durability.

Chinese metallurgical practices evolved in several distinct regions: the northwest in the so-called Ho-hsi corridor of eastern Xinjiang and the immediately contiguous area; between the Yellow River and the Huang-shui River; in the central plains, but really centered in the Yen-shih/Cheng-chou corridor; the lowest reaches of the Yellow River in Shandong; and the southwest, emblematized by the dramatic cultural manifestations of San-hsing-tui.[19] However, the most numerous and earliest bronze artifacts, some 1,500 in comparison with only about 200 from the middle reaches of the Yellow River in Yü-hsi, have been found in the northwest, encompassing Gansu, Qinghai, and Xinjiang, where some small objects strongly resemble external styles.

Early knowledge of metals and metalworking seems to have been widespread but highly limited in actual application during the third millennium BCE. Copper and primitive bronze alloys came into use between 3000 and 2300 BCE, and the Bronze Age seems to have commenced around 2400 to 2000, though assessments vary.[20] In terms of identifiable cultures, only a few early knives have been recovered from Ma-chia-yao (3300–2650) and Ma-ch'ang (2650–2000) cultural sites, while the increased number of artifacts, roughly 130, including axes, knives, daggers, and awls from the Ch'i-chia culture (2200–1800), lying in the intermediate region between the core and the northwest, indicates greater but still sporadic interest in metals. However, the more than 300 copper and bronze objects and the first stone molds discovered at Ssu-pa (1950–1550) cultural sites, said to be the transmission nexus for steppe and thus Western metallurgical knowledge, mark a transition to metal and stone's coexistence, at least in consciousness if not in quantity.

In the incipient stage the greatest strides and most extensive production occurred in Xinjiang and especially Gansu, where the majority of early artifacts have been recovered. Based on fragments of crucibles, knives, axes, awls, ornaments, mirrors, small copper items, and partially refined metallic globules, it is generally claimed that copper was already being mined and smelted along the upper Yellow River during the Yangshao and that copper and bronze manufacturing was being conducted at more than forty sites in this general area by the end of the Ch'i-chia culture.[21]

The first metal alloys have long been identified with Lin-t'ung Chi-ang-chai, a site that was continuously occupied from the Yangshao through Shih-chia-lei, Miao-ti-kou, and late Pan-p'o.[22] Two metal plate fragments dated to about 2700 BCE have been recovered, incontrovertible evidence that metals were already known, but their high zinc content (65 percent copper and 25 percent zinc) makes them somewhat problematic because the knowledge and technology for producing brass objects would not exist for another four millennia due to zinc's volatility, prompting claims that they could not have been produced in China.[23] However, they also have a high sulfur content, indicative of the earliest stages of smelting; the region's copper sources are marked by the presence of high concentrations of other metals, including zinc;[24] and experiments have proven it is possible to produce brass bits identical to those from an awl found in Shandong from locally available, comparatively high-zinc-content ore.[25] Subsequently, Ch'i-chia culture in Gansu and later the early Hsia had access to this material.[26]

Chronologically next in importance would be knife remnants from the definitive Ma-chiao-yao (3400–2000 BCE) cultural site in Gansu, variously dated from 3280 to 2740 BCE but more likely closer to the latter. Hardly primitive, the knife was fabricated from an alloy containing about 6–10 percent tin and cast in a two-part mold indicative of a new orientation to quantity production; it remains the earliest bronze implement yet found. A few casting remnants have also been found that reportedly consist of an imperfectly refined intermixture of iron and copper, evidence of smelting and the achievement of both copper and bronze.[27] Copper knives dating somewhat later have also been discovered in the Ma-kuang, Juo-mu-hung, and K'a-yao cultures.

The forty-five to fifty artifacts recovered from Ch'i-chia (2055–1900 or 2200–1800 BCE) cultural sites in Gansu not only range from pure copper through lead/copper and tin/copper but also show a distinct trend from copper to bronze, prompting the conclusion that both casting and hot forging were being employed by about 2000 BCE.[28] Knowledge of metals and alloys was clearly increasing, but still remained at an intermediate stage. Even though a large bronze spearhead and a forged arrowhead dating to about 2000 BCE have been recovered, small items such as copper knives and awls predominate, copper apparently still being too valuable to waste on casting expendable arrowheads. However, alloys containing from 5 or 6 percent up to 10 percent tin had been achieved, with cold hammering, hot forging, and some casting (such as of a copper knife discovered at Min-hsien) in two-part molds all being employed.

Xinjiang, which has received far less analytical coverage, generally reflected developments in nearby Gansu and the contiguous Andronovo and Sintasha-Petrovka cultures. Small copper objects dating to about 3000 BCE are known, and bronze is well attested by 2000 BCE, because alloys marked by a heavy arsenic component ranging from 8 to 20 percent have been found.[29] Martial items produced over the fifteen centuries from 2000 to 500 BCE include small knives, short daggers, small axes, and arrowheads. Although metallurgical knowledge and practices generally lagged behind Gansu and even the core cultural area, Xinjiang would develop a basic knowledge of iron somewhat earlier, though only to be subsequently surpassed by other areas.

The numerous bronze objects recovered from Ssu-pa cultural sites in Gansu's Ho-hsi corridor have prompted the conclusion that Ssu-pa culture was the final metallurgical precursor to Erh-li-t'ou, even though suggested termination dates of 1600 and 1400 BCE place it well within the Shang horizon. This seems to represent the crucial stage at which metal objects become more common, reportedly as a consequence of increasing class distinction.[30] Rapidly increasing in number and complexity, the items being fabricated soon encompassed axes, knives, and daggers. Slightly more bronze than pure copper objects were produced, alloy formulations multiplied, casting in stone molds began, two-part molds then developed, and the first cast arrowheads appeared.[31] However, production methods were highly varied, ranging from hot forging and casting through cold working and casting with subsequent working.

The three sites that have been extensively excavated show a surprising degree of variation in object preferences and significant differences in alloy composition, the latter no doubt the result of locally available ore. Apart from relatively pure copper, alloys of tin and arsenic predominate, but a few objects were also cast from a combination of copper, tin, and arsenic.[32] (The western sites primarily employed tin, the eastern ones arsenic.) Some residual iron, evidence of incomplete refining, has also been discovered intermixed in many metallic fragments.

Half of the 200 pieces so far discovered at Huo-shao-kou are pure copper, the rest being bronze of varying composition, a few more tin than lead based, but six ternary pieces of copper/tin/lead and a couple with arsenic number among them. Many of the cast pieces underwent subsequent heat and cold treating, and a two-piece stone mold for casting arrowheads has been discovered. In contrast, many of the hot-forged artifacts found at Tung-hui-shan have a heavy arsenic component, whereas those at Kan-ku-yai (dating to about 1900–1600 BCE) apparently include every possible variant, arsenic not excepted, some being hot forged and others molded.

The central plains area, where late Lungshan (roughly 2400–1700 BCE) culture flourished—defined as the western part of Henan, southern Shanxi, and perhaps the southern part of Hebei—generally lagged behind the northwest in metallurgical developments.[33] Unlike the northwest, which experienced a transition from copper to an arsenic alloy, then bronze with tin, the central area progressed from copper to a tin alloy without any intermediate arsenic stage, though this may simply be because the sources were not contaminated. Copper thus tended to dominate, and only a few small fragments have been recovered that might predate the metal container fragments found at the late Lungshan site of Wang-ch'eng-kang.

In Yü-hsi and Chin-nan pre-Hsia culture and in the earliest stages of Erh-li-t'ou culture (1780–1529 BCE) in Henan, Shaanxi, Shanxi, and northern Hebei, but especially Erh-li-t'ou itself, the quantity and variety of bronze items suddenly multiplied. Significant technological advances were realized in Erh-li-t'ou's second period, including the introduction and then widespread use of two-part and multiple-cavity molds, explaining the sudden preponderance of metallic objects in the era's graves; the casting of dagger-axes, axes, and finally arrowheads, reflecting the growing importance of warfare and willingness to employ copper for irrecoverable missiles; and the appearance of extensive decoration.[34]

The contents of slag heaps and crude metallic fragments discovered at expansive workshops confirm that the limitations of pure copper had long been transcended. The beneficial effects of tin and lead in lowering the melting point, the enhanced pouring characteristics of high-lead-content alloys and lead's ability to impart flexibility to weapons, and tin's effect in increasing the overall hardness and thus the sharpness of edged blades had all been fathomed, resulting in a full range of bronze alloys, including the primary formulations of copper/tin, copper/lead, and copper/tin/lead, all being employed by about 2000 BCE.[35] These developments in turn depended on earlier advances in smelting and refining technology that had managed to achieve nearly pure (97.86 percent) copper by this time, as shown by evidence on crucible walls and metal remnants discovered at Mei-shan (2290–1900 BCE).[36]

Under the influence of techniques indigenous to Ch'i-chia culture to the west, two major developments that had begun during the second period came to fruition in Erh-li-t'ou's florid third period. First, cumbersome stone molds that required tedious working and imposed severe limitations on size and complexity were gradually replaced by clay and then heat-fired ceramic molds that could withstand higher casting temperatures. This not only facilitated the multiple replication of smaller objects and weapons but also made possible greater precision in realizing complex designs, initiating an era of intricate detailing and the production of abstract patterns similar to those subsequently seen on symbolic Shang axes and ritual vessels. Not unexpectedly, many of the objects cast by this process were imitations of preexisting ceramic and stone versions. However, ceramic molds were never adopted at peripheral Hsia production centers such as at Tung-hsia-feng.

The second significant advance was the discovery and adoption of the piece-mold casting process, which made possible the larger, more complex ritual vessels that would proliferate in the Shang. Somewhat surprisingly, early piece molds were apparently used only once, even though one of their great advantages should have been multiple employment.

THE SHANG REALIZATION

By the late Shang great progress had been made in recognizing terrain characteristics and plant varieties indicative of likely ore deposits. The development of wood-reinforced shafts and galleries, some of which remain nearly viable today, and techniques to minimize water intrusion and even partially remove pooling water facilitated their increasingly systematic exploitation. Mining efficiency was further improved through specialized tools, both metallic and nonmetallic.

As already noted, China's naturally occurring ores vary greatly in composition. Despite ongoing advances in knowledge and techniques, even identical processing could yield somewhat different raw materials. Furthermore, an examination of an artifact's lead isotope ratios often allows probable sources to be identified, such as the copper employed in Hsin-kan (Wu-ch'eng) and San-hsing-tui bronzes.[37] Somewhat surprisingly, they also reveal that crude metals produced from several geographically distinct sources were frequently intermixed both in the cultural core area and out on the periphery during the Shang, despite locally available quantities being more than sufficient, such as in the southwest.[38] Moreover, changes in Anyang bronze isotope ratios over time indicate a shift in the copper source, whether out of necessity or preference.

Despite being numerous, ancient China's ore sources were widely scattered and characterized by local concentration. Conscious, dedicated effort therefore had to be expended to discover and exploit them. The larger ones mined early on were generally found in the Gansu region around the Ch'i-lien Mountains, Yünnan in the southwest, and Jiangxi and Anhui along the Yangtze River.[39] Somewhat sparser deposits were also accessed in the core Hua-Hsia area of Yen-shih to Cheng-chou and also out in Shandong, explaining how metallurgy could have evolved in both the upper and lower reaches of the Yellow River. Although recoverable copper was frequently found intermixed with other metals such as zinc, iron, lead, and sometimes even silver or gold,[40] tin deposits were dispersed and highly limited, requiring production to be undertaken separately.

The copper mines in Yünnan and the lower Yangtze River area were particularly productive. Their distance from Hsia and Shang administrative centers quickly stimulated the development of several major transport routes that took advantage of China's many interconnected rivers and lakes wherever possible, as well as the dispersion of martial forces, emplacement of strongpoints, and construction of bastions such as P'an-lung-ch'eng. In the lower Yangtze area where both copper and iron are found, the copper content in highly productive mines that operated from the middle Shang through the Warring States period was generally 5–6 percent, with local concentrations sometimes reaching 10 to 20 percent.[41] Massive slag heaps estimated at a staggering 500,000 tons total indicate 100,000 tons of copper may have been extracted over the centuries.[42] Partially processed ore, slag heaps, and other evidence of extensive processing activities have also been found at two smelting sites discovered at Anyang and Wu-ch'eng.

This immense quantity of copper was primarily employed to fabricate crucial emblems of Shang power, ranging from precisely incised drinking vessels to great axes and massive cauldrons designed for ostentatious use, commemorative employment, and ritual performance. The technological achievements embodied in these opulent vessels, being well documented and widely known, need no further elaboration. However, weapons also consumed increasingly large quantities of metal as their types and numbers multiplied in response to warfare's escalating needs, though *ming ch'i* or semblance weapons interred with the deceased soon began proliferating in an obvious effort to conserve expensive, limited resources.[43] Metal was employed for prestige first, then important weapons, and finally expendable weapons.

15.

EARLY
WEAPONS
AND THE AXE

◆ ◆◆ ————————————————————————— ◆◆ ◆

THE LESS THAN unique role played by tools and agricultural implements immensely complicates any attempt to characterize the nature of conflict in preliterate societies. Like contemporary revolutionaries
and subjugated peoples throughout history, early men consciously
adopted tools and other implements for combat purposes and no doubt
instinctively wielded whatever objects might provide an advantage in
exigencies. The late Warring States *Six Secret Teachings* discussed how
to exploit their inherent combat potential:[1]

> The implements for offense and defense are fully found in ordinary
> human activity. Digging sticks serve as *chevaux-de-frise* and caltrops.
> Oxen and horse pulled wagons can be used in the encampment and
> as covering shields. The different hoes can be used as spears and spear-
> tipped dagger-axes. Raincoats of straw and large umbrellas can serve
> as armor and protective shields. Large hoes, spades, axes, saws, mortars
> and pestles are tools for attacking walls. Oxen and horses are the
> means for transporting provisions. Chickens and dogs serve as look-
> outs. The cloth that women weave serves as flags and pennants.
>
> The method that the men use for leveling the fields is the same
> as for attacking walls. The skill needed in spring to cut down grass
> and thickets is the same as needed for fighting against chariots and
> cavalry. The weeding methods used in summer are the same as used
> in battle against foot soldiers. The skills used in repairing the inner
> and outer walls in the spring and fall, in maintaining the moats and

channels, are used to build ramparts and fortifications. Thus the tools for employing the military are completely found in ordinary human activity.

Wood's rapid decay unfortunately causes almost every trace of such basic weapons as clubs, spears, javelins, and staves to vanish, obliterating the evidence necessary to reconstruct their evolution. Lacking fortuitously preserved specimens, the inception of the crude wooden bows and fire-hardened arrows that might push the origins of armed conflict further back into the mists of time can only be inferred from early stone arrowheads. Because few shaft impressions remain from the Shang or even Western Chou, it is extremely difficult to determine the actual length of various weapons, their striking range, and whether they were designed for wielding by one or two hands. In addition, even when they can be readily identified by their light weight, inferior metal, or intricate embellishments, the existence of numerous bronze versions manufactured specifically for ceremonial display or burial with the deceased rather than battlefield use further complicates the process of historical reconstruction.[2]

Archaeological reports often document the recovery of several distinct styles of single weapons such as an axe from individual tombs.[3] Whether this means that weapons from earlier periods were carefully collected, preserved, and employed; earlier styles continued to be copied; or the different regions preserved certain styles by habit or preference and their products circulated to some extent is unknown, but trade and capture by warfare were both extensive, and all three possibilities are likely. The quest for lethality also produced some unusual, even bizarre weapons of unknown or forgotten origins that continued on as anomalies.[4]

As weapons became longer, stronger, and more lethal, they basically evolved from roughly contoured designs laboriously fabricated from natural materials to increasingly precise, forged or cast metallic realizations. Shapes became more complex and dynamic, finishes smoother, and decorations and embellishments more elaborate. However, neither the invention of new weapons nor changes in basic materials necessarily resulted in the latest variants immediately displacing previously popular styles. This phenomenon is readily understandable, if not fully explainable, by remembering that although great energy may be devoted to the unremitting quest for even a minute advantage, an inherent reluctance to change familiar weapons and previously successful tactics has always beset military enterprises. In addition, apart from any antiquarian impulse, ancient weapons invariably required lengthy craft processes to produce and were therefore cherished in cultures that esteemed martial values, including Shang China.

Even when conducive materials such as flint were readily available, highly tedious labor processes were required to transform stone blanks into usable weapons, invariably resulting in slight but noticeably different characteristics, including shape and weight. As part of its emphasis on weapons fabrication, the Hsia embarked on a casting program that did not simply copy the old stone versions but instead embraced new forms and improved designs, initially made possible by copper's malleability, then its ductility. Even though the mining and smelting of ore required a massive labor commitment, the Shang quickly exploited molds to cast uniform axes and arrowheads.

Although it has generally been claimed that these bronze weapons were sharper, stronger, or otherwise vastly superior in some indeterminate way, these assertions should be closely scrutinized because, for example, arrowheads fabricated from flint were often sharper than variants produced in bronze. In addition, even though astonishing amounts of copper were soon being produced, the quantity was not unlimited, and bronze had to be prioritized, the majority being allotted to the production of the ritual vessels essential to manifesting and maintaining power. It can therefore be readily understood why newly created weapons never immediately displaced previous versions, stone axes continued to be important in the Shang, and enormous numbers of bone arrowheads are still found in Western Chou sites.[5]

A detailed history of Chinese weaponry is too complex and encumbered by regional variation to undertake here, but the following simplified analysis based on the work of numerous scholars and archaeologists should prove useful to understanding the combat modes and tactical possibilities prevailing in the Hsia and Shang. Unfortunately, despite a number of overview articles (albeit of limited scope) having appeared over the past thirty years, no comprehensive study has been undertaken for nearly four decades.[6] Nevertheless, by employing these early efforts in conjunction with hundreds of archaeological reports the broad outlines can be clearly discerned, numerous implications drawn, and a few traditionally espoused claims quickly disproven.

Additional insights may be gained by evaluating the combat implications of recovered artifacts against the encyclopedic weapons knowledge and training practices preserved in written manuals and actualized on a daily basis in traditional martial arts schools.[7] Naturally this knowledge must be judiciously employed because many techniques have become highly stylized, designed more for flourishing display than real-world effectiveness. However, since the body's kinesthesiology remains unchanged, insights gleaned from them can aid in understanding how ancient Chinese weapons may have actually been used on the battlefield, as well as providing a sense of their limitations.[8]

Combat with cold weapons is frequently resolved in a few seconds rather than determined by the sort of extended slugfest depicted in contemporary movies. Poor technique, fatigue, weakness, overextension, loss of balance, or a lack of familiarity with the enemy's weapon, even when not decisive, can sufficiently if only momentarily impair a fighter, allowing the enemy to successfully strike. Recovery, even survival, may then prove impossible.

It should never be forgotten that training is the basis of warfare, combat between unskilled fighters is simply a matter of chance, and disorganized groups of warriors can only produce chaos and uncertain results. Every weapon has a unique method of employment, range of effectiveness, required hand placement, ideal arm movement, critical body rotation, and essential leg action, all moderated to achieve the necessary dynamic balance between stability and speed. Soldiers unpracticed in manipulating their weapons pose a danger not just to themselves but to everyone about them.

For every weapon there is also an ideal combat space that allows maximizing its effectiveness while minimizing potentially adverse consequences for the force as a whole. This is one of the crucial differences between single combat on an open field, in which a fighter's wild or bizarre actions may prove surprisingly effective, and military combat between organized contingents on a battlefield, whatever their numbers.

The essence of both weapons training and group fighting is constant, unremitting repetition that makes movement instinctual and response immediate. As the *Art of War* makes clear, warfare is a matter of ruthless efficiency; other factors being reasonably equal, whoever achieves the greatest efficiency in every aspect, including tactics and individual weapons, will prevail. Thus, even though little is known about it, military training must have existed in the ancient period, possibly centered on rudimentary versions of the forms employed in contemporary martial arts practice and discussed in the military classics.

In thus charting the history of weapons and attempting to assess their impact, it should especially be noted that ancient China was populated by several disparate cultures that have only recently begun to be recognized as distinct sources of innovation and technological divergence rather than simply beneficiaries of advanced Hua-Hsia achievements emanating from the Yellow river valley. No longer can it automatically be assumed that a certain weapon such as the dagger-axe originated in the northern plains and then spread by diffusion through trade or conquest throughout the rest of China, each area developing its own more or less imperfect copy. Instead, the myriad weapons designs that have been discovered should be viewed as locally engineered styles or regional variants that embody indigenous cultural characteristics and technological constraints. However, while improving the general understanding of cultural interaction and regional differences, these insights inevitably complicate any attempt to discern functional patterns within the thousands of recovered artifacts.

THE AXE

Because the simplest unimproved stick can deliver a painful, disabling strike by targeting the head, the earliest weapon associated with combat throughout the world has always been the club. Although crushing blows from heavy truncheons can prove fatal, lighter versions are easier to maneuver; however, they suffer from limited impact and therefore require a series of adroit strikes. Nevertheless, being basically amorphous and therefore less restricted than bladed weapons, clubs and short staffs can be employed to attack from almost every position and direction, including sideways or upwards, and still strike nearly every part of the enemy's body. It has been reasonably said that all combat undertaken with short weapons, whether crushing, piercing, or slashing, is necessarily based on the stick's mechanics as well as premised upon forearm movement rather than grandiose arm swings. Depending on the type of head affixed to the shaft—dagger, axe, hammer, knife, or even weighted ball—the arm's natural motion must be constrained and often retrained to wield a compound weapon effectively.

However extensively clubs and staves may have been employed, the bow and arrow and early versions of the axe (but surprisingly not the spear) came to dominate the ever intensifying conflict that plagued China during the Neolithic period. Stone axes represent an important development because the head's weight, being concentrated at the end of an extended lever whose fulcrum is the warrior's elbow (unless the axe is being employed through a rather ineffective "wrist snap"), magnifies the energy that can be delivered to a focal area and thus the destructive impact. Despite still being considered a crushing weapon, the axe's relatively narrow, sharpened edge can also inflict serious internal damage by cutting and severing when wielded in the same overhand mode as a club or truncheon.

The axe assumed many forms in early China, ranging from carefully balanced designs to odd asymmetrical shapes that display remarkable variations in dimensions, materials, and sharpness. Nevertheless, they have traditionally been classified into just two broadly defined categories, the *fu*, which tends to be longer and narrower, and the *yüeh*, which is generally wider and somewhat similar to a Western broadaxe. Both types were similarly edged, sometimes gradually but clearly tapered over the last centimeter or two, sometimes just sharpened right at the tip, with the blade edges always being vertical, oriented parallel to the shaft, rather than horizontal as in a mattock.

Unless they are unusually thin and therefore replica or ritual weapons, axe weights and thicknesses are rarely given in archaeological reports. However, the meager numbers available indicate that apart from a few heavy but purely symbolic *yüeh*, the heads for both were comparatively light, the weight for functional weapons varying from a very low 300 grams to a maximum of about 800, but mostly

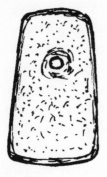

STONE FU   BRONZE FU   BRONZE YÜEH

falling in the 400 to 600 range.[9] Moreover, many large bronze *fu* are actually lighter than most compact versions because their increased size allowed them to be molded with a hollow core that extended throughout the blade's length.

Even though large numbers of axes have been recovered from the Hsia, Shang, and Chou eras, it has generally been held that the axe was not a factor on pre-Ch'in battlefields.[10] Unfortunately, assessing the actual combat role, if any, anciently played by these two axes in all their variants is somewhat problematic because they primarily served as tools for logging, woodworking, and agriculture. Their ready availability almost certainly resulted in them being extemporaneously employed in sudden conflict, but their very ubiquitousness muddies any attribution of a focal combat role. Worldwide, the combat axe has generally been a dedicated weapon, one distinctively shaped to ensure that debilitating blows are inflicted. In comparison with woodcutting, in which repetition and resilience are important, the needs of combat tend to be brief but intense; therefore a certain amount of brittleness can be tolerated in exchange for lethal advantages such as highly sharpened edges.[11]

An additional complication could have been their potential employment as missile weapons at close range. However, despite martial arts movies sometimes depicting secret societies and anti-Ch'ing loyalist groups throwing hatchets as a matter of choice, it is not a traditionally attested mode of combat. Axe throwing also requires considerable practice to master, especially with weapons that have not been properly balanced, suggesting it probably remained a method of last resort.

Finally, excavation reports tend to lack consistency in their classification of individual examples as *fu*, *yüeh*, or *ch'i*, the latter a variant of the *yüeh*. Well-illustrated articles often have identical-looking items differently named, even though the *fu* has traditionally been understood as marked by a longer, narrower shape and the *yüeh* by a much broader blade whose width can even exceed the head length. Justification is rarely provided for identifying an individual artifact as either a *fu* or *yüeh*, and subsequent articles may reclassify previous examples, prompting puzzled comments even from experts.

Despite these vexing aspects, a general trend to more symmetrical shapes, greater consistency, and increased smoothness and sharpness is clearly visible in the Neolithic stone variants and then the bronze versions that appear in the Hsia. However, as with all weapons and metallurgical techniques, significant differences persisted across China, and peripheral areas such as Fujian generally lagged in adopting various advances. Localized variation in design and size also tended to become more pronounced once bronze casting commenced, resulting in unique shapes and bizarre realizations even though interaction through trade and conflict could transmit highly esoteric influences to the most remote regions.[12]

By the Neolithic period the *fu*, which first appeared in uncertain but remote antiquity, had assumed fairly definitive form due to the maturation of the lithic industry. As attested by blunt, relatively long stone precursors that show evidence of heavy use, the *fu* was primarily a utilitarian implement, a tool first and foremost. However, contrary to some claims, it must have played a minor combat role, because a few recovered from comparatively munificent graves were embellished with motifs identical to those found on the accompanying dagger-axes, spears, and *yüeh*. Presumably because they were less expensive to manufacture and bronze had to be conserved for ritual vessels and weapons, stone *fu* persisted into the Shang even though bronze casting techniques had progressed sufficiently to allow multiple molds, hollow blades, effective mounting sockets, and large-scale production.[13]

Traditionally defined as a "large *fu*" by the *Shuo-wen* and other exegetical texts, *yüeh* were generally much broader, thinner, and sharper than most *fu* and therefore more suitable for warfare and severing heads.[14] (The *yüeh* variant known as the *ch'i* seems not to have been distinctive apart from being slightly more compact and thus more easily wielded in combat than an executioner's axe.) Although the earliest examples show signs of wear and are identified as tools, *yüeh* seem to have assumed a combat role virtually from inception. Moreover, being found almost solely with opulent ritual vessels and other weapons in the tombs of obviously prominent people (such as Fu Hao), their possession may have been deliberately confined to "men of power" ranging from clan rulers through tribal kings and battlefield commanders, the latter being derivatively held through deliberate award.[15] With the passage of time more elaborate but paradoxically lighter forms appeared, purely symbolic weapons intended to denote authority.

Later writings envision the *yüeh* as having played a highly symbolic role in the initial years of the Shang and Chou dynasties. For example, the *Shih Chi* states that "T'ang grasped the *yüeh* himself in order to attack the K'un-wo and then Chieh, king of the Hsia."[16] Similarly, King Wu of the Chou reportedly held a yellow *yüeh* in his left hand when his army proceeded against the Shang and employed it to chop off Emperor Hsin's head after the Battle of Mu-yeh.[17] Hsin's execution with a *yüeh* fully accords with the idea that in antiquity "they first employed armor and weapons in major punishments, next *fu* and *yüeh*."[18] Furthermore, presumably as described in the *Liu-t'ao* ritual already reprised, Chou dynasty command authority was bestowed upon a newly appointed commander-in-chief through the symbolic passing of both a *fu* and a *yüeh*.[19]

The *yüeh*'s comparatively broad face also presented an extensive area for elaborate decorations, including abstract patterns and three-dimensional figures such as animal heads highly symbolic of power that project one to two centimeters out from the upper blade. The addition of "incised" (intaglio) embellishments required parts of the blade to be thickened, resulting in otherwise identically shaped blades displaying different profiles when viewed edgewise.

As the result of new grinding techniques, by the middle Neolithic the utilitarian axe or *fu* had already moved beyond the earliest stages of flaking and percussive forming to be fairly well defined and comparatively smooth. Some of the earliest, essentially rectangular P'ei-li-kang examples that date to about 5300–5200 BCE, although still small at only

6 to 12 centimeters in length and simply lashed to a shaft without any binding holes, show extensive signs of use.[20] Although a few specimens of comparable size from this era reached 3.5 centimeters in thickness, most are a rather thin 1.0 to 1.5 centimeters and some have a single hole in the blade to facilitate lashing.[21] Thereafter, even though smaller sizes for specialized purposes and exceptions that attain dimensions comparable to *yüeh* and presumably had a combat use continue to be recovered from individual sites, *fu* gradually became larger, more rectangular, and heavier.[22]

Despite being generally thought of as a comparatively late, regal weapon, *yüeh* already appear in the late Neolithic, especially in the south. A few clearly show evidence of use, but the many characterized by thin, nonfunctional stone blades and complete lack of discernible wear indicate that the *yüeh* must have already assumed a symbolic function even in classic Lungshan cultural manifestations. For example, even though the twelve stone *yüeh* recovered from a Hubei site vary in blade length from 11 to 22 centimeters and in width from 9 to 17.8, their thickness ranges from a mere 1.0 centimeter down to a useless 0.5, with many being about 0.8 centimeter, possibly a compromise between weight and substantiality.[23] (One *yüeh* only 0.6 centimeter thick shows signs of wear, suggesting 0.6 centimeter might have been the lower limit for any sort of functional blade thickness.)[24] The *yüeh* at this site already display three of the basic five shapes: rectangular, a gradually expanding blade, and the pinched waist or slight hourglass shape. All twelve have a large binding hole in the upper third of the blade but no tabs or other lashing slots. At another Hubei site whose *yüeh* has been termed a tool rather than weapon, the blade seems to have been inserted into the shaft before lashing in three directions.[25]

The greatest concentration of late Neolithic *yüeh* blades having been found in the Liang-chu culture, which was centered in Jiangsu province and flourished from about 3000 to 2000 BCE, suggests that developments in the south provided the impetus for the weapon's adoption in the Shang, especially as Fu Hao's *yüeh* (described below) is decorated with a southern tiger motif associated with the indigenous Hu culture. Furthermore, *yüeh* have also been recovered from an incipient Liang-chu cultural site at Ch'ang-shu that has been dated even earlier, somewhere between 3500 and 3000 BCE. Nine of the fourteen graves there, including four of distinctively higher rank, contain a total of twenty-five specimens in four different styles that display little or no signs of use.[26] Because some of the skeletons were incomplete and showed other signs of being casualties of war, the excavators concluded they had been brought back for burial and that the *yüeh* were symbols of martial power. Generally rectangular in shape and still small at only 12 to 14 centimeters in length, the relatively smooth, thin blades still had sharp edges.[27]

Another eleven *yüeh* have recently been discovered amid artifacts dated from 4500 to 3500 BCE at San-hsing-ts'un (not to be confused with San-hsing-tui), also in Jiangsu.[28] Apart from a single jade specimen, they are all smoothly worked stone versions whose blades were affixed by partly inserting the top into a wooden shaft, allowing the unusual addition of a carved bone filial or cap along the shaft just above the blade. All the *yüeh* have medium to large lashing holes in the upper portion of the blade, and the shafts apparently once had end caps carved from bone or teeth attached. Recovered shaft remnants of 45 and 53 centimeters conclusively show that they were easily managed, single-handed weapons designed to be wielded with a well-controlled forearm motion.[29]

Six tombs dating to the late middle phase of Liang-chu culture, located somewhat more westward on the plains in the T'ai-hu area, contain a surprising nine *yüeh* among just thirteen stone objects.[30] Both stone and jade versions were recovered, with the latter generally being more polished and symmetrical in shape than the stone specimens.[31] However, symbolic *yüeh* in both stone and jade have been found even farther afield, both to the north in Liaoning and along the coast in Fujian. A basically square jade specimen dating to the Hungshan culture, recovered in Liaoning, has a well-rounded blade, large center hole, and unusual small double hole with a connecting slot for binding near the top. Just 12.4 centimeters high, 10.5 centimeters wide, and an extremely thin 0.6 centimeter, it has been identified as a purely symbolic martial form that evolved from earlier tools.[32] Neolithic examples recovered in Fujian dating to a distinctively late 2000 BCE are, however, still small and basically similar in style to the *fu* simultaneously discovered, much in keeping with the general trend of imitating Shang bronze weapons such as the dagger-axe in stone.[33]

The latest concentration of Liang-chu stone *yüeh* dates to some-where between 2000 and 1700 BCE and thus falls within the predynastic Shang's horizon.[34] Twenty-eight *fu* and five *yüeh* have been discovered in just twenty-three graves at this Shanghai area site, evidence that they played an important role in this somewhat peripheral manifesta-tion. Perhaps most significant but of uncertain meaning, a youth in one grave was accompanied by two *fu* and three *yüeh*. However, their interment is thought to be an expression of hope for the afterlife, be-cause the inhabitants dwelled in a complex society that integrated agri-culture, warfare, and hunting, one in which the *fu* and *yüeh* were both tools and weapons.[35]

Reconstructing the *yüeh*'s history in bronze is rendered somewhat difficult by the comparative lack of samples, only 200 or so having been recovered from the Shang and earlier eras in comparison with 1,000 spears and perhaps 2,000 dagger-axes, as well as the presence of anom-alies and the persistence of older versions.[36] Nevertheless, perhaps be-cause of their uniqueness, *yüeh* are prominently mentioned in excavation reports, making it possible to discern certain trends in size and com-plexity, though not with any great linearity. The most basic forms were square and rectangular, but variants that gradually expand outward down the whole length of the blade quickly appeared. Further modifi-cations included rounding the top somewhat, imparting curvature to the blade ranging from slight to extreme, reducing the middle portion to produce a sort of hourglass-shaped axe, and various combinations of these developments.[37]

The earliest heads were initially mounted by simply lashing the somewhat ill-defined blades to a shaft, thereafter by partially inserting them into a shaft and lashing with multiple bindings that passed through a two- to three-centimeter hole in the upper third of the blade. However, tabbed and socketed blades also quickly developed, the former utilizing a tab created by reducing the blade's width at the top by about 50 percent to produce a rectangular portion that could be passed through a slotted shaft. As a result the outer portions of the blade pushed against the staff, while the lashing hole, frequently found in the protruding portion of the tab, and two additional binding slots in the upper shoulders en-sured fastness. In some versions flanges provided additional surface area, reducing wobble and preventing push-through. Socketed versions, which developed in the northwest, primarily relied on a tight mechanical fit between the interior of the socket and the shaft, both generally rendered somewhat oval to reduce blade rotation in use, but pegs and early nails were sometimes employed to augment the solidity.

Two rather simple bronze *yüeh* recently recovered from Erh-li-t'ou mark the actual inception of the cast form. The first one discovered, a rectangular blade some 23.5 centimeters long but only 3.1 centimeters wide, was originally (and it would seem correctly) termed a *fu* but has now been reclassified as a *yüeh* or possibly a *ch'i*. However, the second is decidedly less controversial, a sort of rectangle that splays out slightly at the bottom of the blade area, has a decorative band near the top, and would have been lashed through one moderately sized hole. Rather small, with a total length of 13.5 centimeters and a width that tapers inward from 7.6 at the blade edge to 6.1 centimeters at the top, it is marked by a low tin content of 5.7 percent and extreme thinness of 0.5 to 0.6 centimeter, evidence of being a symbolic embodiment.[38]

Although stone versions would continue to be produced throughout the Shang, bronze *yüeh* begin to be noticeable at Yen-shih and Cheng-chou, become somewhat more common after the government's shift to Anyang, and then basically disappear by the end of the Chou. Increasingly an emblem of power, the bronze specimens discovered in the core domain and down at P'an-lung-ch'eng (dated to the upper Erh-li-kang) reflect the era's vastly improved metallurgical techniques, including

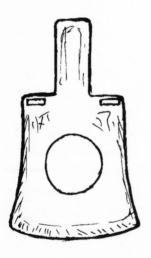

the ability to mold increasingly intricate designs in the walls of ritual vessels and onto *yüeh* blades, particularly in comparison with the more pedestrian *fu* that would always be found in vastly greater numbers.[39]

For example, although one of the three *yüeh* recovered from P'an-lung-ch'eng is plain, the largest *yüeh* discovered to date in China (at 41.4 centimeters long and 26.7 centimeters wide) displays a somewhat classic bell shape that tapers outward toward the blade, an unusually long tab without any holes, two large rectangular binding slots but no flanges, and a rounded blade edge. It incorporates a very large hole in the center of the blade (as shown in outline), and an intaglio design decorates the border and the upper portion below the shoulder.

The second, considerably smaller *yüeh*, at 24.4 centimeters long and 13.3 centimeters wide, has more visibly pinched-in sides, a comparatively short but wide tab without any holes, an even larger center hole, and two binding slots on the shoulders. However, the third *yüeh* is a unique semicircle 22 centimeters high that flares upward into points, and has a huge beveled hole in the center of the blade, narrow shoulders, and a short tab but two large binding slots.

By the early reigns at Yin-hsü the *fu* and *yüeh* had basically realized their final form. However, rather than being produced in large quantities like the *fu*, individual *yüeh* were cast in extremely small numbers, often individually for specific people. They therefore lack the *fu*'s uniform weights and dimensions and embody far more complex, often extravagant decorations ranging from abstract, "incised" or intaglio *t'ao-t'ieh* motifs through raised depictions of animals, faces, and the pernicious grimace molded into China's most famous *yüeh*, shown in outline below.[40] Although trade and other forms of interchange resulted in examples of highly localized styles being disseminated throughout China, few socketed *yüeh* have been recovered at Yin-hsü, the majority (including in the early stage) employing centered tabs for mounting. A large bronze *yüeh* with an iron blade and another bronze *yüeh* decorated with an animal motif have also been recovered at T'ai-hsi.[41]

The four *yüeh* recovered from Fu Hao's tomb, although not the only ones dating to the early Yin-hsü, epitomize the weapon's symbolic nature and confirm its role as the ultimate prestige battlefield implement. The largest two are thick, heavy specimens in a square style that measure 39.5 and a nearly identical 39.3 centimeters high and have blade widths of 37.3 and 38.5 centimeters, respectively. The former has slightly indented sides, a somewhat rounded blade, a wide tab, and two binding slots on the shoulders, and is decorated with two tigers leaping toward a man in the center of the blade itself.[42] Although basically rectangular, one of them being 24.4 centimeters long and 14.8 wide at the blade, the two middle-sized *yüeh* have the deeply indented sides of an hourglass shape, and *t'ao-t'ieh* patterns embellish the upper portion of the blade, but no flanges.[43]

The seven *yüeh* discovered in the tomb of a high-ranking military commander named Ch'ang, apparently the progenitor of the Ch'ang clan, dated to late in the second period at Yin-hsü, well illustrate the tendency to individuality. Not only are the shapes and decorations unusual, but the characters *ya Ch'ang* are also included on one blade.[44] Nearly as large as Fu Hao's *yüeh*, the most massive and interesting specimen has a 40.5-centimeter-high blade, a maximum width of 29.8 centimeters where the blade flares outward, and an extremely heavy weight of 5.95 kilograms. Marked by a somewhat asymmetrical curve at bottom and a top shape that conforms to the molded design of the protruding decorations, it was secured by a large, embellished tab and lashing holes at the top of the blade. Dragons and the characters for the commander's name complete the appearance. Five of the other six are similar, being more rectangular, with a length of about 20.5 centimeters; long tabs; a surprisingly light weight of about 0.67 kilogram; and a combination of stylized circles, triangles, and an animal motif for decoration. However, the last specimen, rather squat at 21.2 centimeters high and 18.7 wide, has a comparatively simple, symmetrical curved edge with a large hole centered in the upper blade, a centered tab, and weight of 0.75 kilogram.

Whether recovered from Anyang or farther afield, the *fu* and *yüeh* dating to the final reigns show continuity with previous styles but a pronounced tendency to be symbolic, as attested by specimens whose thinness precludes any combat utility.[45] Although a few are marked by elaborate decorations including complex *t'ao-t'ieh* patterns or three large triangles, others, probably intended for less distinguished commanders or even for interring with the deceased, display simplified, abstract patterns.[46] However, exceptions and anomalies (such as asymmetrical blade shapes) are not unknown,[47] especially out in the northwest, where sock-eted versions evolved, and local characteristics as well as external influences are strongly evident, such as in a comparatively narrow but long half-moon blade with three large holes mounted lengthwise at the top of a short shaft.[48] Even some tabbed versions came to display unusual features, including a three-dimensional ram's head on both sides.[49]

16.

Knives, Daggers, and Swords

❖ ❖ ❖   ❖ ❖

Rather amorphous, minimally effective knives percussively flaked from rough mineral blanks began to assume consistent, discernible form early in the Neolithic period, initiating a slow evolutionary process that took advantage of the ongoing advances in lithic and eventually metallurgical knowledge to improve their contour and quality.[1] Although it is generally believed that they were employed solely as tools and in hunting, any knife can always be utilized, albeit awkwardly and with considerable difficulty, as a weapon of last resort at close range, as well as for slitting the throats of the unwary, slaying sacrificial victims, and dismembering enemies. Eventually knives would lengthen to become the large *tao* or sabers carried by late Warring States cavalry riders.

Selected characteristics derived from the more deadly but still utilitarian knives created on the periphery during the late Neolithic influenced the shape of Shang knives and daggers. Nevertheless, almost all the specimens recovered from the Hsia and Shang have simple designs featuring integral handles, were clearly intended for tedious applications, and rarely exceed twenty-five centimeters in length.[2] However, the metal embodiments that were initially molded or occasionally hammered from copper or naturally occurring alloys rather than cast from bronze are generally longer and more elegant, similar to contemporary straight razors and some rectangular knives employed in modern Chinese cuisine.

Shang variations other than length and width are seen in the degree of curvature, if any; the type of handle, flat and straight or circular and therefore suitable for wrapping cord; the pointedness of the tip as well as whether it suddenly arcs upward or downward; and the profile of the bottom edge of the blade insofar as it presents a smooth contour, cuts sharply downward, or expands and contracts along the length.[3] Northern influences primarily affected the handle portion, with the animal figures, heavier tabs, and rings marking later Shang knives and daggers all having been derived from the Northern complex.[4]

Debate has arisen over whether Shang fighters deliberately employed their knives as fighting weapons after expending their arrows and closing for combat or viewed them solely as tools. Their recovery in a few early Shang weapon sets has been seen as implying a combat role even though they were probably intended for purely utilitarian purposes such as maintaining the wooden components of their other weapons.[5] It has even been claimed that the knife's size and the type of decorations embellishing the handle reflected the owner's status within the emerging warrior hierarchy.[6]

However, being extremely short and generally lacking a point for piercing, their utilization would have been confined to surprise and to hacking at disabled opponents. Furthermore, because more efficient weapons tend to quickly displace inferior ones on the battlefield, the simultaneous emergence of daggers, dirks, and short swords that were far more suited for fighting purposes almost certainly rules out any dedicated combat role for the knife. (The fact that daggers and knives coexisted in many northern cultures suggests that the former were weapons, the latter tools.) No knife fighter employing slashing motions at close quarters could survive a clash with an opponent armed with a dagger-axe or short spear!

Nevertheless, a few "lethal-looking" knives recovered primarily outside the core Shang domain may have functioned as weapons in desperate circumstances and for delivering the final cuts when slaying an incapacitated or otherwise constrained individual. For example, three knives recovered from P'an-lung-ch'eng could easily be weapons, including the longest at 35.6 centimeters (or about 14 inches), which has a saberlike elongated profile, sharp point, slight downward curvature in the upper edge, and slight curve in the bottom so that the blade bulges toward the middle. Marked by a fairly wide spine, sharp top and bottom edges, very stubby flanges top and bottom, and a basically flat tang continuous with the upper edge and long enough to affix decent wooden pallets to create a handle, it is definitely a slashing weapon whose point could be used for piercing.

The second one is characterized by a straight top edge and a slight inward curve toward the middle on the bottom edge, but the blade retains a basically saberlike appearance. It has a short tang continuous with the top edge, no flanges, and is 30.8 centimeters long. The first two could have been grasped in the hand without additional improvements, but probably had cord or cloth wrapped on the handle to provide a secure grip. However, the third specimen, 30.4 centimeters long, looks more like a throwing knife but has an open filigree handle with a downward hook at the end and would have been unbalanced.

DAGGERS AND SWORDS

Legends of magical daggers and powerful swords began circulating as early as the Spring and Autumn period around Wu and Yüeh, the two states closely identified with the sword's inception, and became part of the lore essential to *wu-hsia* (martial arts) stories by the T'ang dynasty, an era when swords were just assuming symbolic roles in localized folk ceremonies and Taoist rituals.[7] However, despite exaggerated claims and considerable controversy, archaeological discoveries show that rather than dating back to semimythical antiquity, the true sword (which might be simply defined as a thrusting or slashing blade that has a minimal length of at least two feet) did not begin to develop until late in the Spring and Autumn period. This explains why the *Art of War*, which presumably reflects late Spring and Autumn military affairs, never mentions swords when noting the fiscal burden incurred in warfare for the expenditure of bows and crossbows, chariots, helmets, armor, and shields.[8]

Prior to the late Spring and Autumn, warriors may have carried daggers, traditionally termed "short swords," or extemporaneous weapons such as spearheads or dagger-axe blades as their weapons of last resort. Only after infantry forces multiplied and metallurgical techniques advanced would thrusting weapons designed for extremely close combat begin to lengthen and displace the dagger-axe and spear throughout central China, creating the requisite context for the emergence of slashing swords as well.[9] However, all the daggers and proto-swords recovered from Shang, Western Chou, and even Spring and Autumn sites have long been said to be designed for thrusting rather than slashing attacks.[10]

Because swords proved to be not just ineffectual but even a liability in chariot combat, many scholars have attributed their proliferation in the late Warring States and Han to the demise of chariot-based warfare and the cavalry's development.[11] (Spring and Autumn swords would have been too short to impale opponents, while exposing the wielder to spear thrusts and dagger-axe strikes. Early swords were also ill designed to deliver the downward, saberlike slashing blows required of warriors standing in a raised chariot compartment.) However, prejudgments that have tended to constrain historical assessment by tying the sword's initial absence to a chariot-dominated mode of warfare should be avoided.[12] No matter how numerous the chariots or how extensive their role, their numbers were still insignificant compared with the masses of troops that normally congested the battlefield.[13] The sword's thrusting ability was clearly approximated by the short hand spear, essentially a dagger point mounted on a handle, and a preference for traditional weapons coupled with technical difficulties in making strong yet resilient swords more likely retarded the sword's emergence as a critical weapon.

Theories about the dagger's inception in China currently range from assertions that they imitated steppe weapons to claims of totally indigenous development, with or without nonmetallic precursors,[14] including from spearheads or dagger-axes that were elongated and strengthened.[15] Improvements in shape, durability, sharpness, and appearance then rapidly followed in accord with advances in metallurgical knowledge and practice. However, swords with slashing power and significant blade length simply could not be fabricated until the Spring and Autumn period, and even then would not flourish until the late Warring States and Han dynasty.[16]

As the cavalry became a critical battlefield element, a single-edged sword with a ring handle, termed a *tao* or "knife," in turn gradually displaced the long bronze Warring States swords and unwieldy early Han iron variants.[17] Thereafter, further advances in metalworking saw two distinctive trends emerge, one toward high-quality, shorter, functional swords, the other toward purely ceremonial and elaborately decorated symbolic weapons. As a result, with the onset of the Sui and into the T'ang, steel "knives" eventually became the slashing weapon chosen by both infantrymen and cavalrymen.

The numerous daggers and short swords that have been recovered over the past few decades allow a minimal historical reconstruction. Hundreds of individual reports have described sites that contain from one to several swords, and a few synthetic articles have outlined the weapon's history in particular periods.[18] However, the former tend to be marred by the idiosyncratic creation of multiple types and subtypes, and the latter often neglect the fundamental structural changes that affect combat efficiency and tend to study the sword in terms of visual qualities such as handle style, decorations, and overall appearance rather than the functionally critical aspects of blade length, relative dimensions, strength, and resiliency.

Although stylistic issues are important for understanding individual cultures and reconstructing their interactions, for military history purposes it may be said that in China the sword gradually evolved from a fairly blunt, extremely short dagger with a minimally defined handle to a more dynamically contoured, elongated blade-and-hilt combination that was molded as a single unit from various bronze alloys. However, just as with the axe, evolutionary changes are often obscured by the deliberate continuation of previous styles. Whether because bronze weapons were expensive or because inherent conservatism stifled military innovation, daggers from previous generations continued to be esteemed, preserved, and employed,[19] often manufactured from laboriously produced old stone molds. Considerable incentive must have been required to discard functional weapons or consign them to the smelter for reprocessing.

Moreover, although not directly relevant to the Shang, with the onset of the late Spring and Autumn period and the general practice of prominently wearing daggers in court and on the battlefield, swords came to be valued in themselves and the basis of a growing mystique. Some men became known in the Warring States for their skill in appraising not only the martial qualities of contemporary and antique swords, but also their auspiciousness, just like the experts who evaluated horses or physiognomized men. Fragments of books that discuss their essential principles have recently been recovered from Han dynasty border sites in Gansu and elsewhere, attesting to a widespread interest in the practice of evaluating weapons, particularly in dangerous areas, that may well have originated in the Shang.[20]

These daggers or short swords were merely ancillary weapons for self-protection, never the primary choice for close encounters. Unlike in Greece and Rome, ancient Chinese warriors did not fight with swords and shields, slashing and hacking away at each other in open combat, but employed intermediate-range weapons—the dagger-axe and the short spear—and only resorted to their daggers when they failed to prevent the enemy from closing or they lost their primary weapon. This is well attested by individual burials from the Shang through the Spring and Autumn, which, when they contain any weapons at all, generally pair a long weapon with a dagger.[21] A late Spring and Autumn tomb pictograph of riverine combat that portrays shipboard warriors wielding a variety of long weapons clearly shows that although many also carried short swords, they remained fastened to their waists.

As noted, based on their design it has been claimed that the earliest daggers were probably intended to function primarily, if not exclusively, as thrusting weapons, even though the traditional Chinese term "short swords" implies that they were used for slashing and cutting as well as penetrating and piercing. However, throughout history warriors have managed to use weapons in surprising ways, and virtually every weapon can be employed in inefficient but still functional modes. Moreover, when present-day knife-fighting methods are examined, it is obvious that military knives and stiletto-type weapons are frequently used for slashing and that slashing movements are often employed to create opportunities for piercing strikes. Without doubt these early daggers could have been used to sever a limb or slash the neck, even though such movements would require changing the mode of attack from forwardly directed thrusts to more circular swipes.[22] However, further evidence that in antiquity the dagger was conceived primarily as a thrusting weapon may be seen in the absence of the sort of perpendicular guard that would be necessary between the blade and hilt to protect the hand against a slashing blow sliding down a blade's length, as well as a concern with thickening the middle of the blade and sharpening both edges along their entire length.

In at least a few peripheral cultures, the dagger apparently enjoyed a secondary use as a throwing weapon. This is attested by later literary references in the *Hou Han-shu* and other works and further suggested by the unusual practice of wearing a double scabbard in both the northwest and southwest. This custom naturally suggests that either warriors fought with two daggers simultaneously or, as indicated by recorded incidents, resorted to throwing them, presumably as a surprise tactic since the technique would be more effective if unexpected and unobserved. The act of throwing would have been facilitated by the handle's length, which was often nearly half that of the entire dagger, though the necessary balance could also have been achieved by increasing the weight of the hilt with a slightly larger, solid handle or adding a pommel, whether a simple thick knob or the sort of decorative figure frequently seen.[23]

The first stabbing or piercing weapons in the Chinese environs are slender, ill-defined "stickers" that in many instances might be more appropriately termed awls or bodkins than daggers. Fashioned from a single piece of stone or bone but occasionally inserted into a bone or wooden handle, they date back to the Neolithic.[24] Although they proliferated in the third millennium BCE, being common in the upper reaches of the Yellow River in the Ma-chiao-yao culture, stone and bone combinations may also be found as early as 6000 BCE in the northeast.[25] Lengths for these early *pi-shou*—*pi* originally meaning "ladle" or "arrowhead," though *pi-shou* now means "dagger" or "stiletto"—run from about 18 to 35 centimeters, but slightly greater dimensions sometimes appear in Hsia stone variants.

Just like knives, other areas outside China's cultural core possessed daggerlike, edged metal weapons long before the Shang, and the absence of the sword in the Shang and early Western Chou, despite apparent nonmetal and metal precursors from early sites, has led to assertions that China imported the sword from them rather than originated it.[26] However, not all scholars agree that the core Hua-hsia cultural area lacked discernible precursors or that there is any causal connection between the sequenced appearance of northern and Shang types, particularly in the light of the preexistence of dagger-axe and spearhead blades.[27] Moreover, even if it were largely true that the sword was imported, few battlefield implications would be entailed because the steppe people did not possess swords of any appreciable length or employ their daggers in any distinctive or more efficient fighting mode, and Shang dagger-axes and spears would have easily outranged them.

Nevertheless, several Chinese forms visible by the early Western Chou, if not earlier, have been identified with the Northern complex and the northeast, though there are claims that other regions such as the southwest (Pa and Shu) were the inception point. The Northern cultural complex—which was once viewed as sweeping almost virtually unbroken across the broad area just north of a line demarcating various ancient walls and the future Great Wall from Gansu and even Xinjiang in the west through Inner Mongolia, the northern parts of Shaanxi, Shanxi, Hebei, and Liaoning out onto the Korean peninsula—is now perceived as roughly divisible into two areas of influence, the northern zone and the northeast.[28] However, archaeological reports indicate that their nebulous border (which apparently fell somewhat north of the T'ai-hang Mountains) must have been extremely porous because what might be considered definitive weapons are frequently recovered in alien areas, whether as a result of gift giving, trade, or seizure.

The northern zone is clearly the more important of these two, because interaction with the Shang resulted in the early exchange of various bronze objects, including the dagger.[29] As already discussed, Erh-li-kang culture expanded along the Yellow River and T'ai-hang Mountains into the north and westward into the Wei river valley before being forced to retrench about the time the capital was established at Anyang, presumably due to a combination of Shang internal weakness and the growing strength of peripheral peoples. Although early Shang ritual bronzes penetrated all these regions, two specialized northern weapons, daggers and knives, intruded into Shang culture and apparently were deliberately copied, no bronze daggers with integral handles or similar-style knives having ever been recovered from the Erh-li-kang phase of Shang culture. Two styles of dagger were widespread in the north: one very similar to contemporary military or commando knives (such as the Fairbarn), characterized by a visibly protruding, somewhat circular spine, and the other marked by distinctive, often fancy pommels featuring animal heads and an essentially rhomboidal cross-section.[30]

In contrast, the daggers recovered in the northeast, particularly beyond Beijing (which was a sort of pivotal area between the cultures of the northeast, south, and plains), out through Liaoning and in a few cases even down to Shandong, are often marked by a highly unusual appearance, something like a violin or Chinese gourd, created by two wavy bulges protruding from the blade's length.[31] How this dagger was conceived or could have ever been considered a functional weapon remains unclear, especially since the tips of many specimens are so rounded as to be incapable of penetrating the slightest thickness of material. Moreover, the second or larger bulge near the hilt, though certainly enlarging a wound opening, would have encountered increased resistance, thus minimizing any unknown advantages it may have possessed over a more dynamically tapered weapon. (This style didn't really influence China until the middle to late Western Chou, roughly the ninth century BCE, when it spread across the north over the next two centuries but then suddenly disappeared late in the Warring States period.)[32]

Apart from their often complex hilts and pommels, the key feature of daggers was the incorporation of a circular or rodlike spine instead of a flattened or rhomboidal cross-section. This core portion essentially continued down to form an integral, basically circular handle differentiated solely by a minimal hand guard. Despite numerous variations in size and style, overall dimensions in the late Western Chou were still only 31 to 32 centimeters in length and 4.4 to 5.9 centimeters in width, numbers that provide a comparative maximum for earlier Shang achievements. Apart from a slight expansion at the bottom of the handle to insert a staff, at this point many daggers looked exactly like spearheads. However, they would gradually become longer and slimmer in the Spring and Autumn period, with representative samples running 27 to 34 centimeters in length, with less bulbous but still quite expansive widths of 4.6 to 5.1 centimeters.[33]

As already mentioned, in the absence of indigenous Erh-li-kang precursors, foreign origins have been sought for the dagger's sudden appearance in an already fairly advanced, well-contoured form. The preponderance of evidence suggests a fundamental northern influence, yet the number of recognized exceptions continues to increase. For example, a highly unusual, so-called Snake Head Seven Star Sword, which has an overall length of 53 centimeters, a width of 4.5 centimeters, and a minimal hand guard with reverse points obviously designed to catch an enemy blade, has been dated to the Shang.[34] Another distinctive, fully formed Shang weapon, marked by a raised broad spine that ends in an oval handle and with an overall length of 29.4 centimeters, including a 22-centimeter blade, a width of 4 centimeters, and a thickness of 0.5 centimeter, has been recovered in Shandong, an area that would subsequently see the importation of the wavy blade style.[35]

Other than these exceptions, the earliest bronze daggers associated with the Shang dynasty are quite simple, generally having a "willow leaf" appearance. The first dagger of this type, recovered from Ch'ang-an Chang-chia-p'o, is a mere 27 centimeters in total length, including the 18-centimeter edged blade, and has a rhomboidal profile.[36] Having been dated to the eleventh century BCE, it is now identified with the late Shang rather than the Western Chou, its previous attribution.[37] The tab's profile required it to be inserted into a wooden handle and then secured with a nail or peg through a premolded hole. This sort of tapered tab would invariably fail over time as the metal acted against the wood, enlarging the original opening and producing a tendency to wobble in the handle. Thus it is not surprising that one of the first technical improvements witnessed in the early Western Chou would be to change the dagger's design, no doubt in imitation of northern styles, to include a continuously molded handle which, having a fairly small diameter, could be gripped securely with the addition of cord wrapping or wooden slats.

Thereafter the sword would witness innumerable changes in the width, length, contours, general profile, edged portions, and cross-section of the blade, some undoubtedly stimulated by the unremitting quest for longer, sharper, and more durable swords with a secure grip, but others of less certain origin. Many of the changes being very subtle and never experimentally investigated, their likely effects can only be speculated upon within a historical context. However, when the myriad variations and stylistic changes are winnowed out, one basic shape, consisting of a unitary molded handle and elongated blade, can be said to have eventually predominated, although earlier versions, stylistic variations, and different lengths always continued to coexist.

17.

THE *KO*
OR
DAGGER-AXE

◆·◆·◆————————————————◆·◆·◆

T HE DAGGER-AXE OR *ko*, a weapon unique to China, was originally designed to pierce the neck and upper body and thereby maim or slay
by cutting and severing, rather than inflicting the sort of crushing injuries
caused by shock weapons. Even in its earliest form the point was fully
capable of penetrating the era's scanty armor and disabling the victim.
Moreover, just as depicted by oracular characters and later tomb picto-
rials, before the lengthened shaft of subsequent eras came to require
two hands, the short, single-handed versions common in the Shang
were certainly used in conjunction with a shield.[1]

Several characters associated with warfare reflect the primary role
that the dagger-axe played in Shang dynasty combat, including the word
for "attack" (*fa*), which appears in the oracle bones as a man with a *ko*.
The most important character for battle in later ages, *chan*, would consist

of a component that provides the sound and originally meant "great" (itself derived from the primary meaning of a particularly large vessel) conjoined with *ko* once again on the right. The character for "being wary" or "guarding against" something, *chieh*, is formed from two hands holding a dagger-axe in a sort of defensive posture.[2]

Although this sort of speculative interpretation can easily become overly imaginative, it might be noted that the basic character for "martial"—*wu*—is composed of two components that are generally interpreted as a foot and a dagger-axe, the latter sometimes placed above the former rather than on the right, suggesting a warrior on foot with a dagger-axe. However, a purportedly post-battle dialogue dating to 597 BCE provides another interpretation, one that would be cited frequently in imperial court debates when claiming that the only real justification for employing martial power is to effect the cessation of warfare.[3] When the upstart but increasingly powerful southern state of Ch'u gained a significant victory over the antique state of Chin in the north, Ch'u's officers encouraged the commander to raise a victory mound with the enemy's dead so that "the martial (*wu*) achievement would not be forgotten."[4] However, in declining he noted that "*wu* means resting (*chih*) the dagger-axes (*ko*)" and then cited various historical examples (including the Chou conquest of the Shang), wherein the victors had manifestly set aside their arms after extirpating the evildoers.

Even though dagger-axes were fabricated from stone in the late Neolithic,[5] the *ko* is primarily a bronze weapon that first appeared in cast form in the Hsia capital at Erh-li-t'ou, proliferated during the Shang, and continued to function as China's distinctive battlefield implement during the Chou in one form or another. Although several puzzles remain because numerous variations developed across China's ancient expanse before the dagger-axe evolved even further through interaction with indigenous cultures,[6] the thousands already recovered provide an adequate basis for reconstructing the weapon's early history. In the ongoing quest for combat lethality, four major styles emerged by the middle Shang, and though three of them would subsequently be abandoned, the so-called crescent-bladed *ko* dominated Western Chou warfare until it was gradually displaced by the *chi* (spearheaded *ko*) during the Spring and Autumn period.[7]

Artifact-based discussions of the *ko*'s evolution are complicated by a number of factors. First, any single weapon recovered from a burial site, although perhaps a favorite of the interred, may actually be just symbolic, intended to enhance the deceased's status at death or in the afterlife. Second, perhaps because they may have been produced from the same molds, the replica weapons that appear early on at Yin-hsü are visibly identical to functional versions. However, they differ in being characterized by a much higher lead content, a change that facilitated casting while saving copper but rendered them too soft for edging or use.[8]

More stylized, abstract, thinner, obviously simplified replicas begin to multiply by the third period at Yin-hsü. Sometimes extensively embellished, they were increasingly produced from lead alone and gradually became common even in ordinary tombs.[9] During the transition period from highly realistic bronze replicas to these purely symbolic realizations, the exact nature of any individual specimen can be difficult to determine. Nevertheless, it can reasonably be assumed that there is a direct correlation between the number and opulence of the weapons discovered in any particular grave and the occupant's military achievements or prestige.

A study confined to undisturbed graves from the four periods at Yin-hsü has revealed that dagger-axes and spears are the only weapons that were buried with all ranks, and that dagger-axes far exceed spears.[10] Late in the dynasty high-ranking commanders and members of the martial nobility might be interred with several hundred weapons, including *yüeh*, *ko*, spears, arrowheads, and large symbolic *tao* (knives).[11] Middle-ranking officers were accompanied by *ko* together with spears and arrowheads or spears alone, often amounting to a dozen or more; low-ranking officers rated fewer than ten weapons, invariably *ko* combined with spears or arrowheads; and ordinary soldiers were usually limited to just a *ko*, spear, or several arrowheads, never interred with any ritual vessels.[12] However, in chariot burials only *ko* are found, never spears.[13]

Based on the numbers and pervasive distribution, it can be concluded that the dagger-axe was the most important weapon in the Shang, even though commanders were honored with battle-axes and perhaps even employed them on the battlefield. Furthermore, although Shang oracular inscriptions never mention them being awarded for merit, *ko* are known to have been given as a token of recognition over 2,000 times in the Western Chou, not only making them the most frequently bestowed weapon, but also preserving the names of several specific types, including "plain *ko*" (*su ko*).[14]

The dagger-axes found at Erh-li-t'ou simply affixed a short, dagger-like blade near the top of a wooden shaft. The blade's upper and lower edged surfaces were both sharpened and essentially parallel except near the front fifth or sixth, where they tapered down to a relatively well-defined point. However, because daggers had not yet appeared, only short single-edged knives, the "dagger-axe" was not fabricated by affixing pre-existing "daggers" to a shaft but instead derived from long, tapered mattocks, hatchets, or similar choppers and axes employed in woodworking or in the fields, to create a more lethal weapon than an axe alone.[15] The dagger-axe has been likened to the Western halberd and has even been called a halberd, but the halberd's blade is traditionally broad, more like a *yüeh* with a spear affixed on top, a variant known as the *chi* in China.

The earliest Neolithic versions were essentially elongated rectangles fabricated from a variety of stone materials (including jade) that tapered down somewhat over the external third of the blade and were rather tenuously lashed to a shaft. However, the improved variants and the first bronze embodiments secured the blade by inserting it through a slot in the shaft as well as generally binding it through a single hole molded into the protruding portion of the tab.

As may be imagined from the dynamics of its employment, affixing a blade to a shaft in the absence of modern nuts and bolts would have been problematic. Two methods were employed: (1) simply drilling or carving a rectangular opening in the shaft and (2) slotting down from the top (combined with preparing the necessary opening) before attempting to secure the blade by nailing, pegging, or most commonly lashing, whether singly or in combination. (Rectangular openings alone could only accommodate simple, parallel edged blades whose tabs could be inserted through them.)[16] However, slotting and drilling could fatally weaken all but the most durable wooden shafts just where the impulse from striking the enemy is transferred, and bindings alone would probably have been unable to prevent early dagger-axe blades from being pushed through on impact.

Numerous impressions left by long-disintegrated shafts in the compacted soil indicate that Shang era shafts averaged about 85 to 100 centimeters in length but could reach 113 centimeters (roughly 44 inches), and the narrow blade was affixed just about a meter (39 inches) from the butt.[17] Despite its being a single-handed infantry weapon, the length was sufficient to ensure considerable head velocity as the shaft turned through its arc, seriously stressing the bindings and the joint where the blade was attached upon impact. (Even greater forces would be exerted when the shaft was extended to 9 feet and the dagger-axe became a two-handed weapon in the Spring and Autumn period, prompting warnings in the later military writings that even though they look impressive in court, weapons with overly long shafts are unwieldy and break easily.)

The earliest shafts were probably fabricated by shaving down tree branches or saplings, presumably explaining why some Shang dynasty characters show a base with rootlike protuberances. (Speculation that the character depicts an integral base that was employed for standing the weapon is absurd, though a detachable stand would have been feasible.) Naturally the wood species chosen would affect the weapon's overall strength, resiliency, and degree of flexing experienced in use. The Han dynasty *K'ao-kung Chi* mentions that although stiffness is desirable in thrusting weapons such as the spear, some flexibility is necessary for hooking weapons such as the crescent-bladed versions of the dagger-axe and *chi*, both of which evolved in the Shang.

By the Spring and Autumn period, if not earlier, laminated shafts were being fabricated from multiple strips of predimensioned wood and bamboo. This advance greatly facilitated realizing the requisite strength and flexibility within dimensional limitations and allowed the shaft to be lengthened to accommodate two hands in the Chou, as demanded by the exigencies of chariot-based warfare. However, in the Shang and Western Chou the shaft was still short and therefore a weapon for ground troops, even though they have been found interred alongside chariots. (No blow could have been struck across the gulf between the chariot box and an enemy standing outside the wheels.)

To resolve the problem of push-through on impact, the blade's rear portion was reduced to create a rectangular tab, producing a cross-sectional profile with blade portions that effectively butted against the

shaft. The contact area was then further increased by adding a molded flange at the rear of the blade portion (as shown in the illustration). During the Erh-li-kang period at Yen-shih, Cheng-chou, and Lao-niu-p'o, this flange was subsequently extended to form two small protrusions above and below the blade's edges just at the shafting point, though these adjustments did not become popular until the second period at Yin-hsü.[18]

Some early versions of these straight-bladed *ko* with protruding flanges also included a lashing slot or two just in front of the flange area and subsequently the roughened area in the front portion of the tab, where it would be inserted into the shaft during the Yin-hsü period, though these slots must have compromised the blade's integrity and somewhat weakened it. An alternative mounting method consisted of creating a tubular socket by molding a vertical shaft hole where the flange on the tab normally affixed the blade to the handle. This resulted in a somewhat pudgy profile when viewed from the top and considerably greater thickness in the rear portion of the blade and front of the tab, developments that initially required the blade's spine to be broadened and flattened. However, blades with rhomboidal cross-sections (which had already evolved) quickly reappeared.[19]

The mechanical joint created by forcing the shaft up through the socket reduced the wobbling experienced with slot-mounted blades and eliminated the danger of push-through in use as well as slippage under conditions of lower humidity, but coincidentally introduced a tendency to rotate on the shaft. Despite being a problem that could have easily been resolved by molding a small opening on the side and inserting a peg or nail horizontally through the socket into the shaft, oval-shaped sockets and matching shafts were instead employed, and any residual wobbling was remedied by jamming thin pieces of wood into the gaps.

The socketed dagger-axe first appeared in the Erh-li-kang period,[20] but the tubular socket is generally believed to have originated somewhere in the Northern complex.[21] Increasingly found at Anyang beginning with a small number dating to Wu Ting's reign,[22] despite claims to the contrary they seem to have proliferated quite rapidly and displaced the straight-tabbed ko, eventually accounting for the majority of dagger-axes found in the tombs of some high-ranking martial officials in Yin-hsü's third and fourth periods.[23] Nevertheless, the development of the crescent blade catapulted the yu-hu-ko style into prominence, presumably because of its lethality in both infantry combat and chariot encounters, resulting in the virtual disappearance of the socketed ko in the Western Chou, though not before more pronounced rhomboidal blade shapes and a variant incorporating a similar downward extension that increased the solidity of the mounting appeared late in the Shang.[24]

A second early development during the third period at Erh-li-t'ou was the enlargement and elongation of the tab (t'ang) in a downward curve. However, the curved tab style did not proliferate until Yin-hsü and then rapidly disappeared after the Chou conquest.[25] Said to have been similarly inspired by the curved handles found on knives from the Northern complex, the initially simple, flat tabs soon gave way to increasingly intricate decorative motifs coincident with the tendency to more elaborately embellish ritual vessels. Abstract patterns, Chinese characters, and fanciful animals were all employed to enhance prestige, identify the user, and seek divine protection. However, the most imaginative and complex figures appear in nearby peripheral cultures, including one associated with the Hu in the south that depicts a tiger eating a man.[26] Even though their intention was ostentatious display, their primary function overawing others whether on the battlefield or in martial flourishes, these longer, thicker, heavier tabs provided a natural counterbalance to the blade head, thereby improving the battlefield dynamics, while the weight increase augmented the energy at impact.[27]

Surprisingly, these enlarged dagger-axe tabs were never edged, reshaped to form any sort of hammer, or pointed, three improvements that would have allowed their use on the back swing or over the shoulder

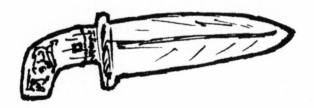

and off to the side in an emergency. However, coincident with the evolution of more complex tab shapes, the *ko*'s overall profile changed somewhat. One of the most obvious alterations was the relocation of the curbed tab to a more upward position so that the top sometimes even formed an essentially continuous line with the blade's upper edge, particularly in the replica weapons that became common late in the Shang.[28] However, these minor stylistic modifications would have had no real effect on the weapon's primary function or effectiveness, unlike the elongated crescent's development.

As early as Erh-li-t'ou's fourth period a slight downward curve at the tip in the straight dagger-axe blades resulted in a slightly longer upper edge and an upper-to-lower length ratio exceeding 1:0.[29] Thereafter, despite preserving the slight downward hook at the front of the blade, during the late Erh-li-kang and early Yin-hsü periods the ratio would reverse as the portion of the lower edge closest to the shaft began to be lengthened in an increasingly discernible arc. Moreover, in the Erh-li-kang period, *ko* blades with sharply tapered, sharpened edges that had initially been flat or characterized by only a slightly protruding spine began to grow wider and evolve the distinctive rhomboidal cross-section that would characterize all subsequent weapons.[30] The greater thickness not only increased the weight and the force at impact, but also strengthened the blade against twisting and breakage.

Although exceptions of considerably greater length have been recovered, the normal length of the blade, including the mounting tab, for functional bronze models in the Shang varied from just under 20 centimeters to nearly 30, with the majority falling between 23 and 26 centimeters or approximately 10 inches, but a few running as high as 38 centimeters. (For lengths of about 15 to 18 centimeters, the ratio for the portion of the blade that extends outward from the shaft to the remainder or tab generally varies from 3:1 to 4:1.) Depending on the blade's thickness, the alloys employed, and whether the tab was a straight rectangle or a heavier, curved, or angled version incorporating complex decorations, the weight could range from a very low 200 to an occasionally hefty 550 grams, with the majority falling between 300 and 450 grams or roughly 10 to 16 ounces.

The large number of dagger-axes that weigh 300, 400, or 450 grams (or just about one pound) suggests that these were considered ideal weights for the particular designs. As with any weapon, extremes tend to result in poor performance. Excessively light *ko* would have been easy to swing and less fatiguing even over the short period of actual combat, but lightening the weapon would have little effect on the arc speed or final velocity. However, because the head's weight significantly contributes to the momentum and thus the impulse or energy available at impact, too light a blade might simply glance off the era's rudimentary body armor or fail to penetrate the body. Conversely, although heavier blades have greater impact, they can become unwieldy and sacrifice precision in striking, accounting for the weight and size constraints suggested by later military writers.

Another crucial issue is the angle at which the dagger-axe blade is affixed to the shaft, because (as some scholars have speculated and our experiments have confirmed) there is a very narrow range of angles that will allow the *ko* to function effectively. Delivering a piercing combat blow in an overhand style requires that the blade arrive more or less perpendicularly to the surface of the target; otherwise, a glancing blow will result that is unlikely to produce a serious wound, if any at all, should the enemy be protected by body armor.

Although a few *ko* have been recovered that employ declinations from horizontal of 20 degrees and ascendant variants of an almost unimaginable 45 degrees, the slight upward angle of approximately 10 degrees that appeared in the early Shang eventually prevailed, no doubt the result of hard-won experience.[31] Any greater angle simply exposes the lower cutting edge, essentially converting it into an extended saber or a precursor of later weapons that mount broad knives at the top of a shaft, such as the Kuan Tao, named after the famous Three Kingdoms general and God of War. Conversely, angles less than perpendicular preclude both cutting and piercing, rendering the weapon useless.

As attested by the large number recovered from Yin-hsü, even without further improvement the dagger-axe had become a substantial weapon with formidable killing power when wielded by practiced warriors. Nevertheless, it continued to evolve, the next major development being the gradual elongation of the bottom edge downward in an increasingly arc-shaped profile. In its incipient Erh-li-kang embodiment this extension did not yet constitute an additional hooking or cutting edge, but instead provided the basis for a somewhat longer flange that further stabilized the blade's mounting while providing an additional lashing slot sufficiently offset below the body of the blade to avoid weakening it. However, even these slight changes must have produced dramatic effects, because the lower portion was quickly extended further downward during the Anyang period, resulting in an essentially crescent-shaped blade that could incorporate additional mounting slots and achieved its final realization in the late third and fourth periods at Yin-hsü, as shown in the illustration overleaf.[32]

Various sizes and basic shapes, including a modified triangular blade, were eventually produced, all generally falling within the overall length and weight limits seen in the straight- and curved-tab models. The triangular- and crescent-bladed *ko* appear to have evolved separately, but it is also claimed that the former influenced the latter. However, tab variants tended to be more dramatic, and a few actually reached lengths approximately equal to the blade itself. Increasingly displaced slightly downward, these tabs also increased in width, providing ample surface for more complex designs.

Because it can be used for hooking and slicing, the crescent or scythe-like blade radically modified the nature of the dagger-axe.[33] Now that it was no longer restricted to fighting in a piercing mode, penetrating strikes were probably relegated to secondary importance, if ever attempted. However, employing the dagger-axe as a hooking and slicing weapon requires a completely different fighting method. Rather than manipulating the point to strike a roughly perpendicular blow, the curved portion of the blade must be employed to catch and pull through the

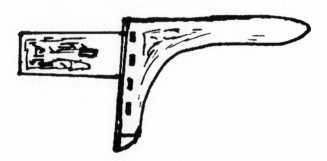

objective. Moreover, as depictions of decapitated bodies and evidence from ritual executions subsequently show, in seeking to sever rather than penetrate, warriors naturally targeted the neck and four limbs rather than the torso. Strikes could also be directed at the enemy's horses, whose speed would actually contribute to the weapon's effectiveness, as well as the chariot's occupants, eventually making it the preferred weapon for chariot combat.

In theory, both the upper and lower edges of the earliest daggerlike *ko*, as well as the upper edge of the hooked or crescent blade form of the *ko*, could have been used as cutting surfaces, but only through rather awkward maneuvers. The lower edge of the earliest piercing weapons can make an effective cut only by striking in a swift downward motion against the shoulder, requiring the wielder to hold the shaft nearly vertical and effect an awkward (and therefore weak) hand orientation before pulling forward or through a contorted horizontal hooking and pulling motion. The mechanics of the crescent blade allow it to be directly brought to bear before the lower edge is far more effectively pulled through as the forearms drop and the hands are rotated downward. On the other hand, despite claims to the contrary, the upper edges of early versions were probably only used "on the rebound" or "in recoil," when an initial swipe missed and the warrior had to abruptly thrust the top edge of the blade back upward in an awkward attempt to strike the enemy in the throat through some sort of reverse blow.

An apparent reference to employing the upper edge of the dagger-axe in this mode appears in the *Tso Chuan*, which states that Lu "defeated the Ti at Hsien and captured a giant called Ch'iao-ju. Fu-fu Chung-sheng

struck his throat with a *ko*, killing him."[34] Commentators have traditionally explained the fatal blow as an upward thrust with the top edge
of the blade, because the enemy's greater height exposed his throat.
(Why he would have been slain after being capturing with a method
appropriate to the battlefield is rather puzzling.) However, they generally
believe that the weapon was a *chi* rather than a simple *ko*, particularly
as the two terms were traditionally used somewhat interchangeably, in
which case the actual implement of death would have been the spear
at the top.[35]

Issues of rotation would not have been as severe for the early or
true "dagger-axe" because, being designed to penetrate, the blow would
necessarily have been directed downward without much sideways angle.
Nevertheless, because of the weapon's unusual design, the early *ko* inherently required highly specialized techniques to wield as an overhead
or overhand weapon and would have had somewhat reduced effectiveness when swung horizontally and even less power when striking upward
in a rising arc from below. However, the arc need not have been wide—
in fact, large sweeps can easily be avoided and there would have been a
tendency for the head to rotate—so the *ko* may have been employed for
short punching blows such as those recently determined to have been
highly effective for similarly shaped Western-style weapons.[36] But in the
case of the hooked version, greater torquing forces would have been
exerted on the hand when the head angle changed from the initial strike
as the blade cut through. In addition, the forces experienced at impact
would have tended to thrust the blade upward on the shaft immediately
upon encountering any resistance, accounting for the ever increasing
length of the crescent blade with its additional lashing slots.

A third, distinctive form of dagger-axe, sometimes referred to as a
k'uei, [37] evolved on the periphery in Shaanxi around Ch'eng-ku.[38] Derived
from precursors traceable back to the Yangshao cultural manifestation
at Pan-p'o and intermediate realizations at K'o-sheng-chuang, it migrated
into the Shang during the Erh-li-kang period and down into Sichuan,
where it persisted as an important style well into the Warring States
period.[39] Basically triangular, it has a relatively broad base but rather
rounded tip and thus somewhat resembles the shape of Neolithic stone
dagger-axes.

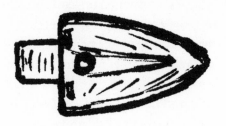

As soon as it began being cast in bronze, this triangular *ko* developed an integral mounting tab that was roughly 50 percent narrower than the blade's width, molded binding slots at the inner edge (but no flanges), holes in the tab, and sometimes even a large hole centered in the blade itself. The preferred form evolved over the centuries, ultimately producing late Shang versions marked by somewhat elongated lower edges, similar to those found on crescent-shaped blades but considerably shorter.

The tab's reduced profile allowed the outer portions of the blade's base to butt directly against the shaft, greatly strengthening the joint, while the binding slots and holes permitted more extensive, tighter lashing, increasing the overall solidity. (Some analysts believe that the noticeable success of the *k'uei*'s lashing holes prompted their addition to the then-evolving crescent-shaped extension on the straight *ko*, and that they may have even preceded the upper and lower flanges that start to become visible during Erh-li-kang.)[40] Although they frequently weigh about 300 grams, triangular "*ko*" tend to have somewhat stubbier blades of about 18 to 20 centimeters, though a few about 22 centimeters long and 9 centimeters wide at the base have been recovered.[41]

By terming the triangular version a "*ko*" rather than a variant of the broadaxe, which it more closely resembles, analysts imply that its mode of employment was similar to the relatively straight-bladed, basic variant. Because it tapers to a comparatively narrow tip, the *k'uei*'s impact area was considerably smaller than that of the average *fu* blade with its wider, more rectangular profile. However, its relative bluntness would still have rendered piercing efforts more difficult, required greater strength to wield successfully, and perhaps transformed it into more of a crushing weapon with somewhat dubious characteristics.[42] The significant popularity it achieved in the late Shang and Western Chou before disappearing by the Spring and Autumn period (because hooking weapons were more appropriate for chariot warfare than the triangular *ko*, an infantryman's weapon) attests to its functional value.[43]

THE *CHI*

Apart from new shapes and advances in construction techniques, the dagger-axe eventually benefited from the somewhat surprising addition of a spearhead at the top of the shaft, thereby uniting two discrete, fully evolved edged weapons into what has commonly been called a *chi* or spear-tipped dagger-axe.[44] (One rare form, called a *kou chi* or "hooked *chi*," affixed a knife to the top of a dagger-axe rather than a spear. This rather puzzling weapon probably attests more to human ingenuity and an inclination to tinkering with weapons than to any improved realization of lethality, because the nearly perpendicular *ko* would interfere with the delivery of slashing knife blows.)[45] Although the earliest known example from the Erh-li-kang period combines a simple straight *ko* with a spearhead, it quickly came to be based on the crescent-bladed *ko* and assumed an appearance somewhat like a Western halberd, even though it remained a hooking rather than crushing or piercing weapon.[46]

The *chi*'s extemporaneous character was lost once unitary molded versions appeared in the late Shang or shortly thereafter.[47] However, as attested by their relative paucity even in tombs and graves with multiple weapons, *chi* remained uncommon until multiplying in the Western Chou, proliferating in the Spring and Autumn, and displacing the spear and *ko* to become the chief weapon in the Warring States period, when there was a resurgence in separately cast pieces.[48] Although originally an infantry weapon (as might be expected in having been derived from the crescent-shaped *ko*), the *chi* is generally viewed as primarily a chariot weapon, especially those with shafts of six feet or more.[49]

The Eastern Chou saw the addition of one or two well-aligned crescent *ko* blades onto these longer shafts to create multiple-bladed *chi*, mimicking a briefly seen development in the *ko* itself.[50] However, the second and third dagger-axes did not have tabs protruding through the shaft, nor did they employ sockets.[51] In addition, these multiple-headed *chi* disappeared in the late Warring States after having flourished in the Spring and Autumn around the Yangtze and Han river areas, including in the states of Ch'u, Wu, and Yüeh.[52] Presumably designed to target the entire space from the tip down to the hands with a single sweep, the resulting weapon must have been too unwieldy even for the strongest infantrymen fighting on firm terrain and should perhaps be considered an oddity with no applicability apart from inspiring terror.

Being a two-part synthetic weapon, the single-bladed *chi* could be used, albeit clumsily, as a spear in forwardly directed thrusts, a vital piercing capability in situations where an overhand rotational attack would be impossible or an arcing strike missed or had been deflected. Thus, when it was necessary to recover from a swing that had carried the head through an arc into a downward position, the spear could simply be angled upward for a reverse strike in a sort of reflexive mode. However, employing the short or single-handed Shang era *chi* as a thrusting weapon probably would have been a secondary use at best, because the unbalanced *ko* head induces considerable awkwardness.[53]

18.

Spears
and
Armor

◆ ◆ ◆ ———————————————— ◆ ◆ ◆

A S ATTESTED BY SPECIMENS dating back 400,000 years, whether thrown or employed in thrusting attacks, the spear has generally been one of the first weapons fabricated throughout the world. Despite assuming many forms, ranging from a sharpened, sometimes fire-hardened length of wood to elaborately cast bronze variants mounted on shafts carved from the rarest timbers, the objective was simply body penetration.[1] However, the addition of a triangular head with sharpened edges, initially of stone but eventually cast from metal, enhanced the spear's capabilities by enabling a new attack mode, slashing and cutting.

Although stone spearheads increased the weapon's lethality, their use entailed multiple problems. Within weight and balance constraints, effective sizes and shapes had to be determined; methods for mounting and fastening developed; and workable minerals sought out, quarried, and prepared. Long, thin spearheads can achieve great penetration but generally cause more limited wounds and, being brittle, are liable to break at every stage from fabrication to impact. Broader blades require greater strength for penetration but make retraction difficult and generally inflict more significant damage. Small heads are light but lack impact; larger, heavier ones transfer greater energy but can shift the center of gravity too far forward, making them cumbersome to use and difficult to control when thrown.

In ancient times the combined length of the shaft and blade could vary greatly, but essentially depended on whether the warriors fought as individuals on a relatively dispersed battlefield or in closely packed, disciplined formations. The maximum length that can be effectively wielded by one hand in combination with a shield in the other has historically averaged about two meters or seven feet. Any longer and the weight of the shaft, coupled with the disproportionate effect of the head at the end, causes the spear to become unmanageable except for the very strongest warriors or through rigorous training. (Contemporary martial arts practice shows that skill sets can be learned that will enable a fighter to single-handedly employ a three-meter spear, but generally only in a very dynamic mode, featuring excessively large swings and considerable body involvement that create indefensible openings for enemy strikes and threaten nearby comrades.)

Because longer shafts generally require two hands to wield and make it impossible to employ a shield, fighters are compelled to depend on agility, body armor, and cooperative action. However, spears also increase the thrusting range, provide a significant advantage against enemies armed with short weapons such as a sword or axe, and make it possible to target chariot and cavalry riders and their horses. Nevertheless, depictions from Mesopotamia, Greece, Egypt, and even Warring States China show that long spears were used single-handedly in a downward-thrusting mode by mounted chariot fighters, contrary to claims that the spear's obsolescence in China resulted from its inherent uselessness in chariot warfare.[2]

Traditional Chinese lore sometimes credits Shao-k'ang's son Chu with inventing the spear, a rather unlikely possibility given the spear's preexistence at the time, but possibly indicative of him having been the first warrior to bind a bronze spearhead to a wooden shaft. Archaeological evidence shows that in China the earliest stone spearheads were mounted by inserting them into a slot at the top of the shaft, then securing them, however precariously, with lashings and bindings. Irregularity in early shapes and the slipperiness of certain minerals that required roughing and grooving must have enormously complicated the task of solidly affixing the head to the shaft. The advent of bronze casting mitigated the problem because metallic spearheads could be molded with an internal cavity or socket extending well into the spearhead. Preshaped, tapered wooden shafts could then be inserted a considerable length, resulting in a tight fit and minimal tendency to rotation when oval and rhomboidal shapes were employed. Other than the *p'i*, all known Shang bronze spears use this method of attachment, a practice that continued right through the Warring States period.[3] The spear's balance was improved by adding a blunt cap at the bottom of the shaft (in contrast with the pointed ones used for dagger-axes).[4]

Unlike the *ko*, whose shaft might snap at weak points along its length from sudden perpendicular (shearing) forces, the spear's vulnerability lay in buckling under compression when the spearhead struck and was thrust into an object. As evident from the *K'ao-kung Chi*'s emphasis on the need for a circular cross-section and no reduction in thickness even at the handhold, consistent thickness and strength were therefore required along the entire length. Unfortunately, apart from one specimen recovered from Ta-ssu-k'ung-ts'un in the Anyang area with an original shaft length of 140 centimeters, nothing more than short remnants and impressions in the sand remain from the Shang. However, a 162-centimeter Warring States shaft constructed from long bamboo strips that had been laminated onto a wooden core and then lacquered black shows that the problem was thoroughly comprehended and illustrates the engineering sophistication eventually achieved.[5]

Even though the stone precursors that have been found in Yangshao and Ta-wen-k'ou sites attest to the spear's employment in the late Neolithic, it is only with the inception of bronze versions that it would assume a focal role. No bronze spearheads attributable to the late Hsia or Erh-li-t'ou, Yen-shih, Cheng-chou, or even Fu Hao's tomb have yet been discovered, and the spear appears to have remained relatively uncommon prior to the late Shang, despite requiring relatively little bronze in comparison with the solidity of axe and dagger-axe blades. In fact, the earliest Shang bronze spearheads appear in peripheral southern culture and the Northern complex, reputedly the twin sources for the modified Shang style that would rapidly proliferate late in their rule from Anyang.[6]

Since few spearheads have been recovered even from Yin-hsü's early years, the spear's history falls into the latter half of the Shang rule from Anyang, when their numbers seem to have rapidly increased.[7] As might be expected given that older bronze spearheads retained an intrinsic value even as newer forms evolved, several styles are visible in the artifacts from Yin-hsü. In addition, just as with the *ko* and *yüeh*, non-functional ritual forms that complicate the assessment of combat designs have also been recovered. Distinguished by the thinness of their blades and the absence of sharpening, they gradually became longer and more elaborate, even occasionally assuming oversized forms similar to the long stone spear found at Liaoning, which dates to the dynasty's end.[8]

The spear actually appears under a number of different names, some simply regional identifications, others derived from a distinctive aspect, such as the *p'i*.[9] However, whether longer or shorter, blunter or smoother, decorated or not, without exception Shang era spears always have just two leaflike blades, presumably in imitation of stone precursors.[10] Even variants with pronounced rhomboidal spines that could have functioned as additional blades with minimal enlargement and sharpening were never transformed into four-edged spearheads, nor did three-bladed models emerge.

Shang era bronze spears have now been divided into three main categories: the southern style, northern style, and a composite embodiment whose derivation remains controversial.[11] Although few dedicated studies have been published, based on numerous recovered specimens it has generally been concluded that what might be termed the southern style was the main source for the Shang bronze spear.[12] Apart from a single, rather primitive-looking specimen recovered out at T'ai-hsi, the very first bronze spears are in fact the three found at two sites at P'an-lung-ch'eng at the Shang's southern extremity, one of which is shown overleaf.[13]

All three have long willowy blades with raised spines, gradually expanding mountings, and a raised rim at the base. Two of the three are further distinguished by the opposing earlike protrusions low on the base, which functioned as lashing holes for securing the head to the shaft rather than for attaching pennants or decorations, as in later periods.[14] Furthermore, two have slightly oval openings, while the third is somewhat rhomboidal.[15]

Although the presence of "ears" has been energetically asserted to be the defining feature for southern tradition spears, sufficient exceptions ranging from single examples to small groups have been recovered in Hubei, including at Hsin-kan, to constitute another type.[16] Proponents of the southern tradition tend to dismiss their importance by noting that they all date to the Yin-hsü's second period or later, thereby implying (without ever specifying) that they are synthetic types that meld Shang influences (which is somewhat problematic given the lack of Shang precursors) with indigenous characteristics. At Hsin-kan, where the spears outnumber the *ko* thirty-five to twenty-eight, twenty-seven of the former do not have the distinctive ears required of the southern type, and another was found at Wu-ch'eng that reportedly exerted a formative influence on Hsin-kan.[17] However, the twenty spears recovered at Ch'eng-ku in Shaanxi (previously mentioned in connection with the southern *yüeh* tradition) all have ears in addition to somewhat pudgy, leaf-shaped blades and decorated bases and are said to be somewhat larger than those at Hsin-kan in comparable styles.[18]

The so-called northern style that appeared about the same time not only lacks "ears" but instead employs wooden pegs inserted into holes in the base to secure the spearhead onto a round shaft, thereby sufficiently augmenting the basically secure mechanical fit to prevent rotation and the head's loss in combat. Generally more elongated than southern-style spears, they are also simpler in appearance and normally lack decoration. Their similarity to spearheads recovered from the Shintashta-Petrova and Andronovo cultures suggests an external origin for this style, but no studies have yet assessed the possibility.

The Shang bronze spears recovered at Yin-hsü are said to be primarily based on the southern style or to integrate both northern and southern features in conjoining some form of ears intended for lashing with a heavier rim, pegging holes, and a rhomboidal socket.[19] However, if pegging and a circular (rather than oval) receptacle are considered definitive northern contributions, enough exceptions have again been discovered

to seriously undermine any claim to a true Shang synthesis, even though pegging would increasingly become the preferred method from the Western Chou onward.[20]

Despite sometimes extreme variations in shape, length, and decoration, late Shang spears assume two primary forms. The more common one, somewhat simplistically identified as the "Shang spear," has a rather dynamic, unbroken contour as its elongated leaf-shaped blade continuously extends down to the rim in a sort of wavelike pattern that first bulges out, then curves inward before finally expanding once again toward the bottom, where lashing holes are usually incorporated. In some variants the blade extends all the way down; in others it cuts inward, leaving a short length of clearly defined shaft that may or may not have a rim.[21]

The other primary form that suddenly appeared in the second period and is also, if confusingly, sometimes identified as the definitive Shang spear, consists of a sort of pudgy-looking, adumbrated, leaf-shaped blade, an obviously protruding spine, short lower shaft, and two ears, as shown overleaf.[22] Thereafter, most Shang spears would be marked by ears in one form or another, though more "earless" versions have been recovered than is generally credited. Both styles initially flourished in the Western Chou, but a shift over the centuries saw a continuous reduction in the blade's waviness, resulting in the longer, thinner, more dynamically contoured and lethal shape that predominated in the Warring States period.

The spear's battlefield role and relative importance in the Shang have long been debated. The number of bronze spearheads recovered from individual Anyang era graves, tombs containing sets of weapons, or collections of internments in a confined area, although cumulatively second only to the dagger-axe, is usually far lower than the latter. It has even been asserted that the simple character for spear (*mao*) doesn't ap-

pear in the oracular inscriptions, though there
are dissenting views.[23] However, the spears recovered from Yin-hsü are almost invariably found
in weapon sets rather than in isolation, including with *yüeh*.[24] It was nevertheless held that
though officers carried *yüeh*, the dagger-axe was
the era's primary combat weapon, and spears
played an uncertain but supplementary role.
However, relatively recent discoveries have cast
considerable doubt on this interpretation, one
being the grave of a presumably high official
with considerable martial authority discovered
in the Anyang area itself that contained 730
spearheads but only 31 *ko*.[25]

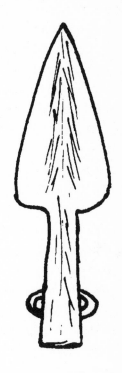

The grave at Hou-chia-chuang consists of
several layers that contain a variety of ritual vessels and weapons indicative of high rank and
military power, and the weapons are arrayed in
a way that suggests they once furnished the arms
for a military contingent. (This hoard would
seem to at least partially confirm the claim that weapons were monopolized by the government, manufactured and stored by Shang authorities, and only dispersed for military activities.) The spearheads were
found clumped together in bundles of ten but distributed into two layers
consisting of 370 in the upper and 360 below. However, the import and
significance of this arrangement remain undetermined. Were the weapons
intended for a unit of 360, the second layer being replacements for the
first 360, whether fully made up with shafts or simply maintained as
spare spearheads to allow battlefield fabrication or replacement as necessary? Or were they intended for a second company of 360, in which
case the total operational unit would closely approximate a modern
army battalion?

The low number of dagger-axes at Hou-chia-chuang has prompted
claims that the spear had already become the army's primary weapon
in the late Shang, in which case commanders for squads of ten were
wielding *ko* and higher-ranking officers carrying *yüeh*. However, other

explanations are possible, including that spears represent a new form of weaponry and were therefore held in central storage, whereas dagger-axes were more widespread, essentially individual possessions. Conversely, the spear is an effective battlefield weapon whose length and mode of action would have easily outranged the era's *ko*, could have been usefully employed against the increasing numbers of horses, and certainly required less combat space in a thrusting mode, making it more suitable for a packed battlefield than a dagger-axe employed in overhand or sidearm strikes.

Minor confirmation that spears were becoming important shortly after Wu Ting's reign appears in the tomb of Commander Ya—who has already been mentioned in connection with the seven unusual *yüeh* that accompanied him into the afterlife—in which slightly more spears (76) than dagger-axes (71) were recovered.[26] Furthermore, the 30 *ko* and 38 spears found among the weapons interred with a high-ranking Ma-wei nobleman dwelling in Anyang in Yin-hsü's fourth period augment the evidence leading to the conclusion that the spear had begun to assume a battlefield role and that the nature of combat was in transition.[27]

Finally, whatever their relative proportions in the Shang, there is no evidence that spears were ever employed as missile weapons in the manner of ancient Greek javelins. This may have been a consequence of the high value placed on bronze, the inconvenience of warriors each having to carry several cumbersome javelins or similar spearlike weapons, or simply the inappropriately short length of slightly less than 150 centimeters. Nor, despite the long tradition of traditional martial arts slashing techniques and the spear's highly esteemed role in the Wushu world, can it be concluded that spears were ever employed other than in a piercing mode.

ARMOR AND SHIELDS

Fairly detailed knowledge of the lamellar armor developed in the Warring States period has been acquired as the result of excavations conducted over recent decades, but even impressions of Shang armor and shields have generally proven elusive because of the rapid degradation of nonmetallic materials. However, this lack of artifacts has not prevented highly speculative discussions and several imaginative attempts at reconstructing armor's inception. Nevertheless, by focusing on the few known impressions and artifacts, some sense of Shang, though not Hsia, armor can be gleaned.[28]

Designs varied imaginatively over the centuries, but priority was always given to the head, then the chest, shoulders, and finally the waist downward. Because percussive blows could be—or more correctly, had to be—simply blocked or warded off, defensive coverings were basically expected to provide protection against arrows and piercing and slashing attacks undertaken with edged weapons. At what stage Chinese fighters began imitating animals and turtles and employed animal skins, furs, or multiple layers of heavy cloth for primitive protection remains unknown, but based on vestiges at Anyang, it certainly predates the Shang.

Perhaps because they lack the gleaming appeal of burnished bronze or the thrill of martial power, shields and armor have rarely been associated with the legendary Sage progenitors of antiquity. (The Spring and Autumn anecdote of penetrating seven layers of armor well illustrates the normal propensity to lionize the bow and arrow while deprecating the role of armor.) However, Shao-k'ang's son Chu, who eventually restored the clan's power amid the internal chaos that beset the early dynastic Hsia, reputedly fabricated the first body armor in order to survive the Yu-ch'iung's deadly arrows.

As attested by vestiges of large leather panels at Anyang, the earliest Shang body armor was apparently a two-piece leather tunic that should have provided some protection against glancing attacks, even though ancient skeletons show that arrows and piercing weapons could penetrate and circumvent it.[29] Decorated with abstract *t'ao-t'ieh* designs in red, yellow, white, and black,[30] Shang armor seems also to have been sometimes partially adorned, rather than layered, with very small bronze pieces. However, although it appears that additional pieces of leather may have eventually been employed for the shoulders, no major improvements can be proven until the early Chou, when more flexible corsets began to be fabricated by employing lamellar construction techniques that linked small leather panels together with hempen cord. Thereafter, it was merely a question of time before metal plates would be substituted, eventually resulting in much of the well-known body armor visible on Ch'in dynasty tomb figures.

Although no examples have yet been recovered, shields constructed from interlaced soft wooden materials, leather, or even leather thongs certainly predate the Shang and would continue to be employed for the next three millennia by peasant forces and in isolated localities. Shang variants were apparently produced in two sizes, a moderate version for infantry warriors confined to the ground and a larger version that is still unattested apart from depictions of shields for chariot fighters intended to block incoming arrows and strikes down the full length of the body and perhaps even cover a compatriot fighting alongside. Materials employed to fabricate these shields seem to have again included thin wooden slats, interlaced bamboo or other reedlike materials, and leather, as well as a leather covering on an underlayer or two of fibrous matter, all stretched and affixed to a simple frame consisting of roughly three-centimeter wooden poles.

From depictions preserved in oracular characters and a body buried with a dagger-axe and shield, it is clear that the moderate-sized, single-handed shield was held in the left hand and used in conjunction with a dagger-axe or short spear in the right. Impressions made by a cache of shields somewhat chaotically deposited in tomb M1003 and a famous reconstruction based on vestiges in tomb M1004 indicate that they were slightly rectangular and had rough dimensions of 70 by 80 centimeters (27.5 by 31.5 inches), or about half a man's height and thus slightly shorter than the era's dagger-axes and single-handed spears.[31] They were held by a vertical handle in the center and had a slight outward bow that should have improved the dynamics of blow deflection while facilitating the warrior's grasp.

Traces in the compacted soil further indicate that leather versions were also sometimes decorated with *t'ao-t'ieh* patterns similarly painted in vivid red, yellow, white, and black and even tigers or dragons, as described and found in later times.[32] The nobility and high-ranking officials apparently affixed small supplemental bronze plates that would certainly, if spottily, have increased the shield's penetration resistance even though they were probably intended mainly as decorative embellishments. One bizarre variant of the infantryman's shield called a *ko-tun* ("dagger shield") even mounted a dagger-axe blade perpendicularly at the top, but its possible utility except as a sort of jabbing distraction in the enemy's visual field is difficult to imagine.[33]

Helmets, presumably of interlaced rattan but possibly leather, also appeared in the Neolithic, and a somewhat lesser-known legendary account appropriately attributes their invention to Ch'ih Yu, even though his aggressive behavior should have prompted others to create them as a defensive measure against his innovative weapons. Whatever form these ancient variants might have taken, no evidence has survived, and the first known metallic helmets appear in the Shang. However, they remain surprisingly sparse in comparison with the vast numbers of weapons that have been recovered and Shang willingness to employ bronze for large ritual vessels, with only one large aggregation having been discovered and a few deposited in the weapons hoards found in scattered high-ranking tombs.[34]

Although several variants can be distinguished, Shang helmets were basically designed to protect the skull from the forehead upward but also extended downward sufficiently to normally, but not invariably, encompass the ears and the back of the neck. Given that some large *yüeh* and shield decorations were molded with bulging eyeholes, they surprisingly did not provide any integrated means of facial defense, a deficiency that may have been remedied with a bronze face mask in exceptional cases.[35] Cast from bronze as a single unit that weighed from 2 to 3 kilograms (a weighty 4.5 to 6.5 pounds) and averaging some 22 or 23 centimeters in height, they were apparently worn with an inner head wrap or intermediate padding that was designed to cushion the effects of blows and protect the skull against

wounds that would invariably have been caused by the interior's roughness. In contrast, the exteriors are all smooth, polished, symmetrical executions of somewhat startling designs that feature stylized projections and protuberances that not only augment a warrior's frightening visage but also strengthen the helmet's structure and increase the distance between the skull and the strike.

19.

ANCIENT
ARCHERY

+·+————————————————————————————+·+

I N CHINA, THE BOW AND ARROW apparently enjoyed at least limited use
by 27,000 BCE, more than twenty millennia before the advent of the
Neolithic civilizations with which our study begins.[1] Thereafter, ancient
sites contain increasingly numerous stone and bone arrowheads coincident with their growing impact in hunting.[2] However, as agriculture
emerged and began to flourish between 6500 and 5000 BCE, other tools
appeared that assumed greater importance, though without diminishing
the number of arrowheads.[3] Even wooden ones dating back to 5000 BCE
have been recovered, virtual throwbacks to the primitive arrows that
were fashioned by sharpening and then heat treating the tip of the shaft
to harden it.[4]

Early arrows were probably multipurpose, but the advent of clan
conflict and tribal warfare stimulated the development of heads specifically designed for military purposes. These variants gradually came to
equal and then quantitatively exceed all other designs, prompting the
further evolution of numerous styles and multiple sizes with increasingly differentiated head characteristics (worthy of a lengthy mono-
graph rather than the cursory examination possible here). The discovery
of copper and ensuing advances in metallurgical techniques then allowed the Shang to produce bronze arrowheads in significant quantities.
Nevertheless, they did not completely displace laboriously fabricated
stone and bone points until late in the Western Chou, despite being ef-
ficiently cast in multiple-cavity molds.

The earliest Chinese bows must have been simple, minimally effective
weapons created by adapting readily available saplings from wood species

with the necessary characteristics of strength and flexibility. Nevertheless, legends claim that either the Yellow Emperor or the Archer Yi invented the bow, and that Yi shot down nine of the ten suns then scorching the earth.[5] However, the *Yi Ching* credits Yao, Shun, and Yü with creating the first bows, but the *Shan-hai Ching* claims that two of the Yellow Emperor's ministers, Hui and Mou Yi, were actually responsible, the former for crafting the first bow and the latter for fabricating the arrows.

In contrast with Western legends that portray knights slaying dragons with magical swords and vanquishing demons with axes, the bow seems to have been valued in China from its inception for its accuracy, power, and ability to destroy enemies at a distance in awesome displays of power. Not surprisingly, when first encountered in Shang oracle inscriptions and archaeological finds, bows and arrows are the weapons of the ruling clan and warrior nobility. In contrast with medieval Europe, where the sword became a highly romanticized, close combat weapon while the bow was condemned for its dastardly ability to kill anonymously and archers were reviled for fighting at a distance,[6] archers and archery have always been highly esteemed in China, as well as in Korea and Japan. King Wu of the Chou even emblematically shot the already dead tyrant Chou (Emperor Hsin) with three arrows before decapitating him with a yellow axe.[7]

Later idealizations of antiquity envision late Neolithic and Shang leaders selecting their officers primarily because they excelled in martial skills vital to the battlefield and the hunt, abilities crucial for surviving amid hostile forces and the untamed environment. Even though these officers might subsequently be entrusted with administrative responsibilities, they were primarily warriors, and it was their "virtue" in wielding a bow rather than a shock weapon that distinguished them.[8] This practice apparently continued in the early Western Chou, with formalized archery competitions intended to reveal both skill and character, the various nobles in attendance being penalized or rewarded based on the performance of their subordinates.

Although these claims need not be accepted in detail, the basic impetus to select men through their martial abilities, later understood as their "character," appears to have been a core element in ancient practices. The *Li Chi* chapter on the archery ceremony states:

Anciently, the son of Heaven chose the feudal lords, the dignitaries who were Great officers, and the officers from their skill in archery. Archery is specially the business of males, and there were added to it the embellishments of ceremonies and music. Hence among the things which may afford the most complete illustration of ceremonies and music, and the frequent performance of which may serve to establish virtue and good conduct, there is nothing equal to archery; and therefore the ancient kings paid much attention to it.[9]

Contemporary bronze inscriptions attest to major archery competitions having being conducted under regal auspices from the Western Chou's inception, implying that they were probably a common practice in the Shang insofar as the Chou adopted many Shang customs. For example, the *Tso Po Kuei* records that the Duke of Tso responded to King Chao's challenge by hitting the target dead on ten times in a row in just such a contest, thereby garnering all ten of the gold pieces that the king had allotted for successful strikes.[10]

In addition to impromptu contests, four formalized archery competitions were routinely held during the Western Chou and perhaps the Shang. (Paradoxically, they would become less frequent despite the evolution of permanent armies and an escalating demand for bowmen.) Moreover, despite almost certainly having had their origins in hunting and purely military contests in which military prowess dominated, with the passage of centuries and the increasing pervasiveness of Confucian thought, these competitions would gradually evolve into a stifling, formalized exercise bereft of martial spirit.

At the end of the sixth century BCE Confucius deemed archery one of the six arts essential to self-cultivation, and bows and arrows were often bestowed as special marks of honor, particularly for military merit. King Wen's appointment as Lord of the West during the late Shang was reportedly sanctified with a bow, arrows, and an axe, and numerous Western Chou bronze inscriptions record the conferring of a bow, often highly decorated and accompanied by a hundred arrows, on various people.[11] The practice continued down into the Spring and Autumn, one *Tso Chuan* entry speaking about bestowing a red bow plus 100 arrows and a black bow with 1,000 arrows,[12] another recording that the king awarded a red bow for martial achievement.[13] Contrary to impressions that these bows may have been fragile and ephemeral, they were treasured, preserved, and frequently transmitted within families as valuable commemorative items. Thus the *Tso Chuan* notes the recovery of the great bow of the state of Lu that had been awarded for martial achievement generations earlier but stolen the previous year.[14]

Archery also played a significant role from the Western Chou onward in ordinary life, including in the Village Archery Ceremony (which will be discussed in detail in *Early Chinese Warfare*). In order to impress his potential wife with his manliness, one suitor in the Spring and Autumn merely shot two arrows, one each to the left and right.[15] In another incident a noble who had received a bow as a gift and was seeking a pretext to hold a private discussion went outside to try it out,[16] which shows that these bows were not intended for idle exhibition and that virtually everyone of rank remained skilled in their use.

DESIGN, POWER, AND ACCURACY OF THE BOW

Lacking reliable evidence for reconstructing the bows employed in the Shang and Western Chou, somewhat ahistorical recourse must be had to depictions preserved in such works as the *Tso Chuan* to infer the bow's likely power, the archer's capabilities, and archery's general impact in combat situations. However, note might first be taken of a pointed debate that has recently arisen over whether arrows were ever effective at a distance, particularly near the limits of their range, where the angle of descent could be a severe 45 degrees or more. It has been forcefully asserted that because these steep incoming angles would have produced glancing rather than perpendicular blows, the arrows would have lacked the impulse necessary to impale, not to mention penetrate, the armor employed during the medieval era in the West.[17] Similar questions may be posed about the comparative efficacy of Chinese arrows and armor for every era in which the age-old clash between offensive and defensive measures, symbolized by the spear and shield (*mao* and *tun*, which form the modern compound for "contradiction"), continued unabated.

Whether wielded to selectively target individuals in the Shang and Western Chou or provide the massed volley fire common in the late

Spring and Autumn and Warring States periods, the extant historical records show Chinese bows were always highly effective. An initial sense of their accuracy and power may be gained by examining the range and the size of the target employed in the Village Archery Ceremony described in the *Yi Li*, a Warring States ritual text. The contestants normally shot at a target hung an even fifty bow lengths (or paces) away, a minimum of 250 feet since the bow's length then approximated a man's height of 64 inches, roughly equivalent to the pace length of 5 feet or 60 inches in both China and the West.[18] Even within the narrow confines of a hall, this must have been a distance easily attained; otherwise it would have been too difficult for the competitors to display the required decorum and master the formalities of ritualized movement while still achieving a respectable result. Although the exact shape and dimensions of the target remain controversial, it basically consisted of a large square marked off into three concentric zones, suspended between extended borders of horizontal material both above and below. To gain full marks an archer would have had to hit a square measuring slightly over 2 feet on edge, a size that realistically approximates the width of an armed soldier.[19]

This nominal ceremonial range of fifty bow lengths comprised but half the distance at which Chinese reflex bows were minimally considered capable of killing an enemy.[20] Thus it is not surprising that archery contests, even highly ceremonial ones such as the Great Archery Ceremony described in the *Yi Li*, were also held at seventy and ninety bow lengths, though certainly outside the hall itself in the surrounding grounds. Somewhat larger targets were reportedly employed for these greater distances, even though maintaining the original fifty-pace target size would have more accurately approximated battlefield requirements.[21] However, tradition also suggests that the targets depicted various wild animals, no doubt part of the heritage of the hunt, in which case they probably would have taken their dimensions from real animals or consisted of actual skins stretched out in a field.

The *Tso Chuan* and later, less reliable writings contain numerous accounts of singular combat in Spring and Autumn battles that apparently provide evidence of the Chinese bow's remarkable power and attest to the skill levels that might be achieved, although these accounts certainly were included because of their exceptional nature and no doubt were dramatically enhanced. (What percentage of ordinary combatants might have achieved such mastery is another question.) For example, at a critical moment during the Battle of Pi, one of the *Tso Chuan*'s six great conflicts, a warrior killed one soldier and then wounded and captured another with just two arrows. At the same battle a warrior successfully shot a deer from a racing chariot before offering it with great bravado to the pursuing enemy.[22] Prior to the engagement two warriors raced forward to challenge the enemy and discharged their arrows, immediately slaying two of their transfixed opponents. In a lesser-known conflict, just three skilled archers managed to hold off an advancing force by killing a large number with their arrows.[23]

By similarly slaying two enemy soldiers with just two arrows at the famous battle of Yen-ling, a single archer reportedly deterred an enemy thrust.[24] The king was also shot in the eye, and another arrow killed one of the enemy.[25] Two arrows loosed by one strong fighter are said to have frightened Ch'u's entire army when they glanced off the harness poles and became embedded in the king's chariot.[26] Furthermore, in an incident that unfolded prior to the battle that has often been inappropriately cited as evidence that traditional China emphasized virtue over martial power, Ch'u's two best archers managed to penetrate seven layers of armor with their arrows. Although this was surely an auspicious precombat omen, their commander surprisingly disparaged the achievement because their efforts inappropriately emphasized might alone.[27] Despite being a static test no doubt conducted against perfectly placed vertical surfaces, it still suggests the bow's enormous penetrating power at short combat ranges. Numerous skeletons from the Shang and earlier with stone arrowheads deeply embedded in thigh and other major bones offer further confirmation of the weapon's lethality.

One warrior proved so accurate at short range that he was able to skillfully shoot a man in the left and right shoulders and thereby persuade him to surrender by simply threatening him with an arrow to the heart, showing that bows were not employed just for open field fighting, but also at close range where they would be even more deadly.[28] Early in the Warring States period the famous military administrator and commander Wu Ch'i suffered an ignominious death in the state of Ch'u when he was pursued through the palace and finally shot to death in the confined quarters of a large room, rather than being stabbed with a dagger or slain with a sword.

In yet another incident an archer killed the ruler with a single shot and then exploited his fearsome prowess to thwart multiple attackers simply by holding two arrows to the bow, ready to fire. Eventually he was killed in an exchange of archery fire with a warrior who arrayed his men behind a protective wall, but despite suffering a wound to his wrist, he still succeeded in slaying the latter before he died.[29] Superlative skill is also visible in numerous tales of surpassing accuracy during hunts and the offhand shooting of smaller targets such as geese and pheasants.[30]

Apart from questions of power and reports of superlative ability in combat situations, there is a real issue of accuracy and its attainment. Exceptionally skilled archers could reputedly hit a flying bird at 200 paces, and superlative archers such as Yang Yu-chi in the Spring and Autumn period reportedly could hit a willow branch at 100 paces, giving rise to the phrase "penetrating a willow at a hundred paces" becoming praise for any extraordinary skill. Similar to the idea of a hundred victories in a hundred engagements, being able to strike a target one hundred times without a miss—*pai fa pai chung*—was another description for surpassing achievement. However, it was also known that these attainments were the result of strength and concentration, with defeat resulting if either wavered.[31]

Conversely, because it was viewed as an achievable skill rather than purely an innate talent, it was thought that high levels of expertise could be realized through study and practice, perhaps explaining the reported enthusiasm of late Chou dynasty students and contestants.[32] Even though Mencius envisioned formal schools dedicated to archery, certain Shang officials were entrusted with the task of instruction, and Shang oracular inscriptions query the appropriateness of one or another individual training men in archery as well as "new archers" being sent to the battlefield,[33] nothing is actually known about Hsia and Shang archery training.

One method that may have been employed to achieve the ideal of one hundred hits in one hundred shots appears in the *Archery Classic*, composed in the T'ang dynasty, predating chapters on archery preserved in such imperial era military compilations as the *Wu-ching Tsung-yao*. According to this text neophyte archers who had been instructed in the proper stance and release methods were to start one *chang* away from the target, and when they could score a hundred hits in a hundred shots at this laughably close range, add another *chang* and continue repeating the procedure until they reached the desired distance of one hundred paces or somewhat more than sixty *chang*. Unfortunately, no estimates of the number that might be expected to advance beyond fifteen or twenty *chang* are ever mentioned.

The classic military writings not only emphasize the effectiveness of bows and arrows for open field combat, but also stress their importance in defensive situations. The *Ssu-ma Fa* asserts that "fast chariots and fleet infantrymen, bows and arrows, and a sturdy defense are what is meant by 'increasing the army'" and adds that "to take advantage of terrain, defend strategic points. Valuing weapons, there are bows and arrows for withstanding attack, maces and spears for defense, and halberds and spear-tipped halberds for support."[34] Engaging aggressors at a distance from an ensconced position with harassing volley fire always proved highly effective in reducing their numbers and was generally advised even if it wouldn't succeed in immediately repulsing them. City walls were to be defended with a hail of arrows and stones;[35] arrows and crossbows had to be prepared for defense.[36]

It might seem that greater physical strength and dexterity would be required to wield a spear or dagger-axe than to shoot a bow. However, even though finger releases were employed as early as the Shang, presumably because of the strength and stamina that would be necessary to repeatedly pull and sustain a fully drawn position, experienced Chinese military thinkers reached the opposite conclusion. Thus the *Wu-tzu* asserted that "the basic rule of warfare that should be taught is that men short in stature should carry spears and spear-tipped halberds, while the tall should carry bows and crossbows."[37] In the Warring States only men who could "fully draw a bow and shoot while racing a horse" were to be chosen for the cavalry.[38] Nevertheless, differences in stature were not completely ignored, because the *K'ao-kung Chi* describes three different bow sizes and the military writings note that the bow's size must match the archer's physique.

The bow was also viewed as an esoteric weapon of power because of its ability to kill suddenly, at a distance, often completely unseen, the very reason it was condemned in the west. Thus Sun Pin commented: "Yi created bows and crossbows and imagized strategic power upon them. How do we know that bows and crossbows constituted (the basis for) strategic power? Released from between the shoulders, they kill a man beyond a hundred paces without him realizing the arrow's path. Thus it is said that bows and crossbows are strategic power."[39]

EARLY CHINESE BOWS

Despite its primal importance in Chinese warfare, there has been a surprising dearth of articles on the bow and arrow. However, from the few attempts,[40] numerous archaeological reports, various comments in the classic military writings, passages such as already cited from the *Tso Chuan* and found in other works including the *Shih Ching* and the late ritual texts known as the *Yi Li* and *Chou Li*, as well as material on early fabrication techniques preserved in the *K'ao-kung Chi*, a sufficient number of basic historical points can be cobbled together to illuminate archery's role in ancient Chinese military activities.[41] Unfortunately, despite extensive evidence pointing to their existence, no bows have ever been recovered from any early cultural site, and even Anyang has yielded only impressions in the dirt.

Nevertheless, close examination of the characters employed in Shang oracle and bronze inscriptions has led to the generally accepted conclusion that late Shang bows were recurved and therefore necessarily composite in construction, as well as quite powerful.[42] (The discovery of numerous water buffalo horns, essential for compressive strength, at Hsiao-t'un provides additional evidence of their composite nature.)[43] Based on jade and bronze end fittings with discernible string notches, the position of various bow components in excavations, impressions left in the compacted soil, statements in the *K'ao-kung Chi*, and recourse to a later realization, Shang and early Western Chou bows are estimated to have been approximately 160 centimeters in length, or slightly less than a man's height of 160 to 165 centimeters at the time.[44] At roughly 60 inches or 5 feet, a bow of this length would have been somewhat shorter than the English longbow as well as far more compact when strung. Unstrung, the tips of the recurved bow were apparently some

65 centimeters apart based on the relative position of end pieces still lying in the ground at Hsiao-t'un.[45]

Idealized descriptions of distinctive bows carried by the king, feudal lords, lesser nobles, and ordinary warriors suggest that several types existed in the Western Chou and perhaps earlier. According to Hsün-tzu, a late Warring States writer, "That the son of Heaven has an engraved bow, the feudal lords have cinnabar bows, and the high officials have black bows (accords) with the forms of propriety (*li*)."[46] In describing the duties of an official bow maker, the *K'ao-kung Chi* states: "In making a bow for the son of Heaven, the criteria is for nine layers to be combined; in making bows for the feudal lords, the criteria call for seven layers to be combined; in making bows for the high officials, the criteria call for five layers to be combined; and in making bows for the *shih* [lower members of the nobility or warriors], the criteria call for three layers to be combined."

Statements such as these reflect an emerging Warring States, Confucian-derived insistence that gradation that should characterize all social and political relationships, projected back onto the early Chou. (More than substance, sumptuary regulations normally constrained the embellishments and quality of materials such as jade and gold that might be chosen for ceremonial or ostentatious manifestation.) If they ever existed, these distinctions must have been limited to ceremonial bows rather than actual field weapons, because it would be unthinkable for the officers and masses to fight with only three or five layers if nine provided superior power.[47] Furthermore, an engraved or otherwise embellished bow that might have been an effective weapon would probably prove useless in real combat when structurally modified to suit the criteria of display.

A far more realistic approach is seen in the *K'ao-kung Chi's* grouping of warriors into three classes based simply on the bow's length, doubtless on the assumption that stronger warriors are able to handle a more powerful pull. "Bows of 6 feet 6 inches should be wielded by superior warriors; bows of 6 feet 3 inches by middle ranking warriors; and bows of 6 feet by the lowest ranking warriors."[48] However, the late Ming dynasty *Wu-pei Chih* concluded that a fully pulled, strong bow is totally inappropriate for real military use because bows need to be fired quickly in order to realize the crucial objective of unexpectedly striking the enemy. Since only the very strongest warriors can pull and hold a powerful bow while awaiting the perfect psychological moment emphasized in ritual competitions (but irrelevant to the battlefield), "softer" bows defined as being well within the capability of the individual archer were deemed essential. Only they could be shot numerous times in the intensity of combat without exhausting the archer, yet also be held fully drawn and poised to fire when necessary to keep an enemy at bay.[49]

Despite ongoing trade and the surprisingly rapid circulation of many technological developments, considerable local variation in bow types existed among the relatively isolated tribes and proto-states of ancient China, some of which even persisted into the Ch'ing dynasty. For example, the *Wu-pei Chih* characterized the unique bows fabricated for employment in the hot, moist south as generally suitable for naval warfare but lacking elasticity and incapable of shooting over a hundred paces. Studies of bows recovered from Warring States sites, especially in the southern state of Ch'u, confirm the existence of strong regional bow-making traditions, and the state of Han was particularly known for the excellence of its bows.[50]

According to the *K'ao-kung Chi*, six basic materials were employed to fabricate effective bows—wood, horn, sinew, glue, thread (or fiber), and lacquer—each of which was believed to furnish a specific attribute. The core or body (*kan*), thought to provide the propellant strength and determine the bow's maximum range, might be constructed from seven different types of wood: silkwood thorn, some sort of privet, wild mulberry, orange wood, quince, thorn, and bamboo.[51] Whatever species might be chosen, in selecting the wood for the body of the bow (or the core in the case of a composite bow) certain irreversible wood characteristics had to be recognized and exploited: heartwood is stiff and can withstand compression but not extension (by being pulled); sapwood is comparatively elastic and can be pulled but not compressed.

Insofar as the *K'ao-kung Chi* reflects Northern Plains and Shandong practices, it is not surprising that bamboo was regarded as the least desirable material for bows, yet (laminated) bamboo bows were common throughout Chinese history in the south where it proliferates, as well as in peripheral areas whose inhabitants could not afford the lengthy fabrication time required to produce a northern style compound weapon.[52] As attested by numerous laminated examples recently recovered from Warring States tombs, Ch'u's skilled archers were probably using bamboo bows.

Although bamboo apparently formed the core or provided one of the laminates for many bows, a single culm could not be used for more than a small child's bow. However, other self-bows—bows generally fashioned from a single piece of wood—have been recovered from numerous places, including Ch'u, again contrary to recorded practices and the expectations that prevailed just a few decades ago. In addition, simple laminate bows made from layers of the same wood or dissimilar woods but not recurved are also known, but played only a minor military role except in peripheral regions and as weapons for training and children.

The tremendous compressive properties of natural horn (*chiao*) made it valuable for the interior of composite recurved bows. Horn has traditionally been said to furnish the arrow's velocity, consigning the bow's wooden core to merely providing a basic structure for laminating. Various types of bovine horns were considered usable, especially water buffalo horns or those from the so-called long-horned cattle found on the western borders.

Sinew or tendon (*chin*), traditionally thought to account for the arrow's ability to penetrate its target, was used to fabricate the exterior or front side of the strung bow because of its relative elasticity and contractive power when stretched. It thus worked in tandem with horn's compressive strength, effectively augmenting the latter by pulling the bow's arms forward just as the horn's power was pushing them outward. Care had to be employed when preparing and gluing it to prevent it from being either too contracted or too stretched and thereby proving useless. The preliminary preparation of sinew was undertaken in the summer, but the actual gluing up—bonding the horn and sinew to the wooden core, which itself would have been composed of strips attached to the central portion—was completed in autumn.

Glue (*chiao*), which provided adhesion, had to thoroughly penetrate the sinew in order to prepare it for bonding to the body of the bow and thus functioned as a plasticizer as well as an adhesive.[53] In the Shang it was probably derived from animals and eventually fish, but the worldwide use of vegetable glues for wood products and their relative ease of preparation suggest they may have been employed in the earlier stages of fabrication despite their greater sensitivity to moisture.

Bindings of silk or bamboo fibers, both of which have great tensile strength (particularly when glued and lacquered), were employed to ensure the adhesion of the individual components. No ancient bow strings have yet been recovered, but depending on the relative availability of indigenous materials, bowstrings were probably made from silk, thin strips of leather, and various plant fibers, especially bamboo, that could be plaited and woven. The *T'ien-kung K'ai-wu* speaks about using a fiber core and a fiber wrap, both twisted, a method consistent with Western practices.

Finally, from the late Shang onward lacquer was applied to protect the finished bow against moisture. Producing and handling naturally derived lacquers must have entailed considerable hardship, and appropriate techniques for applying this difficult material had to be perfected. Because a single heavy coat would result in cracking and crazing the first time the bow was bent for stringing, multiple fine layers and adequate drying time were required for the lacquer to render the bow impervious to moisture, particular care being taken in areas of maximum flex.

Seasonal limitations came to be imposed on sourcing and preparing the component materials during the Chou in order to take maximum advantage of desirable characteristics that predominate in, or may be limited to, different parts of the year due to growth and dormancy cycles.[54] Trees were felled in winter and the wood initially cut and dimensioned when moisture levels were lowest but the wood could still be split without cracking, splintering, or shattering. Horn was soaked and glued in the spring, presumably having been harvested just when it would have been softest and showing new growth. Sinew was prepared in summer. Even then the different strengths and degrees of elasticity inherent to the component materials, essential to the bow's dynamic strength, created forces that constantly tried to delaminate and tear it asunder, and imbalances in any aspects had to be corrected.[55] In the West the bow's upper limb was usually slightly stronger than the lower so that the arrow would be lifted up upon release, but otherwise the bow's components had to be symmetrical.[56]

Successfully fabricating compound reflex bows that would transcend the limitations of a single flexible piece of wood thus came to require a lengthy, meticulous process.[57] A fairly extensive study of traditional techniques still being practiced in the early 1940s found that some three years were required for the wood to be properly seasoned, the other materials prepared, and each stage of the bow's assembly allowed to set and cure correctly in order to avoid inducing fatal stresses or faults.[58] Bowyers therefore had to have numerous bows constantly in process to fulfill even the most basic demand.

Although apparently not a limiting factor in Hsia and Shang warrior culture, the lengthiness of the fabrication time could adversely impact military activities. Even if a multitude of ancient craftsmen could be employed and bows somehow produced in just a year, a shortage of weapons might preclude rashly embarking on expeditionary activities, even render the realm relatively defenseless. When the scope of warfare expanded over subsequent centuries, bows and arrows had to be manufactured rather than built in a craftlike mode, gathered, and stored away in government arsenals well in advance of martial action, just as in medieval England prior to invasions of continental Europe, and officials such as the Ssu Ping (described in the Chou Li), responsible for their disbursement, had to be appointed.

Because a warrior's fate heavily depended on his bow, this manufacturing complexity and the ensuing variation in individual characteristics caused familiar weapons to be highly prized. The Tso Chuan recounts several calamitous incidents caused by strings breaking, but disaster could also result if rain or cold affected the bow's structure, even if with less frequency than in Europe due to the imperviousness of the lacquer. Archaeological and textual evidence suggests that spare bows were sometimes carried, an obvious but cumbersome solution for archers racing across a field with two or three quivers of ten arrows each.

THE ARROW

Although the bow and arrow are inextricably linked, they seem to have advanced in spurts, often jointly but sometimes marked by significant changes in just one or the other. No doubt the dynamic nature of their interrelationship was understood early on, but theoretical contemplation remains scant, primarily preserved within more general discussions of how the bow and archer must be suited to each other. (The late *T'ien-kung K'ai-wu* notes that for a given distance strong archers using powerful bows will be able to penetrate armor, while weaker archers using lower pull bows rely on accuracy for their effect.) Observing that arrows created for the bows of one region invariably fail with bows from another, just a few centuries ago the *Wu-pei Chih* still found it necessary to emphasize that the bow and arrows must be closely matched.[59] As evidence, the compilers noted that large northern arrows used in small southern bows failed to travel more than thirty paces and that southern arrows fitted to northern bows simply snapped.

Recovered artifacts, names passed down in various texts, and discussions in later theoretical manuals indicate that different types of arrows were produced for the divergent purposes of practice, hunting, and warfare as early as the Shang. However, no shafts datable prior to the Spring and Autumn period have survived; therefore, recourse must again be had to the *K'ao-kung Chi*, which, though no doubt somewhat idealized and based on Warring States practices, probably preserves the core of a craft tradition that developed centuries earlier. Fortunately these insights can be supplemented by brief observations preserved in the *T'ien-kung K'ai-wu*, the previously cited report on traditional Chinese bow and arrow making, and knowledge derived from contemporary replication efforts.

The *K'ao-kung Chi* dissects the arrow into four key components: the head (discussed separately below), shaft, feathers, and binding. Shafts might be fashioned from any fairly rigid yet resilient, straight-growing material such as cane, reed, and the smaller bamboos, but the various tree woods, though theoretically possible and found in other cultures and in later eras, generally required too much work to become common. Traditional stories praise administrators who astutely ordered the inhabitants to plant cane because the resulting hedges not only acted as windbreaks, but also furnished materials for crude arrows in times of crisis. Rapidly growing and fairly straight, the hedges thus constituted a latent arsenal.[60]

Replication of traditional Western bows and arrows has shown that numerous woods can be employed, including, but not limited to, splits from pine, ash, and birch and shoots from viburnum, dogwood, hazel, and burning bush. Because arrows vibrate during flight and must flex to clear the body of the bow when released, a combination of hardness and elasticity is necessary. Splits of mature wood tend to retain their linearity and their weight can be closely controlled, but they break more easily than saplings. Conversely, saplings tend to deform, often need repeated straightening, and may vary considerably in density and thus weight despite close growth species of virtually identical dimensions having been chosen.[61]

Even Western replicators emphasize that the raw materials should again be cut in late autumn and winter, because the wood will be easier to dry and less breaking will be experienced.[62] Straightness always being an issue, attempts at heat straightening should commence during drying even before the bark is removed and continue after removal until the wood becomes hard and inflexible, at which point only shaving and sanding can improve the overall dynamics.

Assuming appropriately light candidates can be found, the primary issue thus becomes straightness. In China, particularly in the south, bamboo was the preferred material because, as the *T'ien-kung K'ai-wu* notes, the problems were reduced to selecting the best culms, harvesting in the proper season, and carefully drying to avoid putting a cast into the shaft.[63] Several different varieties were capable of providing the immense quantities required, but one variant, now termed *Pseudosasa japonica*, was so frequently employed that it became known as "arrow" (*chien*) bamboo (*chu*). The best arrows were made from comparatively heavy culms marked by small, closely spaced joints even though they would have to be shaved smooth to reduce air friction.[64]

Although these naturally grown shafts served well for centuries, laminated bamboo arrows laboriously assembled from three long pieces glued together appeared in the Warring States period. Produced by splitting down individual culms and then matching them to achieve the requisite strength and flexibility, this method fully exploited bamboo's desirable qualities while achieving greater solidity in the shaft than naturally hollow bamboo culms, including those less than a centimeter in diameter with closely spaced joints, might provide.

Wooden arrows all have a "down" and an "up" side that results from slight differences in density across the diameter, which may occur because one edge was closer to the heartwood, the other the sapwood, but in bamboo simply derives from differences in relative exposure to sunlight. (Prevailing winds and the additional leafing seen on sunny sides may also cause some inherent curving or lodging in the culms, making it important to choose shafts from the interior and backside of groves.) The *K'ao-kung Chi* thus speaks about determining the *yin* and *yang* sides, no doubt so as to appropriately orient the vanes and head when completing the arrow and thereby avoid undesirable flight tendencies, including the wobble that might occur even in rotating arrows. Shaving and sanding had to be disproportionately applied to the heavier side to improve the cross-sectional balance.

Arrows appropriate for a 160-centimeter Shang bow have been calculated as being 85 to 87 centimeters in length with a nominal (but possibly slightly tapered) diameter of about 1 centimeter. It is well-known that arrows up to a meter long were used in the West for bows of 145 to 165 centimeters in length, and 87-centimeter examples that employed a laminating process have been recovered from a Spring and Autumn tomb in Hebei.[65]

Although arrows, especially short crossbow models, will fly without feathers if the weight is concentrated in the front third, vanes made from feathers impart stability and are normally necessary. Considerable experience and craft are required to choose feathers with the right texture, cut them down to an essential core about 10 to 15 centimeters in length and 2 centimeters in height for the traditional-length Chinese arrow, position them exactly, and secure them permanently with some sort of adhesive augmented by thin lashings of silk. Irrespective of the type of feathers used to produce the vanes, they must all be selected from the same wing.[66]

The best feathers for achieving accurate trajectories were traditionally, though not necessarily in the Shang, thought to come from eagles, hawks, and other birds of prey capable of high soaring flight, found primarily in the north.[67] However, because their numbers were extremely limited and their feathers found mostly in inaccessible locations, comparatively strong geese and duck feathers were commonly employed. Although they were generally shorter, arrows fletched with natural feathers were also used for early crossbows before solid vanes of thin wood and even paper evolved. In both cases the key technical issues were determining the actual length of and appropriate positions for the feathers.

As now conceived, arrows have a center of mass and a center of drag, and the latter must fall significantly behind the former to prevent the arrow from tumbling. This is primarily achieved by having relatively heavy arrowheads and large fletchings, the vanes on well-designed and -constructed arrows also serving to impart the rotation necessary for stable flight. Generally speaking, heavier arrows have more power (momentum), even though lighter ones fly flatter and faster for the same bow. However, lighter arrows are known to be more easily affected by wind, air turbulence, and shooting errors, and air resistance increases with speed.

Under its discussion of the duties of the "officer for arrows," the *K'ao-kung Chi* lists several types of arrows. Even allowing for considerable controversy over textual corruption and the meaning of the various terms, it is clear that early Chinese weapon makers had realized the need for the arrow's center of gravity to fall somewhat forward of middle if effective flight characteristics were to be achieved. According to the text the center of gravity for arrows used by the military and in hunting—*ping shih* and *t'ien shih* respectively—should be 40 percent from the front. (As phrased, the front 40 percent of the shaft with the arrowhead affixed should weigh as much as the rear 60 percent, making it head heavy, as befits military arrows.)

The relative positioning of the arrowhead's points would also be important because early Chinese arrows have only two protruding edges. Being diametrically opposite each other, they would tend to act as wind vanes in flight. Compensation would be provided by the feathers, which, if appropriately positioned, would prevent planing as well as wobble in the tail. However, even though three vanes would be employed in later arrows, the number of fletchings on Shang and earlier arrowheads, possibly only two, remains unknown.

THE ARROWHEAD

Despite significant local variation, the arrowhead's evolution is the most easily charted of the four components because the innumerable recovered artifacts show relatively clear patterns of development. Apart from peregrinations in form, major changes occur in the fabricating materials, the most dramatic being the shift from readily worked natural substances including stone, shell, and bone to cast metals. Bronze first appeared in the late Hsia and early Shang dynasties, but was in turn gradually supplanted by iron in the late Warring States and thereafter.

Just as with the dagger-axe and *yüeh*, in neither case did the new material immediately displace the previous one. Opulent Shang dynasty tombs often contain both bronze and stone arrowheads, whereas stone and bone variants persist in large numbers until near the end of the Western Chou. The oft-repeated claim that changing from stone to metal invariably improved the arrow's lethalness is completely unfounded because many stone arrowheads, especially those fabricated from certain flints and obsidian, possessed razorlike, though brittle, sharpness. Bronze arrowheads did not constitute an improvement but were instead often duller than their mineral predecessors. However, cast metals were synonymous with uniformity, resilience, and ease of production once the laborious work of mining and smelting the ore had been completed.

The arrowhead's history, though still preliminary, has nevertheless identified such a multiplicity of basic types and distinctive substyles as to merit a vast volume rather than a highly abbreviated treatment. However, despite ever more detailed archaeological reports and the specificity of regional variation, the relative advantages of these various types are little understood and thus the reasons for their development, beyond speculating on local fads or the need to adapt to conditions of terrain, objective, and readily available materials, remain opaque. Moreover, whether out of simple conservatism, strong belief in their relative efficacy, the continued convenience of their manufacture, or topographical isolation, certain types continued to be used long after the rest of the realm had changed to more pointed, narrower, or other variations. Nevertheless, pioneering efforts by several scholars, coupled with key archaeological reports and occasional, albeit tentative, overviews, permit the key developments in the slow evolution of styles to be delineated.[68]

Arrowheads can be affixed to a shaft in two basic ways, either by inserting the base into a slot or hole at the top of the shaft or by inserting the shaft into a cavity created in the base of the arrowhead itself. Arrowheads fabricated from natural materials that are worked with difficulty and limited in thickness, including stone and shell, and hammered metal versions were mounted by employing the former method, whereas cast metal variants were also produced with sockets. Arrowheads found in the northwestern part of China, particularly west of the Hsi Ho corridor where socket mounting was commonly employed for spears and dagger-axes, adopted this method more frequently than the east. Whether this was necessary because the region was too cool and dry for bamboo, compelling the use of indigenous woods and solid core cane, or the result of external influences is uncertain.[69]

Subsequent to simply sharpening and probably heat hardening the tip of a shaft just as the *Yi Ching* records, the first arrowheads in China were short yet somewhat elongated stone versions chipped out of flint that date back some 29,000 years.[70] Despite their antiquity, there were few arrowheads until the Neolithic period, when their numbers suddenly began to proliferate because of rising populations in the widely dispersed proto-cultures and the more aggressive hunting activities then being undertaken despite intensive gathering practices. However, by 15,000 BCE triangular arrowheads with lengths of approximately 3 to 4 centimeters had appeared.[71] Their uniformity of size and regularity of surface show that arrow making had emerged as a craft that required specialized skills, significant materials preparation, and a somewhat inviolate sequence of steps. In combination with replica efforts, the ongoing discovery of stone fabrication workshops continues to reveal much about the methodology and complexity of arrowhead and axe production in antiquity.[72]

Bone arrowheads emerged in the Paleolithic and were more commonly employed than stone during the Neolithic because of the comparative ease of carving, filing, and grinding them.[73] (The power of bone arrowheads should not be underestimated, because their superior shape compensates for any inherent softness in the material.)[74] Almost always somewhat longer and more elongated than the stone variants found at individual sites, they also incorporated discernible narrow protuberances or stems at the bottom (*t'ing*) far earlier. However, considerable progress in dynamically contouring the shape of stone arrowheads similarly re-

sulted in noticeable improvements, including
the deliberate thinning of the middle portion
at the bottom of triangular versions to facilitate
inserting into a notch in a wooden shaft (as illustrated) and the tapering of the more elongated or willow leaf style to form a primitive
t'ing for the same purpose.

By the middle Neolithic (roughly 5000 to
3500 BCE) stone arrowheads were being produced by several methods, including chipping
or flaking, percussive hammering, and grinding, as well as a combination of all three with
additional polishing. The fundamental form
remained the relatively stubby, flat, simple tri-
angle with two sharpened edges and a sharp
tip and the rounder, longer, willow-leaf-shaped
variant, as shown to the right and below.[75]

Thereafter, the arrowhead's ongoing evolution resulted in three distinct forms: a basically triangular shape; an essentially rounded
one that tapered into a triangular shape at the
front and narrowed slightly at the base to form
a primitive *t'ing*; and what appears to be a
more fully formed version of the latter, with
a clearly defined *t'ing*.[76] Thereafter, there seems
to have been a gradual progression from a basically rounded but pointed arrowhead to a triangular form with sharply defined spines that
flatten out to the front, two sharpened edges,
and a *t'ing* for mounting. Although this new
design is generally held to have been more
deadly,[77] penetration resistance would have increased as the arrowhead broadened, reducing
the depth of the wound, contrary to the assumption that every evolution increases lethality of design.

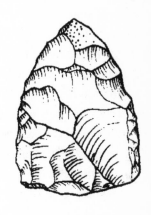

With the advent of the Lungshan culture stone arrowheads suddenly began to multiply, and many excavations yield approximately equal numbers of bone and stone specimens.[78] Nevertheless, variants produced in bone continue to be more elongated, at 5 to 8 centimeters in length, and more precisely formed, marked by a sharp curvature at the bottom and well-defined stems (t'ing).[79] In contrast, although odd shapes frequently appear and a few unusually long specimens have been recovered, the triangular stone embodiments are generally shorter and squatter, at 4 to 5 centimeters, and have somewhat raised spines, but generally lack visible t'ing until late in the period. However, increasingly obvious, functional t'ing evolved in the elongated or willow leaf form, as attested by 8-centimeter heads commonly being found, some including clearly defined t'ing of 3 centimeters, even as the arrowhead tended to lengthen over time.[80]

The process of gradually lengthening the increasingly well-differentiated t'ing is also visible in the peripheral cultures that flourished in the late Neolithic, particularly the Liang-chu culture located in the southeast, long known for the large number of arrowheads recovered from its sites, as might be expected from purported Hsia enemies.[81] However, the best examples tend to be preserved in the northern region, especially Liaoning and Inner Mongolia, where each excavation augments the assemblage of arrowheads that show the t'ing emerging as the preferred mounting method.[82]

Whereas the laborious process of chipping and grinding stone blanks essentially limited stone head shapes to a relatively flat form with two cutting edges and a remnant raised spine, bronze casting allowed an almost unimaginable freedom coupled with virtual perfection in symmetry and balance. Moreover, in contrast to the brittleness of stone arrowheads fabricated out of hard minerals such as flint, which might be extremely sharp but also had to be resharpened after use, bronze arrowheads were virtually indestructible and largely retained their original edges after impact. Nevertheless, stone and bone arrowheads would predominate well into the Shang,

The earliest metal arrowheads yet discovered in China have been recovered at Erh-li-t'ou and Tung-hsia-feng, evidence that the late Hsia had already begun using bronze variants, although in extremely low numbers. (Because the numbers attributable to lower Erh-li-kang sites are equally limited, claims that the Shang enjoyed a substantial technological advantage from their bronze arrowheads obviously lack substantiation.)[83] Highly similar to original bone types, these early bronze arrowheads are characterized by a rounded body with long but undifferentiated *t'ing*, flattened projections, and two clearly defined backward-facing points that would increase the extraction difficulty if they fully penetrate and cause a more disabling wound.

Thereafter, there is a continuous progression from this early bronze form through those excavated at Cheng-chou and then Anyang, the so-called Shang style, primarily consisting of a flattened shape marked by a raised spine and increasingly defined *t'ing*. Yen-shih / Cheng-chou era specimens fall into two categories, a shorter variant some 5 centimeters long that has a rounded *t'ing* and raised spine, and a somewhat longer version with extended rearward projections, highly sharpened points, and a reduced but still obvious *t'ing*. No doubt because multiple-cavity casting was already being employed for bronze arrowheads at Cheng-chou, great uniformity is evident within subgroups.

Although arrowheads rapidly proliferated during Shang rule from Anyang and their sub-styles could be multiplied almost endlessly, a limited number of functionally differentiated forms prevailed in the common mediums of stone and bronze.[84] The short triangular style predominated in the former,[85] but elongated or leaf-shaped variants continued to appear, especially in southeastern peripheral areas. Yin-hsü bronze arrowheads tend to be characterized by a fundamentally triangular shape marked by varying degrees of stubbiness and elongation, a clearly defined *t'ing* of varying

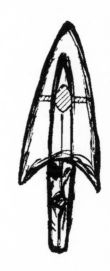

length, and two sharpened blades.[86] Most bronze versions also embody downward-projecting blade tips that could become extreme.[87]

Other aspects that frequently draw attention are the degree to which the core of the arrowhead remained essentially circular, a characteristic also visible on some spearheads, or displayed the more common rhomboidal cross-section. Shang bronze arrowheads invariably have just two blades, even though exaggerating and sharpening the raised rhomboidal portion could have easily produced an arrowhead with four edged surfaces. (In later eras the entire body would be redesigned for three blades rather than the four inherently present, no doubt because matching them with the three vanes that came to be commonly employed ensured better flight characteristics.)

An additional feature that appears in mid- to late Yin-hsü bronze arrowheads is a *kuan* or slight protrusion that emanates from the base just above the *t'ing* (as shown at left), the latter then being reduced in diameter and adumbrated to a length of just a centimeter or two.[88]

However, despite the visibly staggered profile, the design is functionally indistinguishable from its predecessor because only a minor shift has occurred: when the *t'ing* is inserted into the shaft, the top of the shaft now butts against the base of the *kuan* rather than the body of the arrowhead itself. Moreover, the long elongation of the *kuan* and *t'ing* that would be deliberately employed in later periods to add weight at the front of the shaft had not yet materialized. However, as attested by bundles recovered at Yin-hsü, military arrowheads averaged 7 to 8 centimeters in length and weighed from 12 to 14 grams.[89]

Finally, a few long, slightly bulbous arrowheads that completely lack discernible vanes or points are among the anomalous arrowheads excavated from Shang sites. Variously identified as intended for practice or as arrowheads for shooting birds, they seem to be precursors of the many more recovered in the Western Chou and thereafter. Many of the latter have small holes below the tip where some sort of line was probably inserted, evidence that they were employed to shoot flying birds.[90] However, as in the West, the others may have been intended to stun or kill small animals, particularly if the skin or fur was to be undamaged, or even for mock combat, an activity that would have entailed considerable risk for the participants even if their leather armor were directly struck. However, there is no evidence for the existence of this antique equivalent of modern paintball, nor have any scholars suggested it as a possibility.

Beginning with the Western Chou, arrowheads would generally expand in profile, coming to be marked by wider-angled projections and more sharply defined rearward points. Although bronze would finally dominate, stone and bone arrowheads continued to be employed, possibly in an ancillary role such as hunting rather than warfare, though the successful slaying of an animal destined for the dinner table would hardly have been an unimportant event. This tendency to greater width then reversed in the Spring and Autumn with more triangular arrowheads, whose rearward barbs now curved inward and gradually became smaller, in both profile and actual size. This may have been because warfare had become far more extensive, requiring certainty in killing, or because developments in armor necessitated a reduced profile to ensure greater penetration.

20.

THE
CHARIOT
IN CHINA

◆·◆·◆━━━━━━━━━━━━━━━━━━━━━━━━◆·◆

L EGENDS ABOUT THE chariot's inception vary, the most common being
that the Yellow Emperor invented the chariot but that Yü the Great
was the first to employ it. Because he was similarly said to have invented
the cart, the Yellow Emperor has also been known as Hsüan Yüan, a
name whose characters refer to two types of draught pole but in combination designate chariots in general.[1] However, the chariot's creation
is also attributed to Hsi Chung, thought to have been one of the Yellow
Emperor's ministers or to have lived in Yü's era, as well as to Hsiang-
t'u, another of his officials or perhaps the grandson of Ch'i, the Hsia's
first hereditary ruler. Yet another variant combines elements of these
tales by asserting that the Yellow Emperor fabricated the first vehicle
(*ch'e*), Emperor Shao Hao yoked a pair of oxen to it for motive power,
and Hsi Chung widened it and substituted horses, creating a device with
true battlefield potential.[2]

Late writings credit Hsi Chung with the crucial achievement of
bending wood to make curved wheels, harnessing horses and oxen, and
(perhaps reflecting the emergence of bronze tools) being an expert
craftsman.[3] Kuan Chung therefore analogized the workings of the ideal
government by citing his precision in woodworking: "Hsi Chung's skill
lay in chopping and planning.[4] When Hsi Chung fabricated chariot components, the square, round, curved, and straight all accorded with the
compass and lines. Thus the frame and revolving components matched
each other, and when employed it was sturdy and advantageous."[5] How-

ever, another section of the *Kuan-tzu* suggests his skill was more innate than measured, that his spirit naturally resonated with tools and implements such as the axe.[6]

Excavations conducted during the highly troubled middle part of the twentieth century revealed the existence of chariots dating to the late Shang, immediately prompting a few tentative formulations of its history. Subsequent finds have altered the general conclusions but slightly, their chief impact being to augment detailed knowledge of the chariot's construction, moving parts, and harnessing system and reveal a continued emphasis over the centuries on making the chariot lighter but sturdier, capable of withstanding actual field use.[7]

As reconstructed, even the earliest Shang chariots recovered from the tombs and burial pits at Anyang are already complex, well-crafted units whose construction combined lightness and strength. Fabricated from wood, rattan, bamboo, bronze, and leather, they consisted of an essentially rectangular compartment mounted directly over a transverse axle. Powered by two horses connected to a single, centrally mounted draught pole, the chariot was roughly ten feet long from the rear of the chariot compartment to the tip of the shaft. Despite employing fairly compact animals, when the horses' heads; rearward profile of the large chariot wheels; and expanse of the axles, whose hub, fittings, and other projections extended some ten feet or more, are included, even without the blades that were attached in the Warring States period these early chariots occupied the considerable area of approximately ten by eleven feet.

Despite their extensive bronze workshops and surpassing resources, chariots were not unique to the Shang. In addition to the Chou, who similarly devoted considerable effort to building and employing chariots, other peripheral peoples fielded them for warfare purposes, albeit in limited but still comparatively significant numbers.[8] Oracular inscriptions attest to the capture of two chariots late in the Shang,[9] and ongoing archaeological explorations on the periphery have revealed the presence of well-constructed chariots in Shandong that apparently belonged to a smaller regional state and are identical to late Shang models.[10]

The Chou, whose location certainly exposed them to fully formed Western chariots equally early and may have provided the conduit for the chariot's introduction into the Shang, integrated them to a more significant degree. Three hundred were employed as a distinct component force at the Battle of Mu-yeh, and the many hundreds captured from Shang allies in postconquest campaigns indicate that the chariot's adoption had become widespread. Thereafter, despite the expense and length of the manufacturing process, the number found in the Chou and its increasingly independent fiefdoms rapidly increased.

The chariot soon became a symbol of power, and being the army's most visible component, a means for assessing the strength of states. Many restive rulers reputedly could field a thousand vehicles by the end of the Spring and Autumn period, and although massive infantry contingents came to dominate Warring States warfare, the most powerful political entities—Ch'i, Ch'in, and Ch'u—reputedly maintained an astonishing (but still unproven) 10,000 chariots each.[11] Although their organization, training, quality, degree of successful integration, and operational tactics no doubt varied from state to state, they clearly were considered not just effective, but vital.

A sense of the awesomeness invoked by such numbers can be seen in a late Spring and Autumn incident in which a high official from Chin sought to coerce a reluctant state into participating in a conclave that it wanted to convene by saying, "My ruler has 4,000 armored chariots assembled here. Even if he were to employ them contrary to the Tao, they would certainly have to be feared. If he employs them in accord with the Tao, who can be his enemy?"[12] Nevertheless, increasingly supplanted by the rapidly escalating infantry and eventually displaced by the development of effective cavalry contingents in the Han, chariots would eventually resume their original command and transport functions, being employed primarily for distant steppe campaigns.

Even without additional decoration or embellishment with royal insignia, Shang chariots must have been imposing, highly visible symbols of rank and power. Vestiges of elaborate decoration and bronze attachments indicate that the human tendency to display evidence of personal wealth was not absent in the Shang. However, perhaps because of their larger numbers and greater utilization in warfare, despite still being prestigious vehicles and even granted as a sign of imperial patronage, Chou chariots seem to have become more pragmatic. Functional combat versions and less rugged, more stylized vehicles intended for normal transport purposes quickly appeared, initiating a trend that would see highly specialized chariots and other wheeled vehicles being created in the Spring and Autumn and Warring States periods for siege warfare, dedicated assault tasks, and conspicuous display.

Whatever the chariot's prehistory in China, more suitable precursors for the heavy, wheeled platforms that would be required for Warring States overlook towers, battering rams, shielded assault vehicles, movable ladders, multiple-arrow crossbow chariots, and portable shields may have existed in the Shang.[13] In contrast to whatever sort of primitive cart may have been used at Erh-li-t'ou and Yen-shih, it was probably a simple oxen-powered wagon that could be employed for transporting materials and provisions to the battlefront and forwarding tribute from outlying areas.[14] Although none dating to the Shang or Western Chou have yet been found, there are numerous references to provisions wagons being captured in later periods in quantities up to a thousand.[15]

DESIGN AND SPECIFICATIONS

Despite being basically standardized, chariot wheel and compartment dimensions in China differed considerably, not just across time or within an era, but also in the same tomb or chariot pit.[16] (Accordingly, one of Ch'in Shih-huang's accomplishments was mandating a standard gauge in order to remedy the difficulties created by localized preferences, thereby facilitating transport and minimizing the adverse effects on the roadbeds.)[17] The degree of variation, being far greater than might have been expected even for a craft product constructed without blueprints, is surprising. Whether the craftsmen worked from models, had dimensional sketches, or simply constructed the chariots from preexisting examples or experience, just as coopers, wheel makers, and boat builders have done for centuries, is unknown. However, given the chariot's complexity and the preservation of several (often misleading) proportions in the *K'ao-kung Chi*, the existence of basic diagrams or illustrative models seems likely.

Fabricating a chariot required several highly particularized skills, a variety of natural resources, the observation of seasonal constraints,[18] and ultimately the manufacture of hundreds of discrete components that had to be made closely compatible in order to be assembled. Molds had to be made and bronze fixtures cast; glue prepared; leather tanned, treated, and cut to size; wood selected, seasoned, shaped, bent, and dimensioned; and everything assembled by slotting and boring, shaping, lashing, gluing, mortising, and force fitting. It was such a time-consuming process that the *Yi Ching* employed it as an analogy for impossibility by speaking of "bending the wood for a wheel in the morning and wanting to ride in the chariot at night" in its hexagram "Ch'i Fa."

Although the chariot's basic form would remain unchanged, over the centuries there would be a tendency toward stronger, heavier, and swifter vehicles. The various components continued to evolve, reinforcements against breakage and wear were developed, stability was improved, and measures to reduce vibration were implemented. Unfortunately, despite being important for the history of technology and entailing significant implications for military capabilities, chariots require a study in themselves and necessarily fall outside the compass of this book. However, certain aspects are worth considering to provide a basis for pondering the chariot's employment in battle and its possible effectiveness.

From the chariots' initial appearance in China the wheels were already remarkably large in comparison with an average 75 to 90 centimeters for Near Eastern chariots of comparable date. At roughly 120 centimeters to 145 centimeters in diameter, they were approximately the height of the steppe horses of 133 to 144 centimeters being employed for the chariots in the late Shang.[19] Thereafter, they gradually increased slightly over the centuries until late in the Warring States, when they tended to again diminish in diameter. Slightly conical in shape, contrary to Western practices they would generally be "dished" toward the inside rather than outside.

Perhaps because of their considerable size, the wheels incorporated numerous spokes, generally sixteen to twenty in the Shang, with the majority of vehicles recovered to date having eighteen. (One chariot reportedly employed a surprising twenty-six spokes, but it may either be an anomaly or the estimation may simply be incorrect, as frequently suggested.)[20] However, as the wheel maker's art continued to be perfected through the Western and Eastern Chou periods, the spoke count gradually increased, occasionally reaching an astonishing forty. Round to somewhat oval in shape, the wooden spokes on Shang vehicles averaged 3.0 to 4.5 centimeters in diameter, but are somewhat tapered where they were inserted some 2 to 4 centimeters into the hub (nave) and felloes.[21] Even thinner spokes would subsequently be employed, especially for lightweight chariots and carriages intended solely for ordinary transport.[22] Nevertheless, they must have been highly reliable, because the Chou dynasty Yi Ching employed them in an analogy for strength in the hexagram "Ta Chuang."

Being made from wood, without any bronze reinforcements on the exterior or a metal sleeve on the interior, Shang hubs (or naves) had to be relatively long and thick to prevent breaking apart in use. (A well-known Chinese observation asserts that short hubs are advantageous but long ones secure.) Anywhere from 20 to 35 centimeters in total length and about 18 to 28 centimeters thick, the outer portions tapered down somewhat, resulting in a center bulge and an overall profile said to resemble the beads of a traditional wooden abacus. (In contrast, the apparent precursors at Sintashta and Lake Sevan are much thicker, some 40 to 45 centimeters, obviating the need for long spokes.) Decorative metal sleeves would eventually be added to overlay the tapered portions, constraining the wood and reducing the tendency to break apart from cracking, but none dating to the Shang have yet been discovered despite the use of bronze end caps on the axle.

Because the axle was directly inserted into the unlined nave, wood necessarily rotated upon wood, a less than desirable material match.

(Claims that a leather lining may have been employed in the Shang lack substantiation.) Constant lubrication with animal fats or vegetable oils would have been required to reduce the friction being produced, even if the components closely conformed to each other. Despite the lightness of the chariots, the sand and fine plant matter kicked up as they sped down the primitive roads and open fields must have also increased the problem considerably. Not surprisingly, menials responsible for lubricating the wheels are thus mentioned more than once in the *Tso Chuan*,[23] and bronze axle covers and interior rings were being employed on either side of the hub in the Shang that would have held the wheel in place (albeit at the cost of additional friction) and reduced the possibility of foreign material migrating onto the axle at the crucial junction point.

The wheel rim (felloes) consisted of two or three pieces of hard yet relatively pliable wood such as elm, bent after softening in hot water or steam into an appropriate arc, mortised or scarved together, and secured with bronze clips.[24] The felloes were generally thicker than wider, often about 10.0 by 7.5 centimeters respectively, but with considerable variation and even some square versions of 8 by 8 centimeters.[25] Somewhat rounded or tapered toward the inner rim, the wheel presented a flat external profile to the road. No evidence of any sort of metal or leather tire or other external improvement such as studs has yet been discovered.

The wheels rotated on fixed axles rather than being fixed to a rotating axle. Although wheel gauge varied from 215 to 240 centimeters in the Shang, with most being about 215, 225, or 240, the axle averaged slightly over 3 meters in total length with the bronze end caps included.[26] (Only three of the eighteen Shang chariots included in recent tabulations were considerably smaller, at 190, 225, and 235 centimeters; one is reported at 274 centimeters; while the remainder show 294 to about 312 centimeters, the majority of which exceed the 300 mark. This has prompted the thought that the narrower-gauge chariots may have been intended for transporting heavy items,[27] but they could also represent personal-use transport vehicles or simply an earlier or different version of the standard model.) The axle itself, invariably a single round or oval pole rather than two discrete segments, varied considerably in thickness, ranging from 8 to as high as 15 centimeters in diameter, though the majority are 8, 10, or 12 centimeters before tapering at the ends where the axle cap was mounted. (Diameters of 8–12 centimeters must have represented a pragmatic compromise between the lower friction of narrow axles and the higher load-bearing capacity and durability of larger diameters.)

Based on numerous Warring States references to three-man crews, for 2,000 years it has been deemed an unquestioned fact that three warriors standing in a minimally triangular formation manned the earliest chariots.[28] Even though a majority of the burials excavated so far contain only one or two accompanying bodies, every discovery of sites where three bodies are interred in close proximity to a chariot has therefore been seen as confirming this assumption.[29] However, given the compartment's compact size, irrespective of later staffing practices, Shang chariots may well have initially carried only two warriors, in common with Western practices.

Generally rectangular with various degrees of rounding at the front, somewhat greater expanse at the back, and slight distortions in one or another corner, these ancient compartments always emphasized width over depth. Shang chariots averaged about 138 by 96 centimeters,[30] with the vast majority being roughly 130 to 145 centimeters wide, but chariots as small as 94 by 75 (at Ta-ssu-k'ung-ts'un) and as large as 150 by 90 and 170 by 110 (at Kuo-chia-chuang) have been recovered.

In the Western Chou and thereafter the compartment size, particularly the depth, would increase somewhat—perhaps to better accommodate the fighters and allow them to wield longer weapons—but remained considerably smaller than the maximum allowed by the separation of the wheels, including the inner portion of the two hubs. However, irrespective of size, until late in the Warring States the only opening for mounting the chariot was a narrow gap of 25 to 40 centimeters at the back, much in contrast to fully open Western-style backs.[31]

The minimalist frame was constructed from narrow, generally 4- to 6-centimeter poles of rattan, hard cane, wood, or even bamboo, with those framing the bottom being slightly thicker than the 8 to 12 posts employed to erect the walls.[32] The rod forming the top edge of the side wall generally ran horizontally at a maximum height of 45 to 50 centimeters (and thus below the rim of the wheel), though two variants as low as 22 and 30 centimeters have been excavated.[33] (The *K'ao-kung Chi* diagram showing high side walls is simply inapplicable to Shang or later war chariots.) However, the walls were not always uniformly high all around the compartment, because several recovered so far show a tendency to be slightly lower in the front and higher to the rear. For example, at Mei-yüan-chuang the front of the southern chariot in M40 was 39 centimeters high, but the back rail about 50, while the northern chariot is about 30 centimeters in the front and about 40 in the rear.[34] A crossbar or rail that protruded above the side walls to provide a handhold was occasionally added near the front of the compartment.[35]

Fabricated from interlaced material, the compartment's walls would have provided a limited amount of protection against arrows and may even have deterred spear thrusts, depending on the thickness and hardness of the reeds or bamboo. Despite leather being employed for body armor in the Shang and subsequently adopted for chariots in the Spring and Autumn when four horses were commonly harnessed, there is no indication it ever augmented the walls of Shang chariots.[36] (The horses were similarly unprotected by any form of leather armor.) Nevertheless, many chariots presumably intended for transport or martial display were apparently lacquered in red and black and had various marks of insignia or bronze plaques affixed.[37] However, assuming that tombs with accompanying weapons demark a military chariot, little differentiation is seen in the Shang other than such embellishments and perhaps being slightly smaller.[38]

Apart from rare exceptions such as the chariot in M41 at Mei-yüan-chuang, the chariot compartment was normally centered or symmetrically placed over the axle, thereby minimizing the downward load borne by the horses' necks.[39] The frame initially rested directly on the axle and the shaft at four contact points; a somewhat bowed wooden cushioning mount called a "crouching rabbit" (*fu t'u*) may have come to be employed at the end of the era.[40] Leather lashings further secured the compartment frame to the axle and shaft, with another mount sometimes having been employed just over the joint of the shaft and axle.

The shaft always overlay the axle, with both being slotted or in-cut to create a tight joint and present a relatively flat profile across the axle's entire length under the compartment frame.[41] It was originally thought that Shang dynasty chariots employed relatively straight shafts, but advances in reconstructive technology have revealed that they curved upward at various points forward of the chariot compartment, a necessity for the chariot compartment to be level. (Rather than a gradual bend, the curvature seems to have become more radical and pronounced over time, with some chariots simply having a nearly perpendicular upturn at the very end for affixing the crossbar.)

Typical dimensions for two shafts found at Mei-yüan-chuang that date to the third or fourth period of Yin-hsü (and whose chariots show many features previously identified as Chou inceptions) are 280 centimeters for the actual length but only 250 centimeters from the back end to the front tip as measured horizontally, and 265 and 227 centimeters, marked by a 108-degree angle,[42] but only 268 and 261 at Kuo-chia-chuang. The final portion of the shaft at the front also tended to include another curve that placed the tip above the crossbar and well displayed the decorative bronze caps that were employed to embellish both ends. The round to somewhat oval shafts average 15 centimeters in thickness, but taper under the caps.

The two horses were harnessed to the chariot by means of bronze wishbone-shaped yokes suspended from a crosspiece. Rather than directly attached to the shaft, the crosspiece was apparently connected to it with leather straps that could be adjusted for the height of the horses, thereby ensuring that the chariot would not be tilted upward while coincidentally reducing the stiffness of maneuver in lateral and turning motion.[43] Tomb vestiges indicate intervals above the crosspiece from 20 to as high as 40 centimeters, while the crosspieces themselves averaged 110 to 120 centimeters for straight versions and 220 for significantly curved ones, with diameters of 7 to 10 centimeters at the middle but somewhat tapered at the ends, where decorative bronze caps were again attached.

Although claims have long been made that any form of harness that extended down from the upward curved shaft—the so-called throat and girth harness—would have constricted the horses' necks and proved counterproductive under load, the actual method remains uncertain.[44] However, chariots were obviously used with regularity, indicating that this was either not a problem or somehow surmounted. Prior to the inception of Warring States improvements that resolved any residual problems, the bronze yokes suspended from the crossbar would have allowed the use of some sort of chest piece that could have been held in place by lateral lines from the chariot or some sort of girth strapping that transferred the load away from the horses' throats, the latter being made possible by the height of the wheels coupled with the compactness of the horses.[45]

As would be expected for an imported system, even though improvements would continue into the Spring and Autumn period, the bridle, bit, forehead and nose straps, cheek pieces, and reins—the very basis of control—were all essentially complete and functional in the Shang.[46] In the West a variety of materials were employed for the bit, including wood, leather, shell, and metal, and even though the metal bit reportedly did not attain mature form or proliferate until the Spring and Autumn, leather bits were already being displaced by bronze versions by the end of the Shang.[47] For two horses a simple rein system was adequate, but adding an outer pair not directly yoked to the shaft increased the complexity and resulted in the driver holding six lines, a task somewhat facilitated by employing a bronze tube and a so-called bow-shaped bronze fitting affixed to the front of the chariot. Bronze rings and crosspieces were also employed wherever the head ropes, bridle, and reins interconnected, as well as for the harness joints. Various decorative bronze pieces were added to the leather surfaces and a sort of bronze disk sometimes fixed so it would lie just on a horse's forehead.

ORIGINS

Chariots often seem simple, obvious, and mundane in an age accustomed to the complexity of electronic systems and innumerable vehicles. However, wheeled transport required centuries to evolve from the sledges that were generally employed in the West from 7,000 to 4,000 BCE to shift limited amounts of materials in bulk and continued to be used in isolated locations well into the third millennium BCE.[48] The first wagons, generally pulled by slow but powerful oxen, were lumbering, four-wheeled behemoths assembled from rough-hewn logs that relied on solid wheels cut from thick tree trunks. The chariot, a two-wheeled vehicle intended solely for war, eventually evolved but only became truly formidable with the invention of the spoked wheel, fabrication of lightweight compartments, discovery of lubrication methods, mastery of the horse for swiftness, and development of bronze tools capable of more precise woodworking.[49] When coupled with archery, lightweight chariots provided the ultimate military weapon for clashes throughout the Near East starting about 2000 BCE and continuing to 1000 BCE, when the infantry suddenly gained the ascendancy, but China began massively employing them.[50]

In recent decades the origin of the chariots that suddenly appeared in China during King Wu Ting's reign has been the subject of acrimonious debate between those who proclaim the Chinese chariot to be the fruition of purely indigenous developments and opponents who stress the essential continuity of imported design.[51] Previously, the most commonly accepted scenario envisioned the chariot as having been introduced around the fourteenth or thirteenth century BCE through Central Asia, having originated in the Near East. Furthermore, it was believed that the transmission route had been quickly severed because Chinese vehicles displayed unique characteristics but failed to incorporate subsequent Western developments.[52] However, a radically different historical sequence based on new findings and changing interpretations of archaeological data has recently been proposed.[53]

Without getting mired in thorny arguments over the horse's origins or the history of wheeled vehicles, two topics closely entangled with theories about the inception and diffusion of proto-Indo-European, certain discoveries pertinent to the nature of Western precursors should be briefly noted. The crucial developments were the displacement of heavy four-wheeled wagons by lightweight chariots and the domestication of the horse, understood here as the ability to breed the animals in a relatively controlled environment coupled with the knowledge necessary to control them in harness or under mount.

According to one well-argued view, the horse that eventually evolved from among several "horselike" animals primarily emerged in the Pontic-Caspian steppe between the Caucasus and Ural mountains by about 4800 BCE, where it was hunted as a food source.[54] At some stage, possibly 4200–4000 BCE but certainly 3700–3500 as attested by bits found at Botai in north Kazakhstan, the ability to mount and ride a horse, highly useful for controlling even the small numbers being raised for food, reportedly developed, immediately making possible not just previously unimaginable rapid movement but also journeys over a far greater range.[55] Although cavalry contingents would not appear until about 1500 BCE and perhaps only developed as an effective force with the acquisition of the compound reflex bow around 1000 BCE, mounted raiding commenced, dramatically changing the nature of conflict.[56]

The first wheeled vehicles, replacements for the sledges that initially facilitated the transport of moderate loads over limited distances, reportedly appeared sometime between 4000 and 3500 BCE. Wherever they were invented, early forms of four-wheeled wagons dating to 3400–3000 BCE have been found in Mesopotamia, Poland, Germany, and Hungary. Thereafter they rapidly spread in every direction, including out onto the Ukrainian and Russian steppe lands, about 3300–3100 BCE.[57] Meanwhile improvements that occurred in spurts and had highly localized manifestations over the millennium between 3500 and 2500 BCE emphasized making the vehicle more maneuverable and lighter, such as by reducing the weight of the body and adopting tripartite wheels. The next significant stage, their evolution into chariots around 2100 to 1800 BCE, seems to have occurred not in the Near East but in the northern steppes east of the Ural Mountains in settlements identified with the Sintashta culture.

Apart from being heavily fortified with walls and ditches, the approximately twenty sites already found between the upper Ural and upper Tobol rivers were heavily oriented to metallurgical production, primarily from arsenical bronze. Whether they flourished because their casting technology made the aggressive exploitation of weapons possible or simply found themselves immersed in a warlike environment and responded by fabricating weapons, they also developed the first real chariots. Although relatively narrow, with a wheel gauge of around 1.2 meters and axle lengths of 2 meters, being powered by two horses they could easily carry one or two riders into battle.[58] In comparison with chariots from the Near East and Mycenae dating to about 1850 BCE, whose wheels average 75 to 100 centimeters, those from Sintashta are noticeably larger, reportedly 90 to 120 centimeters in diameter.[59] They also had far more spokes, generally 8 to 12 rather than 4 to 8, curved shafts, compartment-centered axles, and cheek pieces.

Somewhat farther to the east the Petrova culture, which flourished from 1900 to 1750 BCE, directly inherited Sintashta's defining aspects, including their focus on metallurgical production (but in tin bronze alloys), use of fortifications, and exploitation of the chariot, prompting scholars to speak of a combined Sintashta-Petrova culture. From here the chariot could have been transmitted as far as the Altai Mountains through the Srobnaya and Andronovo cultures, the latter similarly a tin bronze producer, in the century between 1900 and 1800 BCE. Thereafter it appears to have been another six centuries before the chariot was adopted by the Shang, despite the probable introduction of horses into the precursor Ch'i-chia and Ssu-pa cultures in northwest China sometime between 2000 and 1600 BCE.[60] Meanwhile, the chariot had been spreading through Central Asia (including the area around Lake Sevan) into the Near East and down to India, where it eventually proliferated and assumed a vital role in the indigenous civilizations.

Military chariots that can be employed at speed represent the fruition of centuries of innovation, experimentation, and improvement, not just in materials and structure, but also in the domestication, breeding, training, harnessing, and controlling of horses with bridles, bits, and cheek pins. Advances in knowledge, technology, metallurgy, and craft skills made it possible, but the chariot's successful exploitation as a dynamic system equally depended on a continuous interaction between the driver and the horses. Notwithstanding the irresolvable debate about the nature of technological discovery,[61] mere possession of a physical chariot, however acquired, without the integral manufacturing and equine knowledge would never have constituted a sufficient basis for it to suddenly flourish as a military weapon in China.

Fundamental support for diffusion rather than an indigenous origin for Chinese chariots is also seen in the complete lack of evidence for precursors such as carts or any form of oxen- or horse-pulled wagon despite both oxen and horses already having been domesticated. Vestiges of tracks apparently made by wheeled vehicles repeatedly traveling over the ground have recently been discovered at Erh-li-t'ou, a site where significant roads are visible in the royal quarters. Spaced about 1.0 to 1.2 meters apart and dating to Erh-li-t'ou's second period, roughly 200 years earlier than Yen-shih, they are about 20 to 32 centimeters in width and 2 to 14 centimeters deep.[62] Wheel ruts have also been found at Yen-shih that average 1.2 meters wide and are thus similar in gauge to those at Erh-li-t'ou, but are only half the width of the earliest chariots recovered at Anyang of 2.2 to 2.4 meters.[63] Rather than being carved out by sledges, the ruts were probably made by some sort of small two- or four-wheeled cart that was employed to transport dirt, stone, and other building materials.[64]

Despite traditional claims that sheep and other small animals were hitched to ancient Chinese carts, in the absence of significant horse remains at Erh-li-t'ou and Yen-shih it is more likely that humans, whether pulling or pushing, rather than animals provided the power for these vehicles.[65] Assertions that these wheel ruts obviate any need for technological importation, though intriguing, lack substantiation and thus do not significantly undermine transmission theories, particularly as the carts themselves may have originated in the nearby steppe cultures, where vehicles of similar gauge were common.

More important, many features of Chinese chariots lose their formerly distinctive character when compared with Sintashta-Petrova precursors rather than Near Eastern versions from Mesopotamia, Egypt, or Mycenae.[66] Without denying the essential similarity of chariot construction worldwide, it had been thought that Shang embodiments had far larger wheels with more numerous spokes and a conical shape; that the chariot box was larger, capable of accommodating three men standing in triangular formation, and rectangular rather than rounded; and that it was mounted directly over the axle, in contrast to Western preference for the wheels to be placed increasingly to the rear.[67]

However, most of these definitive features are present, whether fully or incipiently, in the models already noted as having been recovered from Sintashta-Petrova sites that date to at least 700 years earlier. In particular, Sintashta-Petrova chariots employed two horses aside a curved shaft and had multiple spokes, larger wheels, a center-mounted chariot compartment, and a complete Shang-style bridle package, including the bit and cheek pieces essential for controlling the horses. Chariots found near Lake Sevan that date to about 1600 BCE and thus presumably represent somewhat more developed versions provide an even closer match to the Shang manifestations with, for example, twenty-six spokes, and may either have been transmitted back from the Shang or reflect a natural progression of developments. Egyptian versions display several other similarities, such as interwoven walls and leather thong floors, but because they represent localized, nontransmitted developments, they are essentially irrelevant.

21.

The Horse in China

◆◆◆━━━━━━━━━━━━━━━━━━━━━━━━━◆◆◆

DESPITE HAVING HAD A WIDE distribution across the contiguous steppe region and Inner Mongolia for many centuries, horses seem not to have been raised in China until the late Neolithic. Although, as already noted, there is some disagreement about whether these and later Shang horses were larger than those now populating China's northwest region, skeletons suggest a compact animal about 130 to 140 centimeters high, one presumably derived from Przewalski's horse and imported from the steppe rather than native to China. Based on trace archaeological finds at Pan-p'o through Erh-li-t'ou and several early Shang sites, horse breeding gradually moved down into the central Luo River region.[1] Whether these horses were ever employed as pack animals or as draught animals to power carts rather than just eaten before the Shang adapted them to the chariot is unknown, but it is unlikely despite the discovery of wheel traces at Erh-li-t'ou. However, they still matched steppe horses in size before selective breeding increased their overall dimensions and thus their tractive and transport capability.

Even though late historical writings such as the *Shang Shu* assert that the Shang fielded seventy chariots when vanquishing the Hsia, and horses and chariots were virtually inseparable in early China, no chariots or intact skeletons have ever been found at any site predating Wu Ting's reign at Anyang. Moreover, despite vociferous assertions by traditional scholars that riding and hunting on horseback commenced in the Shang following a lengthy period of indigenous development, nothing suggests that horses were being ridden until the Spring and Autumn or even Warring States period, when the cavalry was deliberately created to thwart steppe riders.[2]

Horses suddenly assumed an integral role in martial and royal life in King Wu Ting's reign. They not only powered the small number of chariots employed in hunting and military actions, but also functioned as symbols of prestige and authority. Numerous Shang dynasty inscriptions inquire about the general auspiciousness of horses being sent in as tribute, their suitability for sacrifice, and their prospects for martial employment including hunting. They were designated by the colors white, dark, red, bronze, yellow, and gray and their attributes given special names borrowed from other animals, including deer and wild boar. Whether they would survive or perish (because they might be attacked by tigers or slain on the battlefield) was also a matter of frequent concern.[3] At least one official position within the proto-bureaucracy, the Ma Hsiao-ch'en (Minor Servitor for Horses), was created to oversee equine matters. The raising of horses and improvement of the stock were matters of focal concern,[4] and a number of other menial functionaries were tasked with various duties related to stabling and keeping horses.

Some sixty Shang burial pits containing horses or horses and chariots interred together have been discovered. Nevertheless, horses were generally too valuable to sacrifice except to the highest Shang ancestors or to bury in pairs with a single chariot to honor someone of extremely exalted rank or otherwise distinguished by martial achievement, a practice that would not be abandoned until well into imperial times.[5] They also numbered among the superlative gifts that might be forwarded on important state occasions, provided as tribute, employed as bribes, given to ensure loyalty,[6] or offered as ransom. For example, in the Spring and Autumn period a captured Sung commander was ransomed for 100 chariots and 400 horses;[7] the minor state of Lai managed to halt an invasion by bribing the enemy's chief eunuch with 100 horses and a similar number of oxen;[8] and late in the Shang the Chou secured King Wen's release from prison with a combination of beautiful women, jewels, and horses.

The great value attached to superlative horses is further illustrated by a famous Spring and Autumn incident that provided the basis for the well-known *ch'eng-yü* (aphorism or formulaic phrase) "Having a nearby objective yet making it appear distant," also known as "the Marquis of Chin borrowed a passage through Yü."[9] Subsequently included among the *Thirty-six Stratagems*, it consisted of tempting Yü's ruler with some outstanding horses and a famous jade when requesting permission for Chin's armies to pass through Yü (or "borrow" an access route) and attack the state of Kuo. Naturally Chin's ultimate intent, easily accomplished by the victorious army on its return march two years later, had always been conquering both states. This objective became strategically achievable only after their alliance had been thwarted (as Sun-tzu advised) and one of them vanquished, eliminating any possibility of mutual sustainment. The Marquis of Chin subsequently remarked that although the jade was unaffected by its storage in Yü, the horses had aged.[10]

The horse's importance in the Spring and Autumn and beyond would continue to grow because sedentary China frequently found itself beset by aggressive steppe peoples who raided and plundered the border when not mounting more rapacious invasions. Due to high population ratios and a shortage of agricultural land, the "civilized" heartland would always suffer a severe shortage of horses, placing it at a significant disadvantage when attempting to thwart mounted riders. Moreover, even if arable land were to be devoted to sustaining a herd, the terrain in the interior was viewed as generally unsuitable for their breeding and early training.[11]

TRAINING

It is frequently written that horses are fundamentally shy and that apart from two stallions vigorously contesting a group's leadership, they will flee rather than respond aggressively when threatened. (This tendency is sometimes cited as the reason they normally turn away from solid formations and threatening spears, though their wisdom in not willingly impaling themselves hardly seems a cause for disparagement.) Wild horses are also belittled as useless and stupid in comparison with domesticated variants, which, being free from inbreeding, are reportedly smarter, even though it is more likely that the former are simply untrained and too independent to heed human commands, the reason "coercive" training has often been the norm rather than the exception. However, their inherent gregariousness facilitates employing them as pack animals, cavalry mounts, and chariot teams of two or four, as well as their mass use in warfare.

Horses have to undergo training to make them conformable to use, not to mention reliable in the chaos of the hunt or on the battlefield. Confucius therefore employed them to analogize the universal need for instruction[12] but Chuang-tzu decried the coercively destructive nature of the process:[13]

> A horse's hooves can trample frost and snow, its hair can ward off wind and cold. Gnawing grass and drinking water, rearing up and bucking, this is the horse's true nature. Then Po Le arrives and boasts, "I excel at handling horses!" He then singes them, trims them, shaves their hooves, brands them, entangles them with bridles and leg restraints, and confines them in stables and stalls by which time two or three of every ten horses have died. He makes them hungry and thirsty, gallops and races them, conforms and orders them. Before them they have troublesome bits and cheek pieces, behind them fearsome whips and goads. By then half the horses have died.

Expertise in selecting, training, and employing horses quickly developed, some of it eventually being codified in late Warring States manuals of equine physiognomy. A few men achieved fame for their ability to recognize a horse's innate characteristics, including Po Le, whom Chuang-tzu selected for condemnation because of his renown.[14] Even divination was employed in the Shang to determine a horse's appropriateness for the right side of the chariot,[15] and a few heroes such as Tsao Fu emerged in the Chou who became legendary for their superlative driving skills.

Experienced cavalry riders in the West have frequently commented that the most disciplined horses will still test skillful, even familiar riders whenever an opportunity arises. Greek horses had a reputation for biting and kicking, perhaps the reason Xenophon advised rejecting troublesome horses in his instructions to cavalry commanders, though some tacticians preferred aggressiveness for battlefield employment. From the human standpoint the horses are misbehaving, but from Chuang-tzu's contrarian viewpoint, the fault lies solely with men, who have constrained and contorted their original nature by exploiting them: "Horses dwell on land, eat grass, and drink water. When they are happy they intertwine their necks and nuzzle each other, when angry they turn their backs to each other and kick out. Horses only know this. But when you inflict cross poles and yokes on them and coerce them to conform with bridles, horses then know how to crack the crossbar, twist their heads out from the yoke, resist the harness, thwart the bit, and gnaw the reins.[16] Thus horses acquire knowledge and act like thieves. This is Po Le's offense."

Chariot drivers were confronted by somewhat different problems because they were forced to control two or more horses of less than identical physical capability and personality. In contrast, cavalry riders and their mounts are inescapably bonded by their physical contact to the point that it's often said that they appear as one. This immediacy reportedly allows an accomplished rider to anticipate the horse's behavior just as the horse can reputedly sense the rider's intent even while responding to actual commands. Feedback and anticipation are virtually instantaneous, whereas the chariot driver, who must rely on subtle changes in the reins and any hard-earned rapport with the horses, is invariably, if only slightly, reactive. For the chariot to function effectively, either the horses must be conditioned to absolute obedience, an impossibility, or trust, predictability, and an intuitive synergy in the face of their divergent interests and distinct personalities must be nurtured.

Antiquity recognized that achieving the requisite competence required a long period of dedicated training, any lack of focus in the driver or horses easily resulting in disaster for both.[17] From the Spring and Autumn onward the image of Tsao Fu, the Western Chou charioteer whose superlative skill supposedly allowed King Mu to travel an impossible 1,000 *li* a day and penetrate the distant regions, loomed large. More than a hero to be emulated, he continued to be employed in both common parlance and the military writings as an exemplar of measure and constraint. For example, the somewhat enigmatic statement that "Tsao Fu's skill was not his driving" was explained by saying that "Tsao Fu excelled at 'looking' at his horses, constraining their liquids and food, measuring their strength, and examining their hooves. Therefore he was able to take distant roads without the horses becoming exhausted."[18] It was also said that because he was essentially in resonance with his horses, able to instinctively respond to them, Tsao Fu was able to hunt successfully and race far.[19]

In marked contrast, Yen Hui predicted that the horses of a highly regarded charioteer, Tung Yeh-pi, would soon dissipate their strength. The ruler attributed his remark to mere jealousy, but Yen Hui proved correct when Tung, although capable of arranging a stirring display for the court, exhausted them in actual use. Yen then (perhaps too smugly) commented, "In traversing narrows and going far he exhausted their strength, yet ceaselessly sought more from the horses."[20]

A few well-known incidents in which drivers became disaffected and therefore subverted the mission by deliberately driving into enemy forces have been preserved in the *Tso Chuan*.[21] However, distraction was equally deleterious, as evidenced by a chariot driver at the Battle of Yen-ling (575 BCE), who kept fearfully looking about at his pursuers.[22] The horses could equally affect battlefield operations adversely if they were unfamiliar with the driver or the terrain, as reflected in the following famous incident that unfolded in 645 BCE when a highly motivated Ch'in force invaded Chin in reprisal for several perverse actions on the latter's part:[23]

After Chin suffered three defeats in succession, Ch'in's army reached Han-yüan [the plains of Han]. The duke of Chin said to Ch'ing Cheng, "These brigands have penetrated deeply, what shall we do?"

He replied, "Since my lord has caused this deep penetration, what can we do!"

The duke retorted, "You are insubordinate!"

He then divined who should serve on his right [in the chariot] and found it would be auspicious to employ Ch'ing Cheng. However, he didn't use him. Instead, P'u Yang drove the war chariot and a foot soldier from among his clan forces acted on the right. They were hitching a team presented by the state of Cheng to their war chariot where Ch'ing Cheng commented: "For great affairs the ancients invariably employed native horses. They are nurtured by its water and soil, know the hearts of men there, are settled in their instructions and training, and are thoroughly familiar with the roads. Only when they are used will everything proceed in accord with intentions. But if you now yoke a foreign team to undertake martial affairs, they will prove inconstant when frightened and go against the driver's intent. When their *ch'i* chaotically races and the *yin* components thoroughly arise in their blood,[24] their engorged veins will protrude prominently. They will look strong outside but be dry within. When they cannot advance or retreat, nor turn and wheel about, you will certainly regret it."

Once the battle began, Ch'ing Cheng's analysis proved highly prophetic, because the duke's chariot turned into a muddy patch and was stopped, resulting in the duke being captured when Ch'ing ignored calls to aid him. Although he did dispatch others, who futilely mounted a rescue attempt, he was of course executed after the duke was released. Before citing a number of these passages, the *T'ai-pai Yin-ching* would assert that "martial horses must be accustomed to the water and grass of the places they dwell and their hunger and fullness should be constrained."[25]

The quality of the horses similarly had a significant battlefield impact. For example, one warrior gave his two best horses to his uncle and brother during a conflict, making it impossible for him to escape the enemy with a lesser team, resulting in his being slain after he abandoned his chariot and fled into nearby trees.[26] Although no equine manual comparable to Kikkuli's famous short tract on conditioning or Xenophon's two focal discussions, *The Cavalry Commander* and *The Art of Horsemanship*, ever appeared in China, rules for nurturing horses and employing them for the chariots and eventually the cavalry evolved over the centuries.

Whatever their date of composition, the earliest Chinese passages on the chariot's employment are now preserved in two Warring States compilations, the *Ssu-ma Fa* and the *Wu-tzu*, the latter attributed to the great commander Wu Ch'i. Though they postdate the Shang by nearly a thousand years and thus represent fully articulated views probably not held a millennium earlier, they identify essential operational issues worth contemplating. In view of the need for even well-conditioned horses to have intervals of both brief and extended rest, the *Ssu-ma Fa* emphasized measured control.[27] In addition, in order to prevent chaos on the battlefield the individual chariots had to be synchronized with each other, as well as be coordinated with the infantry forces:

> Campaign armies take measure as their prime concern so that the people's strength will be adequate. Then, even when the blades clash, the infantry will not run and the chariots will not gallop.[28] When pursuing a fleeing enemy the troops will not break formation, thereby avoiding chaos. Campaign armies derive their solidarity from military discipline that maintains order in the formations; not exhausting the strength of the men or horses; and not exceeding the measure of the commands, whether moving slowly or rapidly.[29]

When asked what would ensure victory, Wu Ch'i stressed measure and control:[30]

> Control is foremost. In general, the Tao for commanding an army on the march is to not contravene the proper measures for advancing and stopping; not miss the appropriate time for eating and drinking; and not completely exhaust the strength of the men and horses. These three are the means by which the troops can undertake the orders of their superiors. When the orders of superiors are followed, control is produced.
>
> If advancing and resting are not measured; if drinking and eating are not timely and appropriate; and if, when the horses are tired and the men weary, they are not allowed to relax in the encampment, then they will be unable to put the commander's orders into effect. When the commander's orders are disobeyed, they will be in turmoil when encamped and defeated in battle.

Thorough knowledge of the terrain was vital for avoiding impediments and effectively exploiting the topography so as to reduce the burden on the animals:

> [The commander] should arrange the employment of terrain so that it will be easy for the horses; the horses so that they will easily pull the

chariots; the chariots so that they will easily convey the men; and the men so that they will easily engage in battle. If he is clear about treacherous and easy ground, the terrain will be light for the horses. If they have hay and grain at the proper time, the horses will easily pull the chariots. If the axles are well greased, the chariots will easily convey the men. If the weapons are sharp and armor sturdy the men will easily engage in battle.

Although probably compiled in the middle Warring States period prior to the inception of cavalry,[31] the *Wu-tzu* also preserves insights on equine training and care that would subsequently be incorporated intact by the T'ang dynasty *T'ai-pai Yin-ching* and Sung dynasty *Wu-ching Tsung-yao*:

The horses must be properly settled with appropriate grass and water and correct feeding so as to be neither hungry nor full. In the winter they should have warm stables, in the summer cool sheds. Their manes and hair should be kept trimmed and their hooves properly cared for. Blinders and ear protectors should be used to keep them from being startled and frightened. Practice galloping and pursuit, constrain their advancing and halting. Only after the men and horses become attached to each other can they be employed.

All the equipment for the chariots and cavalry such as saddles, bridles, bits, and reins must be complete and durable. Normally the horses do not receive their injuries near the end of the battle but are invariably injured at the start. Similarly, they are not injured so much by hunger as by being overfed. When the sun is setting and the road long the riders should frequently dismount, for it is better to tire the men than overlabor the horses. You should always direct movements so as to keep some strength in reserve against the enemy suddenly turning upon us. Anyone who is clear about this can traverse the realm without hindrance.[32]

ROLE AND EFFECTS

Horses dramatically affected early civilization and became inextricably entangled in a reciprocal relationship with man. By providing tractive power for vehicles and mounts for riding they immediately expanded the political, economic, and military horizons of ancient settlements. Although their capabilities were not unique, other animals were either too small (donkeys) or large (elephants), with only the ox being comparable. An ancient Chinese saying summarized their respective strengths: "Horses are the means to go far, oxen the means to bear weight."[33]

Observations recorded over the centuries attest to the horse's superiority not just in speed but also in ability to deliver a significantly greater percentage of its load to a designated location within a given time frame. However, far more important for the chariot's utilization in warfare is the horse's ability to trot at seven to nine miles per hour and to race briefly at fourteen to twenty miles per hour depending on the chariot's drag.

Even though horses are said to be shy and shun conflict, troops confronting them are normally disconcerted by their size, irrespective of whether they are being employed as cavalry or are yoked to the front of a chariot. Being highly visible embodiments of temporarily constrained "wild power," their psychological impact, whether in peaceful conditions or the chaos of the battlefield, is great.[34] In measured parades their constrained cadence lends an aura of majesty, and the sound of their pounding hooves as they charge forward augments their impact. Because image can be as important as capability, they were frequently selected for contingents and matched for chariot employment by color, size, and spirit.

Although not as freewheeling and maneuverable as cavalry, chariot forces could still produce immense terror. Particularly when massed in a battlefield charge, their terrifying bulk often shattered formations and scared defenders into breaking and running even after it was historically attested that solid formations that maintained their integrity could withstand such onslaughts. Segmenting and individually racing about, they could cause the chaos and consternation described in Caesar's observations on the British use of chariots.[35]

An incident from the Spring and Autumn period suggests the psychological importance of prestige to the chariot's riders and undermines claims that chariots merely served as transport for dismounted fighters. By then China's sedentary core had already been battling the peripheral peoples for nearly a millennium, dating back to Hsia conflicts with the Tung Yi and others, and some steppe peoples had long fielded forces with chariot components, though others continued to rely solely on infantry. In 541 BCE, when a Chin chariot contingent found itself confronted by a Ti infantry force, the commander ordered his men to dismount and re-form as infantry units. Unwilling to suffer this loss of dignity, one man resisted and was promptly executed, after which Chin scored an unexpected victory.[36]

In some civilizations the horse became the focus of culture and center of existence, even being accorded the status of a spirit or god. In China, apart from being sacrificed to honor (or appease) the ancestors, it early on acquired a number of symbolic and mystic roles. Horses were associated with fire and thought to be most active (or rambunctious) in the summer; therefore, the annual sacrifice to the horse was conducted at this time according to the idealized compilation of Chou rites known as the *Chou Li*.[37] Ritual practices eventually came to integrate a wide range of magical and mystical aspects, including specifying the color of the king's horses so that they would be in accord (or resonate) with the season.

As the horse's prestige and importance increased, methods for discriminating between sturdy and sickly animals evolved and eventually became codified. Incidental aspects also came to be integrated into the vibrant prognosticatory tradition that would characterize China through the ages. Unfortunately, even though the horse is still a prominent member of the twelve animals found in the popular zodiac, little more than remnants of horse prognostication have been preserved, scattered about a few Sung dynasty compilations. Apart from horses reportedly having given birth to humans, three of the more interesting are:[38]

> "When horses eat pebbles and stones, the general is courageous, and the warriors strong, their attack will certainly result in victory."
> "When horses sprout horns the ruler will be defeated."
> "If a colt is born without eyes the ruler will suffer long illness; without a mouth or nose, the ruler will not have any sons; without ears, deaf, or without feet, the ruler will lose his

position; without a tail, troops will arise, the state will be weak, and the ruler will lack posterity."

HORSEPOWER

With allowance for the possibility that not all the horses may have been intended for simultaneous use, the more than sixty Shang and a hundred Chou sites containing chariots and horses already excavated in north and central China (among thousands of graves and tombs) sustain the conclusion that Shang chariots invariably employed just a single pair.[39] In almost every case the remnant chariots had two horses harnessed or placed alongside them irrespective of the grave's opulence, and Shang oracular inscriptions sometimes refer to pairs of horses.[40] Moreover, initial reports of a chariot accompanied by four horses at M20 at Anyang have now been retracted, and no other site that can definitively be dated to the Shang has yet provided any evidence to the contrary.[41] Even the most extensive Chou dynasty tombs rarely include four-horse teams, despite their initial employment at the Chou's inception and escalating employment on Spring and Autumn battlefields.

However, the proto-state of Chou, which may have been the transmission nexus for the chariot and therefore somewhat further along the experiential curve, seems to have initiated the practice of using four-horse teams in warfare, exploiting them to decisive advantage at the Battle of Mu-yeh for the first time.[42] The *Shih Ching*, hardly the most reliable guide to actual practice, always refers to four horses being employed in Western Chou military campaigns, a number confirmed by several interments. Thereafter, whether uncovered in tombs with single or multiple chariots, the ratio varies among two, four, or six horses to one chariot, though four-horse versions would increasingly dominate as chariot-centered warfare proliferated in the Spring and Autumn and became almost universal in the Warring States, just when their importance was diminishing because of the growth of mass infantry contingents.[43] Nevertheless, two- and four-horse chariots are still found in late Warring States sites, and several different basic types and sizes, intended for different purposes, coexisted across the years.

Apart from the complexities of harnessing and control, adding an additional pair of horses complicated management and logistical efforts considerably. The cost of outfitting the chariot soared, training requirements increased, equine behavioral and health problems multiplied, and vulnerability soared with the increased number of horses that might be wounded or suffer injury.

The advantage conveyed by additional horses depends on a number of factors, especially the efficiency of the harness, weight of the chariot, and difficulty of the terrain. If a pair of horses can attain the maximum speed theoretically achievable on a particular terrain, more horses merely diminish the work effort required from each, possibly at too great an overall cost if sustained racing is not tactically required. On rough terrain where increased traction power was necessary or distant battlefields where speed could prove crucial, a four-horse team would allow longer employment under stress and provide a decided advantage. Four horses were also far more prestigious and formidable, clearly favored by anyone bent on ostentatious display or creating awesomeness.

No doubt to curb human proclivity for conspicuous display, sumptuary rules gradually came into effect during the Western Chou that attempted to constrain the types of chariots, carriages, and ornamentation that might be employed, whether in ordinary life or burial. Warring States writers with a propensity for systematization and idealization believed that the number should be a function of rank, with only the emperor being accorded the privilege of driving a team of six horses. Feudal lords would then be allowed four and higher officers two, with most transport carriages being confined to two horses but generally protected by an umbrella. Nevertheless, a few feudal lords usurped imperial prerogative by employing six horses, while many rich and powerful individuals reportedly flaunted their status with four.[44]

The traditionally held belief that the sumptuary regulations exerted a pervasive influence in the Warring States period, their period of systematic formulation, has recently come into increasing question, particularly with respect to whether the Chou king ever drove a team of six.[45] Apart from questions about how they might have been hitched (because only a single shaft was employed until the end of the Warring States period) and their no doubt limited effectiveness (presumably consigning them largely to display), Chou era burials containing six horses matched with a single chariot are isolated and extremely rare. Nevertheless, they do exist, suggesting that they were in fact employed in rare cases, including for the battlefield.[46] There are also Warring States textual references to using six horses, including in the *Yen-tzu Ch'un-ch'iu*, which indicates that the sagacious Yen-tzu found this practice presumptive and therefore offensive.

22.

THE
CHARIOT
IN BATTLE

◆◆────────────────────────────────◆◆

Despite a number of vehicles having been recovered from graves
and sacrificial pits, all aspects of the chariot's employment in the
ancient period pose vexing questions, particularly whether they were
deployed by themselves as discrete operational units or were accompanied by either loosely or closely integrated infantry. Because even the
oracular inscriptions for King Wu Ting's well-documented reign provide few clues, and the tomb paintings recently discovered that date to
the Warring States and thereafter mainly depict hunting scenes and parades, far more is known about the chariot's physical structure than its
utilization. The chariot's essence has always been mobility, but prestige
and displays of conspicuous authority rather than battlefield exploitation
may have been defining factors in the Shang.

Some traditionally oriented scholars continue to assert that chariots
played a significant role in Shang warfare; others deny that they were
ever employed as a combat element.[1] The Shang's reputed employment
of chariots, whether nine or seventy, to vanquish the Hsia is highly improbable given the complete absence of late seventeenth-century BCE
or Erh-li-kang artifacts that might support such claims. However, Warring States writers idealistically ascribed differences in conception and
operational characteristics to the Three Dynasties: "The war chariots
of the Hsia rulers were called "hooked chariots," because they put uprightness first; those of the Shang were called "chariots of the new
moon," because they put speed first; and those of the Chou were called
"the source of weapons," because they put excellence first."[2]

The few figures preserved in Shang dynasty oracular inscriptions, Chou bronze inscriptions, and other comparatively reliable written vestiges indicate that chariots were sparsely employed in Shang and Western Chou martial efforts.[3] The chariot's first recorded participation in Chinese warfare actually occurs about seven to eight hundred years after their initial utilization in the West, ironically just before the Near Eastern states would abandon them as their primary fighting component due to infantry challenges. King Wu Ting's use of a hundred vehicle regiments for expeditionary actions, already discussed, seems to have initiated their operational deployment, though the only concrete reference to Shang chariots (ch'e) appears in the quasi-military context of the hunt.[4]

Chariots must have been extensively employed in the late Jen-fang campaigns, but no numbers have been preserved. Thus the next semi-reliable figure is the universally acknowledged 300 chariots that were employed by King Wu of the Chou to penetrate the Shang's massive troop deployment at the Battle of Mu-yeh, precipitating their collapse. Some accounts suggest that the Chou had another 50 chariots in reserve, while the number fielded by the Shang, strangely unspecified in the traditional histories, could hardly have been less than several hundred. King Wu reportedly had a thousand at his ascension, some no doubt captured from the Shang, though others may have belonged to his allies and merely been numbered among those present for the ceremony. Several hundred were also captured from the Shang's allies in postconquest campaigns, as well as in suppressing the subsequent revolt.

Nevertheless, chariots seem to have been minimal in early Western Chou operational forces. Scattered evidence suggests that field contingents never exceeded several hundred, with as few as a hundred chariots participating in expeditionary campaigns. Although one of their efforts against the Hsien-yün resulted in the capture of 127 chariots from a supposedly "barbarian" or steppe power, King Li's campaign against the Marquis of E seems to have been typical. Despite total enemy casualties being nearly 18,000, inscriptions on the bronze vessel known as the *Hsiao-yü Ting* indicate that a mere 30 chariots were captured in one clash, though a second force of 100 is also mentioned. Somewhat larger numbers were deployed slightly later in campaigns against the Wei-fang, but the maximum figure ever reported for the Western Chou, the 3,000 supposedly dispatched southward against the rising power of Ching/Ch'u in King Hsüan's reign (827–782), is certainly exaggerated despite the king's reputation for having revitalized Chou military affairs, as well as unreliable because it is based solely on an ode known as "Gathering Millet."

The chariot's effectiveness in the Shang, early Chou, and perhaps even beyond must be questioned in the face of the constraints discussed below, the difficulties that will be examined in the next section, and the lessons that can be learned from contemporary experiments with replica vehicles. However, it should be remembered that although numerous reasons can be adduced why chariots could not have functioned as generally imagined, voluminous historical literature, both Western and Asian, energetically speaks about their employment in battle. Ruling groups were still expending vast sums to build, maintain, and employ chariot forces in the Warring States period, and the Han continued to field enormous numbers against steppe enemies, incontrovertible evidence that rather than being historical chimeras or simply artifacts of military conservatism, they continued to be regarded as crucial weapons systems.

Although all the Warring States military writings contain a few brief observations on chariot operations, only two, the *Wu-tzu* and *Liu-t'ao*, preserve significant passages. Primarily important for understanding the nature of the era's conflict, they still furnish vital clues to the chariot's modes of employment and identify a number of inherent limitations that would have inescapably plagued the Shang and Western Chou, long before chariots would explosively multiply to become the operational focus for field forces.

Chariots were considered one of the army's core elements: "Horses, oxen, chariots, weapons, relaxation, and an adequate diet are the army's strength. Fast chariots, fleet infantrymen, bows and arrows, and a strong defense are what is meant by 'augmenting the army.'"[5] Several passages indicate that chariots were viewed as capable of "penetrating enemy formations and defeating strong enemies."[6] Those used in conjunction with large numbers of attached infantry and long weapons were said not only to be able to "penetrate solid formations" but also to "defeat infantry and cavalry."[7] "When the horses and chariots are sturdy and the armor and weapons advantageous, even a light force can penetrate deeply."[8] "Chariots are the feathers and wings of the army, the means to penetrate solid formations, press strong enemies, and cut off their flight."[9] Before the advent of cavalry, they also acted as "fleet observers, the means to pursue defeated armies, sever supply lines, and strike roving forces."[10]

Passages in Sun Pin's *Military Methods* and other works indicate that somewhat specialized chariots evolved in the Warring States, the basic distinction being between faster (or lighter) models and heavier chariots protected by leather armor and designed for assaults. A few of even greater size and dedicated function were thought capable of accomplishing even more: "If the advance of the Three Armies is stopped, then there are the 'Martial Assault Great *Fu-hsü* Chariots.'"[11] "Great *Fu-hsü* Attack Chariots that carry Praying Mantis Martial warriors can attack both horizontal and vertical formations."[12] Variants with a smaller turning ratio, known as "Short-axle, Quick turning Spear and Halberd *Fu-hsü* Chariots," might be successfully employed "to defeat both infantry and cavalry" and "urgently press the attack against invaders and intercept their flight."[13]

Chariots were deemed astonishingly powerful: "Chariots and cavalry are the army's martial weapons. Ten chariots can defeat a thousand men, a hundred chariots can defeat ten thousand men." The *Liu-t'ao's* authors even ventured detailed estimates of the relative effectiveness of chariots and infantry: "After the masses of the Three Armies have been arrayed opposite the enemy, when fighting on easy terrain one chariot is equivalent to eighty infantrymen and eighty infantrymen are equivalent to one chariot. On difficult terrain one chariot is equivalent to forty infantrymen and forty infantrymen are equivalent to one chariot."[14]

These are startling numbers, all the more so for having been penned late in the Warring States period when states still numbered their chariots by the thousands. Even allowing for exaggeration, given that the *Liu-t'ao* generally reflects well-pondered experience and is a veritable compendium of Warring States military science, the era's commanders must have had great confidence in the chariot's capabilities. Nevertheless, it might be noted that the great T'ang dynasty commander Li Ching, upon examining these materials in the light of his own experience at a remove of a thousand years, concluded that the infantry/chariot equivalence should only be three to one.[15]

Chariots were also employed to ensure a measured advance in the Spring and Autumn, Warring States, and later periods when they no longer functioned as the decisive means for penetration. Li Ching's comments about his historically well-known expeditionary campaign against the Turks indicate that even in the T'ang and early Sung they were still considered the means to constrain large force movements: "When I conducted the punitive campaign against the T'u-ch'üeh we traveled westward several thousand *li*. Narrow chariots and deer-horn chariots are essential to the army. They allow controlling the expenditure of energy, provide a defense to the fore, and constrain the regiments and squads of five."[16]

Although certainly not applicable to the Shang, chariots could also be cobbled together to provide a temporary defense, particularly the larger versions equipped with protective roofs.[17] The authors of the great Sung dynasty military compendium, the *Wu-ching Tsung-yao*, after (somewhat surprisingly) commenting that "the essentials of employing chariots are all found in the ancient military methods," concluded that "the methods for chariot warfare can trample fervency, create strong formations, and thwart mobile attacks. When in motion vehicles can transport provisions and armaments, when halted can be circled to create encampment defenses."[18]

Numerous examples of employing chariots as obstacles or for exigent defense are seen as early as the Spring and Autumn period.[19] The later military writings cite several Han dynasty exploitations of "circled wagons" being employed as temporary bastions, including three incidents in which beleaguered commanders expeditiously deployed their chariots much as Jan Ziska would in the West to successfully withstand significantly superior forces.[20] Sometimes the wheels were removed, but generally the chariots were simply maneuvered into a condensed array.

WARRIOR COMPLEMENT AND ACTIONS

Based on burial patterns and traditional texts, it has been strongly, but perhaps erroneously, argued that three warriors manned the Chinese chariot from inception: an archer who normally stood on the left, the driver who controlled the horses from the middle of the compartment, and a warrior on the right who wielded some sort of shock weapon. Nevertheless, graves with only one or two warriors interred with a chariot are common in the Shang, the *Tso Chuan* occasionally refers to a fourth rider as if his presence was unusual but not exceptional, and a few Eastern Chou incidents note only two occupants.[21]

Sun Pin's comment that "those who excel at archery should act as the left, those who excel at driving should act as drivers, and those who lack both skills should act as the right" indicates that driving a chariot was also considered a specialized skill.[22] Archery and charioteering would continue to comprise two of the six essential components of the *chün-tzu* or gentleman's education known as the *liu yi* through the end of the Spring and Autumn period and into the early Warring States, even though Confucius personally disdained charioteering and has been generally perceived as disparaging warfare.

Contrary to possible impression that chariot service imposed lesser physical demands on the occupants than on the average foot soldier, outstanding conditioning and ability seem to have been required: "The rule for selecting chariot warriors is to pick men under forty years of age, seven feet five inches or taller, with the ability to pursue a galloping horse, catch it, mount it, and ride it forward and back, left and right, up and down, all around. They should be able to quickly furl up the flags and pennants, and have the strength to fully draw an eight picul crossbow. They should practice shooting front and back, left and right, until thoroughly skilled."[23] Incidents in the *Tso Chuan* confirm that great strength and courage were necessary for the chariot's occupants to survive the rigors of battle and that constant training was needed to ride in a chariot and simultaneously shoot an arrow.[24]

Apart from acting as an integral component capable of mounting penetrating attacks and engaging other chariots, the mobility provided by chariots had the potential to radically affect the course of battle. However, their actual mode of employment in the Shang remains uncertain despite deeply held traditional explanations. Nevertheless, only three possibilities exist: serving as a command platform for the officers at various levels; providing an elevated platform for observation and archery; and simply transporting the occupants to points on the battlefield, where they fought dismounted as in Western antiquity.

Even though oracular inscriptions have been interpreted as indicating that war chariots were sometimes called up in contingents of a hundred, because of their minimal numbers and novelty most Shang chariots must have been reserved for members of the ruling clan, high-ranking officials, and important officers dispersed across the battlefield. Initially they would have served as transport throughout the several days typically required to reach the battleground and then as mobile command platforms during the engagement.

When the effects of the axle, mounting bar, and box structure are included, the fighters standing in the chariot were elevated at least two and a half feet above the ground. This facilitated observing the battlefield and allowed ancient China's highly skilled archers to shoot over the heads of any accompanying infantry at the enemy. However, their increased visibility also exposed the officers to attack and identified them as prime targets for enemy archers and spearmen.

Whether even the most skilled archers could maintain their vaunted performance levels despite the bouncing, jostling, and instability experienced while standing in a racing chariot seems problematic. Nevertheless, a few Spring and Autumn episodes recount instances of surpassing skill not just in shooting at ground forces, but also in targeting other warriors in rapidly moving vehicles. For example, even though he was fleeing during the Battle of An in 589 BCE, Duke Ch'ing of Ch'i refused to allow his archer to shoot Han Ch'üeh, who was pursuing him in another chariot, because he was a *chün-tzu* (gentleman). The archer therefore shot and killed the two occupants standing on either side of Han Ch'üeh.[25]

Even if these incidents are not highly embellished or outright fabrications, they were probably recorded simply because of their exceptional or infrequent nature, thereby implying warriors on rapidly moving chariots rarely achieved such mastery. Questions remain, and the evidence is simply insufficient for a definitive conclusion, but insofar as the infantry dominated Shang and early Chou battlefields, chariot-mounted bowmen need not have been particularly accurate to strike down enemy soldiers with a flurry of arrows aimed in their general direction.

Considerable textual and archaeological evidence indicates that the chariot's occupants had specialized functions, but the placing of complete weapons sets—a bow and dagger-axe—with both the archer and the warrior on the right side in one tomb at Hsiao-t'un suggests that both occupants may have functioned as archers in the Shang, when the shortness of their piercing weapons would have precluded direct chariot-to-chariot combat. On the other hand, the shields occasionally found in chariot graves, even though probably employed by the warrior on the right in conjunction with his dagger-axe, may have been used to protect the archer during battle, a mode known in the West as well.

After employing missile fire at a distance, closing with the enemy, and becoming caught in the melee, the chariot's occupants would have had to dismount to engage in close combat because their shock weapons only averaged three feet in length, far too short to strike anyone while standing two and a half feet above the ground in a chariot compartment inset from the wheels and isolated by the bulk of the horses. Whether commanders continued to remain aloof or, far more likely, every Shang clansman was a fighter remains unknown. However, one warrior's famous refusal to fight dismounted in the Spring and Autumn suggests that the prestige of being a chariot warrior carried considerable emotional weight and that later eras did not view chariots simply as battle taxis, whatever their function in Shang warfare.

INTEGRATION WITH ACCOMPANYING FORCES

Despite extensive speculation, how the chariot and any accompanying forces may have been coordinated remains murky and confused. Operationally the crucial question is whether the foot soldiers, if any, were nominally attached or were integrated in some sort of close spatial configuration that would enable the commander to direct them in the execution of basic tactics. Vestiges in the oracular records, bronze inscriptions, and traditional historical works note soldier-to-chariot ratios ranging from 10:1 to 300:1, with those preserved from the early Chou commonly being 10:1, a reasonable contingent to have been commanded by a chariot officer.[26] However, significant variation is seen with the passage of time, and standing ratios for states in the late Spring and Autumn and Warring States periods lack consistency. Operational and historical discussions in the military writings describe attached contingents as numbering anywhere from 10 or 25 through 72 (plus 3 officers), even as many as 150, further complicating any assessment of prebattle formations and tactical deployments, both of which varied over time and were subject to local organizational differences, as in Ch'u.[27]

Battle accounts and memorial inscriptions perhaps offer a more realistic picture. When chariots were few in comparison with total troop strength, as in the Shang, the overall battlefield ratio was doubtless much higher. According to Shang oracle inscriptions, 100 chariots were occasionally assigned to a fighting force of 3,000, with the chariots apparently acting as an operational group rather than being dispersed and accompanied by a designated number of infantry. Traditional accounts also indicate that the Chou vanguard that penetrated the Shang lines at the battle of Mu-yeh consisted of 3,000 tiger warriors accompanied by 300 chariots. Although this is a highly realistic 10:1 ratio, given the assault's ferocity the 10 may have simply been running behind their assigned vehicles.

Based on 100 chariots being accompanied by 1,000 men, the famous Yü Ting, dating to the reign of King K'ang of the Chou (1005–978 BCE), similarly indicates a ratio of 10:1. The 100 chariots that had to be dispatched as reinforcements in the conflict with the marquis of E were also basically accompanied by 1,000 men, even though supplemented by another 200 in support. Despite being a Warring States work, the Kuan-tzu indicates that feudal lords were enfeoffed with 100 chariots and 1,000 men and that major lords were expected to supply 200 chariots and 2,000 men for military efforts, minor ones (hsiao hou) only 100 chariots and 1,000 men.[28] However, several other ratios are also recorded, including 300 chariots accompanied by 5,000 men and 800 chariots with 30,000 men;[29] a taxation system specifies seven armored soldiers and five guards for each of the chariot's four horses, yielding an odd number of 48 men per chariot;[30] and the state of Ch'i supposedly had 100,000 armored soldiers but only 5,000 chariots, still an astonishing number, for a 20:1 ratio.[31]

With the inception of specialized vehicles for dedicated purposes, the ratios were apparently revised and troops attached only to certain vehicles such as the attack chariots.[32] Although the configurations described in texts such as the Liu-t'ao may never have been deployed, they incontrovertibly conjoin ground troops with chariots, further supporting the link's historicity.[33] However, even if one or another of these ratios characterized Shang armies, the problem of how the accompanying infantry actually functioned in any era remains unresolved, because the early military writings fail to provide any useful information.

The only evidence to date for the existence of prescribed numbers of men being attached to Shang chariots is an intriguing aggregation of graves at Yin-hsü that has rather controversially been interpreted as an array intended to deliberately depict the composition of a chariot company and therefore prompted claims that the late Shang fully integrated ground troops with chariots in an organized manner.[34] The excavators have categorized the graves into eleven types based on the occupants and accompanying artifacts, ten of which are seen as having direct martial implications. Mainly slain by decapitation, all the occupants appear to have been sacrificed on site rather than brought for interment after dying in battle or other violent circumstances.

Five chariot graves laid out in a roughly T-shaped pattern comprise the core of the contingent. Two are offset at the ends of the T; the other three are arrayed in an essentially vertical line that commences somewhat below, giving the group a south-facing orientation. On the assumption that each grave contains one chariot and a crew of three warriors armed in the traditional manner, the site apparently preserves the first concrete evidence that the five-chariot company, long pronounced as a virtual matter of faith, actually existed.[35]

A horizontal line of five graves, each of which contains five warriors with red-pigmented bones, runs across in front of the chariot group, apparently the contingent's vanguard. Although analysts immediately concluded that they represent the so-called provocateurs employed in later warfare to enrage the enemy and raise the fighting spirit of one's own troops, the sole reference found in the oracle bones to dispatching a small forward contingent shows a unit of thirty horses being employed as an advance element. Therefore, if the practice of deliberate provocation in early Chinese warfare ever existed other than in the imagination of historical writers, its inception should be traced to the Spring and Autumn, not projected back into the Shang, with small units subsequently being employed in Warring States warfare to deliberately probe the enemy.

More or less to the right of the chariot group lie three groups of graves containing a total of 125 young, strong fighters, some complete but others just skulls, variously distinguished by the presence of ritual objects, red pigment on their bones or skulls, and some sort of headband, all deemed indicative of rank. Without becoming entangled by the numerous details engendered by these finds and the several controversies prompted by their imaginative interpretation, it appears that these burials constitute a contingent assigned to the chariot company.

This force apparently was based on the squad of five, the standard number that would essentially underpin Chinese military structure for the next three millennia. Each squad had an officer, but rather than twenty-five squads there were only twenty, giving a base of 100, a number that well coheres with the Shang practices and penchant for units of 100 and 1,000. Because four higher-level commanders are distinguishable, the twenty squads were apparently grouped into units of five. With the addition of the contingent's commander, the total reaches the subsequently sacrosanct number of 125 and would essentially cohere with the *Chou Li* articulation of 100 men to a *tsu*, but only on the assumption that the officers were not encompassed by the century.[36] However, if the officers are excluded and the twenty-five men in the vanguard seen as an integral part of the infantry contingent, the number would again reach the magic 125.

A third aggregation of graves marked by a wide variety of utensils and weapons individually interred with single bodies has been interpreted as representative of the full range of support personnel required by the chariot company. Apart from two officials who seem to have had responsibility for overseeing the food and beverages, they appear to have been divided into groups of five and seven, with the former entrusted with responsibility for the weapons, the latter various other vessels, including the portable stoves and associated utensils. Further confirmation is seen in a grave that contains ten sheep, apparently the contingent's mobile food supply.

It has been concluded that these graves show that Shang contingents were systematically organized around the chariots and that the additional ground forces were merely supplementary or auxiliary.[37] However, even though the burials certainly seem to represent a deliberate array intended to honor the king, a few objections have been raised and a number of questions remain. Most prominent among the latter is whether, as evidenced by a certain arbitrariness in selecting the graves for inclusion, the military organization discussed in Warring States writings is not being projected back onto the Shang.[38] In addition, issues of chronology have apparently been ignored in several instances, because a couple of the graves overlap or intrude upon others, evidence of either sloppiness (which is unlikely given the precision of Shang palace and tomb construction) or ignorance of previous burials, synonymous with the absence of any intent to execute a grand design.

Somewhat broader questions might also be raised, including whether this array is merely an idealization, a deployment appropriate solely for the march rather than any operational utilization just as units paraded in review throughout history, or even a form of organization that only characterized the king's personal force. This would explain the apparent overstaffing with support personnel, an extremely inefficient and unrealistic approach for a field army, even though later traditions suggest a small contingent entrusted with such responsibilities was attached to chariot squads.

How this contingent would operate on the battlefield also remains unknown. Presumably the vanguard of twenty-five protected the chariots during the archery exchange, but if they raced ahead on foot during the move to melee, the speed advantage of the chariots would be lost. If the chariots dispersed during the engagement, these elite ground fighters could have been assigned as accompanying infantry, one squad to a chariot, for protection. But similar questions also plague any understanding of the function of the 125 men deployed on the right, whether they operated in aggregate as close support, as dispersed units of 25, or shed any connection with the chariots in the chaos of battle, the most likely possibility.

Despite these significant issues, with allowance for possible chronological problems and considerable arbitrariness in selecting the graves, it appears that the aggregate preserves the elements of a late Shang military formation. Nevertheless, the foot soldiers constitute the true power, and as the era of chariot-to-chariot combat had not yet dawned, they no doubt bore the brunt of the fighting subsequent to the initial archery exchanges. Rather than being a key fighting element, the dagger-axe warrior on the chariot probably acted as a bodyguard for the archer, while the driver merely controlled the chariot, presumably under the direction of the archer.

Formations for advancing, prebattle methods of deployment, and alignments for combat (such as seen in the *Liu-t'ao*) were eventually developed for the chariots, infantry, and chariots intermixed with infantry, possibly beginning in the early Chou. However, dismissing assertions that King T'ang employed nine chariots in a goose formation, the only pre–Warring States confirmation that chariot formations had begun to evolve is a debate found embedded in the *Tso Chuan* over whether it or the crane formation should be employed for the chariots.[39] Moreover, assertions that the platoons were deployed to the left and right only increase the degree of puzzlement, because in the reality of fervent combat the chariots would have had to shed them to act as a *rapid* penetrating force, enjoy the requisite freedom of maneuver, and pursue fleeing enemies.[40]

Unless the chariots were broadly dispersed, conjoining more than a few men to them would simply have clogged the immediate area. Mobile combat becoming impossible, the chariot would have been reduced to functioning as an archery platform and localized command point. Presumably this is the reason the *Liu-t'ao's* authors subsequently devised a rather dispersed deployment: "For battle on easy terrain five chariots comprise one line. The lines are forty paces apart and the chariots ten paces apart from left to right, with detachments being sixty paces apart. On difficult terrain the chariots must follow the roads, with ten comprising a company, and twenty a regiment. Front to rear spacing should be twenty paces, left to right six paces, with detachments being thirty-six paces apart."[41] This dispersed formation would allow enemy chariots to pass through while providing the necessary operational space for maneuver.

All the military writers considered open terrain to be ideal ground for chariot operations. The great middle Warring States strategist Sun Pin therefore continued to advocate exploiting it with chariots (and cavalry) even when confronted by superior infantry strength:[42]

Suppose our army encounters the enemy and both establish encampments. Our chariots and cavalry are numerous but our men and weapons few. If the enemy's men are ten times ours, how should we attack them?

To attack them carefully avoid ravines and narrows. Break out and lead them, coercing them toward easy terrain. Even though the enemy is ten times [more numerous], [easy terrain] will be conducive to our chariots and cavalry and our Three Armies will be able to attack.

Tso Chuan depictions of Spring and Autumn military clashes frequently reduce the chariot's role to facilitating the exchange of provocative taunts and initiating individualized combat between chivalrous warriors—highly romanticized, unrealistic portraits of bravado that had little impact on the battle's outcome—yet instances of massed chariot charges undertaken across relatively open terrain are also recorded. The ultimate chariot clash occurred in the spring of 632 BCE at the epochal Battle of Ch'eng-p'u, wherein a coalition of older, northern states vanquished imperious Ch'u's initial thrust out of the south into the Chinese heartland. Although the rarity of Shang and steppe chariots precludes such clashes from having arisen in antiquity, it can still be pondered as an example of the potential effectiveness of employing chariot contingents en masse. Chin's commanders secured victory through several unorthodox measures, including targeting the enemy's weakest components:[43]

> When they had finished cutting down the trees, Chin's forces deployed north of Hsin. Hsü Ch'en, in his role as assistant commander for Chin's Lower Army, deployed opposite the forces from Ch'en and Ts'ai [allied with Ch'u].
>
> When Tzu Yü, accompanied by six companies of Juo-ao clan troops, assumed command of Ch'u's Central Army, he said: "Today will certainly see the end of Chin!" Tzu Hsi was in command of Ch'u's Army of the Left and Tzu Shang their Army of the Right.
>
> Hsü Ch'en covered their horses in tiger skins and initiated the engagement by assaulting the troops from Ch'en and Ts'ai. Their troops fled and Ch'u's right wing crumbled.
>
> Hu Mao [of Chin] set out two pennons and withdrew his forces. Meanwhile Luan Chih [of Chin] had his chariots feign a withdrawal,

dragging faggots behind them. When Ch'u's forces raced after them, Yüan Chen and Hsi Chen cut across the battlefield to suddenly strike them with the Duke's own clan forces. Hu Mao and Hu Yen [of Chin] then mounted a pincer attack on Tzu Hsi's army that resulted in Ch'u's left wing being shattered and Ch'u's forces decisively defeated. Tzu Yü gathered his clan forces and desisted from further action, avoiding personal defeat.

Two actions shaped the battle's course and determined its outcome. First, Hsü Ch'en's elite warriors initiated contact with an unexpected, concentrated thrust while the preliminary posturing was probably still under way. Its surprising fervency shattered Ch'u's insecure allies deployed on the right wing, giving Chin's forces unconstricted mobility and exposing the field to flanking attacks. Second, coordinated feigned retreats by Chin's right wing and elements of the left wing that had not participated in the initial thrust, masked by clouds of dust that were deliberately created by the branches that had been cut down and attached to the rear of the chariots, easily drew the overconfident Ch'u armies forward in a disordered attack. Presumably Ch'u's central forces, under Tzu Yü's personal command, also moved forward to exploit Chin's retreat and engage any remaining center forces; otherwise, the *Tso Chuan* account would not have noted that Tzu Yü "gathered his personal forces and desisted."

In the swirl of battle all the actions were obviously performed by chariot forces exploiting their mobility, any infantry forces that were present having been left behind, perhaps as ensconced forces defending fallback positions. Therefore chariot clashes of the type envisioned by many traditional historians, reputedly common in the second millennium BCE in the West, not only could, but did, occur. The *Art of War*, which presumably reflects late Spring and Autumn and early Warring States warfare, specifically speaks of "chariot encounters," implying that these were separate actions that didn't entail infantry participation: "In chariot encounters, when ten or more chariots are captured, reward the first to get one. Change their flags and pennants to ours, intermix and employ them with our own chariots."[44]

Somewhat later, while admonishing his 50,000 troops and 500 chariot commanders prior to engaging Ch'in in a major clash, Wu Ch'i proclaimed:[45] "All the officers must confront, follow, and capture the enemy's chariots, cavalry, and infantry. If the chariots do not make prisoners of the enemy's chariots, the cavalry not make prisoners of the enemy's cavalry, and the infantry not take the enemy's infantry, even if we forge an overwhelming victory no one will be credited with any achievements." His pronouncement would seem to imply that the three component forces targeted their counterparts rather than engaging in a general melee or that the infantry was tied to the chariots.

As their numbers increased in later eras, rather than just being employed as a single mass, chariots were segmented into operational units ranging from a basic three—center, left, and right—to more tactically specific contingents[46] controlled by drums.[47] However, the low number of just 100 per army in the Shang probably precluded any subdividing, though some analysts have suggested groups of 30 may have been employed. The Shang and early Chou probably mark the transitional period when the chariot's role had not yet been defined and individual chariots had not yet been integrated with accompanying infantry, a stage that would require rethinking, apportionment, and the development of training and operating procedures to avoid the battlefield chaos that would inevitably result if the chariots were merely added to the mix of combatants.

23.

CHARIOT
LIMITATIONS AND
DIFFICULTIES

ALTHOUGH CHARIOTS WERE extensively used by Egypt, Mesopotamia, and other states from about 1800 to at least 1200 BCE, then continued on in a more limited role, their combat effectiveness in China and the West has come into question.[1] The chariot embodied power and mobility, but the limited numbers employed in the Shang probably served as command and archery platforms rather than assault vehicles or blocks deployed with overwhelming impact. Nevertheless, certain problems described in the historical and theoretical military writings must have negatively impacted every form of employment, limiting their possible utilization and modes of combat.

Apart from the control issues inherent in employing willful creatures in a synergistic mode, any exploitation of the chariot for martial purposes invariably entailed a number of maintenance, logistical, and environmental problems. Moreover, as chariot operations increased in importance, so did the army's dependency and vulnerability. Horses might be struck by arrows, cut down with hooking and piercing weapons, disabled by pits and traps, enervated by thirst, or slain by poisoning their water supplies.[2] Even when not adversely affected by inclement weather, temperature, excessive humidity, overuse, poor food, bad water, or routine injuries, they required proper provisions, constant care, periodic rest, and especially the disposal of massive amounts of potentially dangerous waste when encamped. Being easily incapacitated both on the march and in battle, their losses could quickly become insurmountable.

Even when the horses were protected with armor and their health otherwise ensured, the chariot's many components, being fabricated from bronze, leather, and wood with dissimilar material characteristics, frequently failed in both routine use and battle. No literary evidence that artisans and engineers accompanied field forces appears until the Warring States period, yet these and other specialists must have assumed a critical role early on, because expeditionary campaigns subjected the chariots to prolonged use under harsh conditions even before their failure rates soared in battle.[3] Low-ranking personnel had to be assigned to ongoing maintenance tasks, whereas skilled craftsmen such as metalsmiths, carpenters, joiners, tanners, wheelwrights, and others were required to undertake the more complicated repairs stemming from component fatigue and catastrophic breakage.[4]

The durability of moving parts, particularly the ability of the wheels to revolve on the axle without binding due to friction, adhesion, grooving, and other forms of damage, was also questionable. Before the invention of ball bearings the areas of contact within the wheel hub at the bottom, where it sustained the axle's weight and where the revolving hub constantly pressed against the cap and interior mounting, must all have been large. Bronze fittings reinforced many other contact points, and lubricants were employed in the hub or nave, but none eliminated the wood-to-wood contact that produced the greatest destructive wear. Furthermore, neither bronze nor brass, probably the best material for moving fittings but unavailable at any time during the chariot's fluorescence, was yet employed inside the nave. Therefore the slightest deviation from the requisite component profiles could quickly doom a Shang chariot.

Axles and spokes were also known to snap, wheels sometimes came off in the heat of combat, glue joints failed, speed and bumpy conditions caused structural breaks, and leather bindings tore. In one case a charioteer boasted that his surpassing skill alone had allowed him to keep the traces intact during the day's combat, a claim that was subsequently supported when they snapped upon simply driving over a piece of wood.[5] In another Spring and Autumn incident, a chariot threw its axle mount, disabling it.[6]

Battlefield encounters resulted in various degrees of irrecoverable damage and destruction. Collisions, whether accidental or the result of deliberate action, inflicted a heavy toll. Crushing blows sometimes sundered major chariot components, attesting to their ferocity as well as to the fragility of wooden structural members,[7] and a crossbar reportedly broke when a soldier was hurled against it during the Battle of Yen-ling.[8] At the end of the Spring and Autumn period Sun-tzu estimated that six-tenths of a state's resources would be consumed in a normal expeditionary campaign[9] and therefore advised seizing and incorporating opposing chariots whenever possible, thereby "conquering the enemy and growing stronger."[10]

Minor variations in the terrain, particularly sharp depressions, pits, and holes, not only caused difficulty, but frequently halted or even disabled chariots. Heavy vegetation and the boundaries between cultivated fields would be expected to impede progress—after vanquishing Ch'i at the Battle of An, Chin insisted that Ch'i should henceforth orient the boundaries between their fields to run from east to west, presumably to facilitate operations as they invaded from the west[11]—but even shrubs, tree roots, and fallen limbs could prove surprisingly problematic. In one Spring and Autumn incident a chariot hit the root of a cassia tree and overturned while the occupants were charging forward and shooting their arrows, resulting in the charioteer being slain.[12] During the Battle of An in 589 BCE, when the Chin commander Hsi K'o complained that he had been badly wounded but was carrying on, his spearman noted that he had frequently gotten down and pushed the chariot on difficult terrain.[13]

An incident in which a chariot got caught in some sort of depression a few years earlier preserves evidence of the taunts and repartee hurled between opposing fighters:[14]

Among the Chin forces there was a chariot that sank down in a depression and was unable to advance. A soldier from Ch'u sarcastically advised him to remove the crossbar for the weapons. The chariot was able to advance somewhat but the horses wanted to turn about. He again sarcastically advised removing the pennant staff and throwing it across, after which they got out. Turning about the Chin charioteer said "I am not as experienced as your great country in frequently fleeing!"

Efforts were therefore sometimes made to prepare the battlefield and smooth the area of an expected clash by at least filling in the largest holes. Water, however, posed an additional problem, and though a sprinkle might improve dusty fields and roadways, significant rainfall would prove inimical:[15]

> Marquis Wu asked: "When it has been continuously raining for a long time so that the horses sink into the mire and the chariots are stuck while we are under enemy attack on all four sides and the Three Armies are terrified, what should I do?"
>
> Wu Ch'i replied: "In general, desist from employing chariots when the weather is rainy and the land wet, but mobilize them when it is hot and dry. Value high terrain, disdain low ground. Whether advancing or halting, when racing your strong chariots you must adhere to the road. If the enemy arises be sure to follow their tracks."

This was a lesson that Sisera and his 900 iron chariots painfully learned in a famous historical battle when a downpour muddied the terrain, hampering their mobility and making them vulnerable to Israelite infantry attacks under Barak.[16]

Being confronted by more permanent bodies of water, including wetlands that might initially seem conducive to mobile operations, required abandoning chariot warfare altogether:

> Marquis Wu inquired: "If we encounter the enemy in a vast, watery marsh where the chariot wheels sink down to the point that the shafts are under water; our chariots and cavalry are floundering; but we haven't prepared any boats or oars so we cannot advance or retreat, what should we do?"
>
> Wu Ch'i replied: "This is referred to as water warfare. Do not employ chariots or cavalry, but have them remain on the side. Mount some nearby height and look all about. You must ascertain the water's condition, learn its expanse, and fathom its depth. Then you can conceive an unorthodox stratagem for victory. If the enemy begins crossing the water, press them when half have crossed."

Sun Pin similarly advised "making the infantry numerous and the chariots few" in aquatic warfare,[17] while the *Liu-t'ao* warned against undertaking operations on wet terrain (in passages already cited). Although it was never explicitly discussed and generally not insurmountable, compelling reluctant horses to cross rivers and streams was another campaign problem frequently encountered.[18]

This fear of wetlands, no doubt derived from extensive experience with unfathomably deep and extensive quagmires, is historically attested by a few Spring and Autumn incidents, including one that occurred at the famous Battle of Yen-ling (575 BCE) between Chin and Ch'u.[19] Because they had initially deployed with a mire or swampy area in front of them, Chin divided their chariots into two groups intended to maneuver around the outer edges. Somehow the commander's chariot failed to clear the muck and sank in, forcing the warrior on the right to dismount and raise it up sufficiently to lurch forward. Fortunately, Chinese chariots were light enough that one man alone could lift them.

As a result of painful experience it was quickly realized that except on the grassy steppe, open plains, well-developed roads, or other easily traversed ground—what the *Art of War* terms "accessible terrain"—chariots would not convey any operational advantage. Furthermore, efforts to surmount the difficulties caused by the ground could only add to the battlefield's inherent chaos, making awareness of inimical terrain paramount. Thus, within the pronounced thrust initiated during the Spring and Autumn to categorize different configurations of terrain and develop tactical measures for exploiting them, outlining the parameters for chariot operations seems to have been focal.[20]

The first known discussion of maps, preserved in the *Kuan-tzu*, states: "Now the army's commander-in-chief must first investigate the maps and thoroughly know the tortuously winding constricted areas, rivers that will overflow chariots (attempting to ford), famous mountains, traversable valleys, key rivers, and where the plains are marked by mounds and hillocks, what areas are heavily vegetated by reeds, grass, and water rushes, whether the road is distant or near, the size of interior and exterior walls, famous and decrepit towns, and difficult and fertile terrain."[21] It was so widely recognized that chariots get entangled and obstructed that in recommending a commander Kuan Chung reportedly said, "In keeping the chariots from being bound up in their tracks or the troops from turning about on their heels [out of fear], beating the drum so that the warriors of the Three Armies regard death like returning home, I am unequal to Wang-tzu Ch'eng-fu. I request that you make him Minister of War [Ta Ssu-ma]."[22]

Chariots require extensive space to maneuver because of their fixed axles and the inability of the horses to move laterally when harnessed to the draught pole. Conversely, noting that chariots are hampered by constricted terrain, the military writings frequently suggested exploiting it whenever infantry forces found themselves confronted by chariot contingents:[23]

> "Suppose our army encounters the enemy and both establish encampments. Our men and weapons are numerous but our chariots and cavalry are few. If the enemy's men are ten times ours, how should we attack them?"
>
> [Sun Pin replied]: "To attack them you should conceal yourselves in the ravines and take the defiles as your base, being careful to avoid broad, easy terrain. This is because easy terrain is advantageous for chariots while ravines are advantageous to infantry. This is the Tao for striking chariots."

The climate in the south, especially in the increasingly semitropical areas, historically posed nearly insurmountable problems for horses, particularly those acclimated to steppe temperatures and humidity. In the imperial period northern invaders such as the Khitan and Jurchen would often find themselves compelled to retreat with the onset of spring or be decimated as their horses fell ill and died. Knowing of this vulnerability, when preparing to conquer the still-obstinate state of Ch'en south of the Yangtze River, the Sui tricked them into misdirecting their defensive resources and acquiring large numbers of horses that quickly weakened and perished.[24]

Not surprisingly, the extremely wet terrain in the southeastern states of Wu and Yüeh generally deterred Wu from adopting the chariot, despite the Duke of Shen's advisory mission undertaken at Chin's behest in 584 BCE. Perhaps somewhat successful in improving Wu's military organization and training, he reportedly failed to convince the leaders to adopt the chariot despite bringing thirty horses because of the problems posed by riverine warfare.[25] Wu and their nearby nemesis Yüeh therefore stressed infantry and naval forces and developed weapons for close combat, particularly swords that became famous throughout the realm and still retain their sharpness and surface qualities when unearthed today.

At the same time the commonly expressed view that the south's wetness completely precluded the chariot's adoption is erroneous because Wu subsequently relied on them to invade Ch'i to the north and Ch'u to the southwest, adroitly ferrying them upriver wherever necessary when mounting the major thrust that carried them to Ch'u's capital in 506 BCE. However, because heavily watered terrain would always convert chariots into a liability, their utilization had to be judiciously planned.

Requiring chariot and cavalry operations to "adhere to the road" not only constrained their freedom of movement but also made their routes predictable. Insofar as roads in the Shang and even early Chou were minimal, often just unimproved narrow trails that meandered along higher terrain, their speed of advance would have been severely curtailed. Only on the broad roadways that gradually developed in the later Western Chou and Spring and Autumn and on the flat, open plains and nearby steppe would the swiftness discussed in the theoretical manuals ever be attained, with somewhat reduced effectiveness inevitably being experienced whenever they crossed fields of millet or other dry crops.

Much in accord with the emphasis on identifying basic topographical features and developing appropriate tactics for them, the *Liu-t'ao* eventually delineated what might be termed ten inimical terrains—actually combinations of environmental factors and tactical situations—"upon which death is likely" for chariot forces. Despite being compiled from a millennium of experience, insofar as they are founded on the idea that "chariots value knowing the terrain's configuration," they provide insights into the operational problems that Shang forces certainly would have encountered:[26]

> Terrain on which there is no way to withdraw after advancing is fatal for chariots. Passing beyond narrow defiles to pursue the enemy some distance is terrain that will exhaust the chariots.

Terrain on which advancing to the front is easy but to the rear
  is treacherous will cause hardship for the chariots. Penetrating
  into narrow and obstructed areas from which escape will be
  difficult is terrain on which the chariots may be cut off.
If the land is collapsing, sinking, and marshy, with black mud stick-
  ing to everything, this is terrain that will labor the chariots.
When the terrain is precipitous on the left but easy on the right
  and there are high mounds and sharp hills, it is terrain contrary
  to the use of chariots. Terrain in which luxuriant grass runs
  through the fields and there are deep watery channels throughout thwarts the use of chariots.
When the chariots are few in number, the land easy, and you
  are not confronted by enemy infantry, this is terrain on which
  the chariots may be defeated.
When water-filled ravines and ditches lie to the rear, deep water
  to the left, and steep hills to the right, it is terrain upon which
  chariots will be destroyed. If it has been raining day and night
  for more than ten days without stopping and the roads have
  collapsed so that it's not possible to advance or escape to the
  rear, it is terrain that will sink the chariots.

Chariot operations were thus known to be severely impeded on
any but ideal terrain unbroken by woods, furrows, cultivation, irrigation
ditches, shrubbery, and other obstacles. Although fewer in number,
certain conditions in the enemy and types of terrain "upon which victory can be achieved" were correspondingly envisioned as highly suitable for chariot utilization.[27] However, as those considered conducive
to victory reflect deficiencies in the enemy's condition rather than op-
timal topographical and environmental conditions, they need not be
reprised here.[28]

The difficulty of coordinating chariots with each other and any accompanying infantry required advancing at a measured pace, just as
mandated by King Wu in his preconquest instructions reputedly preserved in the *Shih Chi* and *Ssu-ma Fa*.[29] Observing these constraints would
have severely tempered an assault's maximum speed and allowed enemy
infantry to surround, overturn, or otherwise obstruct the chariots. Slow

rates of movement necessitated by adverse conditions would have also made them vulnerable to spears being inserted into the wheels, stopping them from turning or causing the spokes to break and the wheels to fail.[30]

As the battle with the peripheral peoples initiated by Shang suppressive campaigns continued through the early Chou into the Spring and Autumn, highly illuminating asymmetrical clashes arose when Chinese chariot contingents encountered purely infantry-based forces. An incident that unfolded in 714 BCE suggests the chariots normally proceeded so slowly that they became vulnerable to being outflanked by infantry, causing commanders to fear them: "When the Northern Jung mounted an incursion into Cheng, the Duke of Cheng actively resisted them. However, troubled by the Jung army he commented, 'it is composed of infantry whereas we are on chariots. I am afraid that they will maneuver around behind us and launch a sudden attack.'"[31]

The commander's belief is really astounding, because speed and maneuverability are defining operational characteristics for chariots and cavalry, not infantry. Ultimately he prevailed in this clash through an imaginative, unorthodox combination of feigned retreat and ambushes.[32] Nevertheless, another famous confrontation between unsupported Chinese chariot forces and a large steppe infantry contingent on constricted terrain in 541 BCE reaffirms this concern:[33]

Hsün Wu of Chin defeated the Wu-chung and several other Ti tribal peoples at T'ai-yüan by stressing foot soldiers. When they were about to engage in battle Wei Shu advised, "They are composed of infantry forces and we will be encountering them in a narrow pass. If they encircle each of our chariots with ten men, they will certainly overcome us. Moreover, since the narrowness of the pass will put us in difficulty, they will again be victorious. I suggest that we all act as foot soldiers, beginning with me."

He then discarded their chariots in order to form ranks of infantry with the staff from every five chariots making up three squads of five. One of Hsün Wu's favorite officers who was unwilling to serve as an infantryman was executed as an example to the others. They then deployed into five dispositions, forming a deer like configuration with two contingents in the front, five to the rear, one at the right point, three at the left, and a narrow force acting as the vanguard in order to entice the enemy. Laughing, the Ti failed to deploy, allowing Hsün Wu to mount a sudden attack and severely defeat them.

Although the meaning and significance of the terms for the various contingents in the deployment would occasion much debate over the centuries, it appears that Hsün Wu came forth with a fast, roving force similar to the large chariot expeditions mounted against the Hsien-yün by the early Chou. Lacking infantry to protect the chariots from disabling multisided attack, they would have been doomed without Wei Shu's suggestion. Since their highly unorthodox deployment was essentially a chariot formation best suited for open terrain, the arrangement, limited numbers, and decision to fight dismounted prompted derisive laughter among their opponents, providing a momentary opportunity for the unexpected Chin assault that swiftly vanquished them.

Without an accompanying contingent of protective infantry, the chariot's occupants were thus seen as susceptible to the piercing and slashing weapons wielded by ground forces. The traditional historical writings preserve a few examples of chariot commanders being slain by spear thrusts[34] and their comrades being struck by arrows, cut down by dagger-axes, or having an arm or leg severed.[35] There are even incidents of foot soldiers grabbing the occupants, wresting them out of the chariot by hand, and throwing them down onto the ground or taking them prisoner.[36] For example, at the Battle of Yen-ling in 575 BCE, one warrior suggested some reconnaissance troops attempt to intercept an enemy fleeing in a chariot so that he could pursue and pull him down from behind, making him a prisoner.[37] To defend against these ground attacks, the warrior on the right, who was entrusted with primary responsibility for wielding the piercing and crushing weapons, seems to have frequently descended to repel attackers. Ancient Western armies sometimes attached skirmishers or runners to the chariots for protection, as well as to dispatch warriors from disabled enemy chariots.

If both sides commit to chariots, they become a weapon of choice rather than just a delivery system, but if one side opts not to become so entangled, the chariot can become a liability. The Liu-t'ao notes that the chariot is useless when not in motion, not even as effective as a single infantryman, presumably because of the difficulty of defending it. Although Warring States military writings attributed surpassing power to the chariot, it was conceded that infantry forces could still prevail by executing appropriately conceived tactics, implementing defensive measures that emphasized solidity, and exploiting constricted terrain:[38]

> When infantry engage chariots and cavalry in battle they must rely on hills and mounds, ravines and defiles.[39] Long weapons and strong crossbows should occupy the fore, short weapons and weak crossbows should occupy the rear, with them firing and resting in turn. Even if large numbers of enemy chariots and cavalry should arrive, they must maintain a solid formation and fight intensely while skilled soldiers and strong crossbowmen prepare against (attacks from) the rear.

Tactics for employing the chariots, particularly the larger specialized vehicles, in defensive situations developed during the Warring States period. Although a complete examination must necessarily be deferred to the tactical portions of a subsequent book, it should be noted that the more common include intercepting the enemy, blocking and thwarting them, and cobbling together temporary ramparts by deploying them without their horses in either a circular or square formation.[40] When operating in mountains, special measures had to be implemented to avoid being trapped high up and cut off from water supplies, but contingents fighting in confined valleys merely had to rely on the solidity of their chariot formations.[41]

The problems posed by natural undulations, pits, and holes prompted the realization that deliberately excavated and well-concealed ditches, holes, and other traps could cause the horses to stumble and break their legs. Although probably neither invented nor deployed before the end of the Spring and Autumn period, several devices designed to impede the passage of chariots and the more mobile cavalry by incapacitating the horses are known to have been employed in the Warring States period. With little effort, pointed devices ranging from the complicated tiger drops described in the *Liu-t'ao* through the easily dispersed caltrops—metal pieces with four points shaped like jacks—could quickly be scattered about the terrain.[42]

COMBAT ISSUES

Fighting from a moving chariot would have been difficult at best, given the bumping and jarring, not to mention the fleeting moment when a shock weapon could be brought to bear against nearby fighters on the ground or used to strike warriors in an oncoming vehicle. Thus the exceptional accomplishments attributed to racing archers may have been preserved precisely because of their uniqueness. Furthermore, even if the chariots merely served as transport to the point of conflict, fighters manning the compartment would have suffered the discomfort of confinement.

Though seemingly spacious, the approximately 32-by-48-inch compartment turns out to be highly limiting when occupied by three warriors bearing weapons and garbed in rudimentary protective leather armor. Experiments conducted over several years with martial artists well trained in such traditional weapons as long- and short-handled halberds, battle axes, daggers, and swords prove that they would have lacked the freedom of maneuver required to fend off, let alone vanquish, attackers. The driver, who faces no threat from the front where the horses block access, is mainly vulnerable to an oblique attack. However, being pinned in the center with the horses and shaft protruding in front of him, he is unable to contribute much to either the attack or defense, whether in motion or at rest. But the other two combatants are exposed from about 45 degrees right around to 180 degrees dead center at the back, where neither shields nor any other form of protection was ever affixed.

If the archer positions himself somewhat laterally on the right side so that his shooting stance puts his arm toward the outside of the chariot rather than to the inside against the driver, he can fire toward the front or out to the sides with little interference.[43] However, swinging around to shoot to the rear is virtually impossible. Conversely, an archer standing on the left, reputedly the normal Shang position, is badly hampered by the driver (even if the driver is kneeling) as he tries to fit an arrow to his bow and fire in any direction. Shots to the rear become possible if he stands laterally facing outward and thus draws his bow on the exterior side of the compartment, in mirror image to an archer positioned on the right side aiming forward.

Wielding the era's preferred shock weapon, a dagger-axe with a three-foot handle, is easily accomplished on the right side, particularly for blows directed to the front or somewhat alongside, but when swinging outward to counterattack perpendicular to the chariot's forward orientation, care has to be taken to avoid striking the archer standing on the opposite side on the backswing. Blows directed to the rear that require swinging around prove impossible without dramatically modifying the motion, as well as fruitless because potential attackers, already at the limit of effective range, can easily dodge any strike.

Even if solitary attackers might be thwarted, multiple attackers, especially those bearing five-foot-long spears, would have been easily able to slay the chariot's occupants without being endangered, unless the archer employed his bow at point-blank range. Whether armed with long or short weapons, multiple attackers create chaos because the heavily confined chariot crew, standing back to back and arm to shoulder, are unable to dodge, bend, or deflect oncoming blows and can only rely on any shields they may have carried or the protection offered by early body armor. Vulnerability would therefore have been especially acute to the rear, though presumably somewhat mitigated by the chariot's forward battlefield motion.

A single occupant wielding a full-length saber or long two-handed weapon fared far better in these admittedly static tests. Two men, though sometimes impinging on each other or even colliding, still had sufficient freedom of maneuver to fight effectively, even if the archer occupied the left side as traditionally portrayed. Three men suffered the difficulties noted; four became an example of "close packing," all four being totally incapable of wielding any sort of crushing weapon.

These problems apparently prompted the development of very long-handled spears and dagger-axes in the Spring and Autumn that were presumably intended for battling similarly equipped warriors in enemy chariots. However, for the three chariot occupants this additional length simply exacerbated the lack of maneuverability, particularly because the weapons tended to be held at least a quarter of the way up the shaft rather than at the very butt. (Grasping with two hands increases the power and control, but at the sacrifice of maneuverability.) Even with these longer weapons, two warriors riding fast-moving, converging chariots would only have had a moment to strike each other—making it not impossible but highly unlikely to significantly contribute to the battle's effort.[44] Rather than as conventionally depicted in contemporary movies, the drivers probably slowed, even halted, to allow the occupants to clash.

Experiments also revealed the height of the compartment to be not just a detrimental factor but also highly puzzling. A horizontal pole or rim that falls somewhere around the middle of the upper thigh provides adequate stabilization for a warrior to maintain a fighting stance and would have prevented falling over in sudden motion, but to provide real functional support the height should rise approximately to a man's waist. However, though not entirely useless, Shang chariot walls would have risen to just above knee level, a height that tended to cause modern fighters to lose their balance and tumble out because the rail effectively acted as a fulcrum.

The axle's high placement in a relatively lightweight vehicle would have resulted in a high center of gravity, making stability a crucial issue for any occupants trying to employ their weapons at speed. In addition, there were no springs or any sort of suspension mounting for the chariot box, even though late Shang models apparently began to employ the cantilevered wooden junction called a "crouching rabbit," which was obviously designed to reduce the effects of the wooden wheels bouncing over the terrain through its tensing and bowing action. The horses loosely coupled to the front shaft and the weight of the three-man crew would have stabilized the vehicle somewhat, but the traditional chariot would certainly have been inherently unstable and rocked jarringly from side to side on the uneven terrain of natural battlefields, just like a modern lightweight SUV.

The straw and moss padding spread on the compartment's wooden floor to provide additional damping proved to be minimally absorptive while inducing further instability, just as sponge padding might on the floor of an open pickup truck. (Comfortable when stationary, sponge-like substances tend to exhibit less desirable properties when the vehicle is in motion or the fighter is active.) In some cases the floors were fabricated by interweaving leather thongs, but their effectiveness in reconstructive experiments was decidedly poor, particularly after they lost their initial tension, and they could even result in the fighter's stance becoming more tenuous. The use of interior straps and efforts to improve the battlefield in the Spring and Autumn period confirm that stability continued to be a problem: the warriors were jostled about as the chariot moved at speed across the terrain.[45]

24.

ANCIENT
LOGISTICS

◆•◆────────────────────────────◆•◆

N O STUDY OF MILITARY AFFAIRS can ignore the crucial issue of logistics, herewith understood in the constricted sense of the art of supplying and sustaining armies both in movement and at rest. (Logistics thus encompasses the acquisition and transport of materials and provisions, feeding of the forces, and movement of armies rather than, as Jomini said, "all aspects ancillary and apart from the 'conduct of war itself'.")[1] Pioneering studies of Western armies provide exemplary models, and the probable requirements of individual soldiers can be projected, even historically employed, to assess the believability of massive campaign forces.[2] However, armies often achieved seemingly impossible tasks, little is known about ancient dietary requirements, and concrete evidence is lacking for most of antiquity. Accordingly, our consideration of logistics is necessarily confined to a few introductory comments, an exploration of the main difficulties identified in Warring States writings and subsequent practice, and a brief overview of the measures and structures that may have characterized supply and support efforts in the Hsia and Shang.

The problems inherent to supplying expeditionary armies in China rarely deterred military initiatives and were never articulated until broached by the classic military writings. The earliest forces, which numbered less than a few thousand, survived by carrying large quantities of foodstuffs at the outset, being supplied (willingly or not) by allies and subject peoples en route, seizing accumulated resources, foraging, and frequently pausing to fish and mount massive hunts in forested areas.[3] Nevertheless, the onerous requirements of campaign sustainment gradually prompted the assignment of supply responsibilities in even the most primitive administrative structures. Thereafter, the passage of time and fielding of increasingly larger forces compelled the development of specialized positions.

For the Chou, Shang, and remote Hsia, the *Chou Li*'s discussion of administrative hierarchies and functional responsibilities is generally acknowledged as an unreliable idealization. Officials noted as having exercised logistical duties may have existed but not necessarily been assigned the indicated role, or the titles may be wrong but the activities correct. Even then, "quartermasters" are not actually described until late in the Warring States period in a chapter of the *Six Secret Teachings* outlining the essential members of a general staff.[4] Among the eighteen basic categories of officials, there should be four "supply officers responsible for calculating the requirements for food and water; preparing the food stocks and supplies and transporting the provisions along the route; and supplying the five grains so as to ensure that the army will not suffer any hardship or shortage."

The army's first priority in the field would have been locating adequate water resources and ensuring they had not been poisoned or contaminated, two pernicious measures that would be practiced from the sixth century BCE onward.[5] Water, especially its denial, is a focal issue in the later military writings, but even Hsia and Shang forces must have been acutely conscious of the need for it, particularly when venturing out against enemies in the semiarid steppe.[6] How much water they carried, what sort of containers they employed (such as gourds), and whether wheeled vehicles—either human or animal powered—were used are all unknown. However, being heavy, bulky, and fluid, water is inconvenient to transport; the countryside was still relatively unpopulated; techniques for digging wells were already known; and potential sources were numerous, especially leading up to the early Hsia and in the Shang from Wu Ting's reign onward, suggesting they depended on ongoing acquisition.

Because the primary nourishment was provided by millet, then wheat, and finally rice, and all three require cooking, firewood had to be gathered and primitive stoves or other cooking arrangements set up. (To secure the allegiance of their troops, Warring States writings advised commanders to emulate famous generals like Wu Ch'i, who never ate or drank until the army's wells had been completed and the fires lit for cooking.)[7] Once they exhausted the local firewood, the inability to boil water and prepare hot food would have immediately increased the army's misery and the likelihood of disease, especially in winter months and the rainy season.

In the Spring and Autumn and Warring States periods, campaign provisions would be increasingly sourced through official confiscations that took several forms, including outright seizure and military impositions (*fu*), while laborers were assembled through broad-based conscription and onerous work levees. Tax obligations might include supplying any of the grains, furnishing an ox or horse, or providing certain types of equipment, and it is claimed that conscripts in the Warring States period were required to report for duty fully equipped and even to sustain themselves for brief periods, though this quickly proved ineffective. Across the centuries monetary impositions would be increasingly imposed, allowing the government greater flexibility in purchasing the requisite equipment and provisions while minimizing the inconvenience and expense of transporting bulky commodities in every conceivable amount around the realm.

By improving trails and demarking regal highways, the Shang initiated an unbroken heritage of increasingly ambitious road projects intended to facilitate administration, communication, and the rapid dispatch of troops to quell unrest or counter peripheral threats. Nevertheless, because oxen rapidly consume a high percentage of their grain loads, and horses were comparatively few and inordinately inexpensive, ordinary soldiers still carried hefty rations.[8] However, the small boats that began to ply China's extensive rivers and lakes in the middle to late Shang could have been employed to transport grain, and oracular inscriptions indicate the king actively contemplated the possibility of moving troops over water. Many centuries later, at the very inception of canal building in the Spring and Autumn, the state of Wu would construct a canal solely to facilitate moving troops and provisions from the southeast into the heartland. Lengthy canals intended for military and dual-use purposes such as supplying interior capitals would also multiply in the Ch'in and subsequent dynasties.

Although the expenses incurred in the Hsia and Shang for military activities are likely to remain unknown, some sense of martial costs and their greater impact can be gleaned from later calculations and comments. The average 100,000-man Warring States operational force, with its numerous chariots and complex siege equipment (but no cavalry), required an extensive supply train, numerous support personnel, and Herculean efforts. The *Art of War*, attributed to Sun-tzu, but probably compiled sometime early in the Warring States period, notes that, "if there are 1,000 four-horse attack chariots, 1,000 leather-armored support chariots, 100,000 mailed troops, and provisions are transported 1,000 *li*, then the domestic and external campaign expenses, the expenditures for advisers and guests, materials such as glue and lacquer, and providing chariots and armor will be 1,000 pieces of gold per day. Only then can an army of 100,000 be mobilized."[9]

A thousand pieces of gold being an almost incalculable expense at that time, Sun-tzu warned that prolonged warfare would not only exhaust the people, but also consume some 60 to 70 percent of the state's resources and entangle seven families for every man who served: "Those inconvenienced and troubled both within and without the border, who are exhausted on the road or unable to pursue their agricultural work, will be 700,000 families."[10] (These numbers may have been derived from rather late, idealized concepts of the well-fielded organization, in which eight families were supposedly allotted individual plots arrayed much like a tic-tac-toe board around a center portion that they farmed in common to sustain the government.)

Seconding Sun-tzu's conclusion that "one bushel of the enemy's foodstuffs is worth twenty of ours, one *picul* of fodder is worth twenty of ours,"[11] most military writers subsequently advised capitalizing on whatever might be acquired en route or seized in enemy territory, including from the enemy themselves in armed clashes. Victorious Spring and Autumn forces occasionally captured "three days of supplies," suggesting that this might have been the minimal reserve for a field force, and larger amounts were sometimes noted.[12] In no doubt characterizing Warring States practices, the *Six Secret Teachings* states that the vanguard should carry three days of "prepared food" to facilitate rapid movement and transitioning to combat as necessary. However, the advance force that preceded the vanguard had a six-day supply, and "the main army set out with a fixed daily ration."[13]

To implement Sun-tzu's belief that "if you forage in the fertile countryside, the Three Armies will have enough to eat," generals across the
ages routinely dispatched contingents tasked with plundering and foraging.[14] However, the prospects for success would depend on the terrain's
accessibility and the existence of warehouses, granaries, animal herds,
and readily harvested crops. Yet armies around the world have managed
to sustain major field efforts for almost unimaginable periods, though
always at a devastating cost to the local populace. (At a minimum the
land is denuded, the infrastructure damaged, heavy collateral casualties
suffered, the local population displaced, and starvation endured, especially when seed crop allotments are confiscated.)

Population and agricultural yields both continued to increase
throughout the Lungshan period, resulting in local surpluses. As previously noted, recently investigated storage pits indicate that surprisingly
large amounts of grain could be accumulated, a situation that probably
persisted in the Hsia and Shang. The production of inebriating liquors,
attested by the proliferation of drinking vessels in the Shang as well as
being a purported cause of their downfall, is generally seen as further
evidence of a grain surplus.[15]

Contradictorily, the early military writings also decried confiscation
policies as counterproductive because they would stiffen enemy resistance.[16] Moreover, even the stupidest commander would have herds
shifted away from projected lines of march, structures dismantled, goods
moved into fortified towns, a scorched earth policy implemented, and
as many provisions as possible acquired before becoming entangled in
a probable siege situation, just as outlined in sections of the Mo-tzu and
the Wei Liao-tzu.

Battlefield experience would stimulate an acute consciousness of the
value of food and its denial as a weapon, both in practice and in the classic
theoretical writings. For example, in stressing the role of measure and
constraint in campaigns, Wu Ch'i said:[17] "If their advancing and resting
are not measured, drinking and eating not timely and appropriate, and
they are not allowed to relax in the encampment when the horses are
tired and the men weary, then they will be unable to put the commander's
orders into effect. When the commander's orders are thus disobeyed,
they will be in turmoil when encamped and will be defeated in battle."

Armies operating in the field for prolonged periods often found themselves either out of supplies or cut off by the enemy, resulting in weakness, starvation, and even death should they be compelled to surrender, as happened to the 400,000 from Chao at Ch'ang-p'ing in the third century BCE. The middle Warring States *Six Secret Teachings* counseled trickery when confronted by a lack of supplies or inability to forage,[18] and even included a five-inch strip for "requesting supplies and additional soldiers" among its set of tallies for secret communications.[19]

The *Wu-tzu*, an early Warring States compilation nominally attributed to the great general Wu Ch'i, included exhaustion of the food supplies and an inability to acquire firewood and fodder as such debilitating conditions for the enemy that they could be attacked without further contemplation or assessment.[20] The critical importance of supplies is also evident from their deliberate abandonment, a desperate measure designed to stimulate death-defying resolve in the troops and impress upon them the finality of their situation as they battled on "fatal terrain."[21]

By the Warring States period well-provisioned cities would be viewed as relatively impregnable to attack, but those who had not made appropriate preparations remained highly vulnerable: "If the six domesticated animals have not been herded in, the five grains not yet harvested, the wealth and materials for use not yet collected, then even though they have resources, they do not have any resources!"[22] Conversely, "collecting all the grain stored outside in the earthen cellars and granaries and the buildings outside the outer walls into the fortifications will force the attackers to expend ten or one hundred times the energy, while the defenders will not expend half."[23]

When sieges commenced in the Spring and Autumn period, aggressive actions against fortified towns sometimes had to be abandoned because the attackers had exhausted their food supplies. Thus, in later periods the problem of sustaining a prolonged siege was sometimes partially solved by assigning a portion of the troops to undertake localized farming efforts, a sort of precursor to deliberately having long-emplaced garrisons simultaneously farm and perform defensive functions along the border from the Han onward. Warring States writers were acutely aware of how intertwined war and agriculture had become, no doubt accounting for Shang Yang's views and the state of Ch'in awarding rank only for achievements in war and agriculture.[24]

Early in the Spring and Autumn period, Ch'in therefore tried to undermine Chin by refusing any aid when the latter was suffering from famine, despite having previously benefited from their largess. Even more perniciously, at the end of the period Yüeh reportedly sought to debilitate its nemesis Wu through an unorthodox biological attack effected by offering them high-yielding but secretly damaged seed for the next year's crop, thereby enticing them into consuming their reserves.

Because victory invariably depends on securing adequate materials and provisions, severing the enemy's supply routes could compel them to segment their forces or dispatch disorganized raiding parties that might then be assaulted in detail, thereby winnowing down their forces while reducing them to a starving, ineffective rabble. Under the unrelenting pressure of severe deficiencies, commanders tended to mount hasty actions that often turned out to be premature or ill-conceived. In addition, people trapped in fortified cities under extended siege not infrequently resorted to cannibalism to survive.

Acquiring accurate knowledge of the enemy's condition, particularly how long they might endure before hunger would totally dispirit or starvation kill them, could prove critical in formulating effective tactics. As early as the Art of War Sun-tzu noted that "those who stand about leaning on their weapons are hungry" while "if those who draw water drink first, they are thirsty."[25] The status of the enemy's provisions— "whether the army is well prepared or suffers from inadequacies, whether there is a surplus or shortage of foodstuffs"—accordingly came to be viewed as a key factor in assessing enemy vulnerability.[26] The late Warring States, Taoist-oriented Three Strategies of Huang-shih Kung somewhat expansively advised:[27]

> The key to using the army is to first investigate the enemy's situation. Look into their granaries and armories, estimate their food stocks, divine their strengths and weaknesses, search out their natural advantages, and seek out their vacuities and fissures. A state that does not have the hardship of an army in the field yet is transporting grain must be suffering from emptiness. If the people have a sickly cast they are impov-

erished. If they are transporting provisions for a thousand *li* the officers will have a hungry look. If they must gather wood and grass before they can eat, the army does not have enough food to pass one night.

Accordingly, anyone who transports provisions a thousand *li* will lack food for a year; two thousand *li*, two years; and three thousand *li*, three years. This is what is referred to as an empty state. When a state is empty the people are impoverished. When the people are impoverished, the government and populace are estranged. While the enemy is attacking from without, the people are stealing from within. This is termed a situation of "inevitable collapse."

Commanders therefore sought to deny this information through increased security, wide buffer zones, outright deception, and other means. For example, General Ho of the Northern Chou created phony grain mounds by heaping grain on top of piles of sand, causing the local people who had been stealthily observing the camp to report the existence of ample supplies to the enemy.[28]

In terms of logistical practices, it must be conceded that the Chinese Neolithic remains unknowable. However, most clashes were localized affairs, basically raids and brief encounters conducted by a few dozen men within a day's march, and the combatants probably carried sufficient provisions to sustain them for a day or two. However, as regional powers such as the Hsia, San Miao, and proto-Shang arose and conflict escalated to involve hundreds, then thousands, of men, some administrative and organizational measures must have been initiated.

Archaeological discoveries provide evidence that the adoption of agriculture and then its rapid expansion in the Lungshan period meant that edible stores ranging from millet to seeds had come into existence, would be found at every village, and could easily be seized by forces in the field. In addition, pigs and other mobile animals that were raised in large numbers greatly increased the potential food supply, but could also entangle the troops in slaying and cooking them, making them vulnerable to the sort of surprise attacks that would be advocated and exploited in later centuries.[29]

Insofar as hunting and the gathering of fruits and other edibles still played a vital role in the Neolithic and even the Shang, small bands of a few hundred could probably find adequate sustenance in the generally uninhabited countryside about them. Large-scale hunts in which several hundred animals might be captured or slain prior to a campaign would also provide significant provisions. Furthermore, since almost all the settlements and villages, the era's most likely targets, were located near rivers and lakes, fishing and trapping offered another, though somewhat more time-consuming, possibility. Some animals such as the sheep found buried with the Shang chariot formation could have been herded along in viable numbers if the armies, proceeding on foot, did not advance too far or too quickly, thereby lessening the protein burdens of hunting.

The relative advantages and disadvantages of cattle and horses apart, the fundamental problem of providing fodder was simply one of weight and bulk. However, the limited forces fielded prior to the Shang were probably not accompanied by animals, and even Shang armies had few chariots and wagons, so their requirements would have been low and probably satisfied simply by letting the horses and any cattle employed as motive power for so-called *ta ch'e* or large vehicles graze in the immediate area. Reconstructed campaigns indicate that Shang armies on campaign rarely remained encamped for any lengthy period and were thus able to avoid exhausting the locally available fodder and provisions or suffering the many other logistical and health problems posed by stalemates and prolonged sieges.[30]

In the Eastern Chou and later periods winter was generally thought to be the proper season for the military activities of punishing and slaying, in accord with the ascendancy of *yin* and the natural characteristics of the correlated elemental phases of metal (autumn) and water (winter). Thus the officials charged with administering punishments appear among those correlated with autumn in the *Chou Li*, and several weapons makers are subsumed under winter. In addition, military activities undertaken at this time could take advantage of the lull in agricultural obligations and live off recently harvested crops. However, the oracular inscriptions show that the Shang initiated military campaigns in response to external stimuli and perceived threats throughout the year. Even though the effects of cold weather had to be endured, autumn and winter campaigns were no doubt facilitated by the maturation of fruits and nuts and the enhanced availability of agricultural reserves, but clearly were not constrained by their existence.

Warehouses and granaries were maintained in the core area from which the initial supplies for military campaigns could be allocated. In addition, the Shang constantly opened new fields on the periphery and converted conquered areas into farmland, particularly to the west. Armies necessarily passing through these areas could take advantage of locally harvested and stored provisions, and it appears that some of the subjugated states also retained foodstuffs and animals for such use rather than forwarding them to the Shang as tribute, thereby reducing not only any initial amount that might have to be allotted, but also the cost and effort of transport.[31] However, large numbers of animals—up to several hundred oxen on at least one occasion—were also received by the Shang, and though many were consumed and used for sacrificial purposes (prior to also being consumed), some would certainly have been available to supply military requirements.

Under Wu Ting the Shang further reduced its military expenses by dispatching subservient states and coercing allies who were responsible for sustaining themselves in the field. They also seem to have provided supplies when necessary. For example, one inscription preserves a query as to whether the proto-state of Yüeh will supply the needs of the *lü* (regiments) on the march.[32] The value of such contributions should not be underestimated. As the *Art of War* notes, "The state is impoverished by the army when it transports provisions far off. When provisions are transported far off, the hundred surnames are impoverished. One who excels in employing the military does not transport provisions a third time."[33]

Even though many Shang campaigns probably required only a few weeks, weapons questions would have still loomed large. Apart from the piercing and crushing weapons carried from the outset, a huge number of arrows had to be supplied as the campaign progressed. For a single engagement the total of twenty arrows carried in each archer's two quivers might have sufficed, but even at a very slow rate of fire of perhaps five shots per minute, a daylong standoff or intense pre-clash archery duel would easily consume a few hundred per man. No archer could have carried that many, as suggested by the large number of bundles often recovered from Shang tombs, so there must have been some system of supply and resupply.

Opinion differs about whether the Hsia's rudimentary administrative structure included officials responsible for weapons or they were simply provided by the individuals themselves or the various Hsia clans.[34] However, as previously noted, it is generally thought that the Shang monopolized the manufacture and bulk storage of weapons, even though a wide variety of dagger-axes, dirks, axes, and bows must have been in the possession of the Shang warrior elite. Conversely, it has been asserted that the early Chou dynasty military proclamation known as the "Fei Shih," included in the *Shang Shu* but probably dating to the early Spring and Autumn, provides evidence that the troops furnished their own weapons in the early Chou and thus, by projection backward, in the Shang.

However, the highly laconic "Fei Shih" is a very indeterminate collection of statements promulgated by an unknown commander prior to a campaign against the Yi located around the Huai River. Even though they are instructed to prepare their armor and weapons by repairing, sharpening, and generally putting everything in order, there is no information on the origin of these weapons. (The troops are also ordered to prepare cooked rations of grain, but whether the materials were supplied by the government is unknown.) Even the official in charge of weapons noted in the *Chou Li*, who clearly provides them to the users (including those about to learn archery), cannot be realistically envisioned as having existed in the Shang.

Different clans seem to have specialized, whether deliberately or through historical accident, in certain types of productive activities. If so, they could be tasked with responsibility for furnishing or otherwise overseeing various categories of essential military goods ranging from weapons through provisions. Various peoples with expertise in horses, oxen, and grazing animals, such as the Chiang, and the specialized officers for dogs and perhaps livestock would also have provided a ready core of competent officials for sustained military operations and probably served on an ad hoc basis throughout the Shang. In particular, responsibility for providing meat in the field seems to have fallen to the *Tuo Ch'üan* or Chief Canine Officer.[35] Standing border forces under these officials and the *shu* also seem to have undertaken local farming for sustenance purposes.[36]

Finally, the nature and degree of road development in the Hsia and Shang would have considerably impacted the transportation of goods and materials, as well as facilitated (or hindered) the army's movement. Although well-tamped roads have been found in many early sites, including Erh-li-t'ou and Yen-shih, all the discoveries to date have been confined to early cities and towns, the natural focus of excavations. Whether the Hsia or, more likely, the Shang had administrative officials entrusted with the task of road improvement and early bridge building is unknown, though the forwarding of tribute and passage of troops would seem to have stimulated the dispatch of work crews responsible for clearing trees and the simple upgrading of well-traveled paths.[37] No doubt there was a reciprocal relationship between the opening of transport routes as a result of trade and administrative necessity, including the passage of large numbers of troops, and the facilitation of commerce and military activities.

25.

MUSINGS
AND
IMPONDERABLES

◆·◆————————————————————◆·◆

DESPITE THE ANALYSES PROFFERED by a growing number of books on primitive or early warfare, the point at which societies shift from being peaceful and rustic to being dominated by martial values in order to survive remains opaque.[1] Whether an idyllic age of tranquility ever existed apart from later imagination and the projections of theorists who embrace the dogma of matriarchal equality might equally be questioned.[2] However, in China many of the classic Warring States writings already envisioned a sudden devolvement from an era of virtue and tranquility, of simplicity and harmony (whether natural or enforced), through a stage when the members concerned themselves with warfare only when threatened, to the final advent of segregation and conflict.

References scattered throughout the wider collection of pre-imperial materials indicate that growing unrest already troubled the highly idealized, semi-legendary Sage progenitors. Sun Pin noted that seven groups didn't obey Yao's mandates, including two among the Yi and four in the central area, and that he pacified the realm with force.[3] Other observers such as Hsün-tzu suggest that Shun found it necessary to subjugate fourteen recalcitrant entities and that he attacked the Miao, then entrusted Yü with the same task and actually perished in a campaign against them. When Yü finally acquired power he found thirty-three groups refused to submit, and he had to "forcefully impose his teachings." Fortunately, the Tung Yi did not support the Miao in their conflict; otherwise, the Hsia's comparatively paltry forces would have been crushed.[4]

However insubstantial and unreliable, these laconic references still accurately reflect an accelerating trend toward localized and global conflict, the failure of vaunted Virtue to prevail, and a purported unwillingness to recognize what would later be termed "Heaven's will." More significantly, particularly for the weaker or less populous areas, alliance building had already emerged as an important factor. Even the neutrality of supposedly natural allies could doom the isolated to defeat, prefiguring the validity of the *Art of War*'s admonition a millennium and a half later to thwart alliances and thereby incapacitate enemies.

Historians have traditionally argued that accumulating wealth, class differentiation, and the evolution of authoritarian structures are the crucial elements underlying the emergence of conflict and early warfare.[5] Although the temptations of material goods and suffering of deprivation no doubt beget greed and rapaciousness, neither hierarchical structure nor class differentiation was necessary for bands of marauders to begin pillaging simply because it was more profitable than farming and hunting. Strife and conflict, once unleashed, stimulate not just a need for defensive measures, but also esteem for the warriors who can preserve life, as well as for those fighters who trample the cowardly and unprepared, with admirers of the latter comprising the growing pool of miscreants.

Specialized weapons with no useful purpose other than attacking and slaying upright human enemies rapidly multiplied at the end of the Yangshao and early Lungshan.[6] Apart from empowering and emboldening their wielders, their growth, coincident with the emergence of fortified population centers, shows how low-intensity warfare can stimulate inventiveness, organization, and authority. The massive defensive walls that appeared in the Lungshan and are considered one of the distinguishing features of the culture have long been recognized as a disproportionately important development in the inexorable evolution of Chinese civilization, a step in the unremitting march toward fulfilling the subsequently articulated idea that *"ch'eng* [walls] were erected in order to protect the ruler and *kuo* [external walls] were constructed in order to preserve the people."[7]

Despite increasing differentiation in dwellings and other evidence of an incipient hierarchical structure such as ancestral temples, the absence of fortified internal quarters indicates the defensive focus was externally directed, toward unknown others, rather than locally oriented and therefore intended to protect emerging power groups that were increasingly claiming authority over others.[8] In the ancient period it is only at the two Yi-luo capitals of Yen-shih and Cheng-chou that inner palace quarters and fully formed encircling fortifications are found. Even the final Hsia and Shang capitals at Erh-li-t'ou and Anyang are famously unprotected by visible fortifications.

As warfare became more complex and its lethality increased in the late Neolithic, the hard lessons derived from ever-accumulating martial experience prompted the realization that topography inherently conveys strategic advantages and disadvantages. In deciding where to locate their settlements the first "urban planners," a term hardly misused here as many fortified towns clearly implemented a preconceived plan, were confronted by vital choices. Foremost among them was relatively high but dry terrain that ranged from hillsides through hillocks and mesas, versus the alternative of well-watered areas situated alongside rivers, lakes, or other bodies of water.

Height and inaccessibility are key factors in deterring and repulsing marauders, but in the absence of extensive wells even minimal quantities of water require laborious effort to acquire and transport for communities located any distance from aquatic resources. Conversely, settlements near streams and rivers, even though significantly protected by natural barriers and generally located amid fertile alluvial plains, must contend with seasonal flooding and moisture-borne diseases. In providing enemies skilled enough to employ rafts and canoes with virtual highways, these same waterways also enhance perversity's mobility.

Early communities therefore tended to inhabit naturally raised terrain alongside streams and rivers, whether slight mounds or relative heights created by the forces of erosion. However, particularly in the central plains, terrace settlements soon sought the enhanced protection of deliberately excavated perimeter ditches, except where deliberately situated amid the confluence of flowing streams or protected by the presence of a lake and perhaps another river. Despite being a simplistic though onerous measure, the ditches' effectiveness is evident from their continued employment well past the Yangshao and Lungshan periods, including to enhance the complex defensive systems deployed at Yen-shih and Erh-li-t'ou.[9]

Although many of these early ditches were dry, a significant number functioned as moats in the rainy season or when deliberately connected to a water source, augmenting their inherent ability to impede and frustrate aggressors. Claims have also been made that they provided protection against flooding, but it is highly unlikely that they would have been effective because even moderately rising river levels would have rapidly exceeded their limited capacity. Nor could they have functioned as drainage canals in alluvial plains marked by minimal declination.

At first the soil that had been removed was employed to raise the village floor's overall height and construct building foundations, but later it was mounded up on the ditch's interior to form primitive walls. Even these low mounds constituted a significantly enhanced challenge for early aggressors who now had to negotiate not just the shallow ditch but also ascend a low embankment, all the while exposed to spears, rocks, and arrows. Although never universal, solidly pounded walls that soared several meters in the air and extensive conjoined moats soon followed that confronted enemies with more formidable challenges.

Agricultural activities would certainly have consumed most, if not all, of the time of the few hundred able-bodied people who populated the typical late Neolithic settlement, leaving little energy for martial endeavors. Unless the inhabitants devoted some effort to self-defense, an aggressive band of forty or so warriors would probably have had sufficient strength to overpower them, seize their property, and carry away prisoners. Therefore static measures—the exploitation of extant water barriers, excavation of ditches and moats, and the building of walls—frequently provided the only means for sedentary communities to thwart raiders. As the Art of War would subsequently state, "Those who cannot be victorious assume a defensive posture, those who can be victorious attack. In such circumstances, by assuming a defensive posture strength will be more than adequate, whereas in offensive actions it will be inadequate."[10]

Although a few Neolithic villages seem to have been burnt to the ground[11] and King Wu Ting dispatched an attack party to prevent the walling of a town (thereby showing that the defensive advantages provided by fortifications were well understood), little evidence yet exists that assaults were undertaken against fortified towns. Moreover, whether they were too costly or simply doomed to fail, sieges did not commence until the Chou. However, human determination ensured that fifty-meter-wide moats and five-meter-high walls would not be insurmountable unless they were stoutly defended. Even then the San Miao's virtual disappearance despite the extensiveness of Shih-chia-ho fortifications shows the most expansive fortifications would not provide an infallible refuge if the combatants engaged in open field battle.

According to traditional sources, the first known failure dates to the legendary period when Yao, also known as T'ang-shih, embarked upon a remarkably aggressive external policy to vanquish a number of peripheral states, including the Hsi Hsia, an entity of uncertain identity to the west. A brief paragraph in the *Yi Chou-shu* notes: "One who doesn't practice the martial will perish. In antiquity the Hsi Hsia were benevolent rather than warlike in nature. Their walls were not maintained and their warriors lacked positions. Instead, they practiced beneficence and loved rewards, but when their material wealth was exhausted they had nothing to use for rewards. When T'ang-shih attacked them their walls were not defended, their martial warriors were not employed, and the Hsi Hsia perished."[12] Having failed to maintain their walls and integrate their warriors into their sociopolitical structure, the Hsi Hsia lacked the means to either mount an active defense or stimulate the requisite behavior at a crucial moment. As Sun Pin subsequently observed, "No one under Heaven can be solid and strong if they mount a defense without anything to rely upon."[13]

Even though history shows that the forces of destruction normally overwhelm constructively oriented efforts, the defensive solidity provided by the earliest walls and moats made possible the gradual accumulation of the goods produced by the weaving and handicraft industries, facilitated the domestication of animals, protected the emergence and expansion of agriculture, and harbored metallurgical workshops. It also fostered social cohesion and nurtured a sense of identity by separating the community from the external realm. By the middle Lungshan period increasing strength and prosperity enabled these early cultures to construct and occupy fortified towns that encompassed dwelling areas of several hundred thousand to even a million square meters.

The transition, however fitful, from isolated settlements to more centrally focused power centers, the repulsion of aggressors, and the execution of successful external campaigns in aggregate allowed, as well as stimulated, the essentially unimpeded evolution of political structures and authority. The incredible effort required over a sustained period to erect these defensive fortifications and accompanying palace foundations provides evidence of not only a new ability to mobilize a vast labor force with a sense of common purpose, but also the emergence of chieftains strong enough to coerce compliance. Fortifications also made power projection possible and thus no doubt contributed to the growth of multi-tiered settlement patterns whose lesser members could similarly be compelled to participate in their construction. Nevertheless, though the inhabitants may have come to feel walls were essential, they apparently lacked full faith in them; otherwise, they would never have been built on such a massive scale or reinforced with ditches and moats.

This shift from simple circular defensive ditches to well-integrated, technologically sophisticated rectangular fortifications can be interpreted as incontrovertible evidence for an unremitting increase in warfare's frequency and lethality from perhaps 3000 BCE. Equally revealing is the rapid proliferation in the types and numbers of weapons, especially bronze variants dating from their true inception in the late Hsia, as well as their widespread production in the Shang in comparison with bronze ritual vessels.[14] This multiplication was accompanied by a comparatively rapid rise in the number and proportion of weapons interred with the deceased in every cultural manifestation, ranging from the much earlier Yangshao through Hung-shan, Ch'ü-chia-ling, Liang-chu, and Lungshan, highly indicative of the emergence of military authority and of a growing esteem for martial achievement and values. (Confirmation of the latter appears as early as the Ta-wen-k'ou in lavishly decorated but empty tombs intended to honor military heroes whose bodies had not been recovered.)[15] More sorrowfully, the number of humans whose skeletons still betray the effects of violence or torture, who must have been sacrificed, wounded, or perished in battle, also rapidly multiplied.[16]

CONQUEST AND DISPLACEMENT

A fundamental question that might be posed about the nature of ancient Chinese warfare is how one group or culture succeeded in dominating, displacing, or extinguishing another. Attention has been largely focused on the growth of significant power centers and the evolution of states, the process by which a few fortified towns began to exceed the common 10,000-square-meter settlement size before a pattern of primary and secondary centers encompassing several hundred thousand and a few 10,000 square meters, respectively, emerged.

Food surpluses, population increases, and improvements in tools and productivity allowed manpower to be allocated to secondary, non-subsistence tasks such as wall building, military training, and ultimately military campaigns that could, contrary to much traditional thought, be highly productive in acquiring possessions and terrain. (In the brutal context of history it would be astonishing if anyone enjoying surpassing military success ever spontaneously desisted from aggressive activities, whether prompted by greed, a desire for power, or simple hatred of others.) However, the forces that actually enabled any particular group to excel, to culturally and politically, if not physically, overwhelm nearby peoples, remain a mystery even though a few charismatic chieftains may have been disproportionately effective leaders and motivators.[17]

The evolution of multiple cultural centers in ancient China,[18] some in close proximity, others dispersed across the greater landscape, ensured a potential for conflict was inherently present. The theory of a dominant central culture whose aggressive expansion acts as an agent of change,[19] whose technology and craft techniques tend to disperse outward, directly or indirectly, though now viewed as outmoded, may yet accurately characterize some aspects of the dynamics of cultural ascension. Even when the central culture dominates militarily and culturally, indigenous elements often continue to comprise the core content in locally produced items, particularly in areas that later experience a resurgence.

Several groups developed the internal strength to physically and culturally resist challenges, even transform the aggressors, but others were conquered and assimilated. Their totems were destroyed, their cultural manifestations suppressed, and their identity largely obliterated. However, a few that lacked the strength to vanquish potential aggressors, either concretely or abstractly, managed to continue on as independent enclaves until being overwhelmed by the next wave that might wash over them, even though the resulting amalgamation might retain a semblance of uniqueness. Still others responded to the ongoing challenges by evolving highly self-sufficient, warrior-oriented organizational structures and augmenting their martial power, ensuring not just their own survival but even their ascension into the ranks of formidable political entities.

Broadly envisioned, ancient China might be divided into five regions populated by ethnically distinct peoples or disparate cultures, the four quarters plus the core, the latter an inescapable concept since Chinese (Hua-Hsia) culture and power are currently identified as Yi-Luo River basin manifestations.[20] According to their identities in traditional literature, the Yi dwelled in the east, Miao in the south, Ti in the west, and Jung in the north, though it is also possible to limit the dichotomization to simply the Hsia, Yi, and Miao before charting the dynamics of their interaction. Even though clashes of Yi with Yi were not unknown and the Hsia's relationship with the Tung Yi was generally strong, the major schism frequently fell along an east-to-west axis.

Topographically, the terrain in the west was higher and problems with flooding were fewer, but agriculture more difficult; the heavily watered south had vastly more natural resources, ranging from mineral through animal and aquatic, including rice and China's virtually ubiquitous bamboo, but was prone to semitropical diseases such as malaria and cholera; the east was susceptible to flooding but had abundant fishing and hunting and could easily sustain agricultural efforts; and the north quickly transitioned into the semiarid steppe and frigidity. Lifestyles and customs varied from region to region, totems were different and religious beliefs dissimilar, languages highly localized and often mutually unintelligible. To the extent that differences spawn antagonism and antagonism breeds conflict, the vectors of warfare were inescapable.

Surprisingly, the cultures with the greatest natural advantages did not invariably dominate increasingly larger areas despite their initial superiority. Even though the south and southeast enjoyed a hospitable climate and reportedly the relative absence of warfare,[21] the millet-based cultures in the center initially proved more forceful. Furthermore, Liang-chu culture in the southeast and Tung Yi culture in Shandong were materially advanced over contemporary central plains manifestations, but as already seen they ultimately perished, raising numerous fundamental questions.[22] Only after the Hsia strengthened did the Tung Yi become relatively quiescent and nominally acknowledge Hsia authority before becoming closely allied with the Shang.

Subsequently, just as the San Miao when they were being pummeled by the Hsia, perhaps because the Tung Yi lacked a sense of self-identity and were too fragmented to initiate the concerted action against the Shang necessary to survive, they were vanquished.[23] Thereafter, apart from an adumbrated threat from Ch'u, it would be the peripheral entities of Chou and Ch'in somewhat to the northwest rather than any southern conglomeration that would ascend to power, lending credence to the idea, however limited in applicability, that environmental challenges stimulate self-reliance, that strong political entities more readily evolve out of deprivation than abundance.

Viewed year by year, ancient Chinese history appears highly static, bereft of monumental events or violent changes, but when the centuries are compressed an energetic swirl becomes visible in which groups grow and disperse, cultures rise and fall. In times of low population density the inhabitants of an entire village could easily move about, select a comparatively advantageous site, and establish themselves without clashing with others. But by the middle Neolithic most of the choice locations had been occupied and resident groups faced challenges not just from marauders and raiders but also from various-sized groups forced to relocate en masse by natural disasters, environmental degradation, or overpopulation.

As productivity nurtured population growth, new fields had to be opened to produce the grains necessary to sustain ever greater numbers, enlarging what might be termed the village's radius of activity. This increased the likelihood of friction between members of settlements originally a full day's walk away and depleted the small game and other wild resources in the intermediate zone. At the same time fields that required more than two hours to access tended to foster the establishment of locally clustered dwelling places that in turned served as the nucleus for new settlements. This further augmented the population density and the potential for clashes with preexisting communities, who were then compelled to fortify their perimeters in order to deter or exclude potential interlopers.

Whether the nature of warfare in any particular civilization is culturally determined, although an important question, is essentially irrelevant at the incipient stage of conflict.[24] Based on the patterns visible in ancient China, it may be that the development of highly fortified permanent communities was a greater stimulus to conflict and warfare than the mere accumulation of targetable wealth because of the inherent tension between the "provocative" sedentary lifestyle and seminomadic variants. Farming, even the slash and burn type that may have commenced about 7000 BCE and quickly exhausted the land, required a comparatively fixed lifestyle. Clearing the terrain for agricultural purposes immediately reduced the number and types of plants and animals that had previously been harvested from the former woodlands and marshes and therefore increased the community's dependence on semi-cultivated and cultivated productivity, including the raising of animals for foodstuffs, which commenced with pigs around 6000 BCE.[25]

The emergence of vital crafts such as ceramics further reduced the otherwise inherent mobility of settlement members. For example, ceramic kilns were not easily disassembled, and the transport of fire bricks and other essential components such as turning wheels to other locations in the absence of vehicles would have been an onerous task. Building even the simplest production facilities would have similarly required enormous effort despite the advances in tools and the emergence of copper and bronze implements. In addition to thus having a major stake in preserving the gains wrought from the environment and protecting their dwellings, including increasingly opulent palace structures whose foundations required thousands of working days to construct, the settlement's members needed to defend the infrastructure and the community's very being.

Several modes of conflict seem to have characterized ancient China: purely localized clashes that may have stemmed from personal challenges or revenge but escalated to involve members of a greater alliance or tribe; a fundamental collision of major cultures; explosive friction between minor peoples and a major group, the latter often but not necessarily sedentary, the victim of opportunistic raiding and plundering; and conquest activities undertaken by a major power in the quest for space and control. After becoming dominant, though not all-powerful, the Hsia and Shang can be characterized as having undertaken increasingly frequent, limited-strength campaigns to discourage and repress minor enemies, as well as occasionally more prolonged efforts to expel them from contiguous areas or annihilate them and annex their lands. Despite the lack of animal-powered transport that would have provided increased mobility and facilitated enlarging the domain of conflict, the expanding Shang also confronted the Hsia at several focused points before vanquishing it with a major expeditionary force, changing the nature of warfare.

Postconquest treatment of the defeated seems to have been largely determined by initial objectives. Late Neolithic warfare was not an idyllic exercise or some form of ritual activity, but very much a battle to the finish, as attested by often dramatic and compelling evidence that one group had subjugated another, such as by inscribing the victor's name on sacred vessels of the vanquished[26] or the sacrifice of prisoners, and numerous graves populated by a high proportion of youthful individuals, including women and children, who had violently perished. Whether in legendary Hsia battles or Shang conflict with the steppe peoples, it was preferable for those facing imminent defeat and remnants of conquered clans to migrate to more remote and even less hospitable terrain than to be absorbed, enslaved, or annihilated.

Perhaps most important of all, focused mental effort was applied to warfare and the problems of survival, coincident with more intelligent approaches to authority, administration, and the efforts of life. Although most clearly witnessed in the increasingly thorough planning of towns and fortifications together with the choice of strategically advantageous terrain, new developments in weapons, tactics, and even rudimentary strategy all resulted. The Hsia's conquest of the San Miao, however long it may have required, was apparently made possible by the latter's failure to adopt a viable guerilla strategy, exploit the advantages of their mountainous and marshy terrain, and capitalize on their superior archery power. Fragmented and isolated from potential allies, the tribal groups must have been defeated in detail by a Hsia field force that probably never exceeded a few thousand men.

Traditional accounts portray the Hsia–San Miao clash as a decisive campaign with a preconceived objective that was conducted at long range, further implying that the stage of strategic planning had been realized. However, archaeological indications and demythologizing reveal that the events of a century or more were probably compressed, immediately suggesting that rather than an epic battle undertaken by massed forces on some sacred plains, their war proceeded as a series of localized clashes along their fluctuating boundary as the Lungshan/Hsia increasingly encroached upon the Miao.

Although the consequences of defeat were dispersal and extinction, traditional accounts fail to acknowledge the brutality of a quest prompted by Hsia expansionist ambition. Instead they focus on the proclaimed justness and inevitability of the victory, one mandated by Heaven but resisted by the "ignorant," "uncivilized," and "unruly" Miao, who failed to fathom the inevitability of history. Yet many of their cities far exceeded those identified with the Hsia in multiple ways, and their culture rivaled it in most material aspects, including the perfection of jade objects. Here "Virtue" had failed, whether because it was insufficient or because virtue always proves insufficient in the face of mutual antagonism and a desire for empire.

Warfare in ancient China thus stimulated innovation, social evolution, material progress, and creativity in general, but also shattered the tranquility and security of myriad settlements whose inhabitants had formerly been absorbed in the task of wresting a living from their often harsh environments.[27] Whether out of a desire for additional space or simply an expression of the will to domination, human perversity and malevolent intent quickly ended any possibility of peaceful coexistence except for isolated villages and a few hermits ensconced in remote mountain hideaways. In the absence of any overarching political authority or other form of unification, conflict with external groups became an inescapable aspect of life. Anyone unprepared to do battle could have their possessions confiscated, have their family members enslaved, or be slain.

By not just exploiting martial skills but also thoroughly integrating warrior values, the Shang, if not the Hsia, can be said to have set China on a trajectory of state building and aggressive activity. Whatever degree of credence may be given to the concept of a dynastic cycle, only the leisure of postconquest tranquility would allow for extolling the civil virtues deemed necessary for a sedentary society. Paradoxically, despite the ever-increasing lethality of essentially unremitting warfare, the disparagement of martial values and deprecation of battlefield prowess that would pervade court views throughout the imperial age would first emerge in the Warring States period. Though often detrimental to framing a viable response to external threats, resulting in China's defeat and subjugation, the impact of this disparagement and deprecation remained superficial, unable to blunt the expansionist intent and numerous expeditionary campaigns witnessed across the centuries in the virtual interplay of *yin* and *yang* already prefigured in the dynamics of ancient conflict.

✦ NOTES ✦

COMMON JOURNAL AND COLLECTION titles used throughout the notes and bibliography are abbreviated as follows:

AM	*Asia Major*
BIHP	*Bulletin of the Institute of History and Philology, Academia Sinica* 中央研究院歷史語言研究所集刊
BMFEA	*Bulletin of the Museum of Far Eastern Antiquities*
BSOAS	*Bulletin of the School of Oriental and African Studies*
CKCHCHS	*Chung-kuo Che-hsüeh yü Che-hsüeh-shih* 中國哲學與哲學史
CKCHS	*Chung-kuo Che-hsüeh-shih* 中國哲學史
CKKTS	*Chung-kuo Ku-tai-shih* 中國古代史
CKSYC	*Chung-kuo-shih Yen-chiu* 中國史研究
EC	*Early China*
HCCHS	*Hsien-Ch'in Ch'in Han Shih* 先秦、秦漢史
HJAS	*Harvard Journal of Asiatic Studies*

HSCLWC	*Hsia Shang Chou K'ao-ku-hsüeh Lun-wen-chi* 夏商周考古學論文集, Tsou Heng, ed.
HSLWC	*Hsia Shang Wen-hua Lun-chi* 夏商文化論集, Ch'en Hsü, ed.
HYCLC	*Hsia Wen-hua Yen-chiu Lun-chi* 夏文化研究論集, Li Hsüeh-ch'in, ed.
JAOS	*Journal of the American Oriental Society*
JAS	*Journal of Asian Studies*
JEAA	*Journal of East Asian Archaeology*
KK	*K'ao-ku* 考古
KKHP	*K'ao-ku Hsüeh-pao* 考古學報
KKWW	*K'ao-ku yü Wen-wu* 考古與文物
LSYC	*Li-shih Yen-chiu* 歷史研究
MS	*Monumenta Serica*
PEW	*Philosophy East and West*
SCKKLC	*Shih-ch'ien K'ao-ku Lun-chi* 史前考古論集
SHYCS	*Chung-kuo She-hui K'e-hsüeh–yüan K'ao-ku Yen-chiu-suo* 中國社會科學院考古研究所
STWMYC	*San-tai Wen-ming Yen-chiu* 三代文明研究 (Beijing: K'o-hsüeh, 1999)
TP	*T'oung Pao*
WW	*Wen-wu* 文物

CHAPTER 1

1. Just as China's "failure" to foster scientific and industrial revolutions has been too readily attributed to the stultifying effects of Confucian doctrine, the complex question of the perceived antagonism between the civil (*wen*) and the martial (*wu*)—frequently posed as an assertion that Confucianism enervated the national will to action, thereby rendering the state impotent in the face of brutal, militant hordes who were, however, vastly outnumbered—has so far been treated simplistically. Disparate power groups exploited doctrine to their own ends, and what might be termed a debased, hypocritical form of Confucianism (as distinguished from the pristine doctrine of Confucius and his early followers) often muddled martial discussions and frequently thwarted the implementation of realistic measures. Conversely, many reigns were marked by a decidedly martial ethos and embraced outwardly directed aggressive actions that vigorously challenged all but the most sincere believers in evolved Confucian doctrine. (Naturally, beyond necessarily acknowledging core concepts such as righteousness and benevolence, "Confucian" doctrine varied greatly over the centuries and assumed many guises, ranging from simple dogma to abstruse Sung formulations.)

2. If anything has been learned from the astonishing archaeological discoveries of the past few decades, including from the so-called tomb texts—early bamboo-strip editions of books entombed millennia ago—it should be that from the Chou onward, even within well-defined schools incredible diversity has always characterized Chinese thought. Some of these "schools" consist of a single vision or particular understanding, others of immense, highly convoluted philosophical structures. Several flourished for centuries; others disappeared in mere decades. Moreover, although records of court debates and modern reconstructions of intellectual history naturally tend to discern organized, patterned activity, the voices were usually multiple. Similarly, virtually every possible viewpoint seems to have been expressed in the martial sphere at one time or another, even if only briefly, and become a motivation or justification for action. (Despite the current penchant for denigrating traditional terms such as "Confucian" and rejecting their applicability, they are retained here for their convenience in charting relative viewpoints and organizing essential concepts.)

3. Recovered from a Han dynasty tomb in 1972, *Sun Pin's Military Methods* (*Sun Pin Ping-fa*) was composed in the last half of the fourth century BCE or slightly later by disciples or descendants of the legendary Sun Pin, whose biography is coupled with Sun-tzu's in the *Shih Chi*. Badly fragmented, the text tends to focus more on tactical matters than does *Sun-tzu's Ping-fa*, generally known as the *Art of War*. (For clarity and the convenience of readers, in lieu of appending extensive footnotes and parenthetical material, our translations are sometimes abridged or slightly amplified. A close translation with extensive notes for this passage may be found in Sawyer, *Sun Pin Military Methods*.)

4. "Audience with King Wei," *Sun Pin Ping-fa*. It would have been foolhardy for him to deny, outright, the possibility that Virtue could affect others since it was already becoming a well-entrenched belief. However, Sun Pin could have mentioned several other conflicts involving lesser or almost unknown early Sage authorities. (For a reappraisal of purported clashes, see Wang Yü-ch'eng, CKSYC 1986:3, 71–84.)

5. "Preparation of Strategic Power."

6. "Military Strategy."

7. "Li Lun" ("Discussion of Ritual").

8. "Shih Chün," *Lü-shih Ch'un-ch'iu*.

9. "Li Lun."

10. "Military Strategy."

11. "Military Strategy," *Huai-nan Tzu*.

12. "Benevolence the Foundation," *Ssu-ma Fa*. (A complete translation of the *Ssu-ma Fa*, parts of which probably predate Sun Pin's *Military Methods*, may be found in Sawyer, *The Seven Military Classics of Ancient China*.)

13. "Audience with King Wei." Sun Pin's attribution of the four fundamental military concepts—formations, strategic power, changes, and strategic imbalance of power—to the ancient cultural heroes credited with creating the essential artifacts of civilization is uncommon.

14. "Inferior Strategy." (A complete translation of the *Three Strategies* may be found in Sawyer, *Seven Military Classics*.) "Huang-shih Kung" means "duke of Yellow Rock."

15. At least according to an incident recorded in his *Shih Chi* biography, which, though dubious, was accepted as genuine throughout the imperial period. In the midst of a crisis he reportedly advised an endangered ruler, "I have heard that in civil affairs there must be martial preparation and that in martial affairs there must be civil preparations." (He is also noted as having asserted that he never studied military affairs, only ritual and ceremonial ones, thereby providing crucial ammunition for antiwar factions. The *Analects* also contains his offhand remark that he never studied military deployment.) Certainly one of the core issues of Chinese military history, it is beyond the scope of the present volume. However, for an interesting defense of Confucianism not being responsible for China's military weakness over the centuries, see Kuo Hung-chi, CKCHS 10 (1994): 65–71; for an overview of Chinese attitudes toward warfare and its causes, see Sawyer, "Chinese Warfare: The Paradox of the Unlearned Lesson," *American Diplomacy Magazine* (Fall 1998); and for contrast, the initial chapters of *Military Technology: Missiles and Sieges*.

16. In addition to the PRC's dedicated effort to establish the historicity (and priority) of ancient Chinese culture, Chinese popular media draw on every aspect for plots and content.

17. Some of the more insightful among the many articles that have appeared in recent decades include Li Yung-hsien, HCCHS 1988:10, 13–20; Li Hsien-teng and Yang Ying, HCCHS 2000:3, 9–19; Huang Huai-hsin, KKWW 1997:4, 33–37; Ting Shan, BIHP 3, 517–536; Liu Fan-ti 1999, 70–74; Wang Wen-kuang and Chai Kuo-ch'iang, 2005:9, 1–8; Li Tsung-t'ung, BIHP 39, 27–39; Chao Shih-ch'ao, HCCHS 1999:2, 43–45; Ch'en Ku-ying, HCCHS 1985:7, 4–16; Liu Chung-hui, CKCHCHS 1 (1997): 11–15; Ch'en Hsü, HSLWC 293–302; and Ho Kao, LSYC 1992:3, 69–84. Articles by Western writers include Charles Le Blanc, 45–63, and Gopal Sukhu, EC 30 (2005–2006), 91–153.

18. This is anachronistic because the "hundred surnames," later a term for the ordinary people but initially a reference to those granted the equivalent of surnames, the nobility, did not exist in this period.

19. Generally taken as atmospheric factors or the *ch'i* (*pneuma* or vapor) of the five quarters: north, south, east, west, and middle. This reflects Warring States five-phase (or -element) correlative thought.

20. Five types of animals are enumerated, including two bears, reflecting the theme of "five" throughout. They are variously interpreted as the symbols or totems for five clans or tribes, although practitioners in China's long martial arts tradition like to believe he trained his warriors in fighting techniques derived from the individual animals.

21. "Wu Ti Pen-chi." Although most scholars assign the Yellow Emperor to the Lungshan period, a few such as Hsü Shun-chan (KKWW 1997:4, 19–26) date his activities as early as the middle Yangshao.

22. "Tao Shih." The clause "the blood flowed for a hundred *li*" is a recurrent literary device used to describe other battles as well, particularly the Chou conquest at Mu-yeh. (A *li* was about a third of a mile.)

23. The "blood was great enough to float a pestle" is another trope often employed to describe the battle at Mu-yeh.

24. *T'ai-p'ing Yü-lan, chüan* 15, citing the *Chih-lin*. Chuo Li is said to have been in the vicinity of Beijing, as a local name implies.

25. *Chüan* 17, "Ta Huang-pei Ching," *Shan-hai Ching.* For a more extensive mythical recounting, see T'ao Yang and Chung Hsiu, 1990, 504–508.

26. Snakes and dragons figure prominently in several legends about the Yellow Emperor and Ch'ih Yu, both men supposedly being descended from snakes on their mothers' side but from the bear and ox respectively on their fathers'. The Yellow Emperor is frequently associated with a white dragon, and one legend has the white dragon battling with either a red or black tiger, presumably Ch'ih Yu's clan, which perishes (T'u Wu-chou, HCCHS 1984:3, 9–14).

27. "Feng Shan Shu," *Shih-chi.* Ssu-ma Ch'ien notes the eight were said to have been established by the T'ai Kung (traditionally recognized as the founder of the state of Ch'i) and that an altar to Ch'ih Yu had been found on Ch'i's western border (which would befit Ch'i's strong military heritage). Somewhat contradictorily, Ch'i is also noted for having esteemed the Yellow Emperor. (See Hu Chia-ts'ung, HCCHS 1991:1, 19–26.) Various dates based on myths, archaeology, and outright assumptions have been suggested for this clash, 2700 to 2600 BCE being the most common.

28. "Ti Shu," *Kuan-tzu.*

29. As will be discussed in the weapons section, nothing more than a dirk existed around 2600 BCE. There are various lists for the five weapons, some of which include chariots and armor.

30. See Hsiao Ping, CKKTS 1994:11, 7–12.

31. Because maple leaves turn red in the fall, the (bronze) shackles that restrained Ch'ih Yu are said to have turned into a forest of maple trees; maple trees continue to be venerated by Miao remnants even today. (Wang Yen-chün, HCCHS 1988:6, 11–12.)

32. For further discussion, see Wang Chih-p'ing, 1999:4, 95–98.

33. For a discussion of the widespread Han admiration for Ch'ih Yu (contrary to the idea that the Han only esteemed Confucian values), see Wang Tzu-chin, HCCHS 2006:6, 70–75.

34. Found at Ma-wang-tui and now included among the collated texts known as the "Huang Ti Ssu-ching."

35. See Chang Ch'i-yün, 1961, vol. 1, 22–25.

36. Directing troops deployed for battle was one of the most formidable problems of antiquity. China early on developed formations and segmentation and control measures that allowed generals to command rather than simply lead from the front. Citing an ancient text, in "Military Combat" the *Art of War* states: "Because they could not hear each other, they made gongs and drums. Because they could not see each other, they made pennants and flags." Drums were particularly emphasized. (For example, see "The Tao of the General," *Wu-tzu*; "Strict Positions," *Ssu-ma Fa*; and "Orders for Restraining the Troops," *Wei Liao-tzu*.)

37. The classic military writings adroitly exploit righteousness as a motivating factor. For example, the *Ssu-ma Fa's* first chapter, "Benevolence the Foundation," elaborates the conditions under which justified campaigns might be mounted.

38. Chang Ch'i-yün believes they crossed during the winter when the Yellow River would have been frozen (which would obviate any need for boats). However, the climate was considerably warmer at this time, and the water's volume probably was greater due to higher rainfall levels, making it unlikely that it would have fully frozen; crossings in later times required placing a rope across to create an ice barrier.

39. Several scholars have noted that stories about important events, particularly those identified with place-names, tend to enjoy localized preservation. Although not primary evidence, some have proven to retain surprising vestiges of ancient events.

40. Lü Wen-yü, HCCHS 2000:1, 10–17. Conversely, Teng Shu-p'ing (KKWW 1999:5, 15–27) identifies the Tung Yi with Ch'ih Yu and sees the conflict as emblematic of the clash between Hua-Hsia cultural predispositions in the middle and upper reaches of the Yellow River and Tung Yi manifestations in Shandong. (Teng's interpretation seems somewhat problematic because the Tung Yi totem was a bird, whereas Ch'ih Yu had an ox head, a difficulty that Teng

somewhat unsatisfactorily deflects by claiming that they subsequently acquired the bird association.) Teng also believes that four jade Tung Yi artifacts discovered in the west in Shanxi and
Shaanxi provide evidence of the severity of this primordial clash, because Ch'ih Yu's clan had
to disperse to the west and south after their defeat.

41. Li Yu-mou, CKKTS 1994:2, 39–45.

42. Wang Yen-chün, HCCHS 1988:6, 11–15.

43. However, note that the Fu and Sui may have been two tribal groups rather than a single
individual. In addition, not everyone agrees that the Yellow Emperor came from the west and
Ch'ih Yu from the east, while arguments about whether Ch'ih Yu should be identified with the
Miao in the south or the Tung Yi continue unabated. For example, Hsiao Ping (CKKTS 1994:11,
7–12) argues for Ch'ih Yu having been one of the great ancestors of the early southern Miao
chieftains, who were in turn descendants of the Nine Li and active around the Yangtze's middle
reaches, especially the vicinity of Tung-t'ing and P'o-yang lakes.

44. Not impossible if they represent the late Shandong Lungshan cultural strata, in which
bronze weapons began to appear, although in minuscule numbers.

45. How much wetter the east would have been is highly questionable. (There has always
been a significant difference in total rainfall between the north and south rather than the east
and west, accounting for rice being a southern staple.) This interpretation would require that
Ch'ih Yu be a representative of lower Yangtze culture and may perhaps be grounds for rethinking
the conflict, as these coexistent cultures were quite dynamic and resilient.

46. Although such claims lack any evidence, this sort of interpretation is frequently found
in popular works on China's military history, such as Chang Hsiu-p'ing's *One Hundred Battles
That Influenced China.*

47. The twenty-seventh century BCE is frequently suggested as the Yellow Emperor's reign
period, whether actual or mythical / symbolic. (See, for example, Teng Shu-p'ing, KKWW
1999:5, 15.)

48. "All under Heaven" is a late term, and this well-known passage is found in the late part
known as the "Shih Ts'u." (The second part may also be translated as "Availing himself of bows
and arrows, he overawed all under Heaven," thereby emphasizing the Yellow Emperor's
aggressiveness.)

49. See Fang Li-chung, HCCHS 1989:3, 21.

50. Widespread belief in his magical powers is found in the Warring States and thereafter.

51. Recorded in some versions, the fog is apparently a later addition that may have been
prompted by Warring States experience in employing smoke screens. (For the history of smoke
and smoke screens in Chinese warfare, see Sawyer, *Fire and Water.*)

52. Joseph Needham, in *Physics and Physical Technology: Mechanical Engineering*, 286–303, has
speculated that some sort of differential gearing may have been employed and considers it the
first homeostatic machine and an initial step in cybernetics. Andre Sleeswyk, in "Reconstruction
of the South-Pointing Chariots of the Northern Sung Dynasty," has provided a further examination of the gearing, and a modern PRC reconstruction has been prominently displayed over
the past decade.

53. See Yang K'uan's extensive discussions, 1941, 65ff. As symbolized by their respective
totems, the Yellow and Red Emperors' tribes are said to have merged through these conflicts,
creating the heritage venerated (and exploited) by Warring States Confucian culture. (See Lin
Hsiang-keng, HCCHS 1984:1, 3–10; Wang Yen-chün, HCCHS 1988:6, 11–15; and Wu Jui and
Cheng Li, CKCHS 1996:3, 4–8.)

54. Lin Hsiang-keng, HCCHS 1984:1, 3–10; Wang Yen-chün, HCCHS 1988:6, 11–15; Wu Jui
and Cheng Li, CKCHS 1996:3, 4–8.

55. Several monographs on the topic have appeared in the past two decades, significantly diminishing the value of H. H. Turney-High's classic discussion, *Primitive War;* they include Jean

Guilaine and Jean Zammit, *The Origins of War: Violence in Prehistory*; Elizabeth N. Arkush and Mark W. Allen, eds., *The Archaeology of Warfare: Prehistories of Raiding and Conquest*; Steven A. LeBlanc, *Constant Battles: The Myth of the Peaceful, Noble Savage*; John Carman and Anthony Harding, eds., *Ancient Warfare*; Anthony Stevens, *The Roots of War*; and Arther Ferrill, *The Origins of War*. Although no more than a hundred men could have been effectively fielded without a minimal administrative hierarchy, modern studies still tend to claim that China's legendary period lacked any form of military organization, placing their forces below the military horizon in accord with Turney-High's conception. (For example, see Liu Chan, 1992, 4 and 20ff.)

56. This is considered another characteristic of so-called primitive warfare. (One of the remarkable, largely unnoticed aspects of later Chinese warfare was the common practice of armies simply reflagging vanquished enemy troops and integrating them en masse, as advocated in "Waging War" in the *Art of War*. How loyal, dedicated, and enthusiastic they historically proved has yet to be examined.)

CHAPTER 2

1. Two examples would be Chi-kung-shan and Wu-chia Ta-p'ing at Wei-ning in Kuei-chou, both dating to about 1300 to 700 BCE (Kuei-chou-sheng WWKK YCS et al., KK 2006:8, 11–27 and 28–39, and Chang Ho-jung and Luo Erh-hu, KK 2006:8, 57–66).

2. A classic Lungshan site of some 140,000 square meters known as "the defensive ancient city" ("Fang-ku-ch'eng"), located in Shandong, furnishes a particularly good example of fortification continuity; it was employed right through the Warring States, when it served as a stronghold on Lu's eastern border. (See Fang-ch'eng K'ao-ku Kung-tso-tui, KK 2005:10, 25–36.)

3. Even Paul Wheatley's erudite but now outdated examination of city growth, *The Pivot of the Four Quarters*, never ponders the craft of wall building. Furthermore, the two volumes in Needham's *Science and Civilisation in China* series that contemplate essential aspects of fortification—*Civil Engineering and Nautics* and *Military Sieges and Technology*—barely mention Neolithic and Shang fortifications.

4. Despite many hundreds of archaeological reports, only a few synthetic overviews such as P'ei An-p'ing, KK 2004:11, 63–76, and Shao Wangping, JEAA 2 (2000), 195–226, have appeared. (See also Shao's "The Interaction Sphere of the Longshan Period.") A few works have discussed city history, including Ch'ü Ying-chieh, *Ku-tai Ch'eng-shih*, 2003; Ning Yüeh-ming et al., *Chung-kuo Ch'eng-shih Fa-chan-shih*, 1994; and Yang K'uan, *Chung-kuo Ku-tai Tu-ch'eng Chih-tu-shih Yen-chiu*. Valuable materials also appear in Li Liu, *The Chinese Neolithic: Trajectories to Early States*, and Chang Kwang-chih and Xu Pingfang, eds., *The Formation of Chinese Civilization*.

5. Other than at Ma-chia-yüan Ku-ch'eng, no evidence has yet been reported for defensive works atop walls, not unexpectedly since erosion has affected almost every wall so far excavated.

6. For the history of aquatic warfare in China, see Ralph Sawyer, *Fire and Water*.

7. In "Military Disposition," Sun-tzu states: "One who cannot be victorious assumes a defensive posture, one who can be victorious attacks. By assuming a defensive posture your strength will be more than adequate." In "Planning Offensives" he further observes: "The highest realization of warfare is to attack the enemy's plans; next is to attack their alliances; next to attack their army; and the lowest is to attack their fortified cities." The passage continues with a sweeping condemnation of the wastefulness of citadel assaults, though not their complete exclusion. (For further discussion and a complete translation of the *Art of War* with historical introduction and textual notes, see Sawyer, *Sun-tzu Art of War*.)

8. The eighty-eight pits at Ts'u-shan near Wu-an in Hebei held varying amounts of desiccated grain, which, when freshly stored, would have weighed over 50 metric tons. (See Jen Shih-nan, KK 1995:1, 38–39.) Similar reserves (such as 120 metric tons at Ho-mu-tu) have been found at

other sites. However, rather than a surplus, they may have been the basic subsistence requirement for the coming year.

9. The Neolithic is generally taken as encompassing 10,000 to 3500 BCE, though others extend it to 2100, the reputed date of the Hsia's inception. Bronze metallurgy emerged during the Hsia; nevertheless, as it remained primarily a stone culture, the first two centuries are sometimes subsumed within the Neolithic as well.

10. "A Small State with Few People" in the traditionally received text. (For a complete translation and contextual discussion, see Sawyer, *The Tao of War*.)

11. See Jen Shih-nan, 37. (Various dates have been derived from the recovered artifacts, resulting in controversy.)

12. Jen Shih-nan, 37–38.

13. Yen Wen-ming, KKWW 1997:2, 35.

14. Site descriptions are based on P'ei An-p'ing, KK 2004:11, 66–67, and Chang Hsüeh-hai, KKWW 1999:1, 36–43. There are numerous difficulties reconstructing the earliest stage and disagreements about how to interpret the archaeological evidence. Some archaeologists have dated the site to nearly 6000 BCE, but the more commonly recognized range is Yen Wen-ming's 5450 to 5100 BCE (cited on page 35 of *The Formation of Chinese Civilization*).

15. Another P'ei-li-kang platform site in Henan was not only surrounded by water on three sides, but also further protected by two shallow ditches about half a meter deep that could have functioned for drainage or demarcation as much as defense. (One of the ditches varied between a functional 1.65 and 5.15 meters in width, but the other was only 0.75 to 1.1 meters wide.) However, the location's desirability is evident from its continuous occupation into the Erh-li-t'ou cultural phase (Chang Sung-lin et al., KK 2008:5, 3–20).

16. For site reports see Kan-su-sheng WWKK YCS, KK 2003:6, 19–31, and Lang Shu-te, KK 2003:6, 83–89. Only minimal information on the ditch has yet been provided.

17. Lin-t'ung Chiang-chai, which was continuously occupied right into the Warring States, was protected by a circular ditch that encompassed approximately 55,000 square meters or slightly more than Pan-p'o. However, as the ditch itself has not yet been analyzed, little can yet be said about its profile or overall significance. (For a basic report, see Honan Sheng Kung-yi-shih Wen-wu Pao-hu Kuan-li-suo, KK 1995:4, 297–304, as well as Ning Yüeh-ming's appraisal, *Ch'eng-shih Fa-chan-shih*, 13.)

18. For a basic discussion of the Yangshao culture that was originally defined by its red-tinged pottery (in contrast to the black pottery of the later Lungshan period), see K. C. Chang, 1986 (which covers Pan-p'o on pp. 112–123) or Yen Wen-ming, *Yang-shao Wen-hua Yen-chiu*. Yangshao culture, generally dated from 5000 to 3000 BCE, was centered in the Kuan-chung area and included Gansu around the T'ien-shui River and the upper reaches of the Ch'ing-shui, the middle and upper reaches of the Luo River, the upper portion of the Han River in Shaanxi, the southern portion of Ching-hsi, and the Yü-hsi area, with the Wei River as the focus (Chang Hung-yen, KKWW 2006:5, 66–70, and WW 2006:9, 62–69, 78). Our discussion of Pan-p'o is primarily based on Ch'ien Yao-p'eng's two articles, KK 1998:2, 45–52, and KK 1999:6, 69–77. However, also see Yen Wen-ming, SCKKLC, 146–153. (Evidence indicates that Lin-t'ung Chiang-chai was occupied well into the Warring States period.)

19. For example, Ch'ien Yao-p'eng believes the site flourished from 4770 to 4190 BCE or almost 600 years, but others such as Ning Yüeh-ming et al., 1994, 12–13, date it as late as 4000 BCE.

20. Ch'ien Yao-p'eng, KK 1999:6, 69.

21. Both Pan-p'o and Lin-t'ung Chiang-chai are marked by raised platforms and well-smoothed earth. It should be noted that Ch'ien Yao-p'eng, 46, believes that simply widening the walls, as was also done at P'eng-t'ou-shan, should be understood as an afterthought to ditch excavation rather than a deliberate attempt to solidify them and increase their height. In contradiction, Chang Hsüeh-hai concludes that the interior mounded wall was actually the result of a later,

more deliberate effort, and notes that there is evidence for a third ditch some ten meters beyond the main one that may have partially furnished the dirt for the inner wall or a no longer visible outer wall. (See Chang Hsüeh-hai, KKWW 1999:1, 41–43.)

22. See Ch'ien Yao-p'eng, KK 1998:2, 48–52.

23. For a recent discussion see Pi Shuo-pen et al., KKWW 2008:1, 9–17. Yen Wen-ming, WW 1990:12, 21–26, also briefly discusses the site's significance, but cites somewhat different measurements, including 160 by 210 meters for a total area of 33,600 square meters. (The settlement's probable appearance is depicted in Chang Kwang-chih and Xu Pingfang, eds., *The Formation of Chinese Civilization*, 68–69.)

24. For three illustrative sites, see SHYCS Nei Meng-ku Ti-yi Kung-tso-tui, KK 2004:7, 3–8.

25. Liu Kuo-hsiang, KKWW 2001:6, 58–67. In contrast, the settlement at Hou-ma Tung-ch'eng-wang identified with the Miao-ti-kou culture has a ditch about 11 meters wide and 2.6 meters deep. (See Kao T'ien-lin, KK 1992:1 62–68, 93.)

26. Hu-pei-sheng WWKK YCS, KK 2008:11, 3–14.

27. For a general discussion of the nature of the Chinese city, see Liu Ch'ing-chu, KK 2007:7, 60–69.

28. Key site reports include Hunan-sheng WWKK YCS et al., WW 1993:12, 19–30, and Hunan-sheng WWKK YCS, WW 1999:6, 4–17. Also useful are Chang Hsü-ch'iu, KK 1994:7, 629–634, and P'ei An-p'ing, KK 2004:11, 69–70. (The diameter to the interior side of the walls is just 310 meters.)

29. The account that follows is based on P'ei An-p'ing, KK 2004:11, 69–70.

30. Discussion of this site is based on Chang Yü-shih et al., WW 1999:7, 4–15; Ch'ien Yao-p'eng, WW 1999:7, 41–45; Li Hsin, KK 2008:1, 72–80; and Jen Shih-nan, KK 1998:1, 1.

31. Contrary to claims of analysts such as Ch'ien Yao-p'eng.

32. Based on Hsi-shan's engineering, Ch'ien Yao-p'eng, WW 99:7, 41–45, claims that the Yangtze area lagged behind the Yellow River in tamped earth techniques. (Pan-p'o is marked by a gradual slope rather than the sharp profile that can only be accomplished with framing.) However, others have claimed that the Yangtze area was more advanced.

33. See Li Hsin, KK 2008:1, 72–80.

34. For a brief overview of Lungshan sites and culture, see Shao Wangping, "The Interaction Sphere of the Longshan Period."

35. For site reports see Ho-nan-sheng WWKK YCS, KK 2000:3, 1–20; Yüan Kuang-k'uo, KK 2000:3, 21–38; and Yüan Kuang-k'uo, KK 2000:3, 39–44. Although Shao Wangping cites a core date of 2300 BCE for the site and most would see a break about 2100, some controversy has arisen over whether it was immediately occupied by Erh-li-t'ou cultural members or there was a hiatus. The site was finally abandoned late in Erh-li-t'ou's fourth period.

36. Yüan Kuang-k'uo believes climatic warming may have speeded the demise of Lungshan civilization, but few others would attribute it directly to increased rainfall, especially since the climatic optimum had long been passed and disagreement exists over the actual pattern. However, the western wall was apparently destroyed by flooding near the end of the Lungshan period but then rebuilt. Yüan identifies Meng-chuang with the legendary Kung Kung, who reputedly helped quell the floods of antiquity (but should have considerably predated this era, being closely identified with Yü), and asserts that the site was abandoned for two centuries after the flood.

37. Shao Wangping, "Interaction Sphere of the Longshan Period," 103. See also Wu Ju-tso, CKKTS 1995:8, 12–20, and Li Liu, *The Chinese Neolithic*, 193–208.

38. Also referred to as Ch'eng-tzu-ya, it is summarily discussed by K. C. Chang in 1986, 248–250. However, his observation that this constitutes the "first erection of a defensive wall by a prehistoric settlement" has been outdated by discoveries over the past twenty years, as will be seen from the examples given in the discussion. (Other useful contextual discussions are found

in Ning Yüeh-ming et al., *Chung-kuo Ch'eng-shih Fa-chan-shih* 27; Chang Hsüeh-hai, WW 1996:12, 41–42; and Jen Shih-nan, KK 1998:1, especially 2. Unfortunately, there are a number of unresolved discrepancies in the dimensions provided for various aspects of the walls, with Jen giving 430 meters east to west and 530 north to south.)

39. However, the *Chung-kuo Ch'eng-shih Fa-chan-shih*, 27, notes the wall's width as 10.6 meters and suggests it suddenly cuts inward just above the broad platform of the foundation.

40. One analyst even believes that the walls were constructed between 1900 and 1700 BCE and concludes that rather than being a prototypical city or Hsia capital, it acted as a *ch'eng-pao*, a fortified town or early castle. (See *Chung-kuo Ch'eng-shih Fa-chan-shih*, 27, where the various arguments about its identification are conveniently summarized.)

41. Jen Shih-nan, KK 1998:1, 2. Chang Hsüeh-hai (1996, 50) cites a similar tree calibrated date of 4565± 130 BP.

42. Discussions of the site may be found in K. C. Chang, 1986, *Archaeology*, 262–267; Ho-nan-sheng WW YCS et al., WW 1983:3, 21–36; Jen Shih-nan, KK 1998:1, 2.

43. Only Jen Shih-nan, KK 1998:1, 2, has described the presence of this moat.

44. Shan-tung-sheng WWKK YCS et al., KK 1997:5, 11–24.

45. Based on Tu Tsai-chung, CKKTS 1995:8, 5–11. Significant nearby sites include Ting-kung at 108,000 square meters interior area, T'ien-wang with 150,000, Wu-lien T'an-t'u at 250,000, and T'eng-chu Yu-lou at 250,000.

46. Following the dates provided by Tu Tsai-chung. However, Shao Wangping dates the inner wall to the late middle Lungshan period and the outer to the early late Lungshan.

47. Ching-chou Po-wu-kuan, KK 1997:5, 1–24.

48. Chang Hsü-ch'iu, KK 1994:7, 629–634. The site descriptions that follow are taken from Chang and others, as individually noted.

49. Ching-chou-shih Po-wu-kuan et al., KK 1998:4, 16–38; Chang Hsü-ch'iu, KK 1994:7, 630.

50. Dimensions from Jen Shih-nan, KK 1998:1, 6.

51. The best site report is Ching-chou Po-wu-kuan and Chia Han-ch'ing, WW 1998:6, 25–29.

52. Dimensions reported for this site vary greatly. (These are taken from Wang Hung-hsing, KK 2003:9, 68.)

53. Wang Hung-hsing, 65–68. (For a brief discussion of Shih-chia-ho culture, see Fan Li, KKWW 1999:4, 50–60.)

54. Chang Hsü-ch'iu, KK 1994:7, 632. (Chang's article does not include Chi-ming-ch'eng.) Other sites include Ch'i-chiao-ch'eng at 150,000 square meters, Chin-men Ma-chia-yüan at 240,000, and Yin-ch'eng T'ao-chia-hu at a vast 670,000.

55. Chang Hsü-ch'iu believes that the conditions and the technology for wall construction were first realized in the middle Yangtze area rather than the Yellow River and that any influence from the latter's cultures was relatively weak.

56. For the initial report see Che-chiang-sheng WWKK YCS, KK 2008:7, 3–10, and for context see Chang Li and Wu Chien-p'ing, WW 2007:2, 74–80. No radiocarbon dates have yet been provided, but Liang-chu culture is generally dated from 3400 or 3200 to 2250 or 2200 BCE.

CHAPTER 3

1. Liu Kuo-hsiang, KKWW 2001:6, 58–67. In contrast, the settlement at Hou-ma Tung-ch'eng-wang identified with the Miao-ti-kou culture has an 11-meter-wide, 2.6-meter-deep ditch (Kao T'ien-lin, KK 1992:1, 62–68, 93).

2. Articles appraising their differences and similarities are beginning to appear. For example, see Wei Chien and Ts'ao Chien-en, WW 1999:2, 57–62, upon which the following discussion is based.

3. Chin Kuei-yün, KKHP 2004:4, 485–505. Huai-ko-erh-ch'i Pai-ts'ao-t'a, the oldest of the Yellow River sites, has been dated to 3000 BCE, roughly comparable to Pan-p'o's fourth stage. Three others are dated to 2700 BCE, and Chai-tzu-shang, the youngest, to 2300. The sites at Tai-hai cluster around 2500 BCE, those at Pao-t'ou around 2700 BCE.

4. Wei-chün, A-shan, Hsi-yüan, Sha-mu-chia, and Hei-ma-pan.

5. In Ching-ch'eng-hsien, Mt. Lao-hu, Hsi-pai-yü, Pan-ch'eng, and Ta-miao-p'o.

6. At Huai-ko-erh-ch'i: Pai-ts'ao-t'a, Chai-tz'u-t'a, Chai-tz'u-shang, and Shao-sha-wan. At Ch'ing-shui-ho-hsien: Ma-lu-t'a and Hou-ch'eng-tsui.

7. Ch'en Kuo-ch'ing and Chang Ch'uan-chao, KK 2008:1, 46–55. Unfortunately, full details of the stone walls visible on the site diagrams have yet to be published. Additional sites continue to be uncovered, such as a complex dating to the lower Hsia-chia-tien composed by adjoining twin stone citadels marked by massive walls constructed from moderate-sized stones. (See Nei-Meng-ku WWKK YCS, KK 2007:7, 17–27.)

8. Ch'en and Chang, KK 2008:1, 48. However, Yen Wen-ming (JEAA 1 [1999]: 143), has surprisingly asserted that it and all Chinese walls were basically tamped earth with stone facings.

9. Rather than solely or directly exerted by the early Shang, these pressures may have indirectly resulted from vanquished Hsia population groups fleeing to safer, if less hospitable, terrain. However, this is not one of the article's conclusions.

10. Liao-ning-sheng WWKK YCS, KK 1992:5, 399–417. Although identified with the lower Hsia-chia-tien and termed a mixed Bronze Age culture, except for a single 13.2-cm. long bronze knife, all the implements so far recovered were fabricated from stone (axes, knives, and arrowheads) or bone (arrowheads). Reportedly the indigenous populace was subsequently displaced by Bronze Age or other unspecified northern peoples.

11. Somewhat divergent dates appear in the literature, ranging from 2800–2000 to 2500–1800.

12. See Wang Yi and Sun Hua, KK 1999:8, 60–73. (The Pao-tun culture heavily influenced San-hsing-tui, but other factors also played a role over the intervening centuries.)

13. For overviews see Tuan Yü and Ch'en Chien, HCCHS 2002:2, 57–62, and Wang Yi, JEAA 5 (2006): 109–148. Tuan and Ch'en believe the homogeneity of the sites suggests common development, evidence of having evolved from the clan stage to a more centralized power. However, the strong defensive character of the fortifications suggests an environment pervaded by threats and aggression, though whether caused by internal strife or external enemies is unclear.

14. Ch'eng-tu-shih Wen-wu K'ao-ku Kung-tso-tui et al., WW 1998:12, 38–56; Jen Shih-nan, KK 1998:1, 9. (It is also referred to as Yü-fu-ch'eng.)

15. In addition to Wang Yi, 122–125, see Ch'eng-tu-shih WWKK Kung-tso-tui, WW 1999:1, 32–42.

16. Wang Yi, 122–125.

17. For Shuang-ho see also Ch'eng-tu-shih WWKK Kung-tso-tui, KK 2002:11, 3–19. (The site was occupied between 2500 and 2000 BCE but there are discrepancies in the reported dimensions. For example, Wang Yi [127–128] gives 270 meters for the north, 420 for the east, and 110 for the south as the dimensions of the inner perimeter, the west wall having been destroyed, and 325 on the north, 500 in the east, and 120 on the south, with walls that were 18 to 30 meters wide at top, for the outer.)

18. The best analytical article to date is Tuan Yü, HCCHS 1993:4, 37–48. There is considerable disagreement over whether the cultural enclaves at these two sites are lineal predecessors of Shu and Pa. (For context see Chang T'ien-en, KKWW 1998:5, 68–77; Chao Tien-tseng, WW 1987:10, 18–23; and Sung Chih-min, KKHP 1999:2, 123–140.)

19. For a concise discussion, see Robert Bagley's "Shang Archaeology," 212–219.

20. For further discussion of San-hsing-tui and Ch'eng-tu, see Tuan Yü, CKKTS 1994:1, 63–70, and HCCHS 2006:5, 16–20; Sung Chih-min, who emphasizes local origination, KKHP 1999:2, 123–140; Ssu-ch'uan-sheng Wen-wu Kuan-li Wei-yüan-hui et al., WW 1989:5, 1–20; Yang Hua,

KKWW 1995:1, 30–43; Huang Chien-hua, HCCHS 2001:6, 21–27; Lothar von Falkenhausen, JEAA 5 (2006): 191–245; Jay Xu, JEAA 5 (2006): 149–190 (who suggests it was contemporary with the late Shang, 1300–1050 BCE); Ch'ü Hsiao-ch'iang, *San-hsing-tui Ch'uan-ch'i* (1999); and Robert Bagley, *Ancient Sichuan.*

21. Tuan Yü, HCCHS 2008:6, 3–9, has suggested that Shu/San-hsing-tui controlled Yünnan's copper, tin, and lead deposits, compelling the Shang to trade with them and that Shu was never a conquest objective or a subordinate state. (There was a shortage of tin in the Shang core domain. Although the Sichuan plains area had ample copper supplies, it lacked tin, but Shu could have acquired it, just like lead, from Yünnan. Tuan concludes that the absence of Shang artifacts in Yünnan indicates the Shang must have employed Shu as an intermediary and therefore left them alone to guarantee continuity of supply. This would explain the presence of Shang ritual vessels in Shu but almost the complete absence of Shang weapons.) Tuan believes San-hsing-tui flourished around the middle of the Shang.

22. See Tuan Yü, CKKTS, 1994:1, 45, for calculations and methods. Various areas for family space have been proposed, ranging from a very small 30 square meters to a spacious 268. Based on later population densities, approximately 155 to 160 square meters per household seems supportable, with each household encompassing five people.

23. Yang Hua, KKWW 1995:1, 30–43.

24. Using the population density at Lin-tzu in the Warring States period as a reference yields 280,000, but if 155 square meters is employed, the total population could have been 484,000. (See Ch'ü Hsiao-ch'iang, *San-hsing-tui—Ku-Shu Wang-kuo te Fa-hsiang*, 1999, 73.)

25. Tuan Yü, HCCHS 1993:4, 37–54. (A useful study of Ch'eng-tu culture is Chiang Chang-hua et al., KKHP 2002:1, 1–22.)

26. For an overview see Zhu Zhangyi et al., JEAA 5 (2006): 247–276. It has been characterized as the defining site for Shih-erh-ch'iao culture and a precursor of the Shang-wang-chia-kuai culture of the strong Sichuan states that developed in the Warring States period.

27. Many contemporary analysts postulate that the development of targetable wealth constitutes the minimum requisite condition (and motivation) for "true" rather than "ritual" warfare. However, when all men exist at the subsistence level, plundering may be the best alternative.

28. "Ching Chu."

29. "Ch'i Su."

30. Chao Kung, thirty-second year. At that time the feudal states under Chin's nominal leadership had been defending the Chou capital at Ch'eng Chou, and the work was prompted when the king of Chou requested that they wall it instead of maintaining (threatening) forces there.

31. A Chou dynasty *chang* or rod was about eight feet.

32. "Mien," Mao # 237 in James Legge's classic translation, *The She King*, 440.

33. Building fires against walls, although not unknown, was primarily done in the Neolithic period for the interior of houses, house foundations, and floors and in the Yangshao period even for clay smeared onto wooden interior beams to increase their hardness and reduce water absorption (Li Nai-sheng et al., KK 2005:10, 76–82).

34. Tu Cheng-sheng, BIHP 58:1, calculated the lower figure; An Chin-huai, "Shih-lun Cheng-chou Shang-tai Ch'eng-chih—Ao-tu," 77, the larger.

35. An Chin-huai, "Ao-tu," 77. A moderately fit individual digging in a pile of soft soil or sand with a modern shovel can remove two to three cubic feet per minute or several cubic yards per hour, easily sustaining a rate of one cubic yard per hour over a full day.

36. An Chin-huai, "Ao-tu," 77. Even though An's estimates for bronze tools seem ridiculously low, his calculations for the walls, based on employing a mixture of bronze and stone tools, may actually be somewhat optimistic. Agricultural tools were almost always fabricated from stone or bone at this early date, bronze being reserved for ritual vessels and weapons. However, given the low excavation rate, it is more likely that the majority of workers—perhaps 8,000 out

of the 10,000—would have been employed in digging, the rest in transporting and pounding the soil.

37. Modern rammed earth construction in the American Southwest reveals that compacting the layers, even with hydraulic tampers, requires a lengthy period. For example, the relatively small area of two feet in width by ten feet in length easily demands an hour to achieve the required consistency.

38. A questionable assumption given the highly varying densities found in various urban situations, although perhaps appropriate since the site is described as having considerable open space. (Even in the absence of multiple-level dwellings, some early settlements in China were so densely packed that each family often occupied a scant forty square meters.) Despite the analysts' earnestness, calculations of this sort venture far beyond speculative.

39. Tsou Heng, HSCLWC, 59, estimates five years; Tu Cheng-sheng, BIHP 58:1, 10, estimates eight years based on a wall width of 20 meters at the base, 5 meters at the top, and an original height of 10 meters. He would apportion the 10,000 workers as 3,000 excavating dirt, 3,000 moving it, 3,500 pounding it, and 500 employed in miscellaneous tasks; and David Keightley, EC 3 (1977): 58, and JAOS 93.4.1973:530–531, estimates 12.5 years.

40. The rate of 0.5 cubic meter/day has been employed to estimate a construction time of fourteen days for the ditch at Pa-shih-tang, assuming that 700 cubic meters had to be excavated and the work was undertaken by 100 people. (See P'ei An-p'ing, KK 2004:11, 63–76.) On an unstated basis Fang Yen-ming (KK 2006:9, 22–23) estimates that 1,000 sturdy men working eight hours per day would have taken fourteen months to construct the fortifications at Wang-ch'eng-kang, with additional personnel having been required for planning and supervisory tasks. He further asserts that this is far more than the site could provide, clear evidence that central authority was capable of coercing the surrounding twenty or so villages to furnish perhaps 100 men each. However, though Wang-ch'eng-kang clearly attests to the growth of centralized authority and coercive power, based on the average living area of 150 square meters per family, a site of 300,000 square meters could have had 2,000 families of five, a number easily capable of diverting 1,000 males from farming and other tasks even if the work were continuously undertaken.

41. For example, a village dated to roughly 2700 BCE located at Yung-lang on the eastern side of the Pearl River near Hong Kong was able to exploit the surrounding hills for protection. It therefore merely added a moat for protection, thus showing regional variation in an already typical pattern of combining walls and moats. (For the site report see Hsiang-kang Ku-wu Ku-chi Pan-shih-ch'u, KK 1997:6, 35–53.)

CHAPTER 4

1. For a discussion of Hsia writing see Ts'ao Ting-yün, KK 2004:12, pp. 76–83; Li Ch'iao, HCCHS 1992:5, 21–26; and Ch'ang Yao-hua, HYCLC, 1996, 252–265. For writing's inception in China see Feng Shih, KK 1994:1, 37–54; Wang Heng-chieh, KK 1991:12, 1119–1120, 1108; or Wang En-t'ien et al., "Chuan-chia P'i-t'an Ting-kung Yi-chih Ch'u-t'u T'ao-wen," KK 1993:4, 344–354, 375. Feng Shih (KKHP 2008:3, 273–290) has recently argued that the two characters on a pottery shard recovered from T'ao-ssu at Hsiang-fen should be interpreted as "wen yi" and therefore evidence for the Hsia capital and proof of the Hsia's existence. Chao Kuang-hsien (HYCLC, 1996, 122–123), among others, sees sufficient proof of the Hsia's existence in early texts.

2. Ch'en Ch'un and Kung Hsin, HCCHS 2004:6, 3–12. See also Ch'en Chih, CKSYC 2004:1, 3–22.

3. For example, see Ho Chien-an, HCCHS 1987:1, 33–46; Chang T'ien-en, KKWW 2000:3, 44–50, 84; Li Wei-ming, HCCHS 2005:5, 40–45; and Chang Te-shui, HYCLC, 1996, 170–175.

4. Tu Cheng-sheng, KK 1991:1, 43–56; Chang Te-shui, HYCLC, 1996, 170–175.

5. For typical expressions see Shen Ch'ang-yün, HCCHS 2005:5, 8–15, or Tu Yung, HCCHS 2006:6, 3–7; for a brief summary of the conflicting viewpoints, see Wang Hsüeh-jung and Hsü Hung, KK 2006:9, 83–90; for Hsin-chai see, for example, Fang Yu-sheng, HCCHS 2003:1, 35–39, or Yao Cheng-ch'üan et al., KK 2007:3, 90–96; and for a concise overview of later literary materials referring to the Hsia, see Chao Kuang-hsien, HYCLC, 1996, 122–123.

6. For a convenient summary of the Warring States textual records, see Ch'en Ku-ying, HCCHS 1985:7, 10–13. Although it subsequently received great impetus from Confucian thinkers and Mo-tzu, the myth of yielding first appeared in the early Western Chou, roughly four hundred years before Confucius. (For the latter see Yu Shen, HCCHS 2006:3, 39–44, and for a general analysis Chiang Ch'ung-yao, HCCHS 2007:1, 41–46, or Ch'ien Yao-p'eng, HCCHS 2001:1, 32–42.)

7. Mencius's discussion in "Wang-chang" may be taken as definitive, but see also Fang Chieh, HHYC 11:1 (1993): 15–28. There is no evidence that Heaven was ever conceived of as an active entity in Yü's time.

8. This was the essential premise of Wittfogel's well-known but now (perhaps too thoroughly) rejected work, *Oriental Despotism: A Comparative Study of Total Power*. (The need to coerce people into building embankments and organize them for the work must have stimulated bureaucratic growth to at least some extent.) For recent discussions of the "hydraulic thesis"—primarily rejections—see Chang Kung, CKKTS 1994:2, 4–18; Chou Tzu-ch'iang, CKKTS 1994:2, 19–30; Liu Hsiu-ming, CKSYC, 1994:2, 10–18; and Yü Shu-sheng, CKSYC, 1994:2, 3–9. Nevertheless, water management is seen as a decidedly important Hsia accomplishment. (See, for example, Li Hsien-teng, HYCLC, 1996, 27–34.)

9. For further discussions see Joseph Needham, *Civil Engineering and Nautics*, 247ff, or the more traditional account in Meng Shih-k'ai, *Hsia Shang Shih-hua*, 149–154. In two different passages (IIIB9 and VIB11), Mencius clearly asserts that Yü accorded with water's natural patterns and removed obstacles to its flow.

10. *Mencius*, IIIA4, "T'eng-wen Kung, Hsia." (Since it recurs while describing other sages in IVB29, "passing one's gate three times" apparently represents Mencius's ideal of self-denial.) The legend of Yü taming the waters dates to the middle of the Western Chou. (See Li Hsüeh-ch'in, HCCHS 2005:5, 6; Tuan Yü, HCCHS 2005:1, 110–116; and Anne Birrell, TP 83 [1997]: 213–259.) Water played an important role in early China's contemplative tradition, including as a focal element in the *Tao Te Ching* and as an image for irrepressible power in the military writings.

11. "Yüan Tao."

12. Variants of this perspective are preserved in military writings as the *Ssu-ma Fa*, *Six Secret Teachings*, and *Three Strategies of Huang Shih-kung*. (For further discussion, see Sawyer, *Tao of War*, or *T'ai Kung Liu-t'ao* [*Six Secret Teachings*].)

13. Apart from the problems invariably posed by seasonal rains, there was a period of maximum flooding from 4000 to 3000 BCE due to increased moisture levels that effectively sundered Hebei. (See Han Chia-ku, KK 2000:5, 57–67.) Miao Ya-chüan (HCCHS 2004:3, 13–19, 26) has even asserted that flooding caused the demise of the Lungshan culture and facilitated the Hsia's rise because their leaders combined warfare with expertise in curbing water's destructive effects.

14. "Hsia Pen-chi," *Shih Chi*. However, see *Hsia Shih Shih-hua*, 149–164, for a more extensive examination of the relevant accounts.

15. Just as in the *Shang Shu* (upon which the *Shih Chi* account is probably based), the chapter continues with a lengthy description of Yü's accomplishments and enumerates the chief characteristics of the Nine Provinces he demarked. These descriptions represent early attempts to compile topographical knowledge for administrative and military purposes.

16. Every conceivable intellectual discipline has been employed to demythologize tales about the ancient sages. (See, for example, Yi Mou-yüan, HCCHS 1991:2, 4–12; Huang Hsin-chia, HCCHS 1993:11, 25–32; and Feng T'ien-yü, HCCHS 1984:11, 5–14.)

17. Yen Wen-ming, WW 1992:1, 25, 40–49.

18. Yen Wen-ming, WW 1992:1, 25, 40–49.

19. See, for example, Hu Chia-ts'ung, HCCHS 1991:1, 19–26; Li Hsüeh-ch'in, HCCHS 2005:5, 5; and Wang Hui, KKHP 2007:1, 1–28. More compressed dates have also been suggested for the semilegendary Sages, such as 2400 to 2000 BCE. (For example, T'ien Chi-chou, HCCHS 1985:9, 25–32.)

20. See, among many, Ch'eng P'ing-shan, HCCHS 2004:5, 10–21, and P'an Chi-an, KKWW 2007:1, 55–61. (P'an believes that T'ao-ssu, which has been suggested as Yao's capital of P'ing-yang, served as the Yellow Emperor's capital.)

21. These are the dates suggested by Chao En-yü, HCCHS 1985:11, 17–19, who opts for eras rather than realistic life spans. Based on astronomical data, he also claims that Yü's reign had to commence in either 2221 or 2161 BCE and that it lasted for thirty-three years. (However, Chao contradicts his own astronomical dating in concluding that Yü ascended the throne in 2227 and ruled for thirty-nine years.)

22. Stimulated by David Nivison, a series of articles by David Pankenier, Edward Shaughnessy, Kevin Pang, and others two decades ago argued whether the data found in traditional accounts are original or the result of later accretions and reconstructions; whether the phenomena would have been observable or were just extrapolated from other observations; the resolution of various discrepancies; and which records might be deemed authoritative. Based on a passage in the Mo-tzu and a five-planet conjunction, David Pankenier, concluded that Shun's fourteenth year—1953 BCE—was Yü the Great's first de jure year as the Hsia's progenitor (EC 9–10 [1983–1985]: 175–183, and EC 7 [1981–1982]: 2–37). Other critical articles, some of which focus on the broader issue of the reliability of the old and new text versions of the Bamboo Annals, include E. L. Shaughnessy, HJAS 46, no. 1 (1986): 149–180, also reprinted in Before Confucius, and his important article in EC 11–12 (1985–1987): 33–60; and David S. Nivison and Kevin D. Pang, EC 15 (1990): 86–95, with additional discussion and responses, 97–196. (For useful discussions of the Old Text/New Text controversy, see Michael Nylan, TP 80:1–3 [1994], 83–145 and TP 81:1–3 [1995], 22–50, and Hans Van Ess, TP 80:1–3 [1994]: 149–170.)

23. For example, see Chao Chih-ch'üan, KKWW 1999:2, 23–29.

24. The late K. C. Chang is most prominently associated with this debate, but for concise versions of single origination see Cheng Kuang, KKWW 2000:3, 33–43; T'ien Chi-chou, HCCHS 1985:9, 25–32; and Yü Feng-ch'un (who examines the Shih Chi's depiction), 2007:2, 21–34.

25. Ho Chien-an, HCCHS 1986:6, 33–46; T'ien Chi-chou, HCCHS 1985:9, 25–32; Li Min, HCCHS 2005:3, 6–8, 13; and Hsü Shun-chan, HYCLC, 1996, 128–135.

26. His failure to yield, a topic of heated argument over the centuries, continues to be an issue. (For a recent example, see Fang Chieh, HHYC 11:1 [1993], 15–28.)

27. For an expression of this view see T'ien Chi-chou, HCCHS 1985:9, 25–32.

28. For an overview that charts the period of greatest eastern influence see Luan Feng-shih, KK 1996:4, 45–58.

29. For example, see Tsou Heng, KKWW 1999:5, 50–54.

30. For one formulation of the amalgamated view, see Wang Hsün, KKWW 1997:3, 61–68. For a useful discussion of Yüeh-shih culture, see Tsou Heng, HSCLWC, 64–83. (Note that Tsou cites dates of 1765 to 1490 BCE, far too late to have contributed to the predynastic Hsia.)

31. Among many, see Tu Cheng-sheng, KK 1991:1, 43–56.

32. For example, see Wang Ch'ing, CKSYC 1996:2, 125–132.

33. Fang Yu-sheng, HCCHS 1996:6, 33–39, and Shen Ch'eng-yün, CKSYC 1994:3, 113–122.

34. Wu Ju-tso, CKKTS 1995:8, 12–20.

35. For expressions of this thesis, see An Chin-huai, KKWW 1997:3, 54–60; Chu Kuang-hua, KKWW 2002:4, 19–26; and Wei Ch'ung-wen, HCCHS 1991:6, 29–31. However, Ho Chien-an, HCCHS 1986:6, 33–34, believes that the Lungshan Wang-wan manifestation found there and in the eastern part of Yü-hsi around the Loyang plains and in the Yü-hsien to Cheng-chou

corridor would have had to pass through the Mei-shan stage before possibly expanding to transform to Erh-li-t'ou culture.

36. For example, Wei Ch'ung-wen, HCCHS 1991:6, 29–31, believes T'ao-ssu was probably the focal location for Yao, Shun, and Yü, while Feng Shih, KKHP 2008:3, 273–290, has concluded that the Hsia should be identified with T'ao-ssu culture.

37. Just like bronze in the Shang, jade was the material of privileged artifacts in the Hsia (Wen Hui-fang, HCCHS 2001:5, 61–68).

38. Ch'en Sheng-yung, HCCHS 1991:5, 15–36. These assertions raise more questions than they answer—did the Hsia prevail through warfare, cultural power, or some other factor that allowed them to absorb the Liang-chu manifestation? (Some historians have suggested that Liang-chu culture was essentially contemporaneous.)

39. Li Liu and Hung Xu, *Antiquity* 81 (2007): 893–894, and WW 2008:1, 43–52, claim that the Hsia (in its Erh-li-t'ou manifestation) was populated by multiple groups rather than a single clan that emigrated into the area and that it had precursors in Yangshao and Lungshan cultures.

40. Various dates (such as 3200 BCE) have been suggested for the inception of the simple chiefdoms that mark a transition from (Marxist-postulated) matriarchical societies to patriarchical ones. (For example, see Chang Chung-p'ei, HCCHS 2000:4, 2–24.) The power to sacrifice or punitively slay others clearly existed in the Hsia and apparently the late Lungshan as well, though decapitated and contorted bodies pose the different problem of distinguishing sacrificial and battle victims. (An example would be the three recovered at Shaanxi Ch'ang-an K'o-sheng-chuang, for which see Chang Chih-heng, HYCLC, 1996, 109–112.)

41. Wei Chi-yin, KKWW 2007:6, 44–50.

42. Wang Wei, KK 2004:1, 67–77.

43. Suspicion has often been inappropriately cast on these and other efforts because they are largely being pursued under a cultural manifest from the central government and are thus viewed simply as another nationalistic manifestation.

44. Chang Chih-heng, 109–112; Chang Li-tung, 113–118; Yü T'ai-shan, 176–196; and Fang Hsiao-lien, 266–273, all in HYCLC, 1996.

45. Chang Li-tung, HYCLC, 1996, 113–118.

46. One of the sites tenuously identified with Yao and pre-Hsia culture is P'ing-yang, noted in traditional historical accounts as Yao's capital. Neolithic tools and weapons (including jade axes and stone and bone arrowheads) have been recovered from this former T'ao-ssu settlement that apparently was occupied in 2600–2000 BCE, Yao's reign being projected as 2600–2400. (See Shan-hsi-sheng Lin-fen Hsing-shu Wen-hua-chü, KKHP 1999:4, 459–486.) Just like Wang-ch'eng-kang, it has also been termed the first Chinese city (Ma Shih-chih, HYCLC, 1996, 103–108).

47. Wei Ch'ung-wen, HCCHS 1991:6, 29–31. Calculating Yao's era as 2300 to 2200 and the site as 2600 to 2100 BCE, Li Hsüeh-ch'in has suggested that many aspects of the site indicate it might have been Yao's capital of P'ing-yang. (See HCCHS 2005:5, 3–7.) However, P'an Chi-an, KKWW 2007:1, 56–61, who dates Yao and Shun to the twenty-second century BCE, though concurring in P'ing-yang's identification, ascribes it to the era of the Yellow Emperor and Ku.

48. Tu Cheng-sheng, KK 1991:1, 43–56; Shan-hsi-sheng Lin-fen Hsing-shu Wen-hua-chü, KKHP 1999:4, 459–486.

49. Chang Chih-heng, HYCLC, 1996, 109–112.

50. Feng Shih, KKHP 2008:3, 279–283. Feng holds the unusual view that powerful rulers didn't require walls, explaining the failure to refurbish them at Hsiang-fen and their absence at Anyang.

51. For the inner citadel see Ho-nan-sheng WW YCS, WW 1983:3, 8–20. Subsequent reports include Fang Yu-sheng, KK 1995:2, 160–169, and KKWW 2001:4, 29–35; Pei-ching Ta-hsüeh K'ao-ku Wen-po Hsüeh-yüan, KK 2006:9, 3–15; and P'ei Ming-hsiang, HYCLC, 1996, 60–65.

52. From the eastern section only 30 meters of the southern part of the common wall and 65 meters in the western part of the south wall remain. The western section, which is marked by a slight tilt of 5 degrees, is defined by foundation remnants of 92 and 82.4 meters on the western and southern sides respectively and 29 meters of wall on the north.

53. An Chin-huai, "The Shang City at Cheng-chou and Related Problems," 30, cites a date of 4010 ± 85 BP or 4415 ± 140 BP after calibration, which he concludes places it within the Hsia dynasty. However, 2415 BCE—at the extreme limit—would have to be considered pre-Hsia at best if the Hsia is dated as 2100 to 1600 BCE. Chang Chih-heng, HYCLC, 1996, 109–112, asserts it clearly postdates T'ao-ssu.

54. For an overview see Fang Yen-ming, KK 2006:9, 16–23. Fang Yu-sheng, KK 1995:2, 164, and KKWW 2001:4, 29–35, is among those identifying the site as Yü's capital of Yang-ch'eng. (See also Wei Ch'ung-wen, HCCHS 1991:6, 29–31. Prior to the discovery of the outer walls, scholars such as Tu Cheng-sheng, KK 1991:1, 43–56, had felt that Wang-ch'eng-kang was too small for a great chief such as Yü.)

55. Based on its size, location, dating, and a reference in the *Mu T'ien-tzu Ch'uan*, it has been proposed as a good candidate for Ch'i's capital. (For site reports see Hsü Shun-chan, HCCHS 2004:6, 13–17; Ma Shih-chih, KKWW 2007:3, 54–58; Ch'eng P'ing-shan, KKWW 2007:3, 59–63; SHYCS Ho-nan Hsin-chai-tui, KK 2009:2, 3–15, and KK 2009:2, 16–31.)

56. Chang Hsüeh-lien et al., KK 2007:8, 74–89; Chao Chih-ch'üan, KKHP 2003:4, 459–482; and SHYCS Ho-nan Hsin-chai-tui, KK 2009:2, especially 15.

57. Although recent reports are increasingly detailed, they cover only small, scientifically excavated sections, making it difficult to accurately estimate the remaining dimensions.

58. See Cheng-chou-shih WWKK YCS, KK 2005:6, 3–6.

59. Ch'en Hsü, HSLWC, 8–15. Useful site analyses include T'ung Chu-ch'en, WW 1975:6, 29–33, 84; Fang Yu-sheng, KK 1995:2, 160–169, 185; Hsü Hung, KK 2004:11, 32–38; and Fang Yu-sheng, HYCLC, 1996, 81–91. (See also Robert L. Thorp, EC 16 [1991]: 1–38.)

60. See, for example, Li Po-ch'ien, HCCHS 2003:3, 2023, and Li Liu and Hung Xu, *Antiquity* 81 (2007): 886–901, or WW 2008:1, 43–52.

61. The foundations for the large, palatial structures discovered in the royal quarters average one to three meters thick and consist of highly uniform layers of four to six cm. reinforced by interspersed pebbles.

62. For summaries of recent discoveries, including the second wall described in the next paragraph, see SHYCS Erh-li-t'ou Kung-tso-tui, KK 2004:11, 3–13, and KK 2005:7, 15–20.

63. Tu Cheng-sheng, KK 1991:1, 43–56, for example, has concluded it was not the Hsia capital. Similarly, based on a brief reference in *Mo-tzu* to King T'ang having destroyed the walls, Chang K'ai-sheng, HYCLC, 92–102, believes that Erh-li-t'ou could not have been the last royal city and opts for Yen-shih instead.

64. Important site reports and assessments include Cheng-chou-shih WWKK YCS, WW 2004:11, 4–18; Wang Wen-hua et al., WW 2004:11, 61–64; Li Feng, 2006:2, 67–72, and KKWW 2007:1, 62–66; and Hsü Chao-feng and Yang Yüan, KKWW 2008:5, 26–30.

65. Wang Wen-hua et al., WW 2004:11, 64.

66. Their exact locations are a matter of ongoing disagreement, with Ta-shih-ku itself sometimes being suggested as having been either the satellite state of Ku or Wei. (Based on the site's date and degree of destruction, Li Feng, 2006:2, believes it may have been the fabled Shang capital of Po.)

67. Cheng-chou-shih WWKK YCS, WW 2004:11, 14–15.

68. For example, see Li Liu and Chen Xingcan's *State Formation in Early China*.

69. For a discussion, see Liu and Chen, *State Formation*, 69–73. Erh-li-t'ou's remains have been dated to between 1900 and 1600 BCE.

70. Ch'in Hsiao-li, KKWW 2000:4, 46–57, especially 55–56.

CHAPTER 5

1. The *Bamboo Annals* state that in his eighth year, when Yü assembled the feudal lords at Kuai-chi, he killed the clan leader of the Fang-feng, an act that commentators try to justify by claiming he had arrived late. Some analysts date the Hsia's inception to Yü's conquest of the San Miao. (See, for example, Han Chien-yeh, HCCHS 1998:1, 44–49.)

2. In contrast to the Chinese approach that early on stressed molding and quickly developed multiple-cavity molds for the efficient production of small items such as axes, knives, and arrowheads that would emerge late in the period, hammering was the fundamental Western technique.

3. Our account basically follows Yang Hsin-kai and Han Chien-yeh, CKKTS 1995:8, 32–41, and HCCHS 1997:2, 25–30.

4. See Yang and Han, CKKTS 1995:8, 32–41. The San Miao had bird totems.

5. For example, in the area that would become Ching/Ch'u.

6. The *Chan-kuo Ts'e* ("Wei Ts'e, 2") states that when Yü attacked the San Miao, the Tung Yi didn't move.

7. Kung Wei-ying, HCCHS 1988:9, 40–41.

8. An unusual view of the conflict has been offered by Chao Kuang-hsien (LSYC 1989:5, 24–34), who believes it arose in the Yellow river area rather than the south and that the San Miao remnants, despite having been pushed out of the Yellow River valley, remained defiant, provoking further clashes.

9. Appearing among the so-called *Ku-wen* chapters of the *Shang Shu* that are generally acknowledged as having been fabricated centuries later than the Spring and Autumn and Warring States portions, such passages reflect post-Han Confucian concepts more than historical events. (The concepts of a "campaign of rectification" and "five phases" postdate the early Hsia by more than a millennium.) However, scholars have traditionally cited the appearance of common passages and other references in *Mencius* and *Mo-tzu*, both Warring States works, as evidence of the chapter's early origins and presumed authenticity. (In actuality, the *Shang Shu* chapter was presumably created on the basis of these earlier passages, perhaps on the basis of some common text, and simply incorporated them for authenticity. Nevertheless, its authoritativeness remained unquestioned until well into the Ch'ing dynasty.)

10. And having enormous difficulty with his parents, who reputedly even tried to kill him. (The travails that Shun underwent to prove his filiality became a defining characteristic for Mencius and subsequent Confucians.)

11. Although the concept of "returning the army" seems inappropriate here, this is the understanding offered by the traditional commentators for the well-known term *chen lü*.

12. "Fei-kung, Hsia," *Mo-tzu*.

13. "Chao Lei," *Lü-shih Ch'un-ch'iu*. See also Luo K'un, HYCLC, 1996, 197–204.

14. "Yao Tien," *Shang Shu*, also found in "Hsiu-wu" in the *Huai-nan Tzu*.

15. *Bamboo Annals*.

16. See Fan Li, KKWW 1999:4, 50–61; Liu Yü-t'ang, HCCHS 2001:4, 53–55; and Yang Hsin-kai and Han Chien-yeh, CKKTS 1995:8, 32–40.

17. Liu Hsü, HCCHS 1989:7, 21.

18. The battle reportedly unfolded on the bank of the Kan River, but other places, such as southwest of Loyang, have also been suggested. (The *Bamboo Annals* record the clash under Ch'i's second year, but other accounts place it in his third year.) The much-quoted oath is certainly a Warring States fabrication, but many contemporary scholars still believe the "Kan Shih" preserves authentic material, including Chin Ching-fang and Lü Shao-kang, HCCHS 1993:5, 13–17; Li Min, CKSYC 1980:2, 157–161; and even Yang Sheng-nan, CKSYC 1980:2, 161–163.

19. Variously interpreted as calendrical referents or Heaven, Earth, and Man. (References to the "five elements" and seeking Heavenly justification indicate the oath's anachronistic nature.)

20. That is, his mandate to rule, but also his life. (The concept of the mandate of Heaven does not appear until the late Shang or early Chou.)

21. "Obligations of the Son of Heaven," *Seven Military Classics*, selected from pages 130 to 132.

22. Yang Sheng-nan, HCCHS 1991:9, 46, is basically inclined to recognize the existence of chariots in the Hsia despite, as he acknowledges, the lack of evidence.

23. *Bamboo Annals*, Emperor Ch'i, eleventh and fifteenth years. Wu Kuan's fate is not specified.

24. Yang Po-chün, 1990, 936, locates it south of Loyang.

25. Reading *chung* as troops rather than simply laborers or retainers.

26. Also pronounced *"chiao."*

27. *Tso Chuan*, Duke Hsiang, fourth year.

28. *Tso Chuan*, Duke Ai, first year. A virtually identical passage is preserved in the *Shih Chi*'s "Wu T'ai-po Shih-chia." Vanquishing Yi was no mean feat, as he was said to have such great strength that he could push a boat on land ("Hsia Pen-chi," *Shih Chi*).

29. Note the presence of the *Ch'üan* or Dog people, who will become important in subsequent periods. (Their name is written with the radical for field next to "dog.")

30. The Hsia campaign (seen in the "Yin Cheng" and also the *Shih Chi*'s "Hsia Pen-chi") has traditionally been assigned to Chung-k'ang's reign but more recently demythologized as probably dating to Shao-k'ang's restoration. (See Hsü Chao-ch'ang, HCCHS 2004:4, 22–27.)

31. "Chieh-ts'ang, Hsia." Some of these myths are obviously conflated because Shun is also said to have perished in a conflict with the Tung Yi.

32. These and other conflicts with the Yi listed in the *Bamboo Annals* suggest that their submission was nominal rather than total.

33. According to the *Shih Chi*'s "Hsia Pen-chi": "Chieh did not concentrate upon Virtue but on the martial, thereby harming the hundred surnames who no longer sustained him."

34. Although the Hsia populace seems to have primarily dispersed via the upper Yellow River to Shaanxi, Gansu, Ching, and the northwest, scattered Hsia elements have been found in Shandong, Jiangsu, and Anhui, especially in the ancient areas of Wu and Yüeh, as well as Ching/Ch'u. (See Wang K'e-lin, KKWW 2001:2, 48–53.) It is commonly thought that they were the ancestors of numerous steppe peoples, including the T'u-fang, Chiang, and Hsiung-nu. (For a contrary argument see Ch'en Li-chu, LSYC 1997:4, 18–35.)

35. The plan is enunciated in the *Shang Shu* and incorporated in the *Shih Chi*'s "Hsia Pen-chi." The theory of administrative domains known as the *"wu fu"* received its paradigm expression in the late *Chou Li*, but the concept evolved during the Chou. Although a creative idealization (if not absolute nonsense), it has been interpreted as providing a possible framework for understanding the Hsia's relationship with Lungshan and other cultural groups. (See Chao Ch'un-ch'ing, HCCHS 2007:1, 9–19. Other important discussions include T'ien Chi-chou, HCCHS 1985:9, 25–32, and Ch'en Lien-ch'ing, 1991, 863–891. For a general discussion of the utility of these classical texts, see Ts'ui Ta-hua, CKCHS 1995:1, 55–63.)

36. The character *tien* has long been understood as designating the "imperial domain," with *tien* basically encompassing all the territory within 500 *li*.

37. Contemporary enunciations even occasionally appear in PRC theoretical publications, embedded in articles advocating a new world order based on revitalizing this ancient outline.

38. Unfortunately, every characterization of specialized assignments and identification of correlative titles is invariably based on statements in the *Shang Shu* and later works, at best Western Chou and Warring States writings. Even allowing for the possibility of institutional continuity and accepting the presumptuousness of categorically denying that these historical vestiges could accurately depict early practices, even the broadest claims lack substantiation.

39. For analyses that envision a complex structure already in existence, see Ko Sheng-hua, HCCHS 1992:11, 13–18, or Ch'ao Fu-lin, HYCLC, 1996, 136–142.

40. Yang Sheng-nan, HCCHS 1991:9, 45.

41. Analysts such as Ch'ao Fu-lin envision a key transition from a simple chiefdom style—rule by one man, through personal charisma—to the recognition that a certain clan or even family has the right to rule. (See HCCHS 1996:6, 23–32.)

42. For a contrary view, see Ko Sheng-hua, HCCHS 1992:11, 13.

43. "T'eng-wen Kung," *Mencius.*

44. This statement has also been employed to claim that population records were already being kept and then undertake population estimates for the Hsia and subsequent periods, a seemingly flawed approach given the many quantitative unknowns. (See Sung Chen-hao, LSYC 1991:4, 92.)

CHAPTER 6

1. Until recently oracle bone discoveries had been confined to Anyang, but a number of small finds have now been made in other, even peripheral areas, including ancestral Chou locations and at the pre-Shang sites of other cultures, though most of them lack inscriptions. (For example, see Shan-tung Ta-hsüeh Dung-fang KK YCS, KK 2003:6, 3–6, and Sun Ya-ping and Sung Chen-hao, KK 2004:2, 66–75.)

2. Primarily by Tung Tso-pin, Shima Kunio, Kuo Mo-jo, Ch'en Meng-chia, and others in China, as well as David E. Keightley and Paul Serruys in the West. It should be noted that the nature of these inscriptions, whether they represent reports, charges, or entreaties to the ancestors or some other transcendent entity, remains uncertain. (For example, see Qiu Xigui, EC 14 [1989]: 77–172.)

3. For a brief discussion of the dynasty's name see Chang Kwang-chih, EC 20 (1995): 69–78.

4. In addition to the essential materials found in *The Cambridge History of Ancient China*, Bruce G. Trigger, JEAA 1 (1999): 43–62, has advanced an interesting analysis of the Shang state. Among the key issues is whether the Shang should be considered a city or territorial state (even though neither may be applicable). The presence of non-Shang clans in the core domain and their degree of participation in Shang power raises further questions.

5. Articles arguing one or another cause constantly appear. However, the often seen claim that the ruler initiated the move to curb abuses and restore monarchial authority is particularly puzzling, because any tenuousness in his authority would have precluded having the power to compel such a move.

6. The Shang's theocratic nature has been particularly emphasized over the past two decades. For example, see Yang Sheng-nan, CKSYC 1997:4, 16–23; Wang Hui, HCCHS 2000:6, 36–41; Li Shao-lien, STWMYC, 304–312; and David N. Keightley, *History of Religions* 17 (1978): 211–224, and PEW 38 (1973): 367–397.

7. From the oracular inscriptions it is obvious that Ti was held in awe because he had the power to inflict misery, defeat, and disaster on the Shang. Although probably not an anthropomorphic entity, it is unclear whether Ti was conceived of as a titular deity, the spirit of an immediately preceding ruler or ancestor, or some collective but numinous entity. Ti's sanctification was deemed necessary before military campaigns could be undertaken, explaining the importance of announcing them in the ancestral temple, a martial practice that would persist well into imperial times. (For contending viewpoints on Ti's nature and role see, for example, Robert Eno, EC 15 [1990]: 1–26; Chou Chi-hsü, HCCHS 2008:1, 3–11; Yang Hsi-mei, CKSYC 1992:3, 36–40; Fu Pei-jung, *Chinese Culture* 26, no. 3 [1985]: 23–39; and David N. Keightley, JEAA 1, nos. 1–4 [1999]: 207–230. It should be noted that repeated inquiries about undertaking particular campaigns suggest the decision had already been made, that the divinatory process was a mere formality or psychological ploy.)

8. The archaeological records are replete with descriptions of sacrificial burial pits. Family members and retainers followed their lords and masters into death, but prisoners of war were

slain in more general rites, often in exceedingly brutal fashion and numbers that could exceed a hundred. However, whether because productive labor assumed greater importance as the economy burgeoned or there was a shift in religious attitude, the totals diminish as the era progresses. (For overviews see Tuan Chen-mei, 1991, 182–191; Loewe and Shaughnessy, *The Cambridge History of Ancient China*, 192–193 and 266–267; Li Hu, 1984, 130–136; Yang Sheng-nan, LSYC 1988:1, 134–146; and the discussion on the Ch'iang that follows in the focal military section. For just one example of a major sacrificial burial site, see Shan-tung-sheng Po-wu-kuan, WW 1972:8, 17–30. Dramatic statues of sacrificial victims have also been recovered farther afield in Sichuan. See WW 2004:4, 56 and 57, for illustrations.)

9. Even the expansive *Cambridge History of Ancient China* avoids the issue by merely dealing with the artifacts themselves and commencing its account with the Anyang period. Although, as shown by numerous articles in *Early China* and the *Journal of East Asian Archaeology*, interest in the Shang has burgeoned in recent decades, apart from Robert Thorp's *China in the Early Bronze Age: Shang Civilization*, the main Western-language books on the Shang remain three works from the early 1980s: *Studies of Shang Archeology, The Origins of Chinese Civilization,* and the comprehensive but increasingly outdated *Archeology of Ancient China.*

10. For an analysis of the "Yin Pen-chi" see Ch'ü Wan-li, 1965, 87–118.

11. For a recent example, see Wang Chen-chung, HCCHS 2005:1, 3–6, 64.

12. Certain references to Yin in later, supposedly transmitted works nevertheless have prompted considerable discussion of this topic as well. (For typical analyses see Chang Fu-hsiang, HCCHS 2001:5, 57–60; Lien Shao-ming, HCCHS 2000:1, 27–31; and Ch'en Chieh, HCCHS 2001:4, 46–53.)

13. In contrast, based on the name "Yin" another interesting variant concludes that the Shang originated in the upper Huai River area before moving to Yen-shih. (See Ching San-lin, HCCHS 1986:5, 37–46.)

14. How the final location of a remnant people whose destiny was dictated by their conquerors can be interpreted as evidence for the dynasty's founding site is puzzling in the extreme.

15. K. C. Chang's section on the Shang in his *Archaeology of Ancient China* and overview in *Shang Civilization*, 335–355, extensively discuss these and similar issues. Based on the coinciding number of fourteen rulers in the Hsia and predynastic Shang and their lengthy period of intermixed dwelling in the east, in 1936 Ch'en Meng-chia (*Ku-shih-pien*, Vol. 7B, 330–333) concluded that the Hsia and Shang were members of the same *tsu* (clan). However, acrimonious disagreement marks discussions of predynastic Shang culture. (For example, see Li Wei-ming, KKWW 2000:3, 51–55, and STWMYC, 208–213; Yang Sheng-nan, HYCLC, 1996, 143–148; Tu Chin-p'eng, HYCLC, 1996, 160–164; Chang Kuo-shuo, STWMYC, 280–285; Tsou Heng, HSCLWC (reprint of 1993), 221–226; and Tuan Hung-chen, STWMYC, 213–222.)

16. A particularly useful article by Ch'en Ch'ang-yüan, LSYC 1987:1, 136–144, summarizes the various views. However, for representative theories and opinions see Han K'ang-hsin and P'an Ch'i-feng, LSYC 1980:2, 89–98; Wang Yü-che, LSYC 1984:1, 61–77; Ching San-lin, HCCHS 1986:5, 37–46; Kan Chih-keng et al., LSYC 1985:5, 21–34; and Yen Wen-ming, SCK-KLC, 227–247. Cheng Hui-sheng, 1998, 33–34, suggests the name may have been derived from an unusual height where the clan initially dwelled. Chang Ts'ui-lien, KKWW 2001:2, 36–47, argues that the Shang emerged out of the Hsia-ch'i-yüan because the Shang-ch'iu area was already inhabited by the Yüeh-shih prior to the appearance of Shang/Erh-li-kang culture and that Yüeh-shih (or Tung Yi) didn't evolve into Erh-li-kang. Others, noting the presence of numerous Yi and Yüeh cultural elements, have concluded that the Shang may have been an amalgamation of cultures and clans. For example, based on an early Shang site just west of Hsing-t'ai-shih in southern Hebei, Chia Chin-piao KK 2005:2, 71–78, believes that the early Shang benefited from a multiplicity of cultural influences and Hsia-ch'i-yüan can't be the sole defining representation.

17. K. C. Chang, *Archaeology of Ancient China*, 326–329; Chang Ch'ang-shou and Chang Kuang-chih, KK 1997:4, 24–31.

18. See Ch'en Ch'ang-yüan's summary, LSYC 1987:1, 136–144.

19. See, for example, Liu Hsü, HCCHS 1989:7, 21–26, and Cheng Shao-tsung, STWMYC, 423–428.

20. For example, Ch'en Ch'ang-yüan and Ch'en Lung-wen, HCCHS 2004:4, 15–27, argue that the Shang originated in the middle southern part of Shanxi, then moved eastward, evolving into the Chang-ho manifestation before secondarily developing as Hui-wei. (There was a basic shift from mid-Hebei to southern Hebei and settlement along the Chang River.) T'ang therefore should be identified with Yüan-ch'ü in Shanxi.

21. K. C. Chang advanced this idea; for a more recent analysis see Chang Wei-lien, HCCHS 2006:4, 23–32. (For further information on pre-Shang climate and crops, especially the Ta-wen-k'ou to Yüeh-shih cultures in Shandong's Shu river valley, see Ch'i Wu-yün, KK 2006:12, 78–84.)

22. A number of articles have discussed the effects of climate change in the late Neolithic, Lungshan, Erh-li-t'ou, and even Shang eras. Commencing about 3000 BCE, the average temperature apparently dropped about 3 degrees C from the climatic maximum, rainfall amounts decreased, herbs and grasses proliferated, woodlands and marshes shrank, and small-animal wildlife populations diminished (making it necessary to raise more animals), all of which could have prompted Hsia aggressiveness even though they are noted for battling floods, not drought. (See, for example, An-hui-sheng WWKK YCS, KK 1992:3, 253–262; Yüan Ching, KKHP 1999:1, 1–22; Sung Yü-ch'in et al., KK 2002:12, 75–79; and Wang Wei, KK 2004:1, 67–77.)

23. Ancient China developed an extensive psychology of *ch'i* (spirit or the will to fight), and the early military classics include important passages on stimulating and manipulating it. For example, in a purported historical overview the *Ssu-ma Fa* ("Obligations of the Son of Heaven") notes that "Shang rulers swore their oaths outside the gate to the encampment for they wanted the people to first fix their intentions and await the conflict." Further discussion of the concept and psychology of *ch'i* may be found in Sawyer, "Martial *Qi* in China: Courage and Spirit in Thought and Military Practice."

24. See *"Shang Shu"* in M. Loewe, ed., *Early Chinese Texts*, M. Loewe, 378.

25. Note that Karlgren, "Glosses on the Book of Documents," 172–174, interprets the lines as Hsia complaints that T'ang is reciting to rationalize the attack on Chieh. However, since T'ang justifies the assault in the next paragraph as a response to their complaints and answers the even more pointed question as to Chieh's impact on them, the translation follows the common understanding that the words were uttered by the Shang populace. (The *Shih Chi's* "Yin Pen-chi" integrates the oath into its explanation of King T'ang's actions.)

26. Irrespective of whether the historical sequences are realistic or inspiring leaders necessary, the Chou conquest of the Shang so closely replicated the Shang's overthrow of the Hsia that the authenticity of the account and achievements of the key actors have long been derisively dismissed by cynics.

27. "Chung Hui chih Kao," *Shang Shu* (*Shang-shu Chu-shu chi Pu-cheng*, 8:14b). Being from the so-called old text materials, this section probably dates to the fourth century CE, just like the "'T'ang Kao" that follows it. (See Loewe, *Early Chinese Texts*, 377–383.)

28. "Chung Hui," *Shang-shu Chu-shu chi Pu-cheng*, 8:15a.

29. The best account remains Sun Sen's *Hsia Shang Shih-kao*, 1987, 303–319, but similar versions are found in the massive *Chung-kuo Li-tai Chan-cheng-shih*, Vol. 1, 45–54, as well as overviews by historians such as Wang Yü-hsin, HCCHS 2007:5, 14–20.

30. Historians such as Sun Sen, 303–304, believe that the Shang actually moved south, closer to the Hsia, to position themselves to launch an attack that would require less than a full day's march. However, they presuppose the Hsia either sanctioned or lacked the might to oppose such a move. (Although there is great disagreement over the site's identity, most accounts emphasize

that the Shang sought a remote, isolated location [presumably east of the T'ai-hang Mountains] in order to avoid the Hsia's perversity and allow their incipient power to grow unobserved.)

31. See, for example, Hsü Chao-feng, KKWW 1999:3, 43–48.

32. Yi Yin's life and career continue to prompt articles that employ a broad range of traditional and oracular sources. One of the most extensive is Cheng Hui-sheng's "Yi Yin Lun" (1998, 184–208), which points out that Yi was especially mentioned as a recipient for sacrifices after Wu Ting's reign, had exalted status comparable to all the ancestors apart from the Shang's progenitor, and was apparently viewed as chief of the officials. (For inscriptional examples, see HJ27656, HJ27653, HJ27655, HJ30451, HJ27661, HJ27057, HJ26955, and HJ507.) Based on a detailed examination of numerous traditional sources conjoined with materials winnowed from the oracle bones, Cheng affirms the traditional (rather romanticized) view that Yi was originally an orphan of low, even slave status who derived his name from the Yi River, where he had been found, and was brought up by a kitchen helper. Because all the other Shang administrators were members of the nobility, he then reaches the novel conclusion that Yi was the first professional bureaucrat (207–208).

33. *Mencius*, VA7.

34. Hsü Chao-feng and Yang Yüan, KKWW 2008:5, 28.

35. According to the "Hsia Pen-chi," Chieh later expressed regret at failing to execute him.

36. Hsü Chao-feng and Yang Yüan, KKWW 2008:5, 28.

37. The parents of the slain youth, rather than all the common people slain by the earl's troops. However, it may also be understood as "ordinary men and women."

38. *Mencius*, IIIB5.

39. It is also emphasized in the late Warring States *Six Secret Teachings* and *Three Strategies of Huang-shih Kung*.

40. "Employing Spies," *Art of War*.

41. The preface to the "T'ang Shih" spuriously attributed to Confucius summarizes Yi Yin's role simply as having "acted as a minister in T'ang's attack on Chieh." (Since his purported actions are fully analyzed in Sawyer, *Tao of Spycraft*, only a few key points need be noted here.)

42. "Shen-ta Lan," *Lü-shih Ch'un-ch'iu*. (Note that Chieh's fate is rather different in this account.)

43. "Ch'üan Mou."

44. The relationship between the Hsia and the Nine Yi seems to have fluctuated over the Hsia's existence but generally to have been nonantagonistic despite the Hsia's early conquest of the San Miao.

CHAPTER 7

1. The location of Po, the reputed first capital and residence of King T'ang prior to the conquest, is particularly disputed, with claims even being made that it was located in the mountains. (See Ch'en Li-chu, HCCHS 2004:4, 28–37, who believes Po continued on as the permanent ritual center.)

2. These topics are all well covered in the chapters on the Shang in *The Cambridge History of Ancient China*, K. C. Chang's volumes on Shang China, and innumerable articles published over the past several decades. Without doubt the Shang was agriculturally based; many inscriptions inquire about the harvest, opening lands, and other productive concerns (Chang Ping-ch'üan, BIHP 42:2 [1970], 267–336).

3. For site analyses, see Yüan Kuang-k'uo and Ch'in Hsiao-li, KKHP 2000:4, 501–536, and Ch'eng Feng, HCCHS 2004:2, 24–26, 35. The sudden intrusion of Shang (Erh-li-kang) cultural elements with the erection of the walls around the end of the third or beginning of the fourth Erh-li-t'ou period is interpreted as evidence of Hsia elements having been forcibly displaced. Whether this was just prior to, or coincident with, their final conquest of the Hsia might be

questioned. However, to the extent that dating presently allows, the bastion well fits a probable sequence of Shang expansion.

Ke-chia-chuang in southern Hebei west of Hsing-t'ai-shih, at a vital crossroads, is another culturally complex preconquest site. Initially occupied around Erh-li-t'ou's late second or early third phase, it shows the forceful intrusion of lower Erh-li-kang culture at the interstice between third and fourth Erh-li-t'ou or just about the time the Shang moved to vanquish the Hsia. (For a preliminary report see Chia Chin-piao et al., KK 2005:2, 71–78.)

4. Lengths average some 280 meters, remnant widths 4 to 8 meters except on the south, where a 14-meter-wide section is visible. The western and northern segments still retain a height of 2 to 3 meters.

5. See Ch'eng Feng, HCCHS 2004:2, 24–26. However, this sort of explanation is not fully convincing because capital expenditures had already been made and threats were arising from this quarter.

6. The following are informative: Wang Hsüeh-jung, KK 1996:5, 51–60; Tung Ch'i, KKWW 1996:1, 27–31; Chang Kuo-shuo, KKWW 1996:1, 31–38; Tu Chin-p'eng et al., KK 1998:6, 9–13, 38; SHYCS Honan Ti-erh Kung-tso-tui, KK 1998:6, 1–8; An Chin-huai, HCCHS, 1993:11, 332–338; An Chin-huai and Yang Yü-pin, KK 1998:6, 14–19; Fang Yu-sheng, KKWW, 1999:3, 39–42; Chao Chih-ch'üan, HCCHS 2000:1, 18–27; Tu Chin-p'eng, HCCHS 1999:5, 38–40, and WW 2005:6, 62–71; Tu Chin-p'eng and Wang Hsüeh-jung, KK 2004:12, 3–12; and SHYCS Honan Ti-erh Kung-tso-tui, KK 2006:6, 13–31 and 32–42.

7. Tsou Heng has long argued that Yen-shih is the detached T'ung palace to which T'ai Chia was exiled. (See HSCLWC, 120–122, 123–158, and 166–168, as well as JEAA 1, nos. 1–4 [1999]: 195–205.)

8. Residual markings show the framing boards ranged in height from 0.3 to 0.7 meter and each ascending layer was set in about 10 cm. In addition, there is a massive inner waist wall, with a slight 15 degree pitch that extends some 13 meters to the interior from a point about 0.7 meter high up on the core wall. Moreover, the walls were not only constructed on an excavated foundation ditch with an unusual profile (because the terrain to the interior is approximately 1 meter higher than the exterior), but also extend out onto unprepared ground. (This may be partly the result of excavating the soil for the raised interior platform from the immediate vicinity of the core wall, thereby leveling the ground between the wall and the moat.) The foundation ditch has an average top opening of 18.6 meters, and both sides slant inward toward the bottom.

9. See, for example, Wang Hsüeh-jung's lengthy analysis of textual materials coupled with his own on-site investigations, KK 1996:5, 51–60.

10. Exact dimensions are reported as 233 meters for the western wall, 230 for the eastern, 213 for the southern, and a surprisingly short, presumably northern wall remnant of 176 meters.

11. Yen-shih and the later capital of Huan-pei are seen as evidencing a transition to the regularized layout of capitals described in the *K'ao-kung Chi*: three concentric segmented rectangles all demarked by walls of various solidity, arrayed along the same axis, that functioned as the royal quarters, inner city, and outer city. (For a typical discussion see Li Tzu-chih, KKWW 2004:4, 33–42, or Liu Ch'ing-chu, KKHP 2006:3, 296–297. However, Liu doesn't see the city as being fully developed until the imperial age.)

12. SHYCS Honan Ti-erh Kung-tso-tui, KK 2006:6, 13–31. Artificial pools with outflow ditches for irrigation or water supply have been found in the palace complexes of both Yen-shih and Cheng-chou and may have been a regular feature of Shang city construction (Tu Chin-p'eng, KK 2006:11, 55–65).

13. This is the view of analysts such as Chang Kuo-shuo, who emphasize the austere character and martial aspects of this first capital. (See also Chao Chih-ch'uan, KKWW 2000:3, 28–32.)

14. An observation made by many, including Li Tzu-chih, KKWW 2002:6, 43–50.

15. However, as the number of radiocarbon-dated artifacts increases, several other dates,

generally centering on 1600 to 1525 BCE, have been suggested as critical to determining the actual age of the site's walls. For example, An Chin-huai and Yang Yü-pin, 15–18, cite 3395 and 3380 BP, which, when calibrated, give 3650 and 3630 BP (± 125–130 years) respectively. (An and Yang point out that all the artifacts come from Erh-li-t'ou's first three phases, that the walls and building foundations are therefore earlier than lower Erh-li-kang, and therefore, at the latest, they date from the fourth phase. Moreover, since Cheng-chou's walls are built over a lower Erh-li-kang layer, they must postdate Yen-shih.) Separately, Yang Yü-pin, KK 2004:9, 87–92, ascribes radiocarbon dates of 1610–1560 BCE to Yen-shih and 1509–1465 to Cheng-chou. Tung Ch'i, KKWW 1996:1, 30, believes the walls all date to the lower Erh-li-kang period, 1600 BCE or later, and Chang Kuo-shuo, KKWW 1996:1, 34, also argues that the walls at both Yen-shih and Cheng-chou were in existence during the lower Erh-li-kang. Assuming a core date of 3570 BP for Cheng-chou and about 3650 BP for the palace foundations at Yen-shih, An and Yang note a difference of about eighty years. (The most detailed analysis to date is provided by Chang Hsüeh-lien et al., KK 2007:8, 74–89. Other useful articles include Kao Wei et al., KK 1998:10, 66–79, and Ch'en Hsü, HSLWC [reprint of 1985], 119–128.) Despite these potentially significant quibbles, given the radiocarbon deviation allowance being at least a century, ascribing a very early sixteenth-century date to Yen-shih is certainly justified.

16. For example, see Wang Hsüeh-jung, KK 1996.5: 58–59; Tung Ch'i, KKWW 1996:1, 30–31, who believes the walls date from 1600 BCE or slightly later; and An Chin-huai and Yang Yü-pin, 6 (1998): 16–18. Tu Chin-p'eng, KK 2005:4, 69–77, makes the interesting point that there is no real evidence that T'ang conquered Hsia from Po and then returned to Po. (See also Liu Hsü, HYCLC, 1996, 38–41; Ch'en Hsü, HSLWC, 119–128; and Kao Wei et al., KK 1998:10, 66–79.)

17. See Chang Kuo-shuo, KKWW 1996:1, 36–37.

18. Cheng-chou has been the subject of many reports of varying quality and argumentative character, including An Chin-huai, WW 1961:5, 73–80; An Chih-min, KK 1961:8, 448–450; Liu Ch'i-yi, WW 1961:10, 39–40; Honan-sheng Po-wu-kuan, WW 1977:1, 21–31; Cheng-chou-shih Wen-wu Kung-tso-tui, KKHP 1996:1, 111–42; Chang Kuo-shuo, KKWW 1996:1, 31–38; Fang Yu-sheng, KKWW 1999:3, 39–42; An Chin-huai, HCCHS 1993:11, 32–38; Chang Wei-hua, HCCHS 1993:11, 48–56; Hsü Chao-feng, KKWW, 1999:3, 43–48; An Chin-huai, 1986, 15–48; and Louisa G. F. Huber, EC 13 (1988): 46–77.

19. Tsou Heng and Ch'en Hsü have both been strong proponents of the Po identification. (See HSCLWC, 97–100, 101–106, 117–119, and 173–188; and HSLWC, 8–15, 23–35, 36–44, 45–54, 64–72, 73–84, 85–95, 96–103, 104–110, and 115–118 respectively.)

20. This is An Chin-huai and Yang Yü-pin's conclusion based on radiocarbon dates of 1650 or 1630 for Yen-shih's walls and roughly 1570 BCE for Cheng-chou. In addition (following others before them, including An Chin-huai's 1993 article on Cheng-chou), they note that Yen-shih's walls are no later than Erh-li-t'ou fourth period and pre-Erh-li-kang lower cultural layers, whereas the walls at Cheng-chou are built over an Erh-li-kang lower culture foundation. Chao Chih-ch'üan, KK 2003:9, 85–92, while identifying Po with Yen-shih and concluding that it demarks Hsia and Shang interaction in this area, observes that the main features of the palaces and walls show a Yen-shih to Cheng-chou sequence but that the artifacts tend to be opposite.

21. Tsou Heng envisions its construction and occupation as having been contemporary with Yen-shih despite radiocarbon dates cited by others to the contrary. (For example, see HSCLWC, 97–106 and 117–119.)

22. Chang Kuo-shuo, KKWW 1996:1, 35–36, argues that Po was not the name of a single site but a general term for the early Shang capitals and that both Cheng-chou and Yen-shih were necessary to control the Hsia, thus relegating Yen-shih to a sort of secondary status. Li Min, HCCHS 1996:2, 44–46, identifies it with Ao and (contrary to most views that it was fully abandoned) believes that it continued to serve as a military bastion during the Anyang period. (See also Chu Yen-min, STWMYC, 296–299.)

23. An Chin-huai (1986, 43) and others see the location of bronze-casting workshops outside the city as evidence that slaveholders dwelled within the walls, slaves outside them.

24. Chang Kuo-shuo, KKWW 1996:1, 36. An Chin-huai and Fang Yu-sheng, KKWW 1999:3, 39–42, claim that the bronzes discovered in three large hoards at Cheng-chou are evidence of a sophisticated stage of Shang culture; therefore Cheng-chou must be Chung Ting's capital of Ao. (Fang cites the *Ku-pen Bamboo Annals* entry: "After Chung Ting ascended the throne, he moved the capital from Po to Ao in his inaugural year [and] conducted a punitive expedition against the Lan Yi.") Conversely, scholars such as Chang Wei-hua, HCCHS 1993:11, 56, believe that Cheng-chou's development, wealth, and extensiveness argue against it having been so early a capital.

25. The best descriptions of the walls remain those of the Honan Provincial Museum, WW 1977:1, and An Chin-huai's "The Shang City at Cheng-chou," 22–26.

26. Ch'ü Ying-chieh, 2003, 42–44.

27. Ho-nan-sheng WWKKYCS, KK 2000:2, 40–60, and KK 2004:3, 40–50; Yüan Kuang-k'uo and Tseng Hsiao-min, KK 2004:3, 59–67.

28. Chang Kuo-shuo, KKWW 1996:1, 37.

29. Chang Kuo-shuo, KKWW 1996:1, 37.

30. For comparative dates and Cheng-chou's occupation before the conquest, see Yüan Kuang-k'uo and Hou Yi, WW 2007:12, 73–76, and Chang Hsüeh-lien and Ch'iu Shih-hua, KK 2006:2, 81–89.

31. For dissenting views, see Chang Wei-hua, HCCHS 1993:11, 49–55, and Hsü Chao-feng, KKWW 1999:3, 43–48. Chang concludes that King T'ang built Yen-shih immediately after the conquest but only dwelled there briefly because five years of drought—Heaven's punishment for overthrowing the legitimate ruling house—immediately ensued, forcing the Shang to precipitously move eastward. Hsü, however, stresses that Cheng-chou was needed to dominate the east.

32. P'ang Ming-chüan, KK 2008:2, 55–63.

33. For a representative analysis that synthesizes the archaeological discoveries at Erh-li-t'ou with traditional historical accounts, see Chao Chih-ch'üan, KKWW 1999:2, 23–29.

34. Lower Ch'i-tan Shang culture, which predates Erh-li-kang, was primarily centered in the Chi-pei, Yü-nan, and Yü-tung areas (Chao Chih-ch'üan, KKWW 1999:2, 24, and KKWW 2000:3, 28–32).

35. For example, see Chao Chih-ch'üan, KKWW 2001:4, 36–40.

36. Li Liu and Hung Xu, *Antiquity* 81 (2007): 886–901, or WW 2008:1, 43–52. Liu and Xu emphasize the essential continuity in bronzes between ELT and ELK and date Yen-shih to 1600–1400 BCE, the initial construction having occurred during Erh-li-t'ou's fourth phase.

37. Although it evolved out of Henan Lungshan culture, the Hsia's immediate precursor was Hsin-chai-ch'i culture.

38. Despite the newly mandated chronology, numerous Erh-li-t'ou alternatives have been offered, including by T'ien Ch'ang-wu, HCCHS 1987:12, 12–16, who concluded that Erh-li-t'ou's four cultural stages each lasted about a century and that the Shang conquered the Hsia in the midst of the third period, somewhat after 1700 BCE, earlier than most analysts.

39. In addition to Chao Chih-ch'uan, KKWW 2000:3, 28–32, see Yang Yü-pin, KK 2004:9, 87–92, who concludes that Yen-shih was unquestionably Po.

40. Chao Chih-ch'uan, KKWW 2000:3, 26.

41. Shang technologies and craft techniques tended to disperse outward, but local cultural elements often constituted the core content. Interaction with the Yi resulted in numerous shared cultural elements and practices, whatever the direction of transmission.

42. For example, see the extensive discussion in *The Cambridge History of Ancient China*, 124ff., for a comprehensive reflection of the revised view of Chinese history and culture that has

evolved over the past few decades, displacing the former belief that so-called high civilizations define the benchmarks and comparatively determine progress toward some extrapolated ideal.

43. Jui Kuo-yao and Shen Yüeh-ming, KK 1992:11, 1039–1044.

44. For an example of a subordinate state that sent in tribute, could be deputed on warfare tasks, and whose rulers may have been related to the king's clan by marriage, see Wang Yung-p'o, HCCHS 1992:4, 31–40. Many clans and proto-state names are known only from emblems and characters on bronze vessels. (For a concise enumeration of the latter, see Liu Chao-ying and P'ei Shu-lan, STWMYC, 365–372.)

45. For the origins of the plastrons employed in Shang divination, see Sung Chen-hao, STWMYC, 392–398.

46. Yang Sheng-nan, CKSYC 1991:4, 47–59, believes that the king had ultimate control over land allotments and rescissions and that grants were even made to other clans and proto-states outside Shang bounds (constituting a sort of de facto recognition for the purpose of symbolic integration).

47. Wang Kuan-ying, LSYC 1984:5, 80–99. It appears that the Shang was not a centralized state that systematically awarded fiefs but instead the most powerful group among a number of entities characterized by varying degrees of strength and independence. (For example, see Li Sheng, *Chung-kuo Pien-chiang Shih-ti Yen-chiu* 2006:9, 1–8; for a recent overview see Li Hsüeh-shan, HCCHS 2005:1, 29–34, who concludes that some thirty-six states concentrated in central Henan, southern central Shanxi, and around the capital had a close relationship with the Shang without ever being formally enfeoffed, and that they formed a defensive bulwark late in the dynasty that effectively blunted encroachments from the northwest.)

48. The core domain was apparently more heterogeneous than traditionally believed; evidence of non-clan names and even foreign peoples, including Chiang, Chi, Yün, Jen, and Ch'ang, has been uncovered. (See Ch'en Chieh, HCCHS 2003:2, 15–22. Evidence for the Ch'ang clan is seen in an opulent tomb; see SHYCS An-yang Kung-tso-tui, KK 2004:1, 7–19.)

49. For recent overviews see Yüeh Lien-chien, HCCHS, 1993:10, 29–40; T'ang Chi-ken, KKHP 1999:4, 393–420; and Pei-ching Ta-hsüeh K'ao-ku-hsi, KK 2005:6, 17–31.

50. See Yüeh Lien-chien, HCCHS, 1993:10, 29–40, and Sung Hsin-ch'ao, CKSYC, 1991:1, 53–63.

51. Lei Kuang-shan, KKWW 2000:2, 28–34; Chang T'ien-en, KK 2001:9, 13–21; and Li Hai-jung, KKWW 2000:2, 35–47. As Li's analysis shows, Kuan-chung was an extremely complex area that integrated multiple influences from the Shang, Shu, and northern steppes with numerous local factors. (For example, Cheng-chia-pao had better metallurgy and eventually absorbed the nearby Liu-chia before their amalgamated indigenous culture displaced the Shang during the fourth phase of Yin-hsü.)

52. Sung Hsin-ch'ao, CKSYC 1991:1, 53–63. For example, a dagger-axe with the name of a powerful group apparently closely related to the royal clan, noted in oracular inscriptions such as HJ33002 as having been dispatched in a military command capacity and also found on bronze vessels, has been recovered at an apparent Shaanxi outpost (Chang Mao-jung, KKWW 1997:4, 38–41, 49).

53. Chang T'ien-en, KK 2001:9, 13–21.

54. T'ang Chi-ken, KKHP 1999:4, 393–420; Yüeh Lien-chien, HCCHS, 1993:10, 29–40.

55. Sung Hsin-ch'ao, CKSYC, 1991:1, 53–63; Liu Shih-o and Yüeh Lien-chien, HCCHS 1991:10, 15–19. A strongly fortified Shang citadel, it also had smelting and pottery production facilities. (There is some disagreement as to the date of its loss to the Shang, with Liu and Yüeh arguing that it was the stronghold of the state of Sui, a staunch Shang supporter known to have been vanquished by the Chou prior to their conquest of the Shang.)

56. Yüeh Lien-chien, HCCHS, 1993:10, 29–40.

57. For an explication of the site's martial aspects and their implications (which are generally followed here), see Wang Jui, KK 1998:8, 81–91. Additional analysis may be found in Ts'ao Ping-

wu, KK 1997:12, 85–89, and Wang Yüeh-ch'ien and T'ung Wei-hua, KK 2005:11, 3–17. Radio-carbon dates for the initial site of as early as 2000 BCE, clearly pre-Shang, have been reported. In addition to being a military bastion, it appears to have been the center of the later state of Yüan (Tsou Heng, HSCLWC, undated, 204–218).

58. See Wang Jui, KK 1998:8, 90.

59. See Wang Jui, KK 1998:8, 89–90, and Ts'ao Ping-wu, WW 1997: 12, 85–88. Although ancient copper mines have been discovered in the general region, the site lacks any evidence of a direct relationship, and during the period of Shang fluorescence the most important deposits were located in the south. (See Wang Jui, 90, and Wang Yüeh-ch'ien, KK 2005:11, 16.) However, salt was crucial to every dynasty, and the Yün-ch'eng basin around Yüan-ch'ü was one of the chief sources in antiquity, the salt recovered from the nearby salt lakes being shipped down to Cheng-chou. (It has been suggested that Yüan-ch'ü was T'ang's early capital of Po, even though radiocarbon dating precludes this possibility, and that it was the center of the enemy state known as Yüan in Wu Ting's time and thereafter. However, though geographically possible, the recovered artifacts predate Wu Ting's reign and indicate that it was a purely Shang enclave, uncontaminated by the intrusion of other cultural elements.)

60. Ch'en Hsüeh-hsiang and Chin Han-p'o, KK 2007:5, 84–85.

61. Yüeh Lien-chien, HCCHS, 1993:10, 29–40. (Dates cited for late T'ai-hsi center on 1300 BCE or a century before Wu Ting's reign, but some artifacts corresponding to early Yin-hsü man-ifestations have also been recovered.)

62. Ho-pei-sheng WWYCS, KK 2007:11, 26–35.

63. Sung Hsin-ch'ao CKSYC 1991:1, 55.

64. Sung Hsin-ch'ao, 55. As attested by the character "ya" on the various grave goods, tomb number 1 clearly belonged to a Shang commander, probably a member of the royal clan (Ch'en Hsüeh-hsiang and Chin Han-p'o, KK 2007:5, 87).

65. Chiang Kang, KKWW 2008:1, 35–46, but especially 44–45. Initial site reports indicate that Wang-chia-shan consisted of a three-tiered, triangular earthen platform with dimensions of 80 by 180 meters.

66. Sung Hsin-ch'ao, CKSYC, 1991:1, 53–63.

67. The site reportedly shows the extent of Shang power over distant cultures. (A useful early report is Hu-pei-sheng Po-wu-kuan, WW 1976:2, 5–15, and Chiang Kang has examined some of the military aspects in KKWW 2008:1, 35–46, but the most comprehensive evaluation to date is the extensively illustrated P'an-lung-ch'eng Ch'ing-t'ung Wen-hua. Brief evaluations are also found in Cambridge History of Ancient China, 168–171, and State Formation in Early China, 75–78. Li Chien-min [KK 2001:5, 60–69] believes that Pan-lung-ch'eng was the capital of the Shang enfeoffed southern state of Ching-Ch'u [64], though this seems unlikely in view of the absence of feudal enfeoffments and subsequent Shang campaigns to the south.)

68. In contrast to the bronzes, the strong indigenous character of the ceramics has been in-terpreted as indicating the relatively sudden imposition of an external cultural type. (Ceramics are generally held to express local factors, bronze stylistic influences imposed by—or copied from—occupying powers. Although the 159 bronze objects consisting of ritual cauldrons, weapons, and a few tools were locally produced rather than imported [Nan P'u-heng et al., WW 2008:8, 77–82] and include a few distinctive types, they are virtually identical to lower Erh-li-kang Cheng-chou artifacts.) However, contrary to recent interpretations that emphasize resource acquisition, Chiang Kang, 2008:1, 46–48, doesn't believe P'an-lung-ch'eng was engaged in mineral activities prior to the Shang.

69. For a reconstruction of ancient transport routes, the distribution of natural resources in the P'an-lung-ch'eng area, and a discussion of the importance of tribute and redistribution in the Shang's ritually based monopolization of authority, see State Formation in Early China.

(Apparently there were three water routes and one land road for forwarding minerals to the capital, all controllable from P'an-lung-ch'eng. See Chiang Kang, KKWW 2008:1, 44.)

70. Yüeh Lien-chien, HCCHS 1993:10, 34–37, argues for it having been a Shang enclave.

71. Important reports include T'ang Lan, WW 1975:7, 72–76; Chiang-hsi-sheng Po-wu-kuan, WW 7 (1975): 51–71; and Chan K'ai-sun, KK 1 (1995): 36, 63–74. It is also briefly discussed in K. C. Chang's *The Archaeology of Ancient China*, 389–394, and by Robert Bagley in the *Cambridge History of Ancient China*, 171–175.

72. K. C. Chang, *The Archaeology of Ancient China*, 389.

73. "Sun-tzu Wu Ch'i Lieh-chuan," *Shih Chi*. As T'ang Lan points out (1975:7, 73), a slightly expanded, possibly original version is found in "Wei Ts'e" in the *Chan Kuo-ts'e*, which states that "relying on the terrain's difficulty, they did not practice good government."

74. For site reports see SHYCS, KKHP 1983:1, 55–92, and Tung-hsia-feng K'ao-ku-tui, KK 1980:2, 97–107.

CHAPTER 8

1. T'ang Chi-ken, KKHP 1999:4, 410–413. Although cut off from direct contact with the Shang, rather than displacing Shang material culture the local resurgence was based on it.

2. See Yüeh Lien-chien, HCCHS 1993:10, 29–40.

3. It has also been suggested that the Hsia populace had been fully integrated, even assimilated, making the citadel's abandonment possible. See Ts'ao Ping-wu, KK 1997:12, 86–87.

4. Other sequences substitute Pi and Po. Significant discrepancies in Warring States accounts have long prompted rather futile attempts to match them to likely sites. (For convenient summaries see Wang Li-chih, KKWW 2003:4, 41–42, or Wang Chen-chung, KKWW 2006:1, 44–49.)

5. According to the *Ku-pen Bamboo Annals*.

6. For reports and discussion see Ch'en Hsü's five HSLWC articles, 137–144, 145–154, 155–158, 159–162, and 163–170; Ch'en's KKWW 1 (2000): 33–38; and Fang Yu-sheng, KKWW 2000:1, 39–41.

7. Ch'en Hsü, KKWW 2000:1 36–37. Tsou Heng, KKWW 1998:4, 24–27, believes that Hsiao-shuang-ch'iao was the site of Chung Ting's capital of Ao, Cheng-chou having been abandoned, whereas Fang Yu-sheng, HCCHS 1998:1, 58–63, KKWW 2000:1, 39–41, and KK 2002:8, 81–86, has repeatedly argued that Cheng-chou continued to flourish in the Pai-chia-chuang phase as the capital known as Ao and dismisses Hsiao-shuang-ch'iao as a secondary ritual center.

8. Tsou Heng, KKWW 1998:4, 26; Ch'en Hsü, HSWHLC, 161–162, and KKWW 2000:1, 36–37.

9. Ch'en Hsü, HSWHLC, 163–170, and KKWW 2000:1, 36–37.

10. For an overview of their relationship during the Yen-shih/Cheng-chou period, see Ch'en Hsü, HSLWC, 104–110. Considerably earlier than most archaeologists, Ch'en discerns a shift from a strong intermixing of Yüeh-shih elements prior to Cheng-chou to their virtual elimination post–Cheng-chou, not just in central China, but also in Shangdong, coincident with the Pai-chia-chuang phase of Shang culture.

11. Chia Chin-piao et al., KK 2005:2, 71–78; Chin Wen-sheng, STWMYC, 133–136; and Tsou Heng, STWMYC, 42–44. All the cultural elements clearly postdate ELK Pai-chia-chuang but are equally pre-Yin-hsü.

12. For a site report, see Tung-hsien-hsien K'ao-ku-tui, KK 2003:11, 27–40.

13. For a site report and analysis, see Ho Kuang-yüeh, CKKTS, 1995:5, 32–36.

14. For example, based on a comparative reading of the *Bamboo Annals*, *Shih Chi*, and other early writings, Ch'ao Fu-lin concluded that the clause indicating that "the Shang did not move their capital thereafter" is erroneous and that the *Shih Chi* is correct in stating that in P'an Keng's

time their capital was north of the river and that he crossed to the south and again dwelt in "Ch'eng T'ang's old dwelling," in other words the old Po capital at Cheng-chou. He further believes some collateral activity occurred at Yin-hsü prior to Wu Ting's ascension. (See CKSYC 1989:1, 57–67, and for further discussion Yen Yi-p'ing, 1989, Vol. 2, 157–173.)

15. Even this claim does not go unchallenged. Apart from the question of whether the storage pits simply haven't been discovered, the reclassification of certain diviner group inscriptions has resulted in a few being attributed to the three reigns immediately preceding Wu Ting. (See Yang Pao-ch'eng, KK 2000:4, 74–80, and Ts'ao Ting-yün, HCCHS 2007:5, 21–29.)

16. For a basic site report, see SHYCS An-yang Kung-tsuo-tui, KK 2003:5, 3–16.

17. For example, Chu Kuang-hua, KKWW 2006:2, 31–35.

18. Wang Chen-chung, KKWW 2006:1, 48.

19. As suggested by Chu Kuang-hua, KKWW 2006:2, 31–35. (An alternative explanation—that Huan-pei may have been Yen and P'an Keng then simply moved across the river—seems not to have arisen.) Various dates have been suggested for the move to Anyang, such as Hsü Po-hung's rather early 1350 BCE. (See HCCHS 1998:4, 29–36. Note that Hsü dates the Chou conquest to about 1075 BCE.) However, the Xia Shang Zhou Chronology Project has proposed a date of 1298 BCE, with Wu Ting's ascension then occurring in 1251 BCE. The question is not independent of Cheng-chou's abandonment.

20. Li Chi has recounted the myriad problems that beset early excavation work at Anyang in his classic *Anyang*. Apart from the numerous site reports and the fundamental information contained in *Cambridge History of Ancient China*, Chang Kwang-chih's discussion in *Shang Civilization*, 69–135, and An Chin-huai's "The Shang City at Cheng-chou and Related Problems" retain value. Chu Yen-min (1999) provides an extensive survey and analysis, but see also Shih Chang-ju, KKHP 2 (1947): 1–81.

21. For some evidence of the earlier occupation, see Liu Yi-man, STWMYC, 148–161.

22. Li Min, LSYC 1991:1, 111–120, believes the old capital suffered some sort of disaster, probably flooding, but that conditions there had also been exacerbated by environmental degradation. In contrast, the fertilization practices adopted at Yin-hsü preserved the environment by disposing of human and animal waste. However, for another view see Yang Hsi-chang and T'ang Chi-ken, STWMYC, 248–256.

23. For an overview, see Chu Yen-min, 1999, 100–114. No doubt under the influence of Confucianism, later ages envisioned the shift to Anyang as a manifestation of P'an's quest to return the people to Virtue, an imaginative view clearly reflected in the *Shang Shu*.

24. See, for example, Tung Ch'i, WW 2006:6, 56–60, 87. Generally speaking, as indicated by discussions in *Mencius*, *Kuan-tzu*, and other pre-Han writings, every significant city was expected to have inner and outer walls, known as *ch'eng* and *kuo*, as well as moats and segregated quarters. (For a brief overview see Liu Ch'ing-chu, WW 1998:3, 49–57.)

25. Liu Ch'ing-chu, KKHP 2006:3, 283, and others believe the moat, in conjunction with the rivers, furnished adequate perimeter protection. However, Li Min, LSYC 1991:1, thinks the moat was intended to reduce the Huan River's level during floods, thereby sparing Anyang. A small interior ditch apparently intended to protect a sacrificial area, datable to the third phase of Yin-hsü, has also been found at Anyang. Only about 30 to 35 meters in length, it has a maximum width of 2.9 meters and a depth of nearly a meter, but considerably less in some sections. Hardly an effective deterrent, it probably served more as a marker. (See Yin-hsü Hsiao-min-t'un K'ao-ku-tui, KK 2007:1, 37–40.)

26. As argued, for example, by Chang Kuo-shuo, KKWW 2000:1, 42–45.

27. For a complete translation of the *Wu-tzu*, the book attributed to Wu Ch'i, see Sawyer, *Seven Military Classics of Ancient China*. This passage appears in Wu Ch'i's biography in the *Shih Chi*.

28. For a complete translation, see Crump, *Chan-kuo Ts'e*, 374.

29. "Responding to Change," *Wu-tzu*, attributed to Wu Ch'i.

30. "Military Combat."

31. "Maneuvering the Army."

32. "Configurations of Terrain."

33. Sun Pin, "Male and Female Cities." (For a complete translation and discussion of the various configurations, see Sawyer, *Sun Pin Military Methods*.)

34. Yin Chün-k'o, 1994, 114ff.; Li Min, HCCHS 1988:4, 41–48; and Chu Chen, HCCHS 1989:8, 3–10.

35. Something might perhaps be learned from the movement of the Japanese capital from Nara to Kyoto and then the subsequent power shift to Edo where compulsory, burdensome attendance at the shogun's court proved a significant factor in eroding vassal independence and power.

36. Ch'iao Teng-yün and Chang Yüan, STWMYC, 162–174.

37. Both Li Min HCCHS 1988:4, 46–48, and Chu Chen, HCCHS 1989:8, 7–10, discuss the generally neglected subject of Chao-ko.

38. Despite various claims that writing's incipient origins can be discerned in Yangshao cultural manifestations such as Ta-wen-k'ou or Pan-p'o, the evidence cited consists mostly of individual symbols that may represent the earliest form of written expression or may simply be clan markers or totemistic symbols. (See, for example, Chang Mao-jung, 2002, 20–23, or Chiang Lin-ch'ang, HCCHS 2006:4 for more recent summaries of the manifestations and interpretations.)

39. For further discussion, see David Keightley, *Sources of Shang History*, 134–146.

40. This exploration of Shang martial activities is heavily indebted to numerous pioneering examinations of oracular inscriptions by Chinese, Japanese, and Western scholars, such as Kuo Mo-jo, Tung Tso-pin, Ch'en Meng-chia, Li Hsüeh-ch'in, Wang Kuo-wei, Shima Kunio, Chang Ping-hsüeh-ping, David Keightley, and Edward L. Shaughnessy. In addition to Tung Tso-pin's chronologies, our discussion of Shang dynasty martial activities is primarily based on the following important studies: Wang Yü-hsin, "Wu-Ting-ch'i Chan-cheng Pu-ts'u Fen-ch'i te Ch'ang-shih"; Fan Yü-chou, "Yin-tai Wu Ting Shih-ch'i te Chan-cheng" and "Military Campaign Inscriptions from YH127"; Ch'en Meng-chia, "Wu Ting Shih-tai te Tuo-fang," "Wu-Ting-hou te Tuo-fang," and "Yi-Hsin shih-tai Suo-cheng te Jen-fang, Yü-fang"; Lin Hsiao-an, "Yin Wu Ting Ch'en-shu Cheng-fa yü Hsing-chi K'ao"; P'eng Yü-shang, *Yin-hsü Chia-ku Tuan-tai*; and others as individually noted. All these studies, although founded on the same collections of oracle bones (including those in *Chia-ku-wen Ho-chi*, hereafter abbreviated as HJ), tend to selectively emphasize certain aspects of Shang martial activity, resulting in slightly different depictions and divergent conclusions. Citations will indicate which view is being followed, while inscription numbers will signify materials either abstracted from the above because of their particular relevance or developed in the course of our study.

41. Claims have been made that anywhere from 90 significant cities or states to a maximum of 800 place-names (cited by Li Hsüeh-ch'in, HCCHS 2005:5, 3–7) can be identified.

42. For an early report on two bronze Yin-hsü foundries, see SHYCS, *Yin-hsü Fa-chüeh Pao-kao 1958–1961*.

43. The ratio of prisoners to casualties is sometimes astounding. For example, according to *Hsü-ts'un hsia* 915, in a campaign against the Wei, twenty-four prisoners and a clan leader (who was later sacrificed) were captured, but an astonishing 1,570 were slain. Various weapons and other items were also seized, including two chariots. (For an analysis of this inscription see Wang Yü-hsin, CKSYC 1980:1, 106.)

CHAPTER 9

1. Among many, see Chang P'ei-yü, KKWW 1999:4, 62–65. The dates 1239–1181 and 1250–1192 BCE are based on eclipse calculations; the PRC dating project has stipulated 1250–1192;

and *Cambridge History of Ancient China* begins its chronology with an end date of 1189 (1198–1181 is clearly too short for his many accomplishments). Many of the warfare inscriptions are found among HJ 6057–7771.

2. The nature of these entities—extended clans, tribes, chiefdoms, or proto-states—remains unclear. (For a characterization see T'ung Chu-ch'en, KK 1991:11, 1003–1018, 1031; for a survey of the peripheral proto-states see Lu Lien-ch'eng, CKKTS 1995:4,30–56.)

3. This is the scheme adopted by historians such as Meng Shih-k'ai, Lin Hsiao-an, Fan Yü-chou, and P'eng Yü-shang. It should be differentiated from Tung Tso-pin's five-era periodization of Anyang oracle bones, whether as originally formulated or subsequently modified, and Wang Yü-hsin's somewhat nebulous two-part division. (For a discussion of periodization see Keightley, *Sources of Shang History*, 91–100.)

4. Ch'en Hsü, HSLWC, 246.

5. The "rabbit" rather than "earth" people, who were a major threat in the second period.

6. Apart from those noted in the discussion below, see Fan Yü-chou, 1991, 201.

7. According to Ch'en Meng-chia, 1988, 297–298.

8. Ch'en Meng-chia, 296–297. (For relevant inscriptions, see Chang Ping-ch'üan, 1988, 333.)

9. For inscriptions (including HJ6993 and HJ6991) see Lin Hsiao-an, 229, and Fan Yü-chou, 1991, 196.

10. For inscriptions (including *Hsü Ts'un* 1, 609, and HJ6983), see Chang Ping-ch'üan, 491, or Lin Hsiao-an, 230.

11. *Yi* 4693.

12. For a study of the concept of space in the Shang, see David N. Keightley, *The Ancestral Landscape*.

13. Accounts of the campaign may be found in Fan Yü-chou, 1991, 186–191; Luo K'un's *Hsia Shang Hsi-Chou Chün-shih-shih*, 1991, 173–177; Liu Yü-t'ang, HCCHS 2001:4, 59–60; and Liu Huan, CKSYC 2002:4, 3–9, with additional references to the "south" listed in Chang Ping-ch'üan, 316. As usual, readings and interpretations vary, and the pronunciation of the proto-states is highly tentative.

14. The formation of alliances is implied by oracular references to the *nan pang fang* or "southern allied states" (*Chia* 2902, HJ20576), and this repression of the rebellious generally derived from *Shih Ching* commentators (Ch'en Hsü, HSWHLC, 243).

15. "Li shih yü nan" (*duo* 2.62), meaning "to take responsibility/handle affairs in the south," is understood as evidence that the king personally supervised the campaign, perhaps even the battle action. He also performed prognostications en route and in the south. Ch'üeh's participation shows the conflict cannot have been too early in the king's first period, because he was originally an enemy who subsequently became a trusted ally.

16. HJ5504.

17. HJ20576, possibly HJ19946.

18. Not the same state as the "Ghost" state.

19. HJ19946 and HJ20576 convey a sense of urgency because the king is inquiring whether his generals will suffer harm while repeatedly asking if they will receive blessings for the effort—in other words, if the time is right for an action that the king has already decided upon.

20. HJ6667.

21. The severity of his illness so perturbed the king that he felt compelled to offer sacrifice for his recovery.

22. According to Liu Huan, CKSYC 2002:4, 8. (Liu, 6–7, generally derives more sweeping conclusions than other analysts, including that the king retained extensive control over his field troops and engaged in public relations or propaganda pronouncements.)

23. See P'eng Ming-han, CKSYC 1995:3, 101–108. (P'eng suggests that there were two Hu peoples, the Hu in the main Shang domain and a separate Hu-fang, but fails to substantiate his claim.)

24. The term *t'u* (seen in HJ6667) is generally understood as meaning something like "punitively attack" but might merely signify going out to resist them or manifest awesome power. (The Chü is a tributary of the Han River.)

25. P'eng Ming-han understands these three as "war spirits," but they need not be so.

26. As claimed by P'eng Ming-han, CKSYC 1995:3, 101.

27. Although massive battles would not be unknown, Chinese military writers noted that the later (horse-mounted) steppe peoples avoided fixed confrontations. However, it would be erroneous to project similar combat tendencies back onto the essentially sedentary steppe peoples of the Shang period.

28. Our account basically follows Fan Yü-chou's dating and campaign chronology, 1991, 205–207. (Lin Hsiao-an, 236–239, dates it to Wu Ting's first period and Wang Yü-hsin also places it early.) Early Hsüan participation in Shang theocracy argues for the correctness of Fan's dating. (See also P'eng Yü-shang, 1994, 97–103, who similarly assigns it to the king's middle period.) The Hsüan were apparently located to the west of the Shang; Ch'en Meng-chia, 1988, 276, suggests they may have dwelled in Kuei-fang territory.

29. HJ9705. Reconstructions of the campaign may be found in Fan Yü-chou, Vol. 3, 205–206, and BSOAS 539; Lin Hsiao-an, 236–238; and *Hsia Shang Hsi-Chou Chün-shih Shih*, 165–166.

30. *Ping-pien* 307.

31. HJ6937. The problem raised by Lin having damaged the Hsüan in the seventh month (requiring that the Hsüan and T'an be contiguous states) can be resolved if these are inquiries, not necessarily events, and he may have been sent against the Hsüan rather than the T'an, or the reverse.

32. See HJ6924–6939.

33. *Ping-pien* 249, cited by Fan Yü-chou, BSOAS, 539.

34. Fan Yü-chou, 1991, 205, argues that *Ping-pien* 307, which asks whether the king should attack the T'an in the sixth month or go west, shows that the T'an were located in the west. However, the inscription seems to be posing paired alternatives: Should the king campaign in the east, attacking the T'an, or devote his efforts to resolving problems in the west?

35. HJ6942, querying whether the T'an will harm the Ts'ao.

36. For this view see *Hsia Shang Hsi-Chou Chün-shih Shih*, 165–166.

37. Ch'en Meng-chia, 1988 298, 276. (Ch'en Meng-chia provides a useful iteration of the main commanders on 275–276, but for a summary of the most important inscriptions see Chang Ping-ch'üan, 1988, 492; Lin Hsiao-an, 236–239; and Wang Yü-hsin, 1991, 155–158.)

38. Lin Hsiao-an, 237, interprets the simultaneous incursions of the T'an, Hsüan, and Jung as evidence of deliberate, coordinated action.

39. *Ping* 306; Lin Hsiao-an, 237.

40. See, for example, HJ6958, which asks whether Ch'üeh will "pummel" the Hsüan.

41. HJ6946.

42. HJ6959; *Yi* 4380, 4919, 5163, and 5193.

43. *Yi* 4693.

44. For example, see HJ9947, HJ6939, HJ6947, HJ6948, HJ6952, HJ6954, HJ6958, HJ6959, HJ20383.

45. Attested by HJ20393.

46. HJ6205, HJ6567, HJ6240; Fan Yü-chou, 1991, 205 and 207; Wang Yü-hsin, 1991, 157. The Hsüan are recorded as frequently sending in scapula for divination, once in the staggering amount of 1,000. (See Fan, 205–206.)

47. Reconstructions are provided by Edward L. Shaughnessy in "Micro-periodization," 58–82, and Fan Yü-chou in "Military Campaign Inscriptions," 1989, 535–48, and *Yin-tai Wu Ting*, 1991, 202–205. Several inquiries fortuitously include confirmation of the results. (Inscription summaries will also be found in Chang Ping-ch'üan, 491–492, and Ch'en Meng-chia, 287–288.

Also see page 72 and note 28 of "Micro-periodization" for Shaughnessy's determination of the campaign date as 1211–1210 BCE.) Significant differences exist between Shaughnessy's reconstruction (which is essentially followed here, apart from transcribing the state name as P'ei instead of Pu), based on the adroit employment of an intercalary month, and that offered by Fan Yü-chou in his two articles (1989 and 2006). In fact, the divergences illustrate the difficulty of assembling highly disparate pronouncements into a coherent patchwork of probable events.

48. A crucial issue is whether "Fou" should be understood as the name of another state or the head of the Chi-fang, as only Shaughnessy ("Micro-periodization," 76, note 14) concludes. In contrast, Ch'en Meng-chia, Fan Yü-chou (204–205), and others assume Fou was a second state located in the same general western vicinity. (Shaughnessy argues that the actions being performed by Fou—visiting the king, hunting, and dying—could only pertain to an individual, but they could simply be a personified way of referring to the head of the clan or state, like Ch'üeh, Chih, or Hsüan. However, evidence for the correctness of his view may be seen in the virtual impossibility that every inscription in which "Chi-fang Fou" appears would have required identical actions by commanders against two discrete enemies irrespective of their proximity. In addition, it is highly unlikely that two proto-states would have simultaneously undertaken the construction of defensive fortifications despite having come under Shang onslaughts.)

49. See HJ6834a, a well-known inscription related to inquires about the advisability of attacking P'ei and the Chi-fang as well. As noted by Fan Yü-chou, Ping-pien 558 and 1, asking about the prospects for (imminent) conquest, corroborate Hsi's likely defeat.

50. See BSOAS, 1989 537.

51. HJ6577 questions whether Prince Shang will capture the Chi-fang; Ping-pien 1 (HJ6834) inquires whether Ch'üeh, the tuo ch'en, or the shih will capture or conquer Fou; and several other inscriptions such as Ping-pien 171 and HJ6934 show the Shang intended to severely damage or destroy them.

52. HJ13514a. Shaughnessy, "Micro-periodization," 66, translates as "Crack-making on [hsin]-mao (day 28), K'o divining: 'We ought not ?? Chi-fang Fou's building a wall, for Prince Shang will harm (him).' Fourth month."

53. See HJ6573; Ping-pien 302.

54. According to the Shih Ching, "Yin Wu," Mao #305. Being employed by the Shang to refer to any of the external, mainly steppe peoples who were characterized by distinctive cultural practices and presumably spoke different languages, the term ch'iang also has a broader meaning and somewhat derogatory connotations.

55. Meng Shih-k'ai, 1986, 206–207; Ch'en Meng-chia, 281. (Ch'en notes that most of the battles with the Shang occurred in Chin-nan and Ho-nei, around the T'ai-hang Mountains.)

56. Lin Hsiao-an, 241, 260.

57. For a typical expression of this less than universally accepted view, see Wang Shen-hsing, 1992, 116–117. The relationship of the Hsia populace to later steppe groups, including the Ch'iang and Hsiung-nü, has long been a subject of debate. (For example, see Hsü Chung-shu, LSYC 1983:1, 60–61.)

58. Meng Shih-k'ai, 1986, 207–208. (For an enumeration of relevant inscriptions, see Ch'en Meng-chia, 279–281.) Twelve of the fifteen ways employed by the Shang to slay sacrificial victims were used against the Ch'iang (Wang Shen-hsing, 125–130). Liu Feng-hua, KKWW 2007:4, 22–26, notes that T'un-nan refers to beheading forty Ch'iang prisoners, showing that they were slain with impunity.

59. See, among many, Ch'en Meng-chia, 279–281; Meng Shih-k'ai, 207; and Wang Yü-hsin, 1991, 170 and 173. Ch'iang were captured in military expeditions and as an aspect of hunting expeditions, as well as while simply attending to pasturing and farming, and even forwarded to the Shang by others.

60. For a general discussion see Luo K'un, 1991, 405–426.

61. As exemplified by Chiang T'ai Kung, the legendary Chou tactician and commander. It would be frequently claimed that the Chiang were descended from Yen Ti, the Red Emperor.

62. For further discussion, see Ch'en Meng-chia, 282.

63. Wang Shen-hsing (1992, 133–140) notes that they frequently proved troublesome prisoners, rebelling, fleeing, and resisting recapture.

64. See Lin Hsiao-an, 241–242, and Fan Yü-chou, 1991, 191–193, as well as such inscriptions as Ts'ui 1300, HJ6600, HJ6601, HJ6603, HJ6604, HJ6978, and finally HJ6599, which shows that they were perceived as dangerous. (HJ6492 indicates Kuang took fifty prisoners, hardly a large number considering the apparent scope of the conflict, but not insignificant.)

65. Per Lin Hsiao-an's chronology, 258–261. (Lin also asserts, perhaps dubiously, that Lung-fang forces that had just been vanquished by the Shang also participated as battlefield allies.)

66. The inscriptions for the various Ch'iang campaigns are summarized by Wang Yü-hsin, 1991, 170–171; Fan Yü-chou, 1991, 207–209; Liu Hsiao-an, 258–261; Ch'en Meng-chia, 276–279; and Chang Ping-ch'üan, 1988, 492. Despite frequent references to the Ch'iang in Wu Ting's period, Luo K'un believes they were neither a significant threat nor a major warfare objective in his era, though the capture of "barbarian" prisoners was important. (Luo K'un, 1991, 405–426, and Hsia Shang Hsi-Chou Chün-shih Shih, 185–187. Luo bases his view on distinguishing between the specific tribe or proto-state called Ch'iang and the broad use of the term to designate any of several steppe peoples.) However, the fact that a total of some 2,000 victims were ritually slain according to several inquiries preserved on just one bone would certainly seem to argue for large-scale, frequent warfare and against Luo's assertion.

67. K'u 130, sometimes misstated as 310. However, there is considerable disagreement whether this badly corrupted inscription discusses a force targeting the Ch'iang (since "ch'iang" is interpolated) or the T'u-fang. (See, for example, Luo K'un, 186.) Lin Hsiao-an, 259–260, argues (albeit from somewhat later materials) that military organization and prerogatives at the time would have precluded Fu Hao from commanding such a large force. Nevertheless, since Fu Hao did exercise power in such important court functions as sacrifice and prognostication, commanding in the king's stead would not have been beyond the realm of possibility.

68. P'eng Yü-shang, 1994, 145. (Luo, 412, denies that Fu Hao campaigned against the Ch'iang.)

69. Lin Hsiao-an, 260–261. (See especially HJ6630 and HJ6636.)

70. Lin Hsiao-an, 261.

71. Whether this conflict falls at the end of Wu Ting's middle period or the beginning of his third is somewhat nebulous. Fan Yü-chou, dissatisfied with Tung Tso-pin's pioneering work (which assigns the campaign to the third month of Wu Ting's twenty-ninth year), has provided a brief revamped calendar for the clash with the Hsia-wei in Yin-tai Wu-Ting Shih-ch'i te Chancheng, 1991, 213–214, grouping it with other conflicts in the final period. However, in his "Military Campaign Inscriptions from YH127," 539, Fan views it as a late middle period clash because it is the last campaign attested by inscriptions found in pit YH127. In contrast, Wang Yü-hsin, 1991, 150, ascribes it to the first half of Wu Ting's reign and Lin Hsiao-an, 251–253, to the middle period. Overall, although there were clashes early on, the main conflict seems to have unfolded during the final years of the middle period.

72. HJ6477a (Ping-pien 311), for example, preserves simultaneous queries whether the king should attack the Pa-fang with Hsi or the Hsia-wei with Wang Ch'eng. HJ6413 similarly shows campaigns being contemplated against the T'u-fang, Hsia-wei, Lung-fang, and others. HJ6417 records consecutive queries about the Hsia-wei and T'u-fang. (See also Wang Yü-hsin, 149–152, and Lin Hsiao-an, 252–253.)

73. HJ6525.

74. HJ6523, Hsü 137.1.

75. HJ6496. Also note Ping-pien 24 and Ching 1266.

76. Attested by HJ6477, HJ6487, HJ6496, and others.

77. HJ6530a, *Ho* 151.

78. HJ6487.

79. Chang Ping-ch'üan, citing *Chin* 25, 1988, 489.

80. *Yi* 6382.

81. HJ6527.

82. See, for example, HJ6451, HJ6459, and HJ6480.

83. See Wang Yü-hsin, 1991, 149–152.

84. Lin Hsiao-an, 257.

85. It is thought that the Lung were one of the ancestors of the powerful Hsiung-nu, who were active in the late Warring States and early Han. (See, for example, Ch'en Meng-chia, 1988, 283.)

86. HJ6476, *Yi* 5340. (Convenient listings of Lung campaign inscriptions may be found in Ch'en Meng-chia, 283; Chang Ping-ch'üan, 1988, 316–317 and 489–490; Lin Hsiao-an, 257–258; and Wang Yü-hsin, 1991, 164–165.) As apportioned by Fan Yü-chou (230), the commissioning of a few early period generals argues for relatively early clashes, but the first appearance of Fu Ching in a command role implies that they date to Wu Ting's late middle period or later.

87. HJ6587, HJ6590.

88. *Yi* 5340.

89. HJ6585 / *Hsü* 4.26.3 (targeting the Lung and Ch'iang), HJ6584 (targeting the Lung and an unknown enemy but presumably the Ch'iang).

90. HJ6633, HJ6593, HJ6594, HJ6636.

91. HJ6584, HJ6585, HJ6633; *T'ieh* 105.3; *Shih* 5.5.

92. According to Fan Yü-chou, 1991, 212, evidence that they were about to be defeated is seen in HJ6630, HJ6631, HJ6632, HJ6634, HJ6635, HJ6636, and HJ6638, as well as *Ching-jen* 343, *Yi* 462, and *Ts'un* 2.302. In addition, HJ6587 and H6590 inquire whether Shih Pan will "seize" the Lung.

93. HJ8593.

94. HJ6664b.

95. Ho 626 and 630; *Nei-pien* 49, 52, and 132.

96. HJ6630, HJ6631.

97. The problem of relative dating is particularly visible in determining the date of the T'u-fang campaign. Wang Yü-hsin, 1991, 147, who simply divides Wu Ting's reign into two periods, places it toward the end of the first part, but others who employ a tri-partite division (such as Lin Hsiao-an, 262–263) assign it to late in the middle period. (Tung Tso-pin's reconstruction in his *Yin-li P'u* has it commencing in the king's twenty-eighth year, though its accuracy has been questioned by Fan Yü-chou, 1991, 214, and others.)

98. Discussions of critical inscriptions are found in Ch'en Meng-chia, 1988, 272–273; Chang Ping-ch'üan, 1988, 492–493; Lin Hsiao-an, 261–266; Wang Yü-hsin, 145–147; Fan Yü-chou, 214–219; P'eng Yü-shang, 143–145; and Hu Hou-hsüan, HCCHS 1991:2, 13–20.

99. See Lin Hsiao-an (265), who cites this estimate of marching time from Kuo Mo-juo. (Kuo apparently concluded the T'u-fang were therefore located about 1,000 *li* from the Shang center, but in contrast to the maximum average sustained rate of 25 *li* known from historical records, his estimate was based on an impossible daily march of 80 *li*.) Others, such as P'eng Yü-shang, 144, conclude that the T'u-fang were located to the northwest or west of the Shang and actually quite close, which would agree with twelve days' march at 25 *li* per day, or only 300 *li* in total. Although most historians and maps have traditionally placed them to the north, Chao Ch'eng, 2000, 4, locates the T'u-fang east of the Kung-fang, who were active in northern Shaanxi; Hu Hou-hsüan would add southern Inner Mongolia to the T'u-fang's domain, requiring the main Shang campaigns against them to proceed northwest.

100. Whether the T'u-fang were direct descendants of the Hsia has long attracted considerable speculation, most of it based on detailed but highly imaginative interpretations of

vestiges perceived in traditional historical writings. (For example, see Hu Hou-hsüan, HCCHS 1991:2, 13–20.)

101. See, for example, HJ6381, HJ649, HJ6440, and various inscriptions appealing to the spirits for aid against them, including HJ6384, HJ6385, HJ6386, HJ6388, and HJ39879.

102. Several different terms are used for the Shang response, ranging from "attack" to "destroy." See Hu Hou-hsüan, HCCHS 1991:2, 16–17, for an enumerative listing.

103. P'eng Yü-shang, 144, and Hu Hou-hsüan, 16–17. For examples, see HJ6087, HJ6354, HJ6389, HJ6391, HJ6392, HJ6393, HJ6396, HJ6400, and HJ39880.

104. For example, see HJ5412, HJ6087, K'u 237.

105. HJ6452.

106. HJ39889.

107. For example, HJ6401, HJ6402, HJ6403, HJ6415, HJ6416, HJ6417, HJ6419, HJ6420, HJ6421, HJ39885, HJ39887, HJ39888, and HJ6438 (with the San-tsu).

108. In addition to the levies listed in the discussion, undated strips—which may well be redundant—at least twice list call-ups of 3,000 men. (See HJ6438, HJ6407, and HJ6410.) HJ6409 records a levy of 5,000 men to "cheng" (rectify) the T'u-fang.

109. Based on his analysis of traditional sources, Ch'en Meng-chia, 198, 272–273, concludes that all the campaign inscriptions date to Wu Ting's reign and that he successfully extirpated them.

110. This reconstructed chronology is based on Lin Hsiao-an, 261–266, and Fan Yü-chou, 1991, 214–219, slightly modified by recourse to the original inscriptions. Though essentially agreeing on the course of the campaign, Lin and Fan disagree on the years of some strips, somewhat altering the sequence of events. (For the major inscriptions see Hu Hou-hsüan, HCCHS 1991:2, 13–20.)

111. HJ6413, Hsü 3.8.9 (which makes no mention of conscription. Wai 314 also dates to the eleventh month.) Numerous strips inquiring about undertaking actions against the Hsia-wei and T'u-fang or against the T'u-fang and Kung-fang show their simultaneous nature. (See HJ6413, HJ6427, and HJ39884).

112. HJ6438.

113. HJ6452, assuming it doesn't belong to the subsequent year. (See also HJ6541 per Fan Yü-chou; Lin Hsiao-an, 226; and Hou-hsia 37.6.)

114. HJ6409, per Fan Yü-chou. (Lin Hsiao-an, 264, dates this inscription to the succeeding year.)

115. HJ6420, per Fan Yü-chou. See also HJ6417a, HJ6087, and Wang Yü-hsin's account, 1991, 146–147. (Lin Hsiao-an notes HJ6354 indicates the king mounted an attack in the fourth month, but there is no mention of Chih Kuo.)

116. HJ6385a.

117. HJ6412, HJ6417. (However, Wang Yü-hsin, 1991, 149, attributes HJ6412 to the first part of Wu Ting's reign.) See also HJ6087, HJ6416.

118. HJ6414.

119. Lin 2.7.9 per Lin Hsiao-an.

120. HJ6439 per Fan Yü-chou.

121. HJ6057. Note that his report uses the term cheng, translated as "punitive attack," even though it supposedly never was employed to refer to attacks mounted by "barbarians," only the Shang (and later states) going forth to punitively attack or "rectify" offenders. (Additional counterexamples, such as HJ20440 and HJ20441, are easily found.)

122. HJ6057a. See also HJ6354 for a similar report. According to the Yin-li P'u, this was in Wu Ting's twenty-ninth year.

123. Hou-shang 31.6, per Lin Hsiao-an.

124. HJ6057.

125. HJ6087.

126. HJ6454; Hou-hsia 37.6. See also Lin Hsiao-an, 266.

127. Hu Hou-hsüan, HCCHS 1991:2, 19.

128. According to Chang Ping-ch'üan, 1988, 350.

129. According to Fan Yü-chou, 1991, 209, based on *Kuei* 2.15.18.

130. See *Ch'ien-pien*, 4.46.4, for reports of Ma-fang activities; also *Ping-pien* 114 (first month), *Ching* 1681, *Yi* 5408, and HJ6664 (eleventh month) for whether they will receive blessings. (Relevant inscriptions are noted by Wang Yü-hsin, 1991, 165; Ch'en Meng-chia, 1988, 283–284; and Chang Ping-ch'üan, 1988, 492.)

131. Respectively *Ping-pien* 301 and *Chia* 1, per Ch'en Meng-chia.

132. For inscriptions see Chang Ping-ch'üan, 1988, 489, and Fan Yü-chou, 1991, 210–212.

133. Artifacts suggesting that the Shang conducted campaigns against Pa or Shu have also been found in P'eng-chu-hsien in Sichuan. (Ch'en Hsü, 2000, 242–243.) After the Chou conquest of the Shang, Pa groups apparently came up the Han River and were enfeoffed with the state of Yü in Shaanxi as a reward for participating in the coalition action against the Shang. (See Pao-chi-shih Yen-chiu-hui, WW 2007:8, 28–47.)

134. HJ6461, *Nei-pien* 267.

135. HJ6468 (as sometimes interpreted).

136. HJ6473; *Nei-pien* 25, 26, 32, 34; *Yi* 3787.

137. *Nei-pien* 313.

138. *Nei-pien* 159 and 311.

139. HJ8411.

140. *Yi* 2213; *Yi* 8171.

CHAPTER 10

1. Fan Yü-chou, 1991, 227. However, see Ch'en Meng-chia, 1988, 270–272.

2. Convenient summaries of key inscriptions and discussions of their unfolding relationship are found in Ch'en Meng-chia, 270–272; Fan Yü-chou, 226–228; Lin Hsiao-an, 234–235; and Wang Yü-hsin, 1991, 168–170. A brief account will also be found in Luo Kun's *Hsia Shang Hsi-Chou Chün-shih Shih*, 180–181.

3. Attested by HJ6773, HJ6783, HJ6788, and HJ6790.

4. See, for example, *Yi* 2287 and *Yi* 2347 (inquiring about the fate of the Shang army under the *li*, later known as "lictors").

5. *Chia* 3066.

6. HJ6689 through HJ6696 and HJ6724.

7. *Hsü* 6.9.6 (pummel the Chien).

8. *Hsü* 5.8.1, *Ching* 3, *Ching* 5.

9. HJ6783, HJ6786, HJ6788, and HJ6790.

10. HJ6759, HJ6761. (Somewhat inexplicably, the *Hsia Shang Hsi-Chou Chün-shih Shih* compilers, 181, take the character *yü*, normally understood as "to mount a defense" or "defend against," as indicating some sort of mop-up effort subsequent to victory.)

11. HJ6754, *Hsü* 5.28.2, *Hsü* 6.21.11, *Yi* 2287, *Yi* 7764.

12. HJ6702, *Ch'ien* 6.3.54, HJ6704a, all dating to the same year, according to Fan Yü-chou (1991, 227). (Many other strips [HJ6689–HJ6724], some dating to the fourth month, also speak about great Fang uprisings.)

13. HJ6782, HJ6466, and HJ6781 respectively. (See also HJ6778 and HJ6784.)

14. HJ6737, HJ6733, and others.

15. *Hsi* 386.

16. Or at least was in danger of it. (HJ6771a, *Yi* 2287, *Yi* 7764.)

17. HJ6754.

18. *Chia* 243 records the king ordering K'eng to pursue the Fang, implying they had been vanquished and were in retreat.

19. HJ6768, HJ6769.

20. Ch'en Meng-chia, 1988, 273–274.

21. See, for example, Lin Hsiao-an's citations, 273–275, as well as those in Ch'en Meng-chia, 273. Relevant inscriptions include HJ6057, HJ6063, HJ6069, HJ6079, HJ6178, HJ6347, and HJ6359.

22. For example, HJ6057.

23. Wang Yü-hsin, 1991, 160, citing HJ6087a. (See also Wang, 147.)

24. For example, see HJ6354a, HJ6209, HJ6404a, as well as the inscriptions cited by P'eng Yü-shang, 140–142, who emphasizes the king's active role. Alone among historians, P'eng (153) ascribes the conflict to Wu Ting's middle period. The absence of any reference to Fu Hao, except to call upon her for protection, suggests she had already died, and Wang Yü-hsin (148) employs her demise, as attested by sacrifices being offered to her, as one of his defining chronological points (though he also seems to simultaneously hold a slightly contradictory view [163]). Lin Hsiao-an (273) concurs that she was deceased and had become the recipient of prayers for the campaign's success.

25. As Lin Hsiao-an notes, 265.

26. In addition to the selected oracle references provided for each commander in the list that follows, see Ch'en Meng-chia, 273–274.

27. See HJ6344 and HJ8991.

28. See HJ6297, HJ6299, and HJ24145.

29. HJ6083. (Wang apparently was not involved in the final stages of the conflict [Wang Yü-hsin, 1991, 162].)

30. HJ6135, HJ6161. (Chih Kuo seems to have concentrated on the T'u-fang.)

31. *Kuei* 2.8.12.

32. HJ6178, *Yi* 51 (further confirming the formidable nature of the Kung threat).

33. HJ6376.

34. HJ6371.

35. HJ6196.

36. HJ6209.

37. HJ6072.

38. HJ5785, HJ6209, HJ6272, HJ6335, and *Ch'ien* 6.58.4. See also Lin Hsiao-an, 276–277.

39. *Ch'ien* 4.31.3.

40. See Fan Yü-chou, 221–222.

41. Both Tung Tso-pin and Fan Yü-Chou (who essentially rejects Tung's reconstruction) have offered chronologies. (See Fan Yü-chou, 217–224; his criticism of Tung, not unlike that raised by Ch'en Meng-chia, appears on 214. [Tung charts a 4.5-year campaign stretching from the seventh month of Wu Ting's twenty-eighth year through the twelfth month of his thirty-second year.]) For further discussion of the Kung campaign, see Lin Hsiao-an, 264–265 and 272–279; Wang Yü-hsin, 146–148 and 160–164; and P'eng Yü-sheng, 138. (P'eng also provides an interesting campaign route for the king's final effort on 198–199.)

42. HJ6063a.

43. *Ch'ien* 5.13.5.

44. *Ching* 1229, HJ6112.

45. HJ6057, HJ6060. (See also HJ6354.) Numerous scapulae record both T'u-fang and Kung-fang aggressiveness together with actions being undertaken or contemplated against them. Their physical presence on a single prognosticatory medium, whether scapula or plastron, is of course evidence that they occurred concurrently or within a few days. See, for example, HJ6087.

46. HJ6057a.

47. See Fan Yü-chou, 222, and *Hsia Shang Hsi Chou Chün-shih Shih*, 181–182.

48. See Fan Yü-chou, 219, for numerous references.

49. See Fan Yü-chou, 219–220. (Other, undated inscriptions show the king seeking spiritual aid through sacrifices and prayer, while HJ6347 suggests that these efforts began as early as the seventh month.)

50. HJ6732. Note that HJ6371 queries whether the Kung would destroy Yüeh in the tenth month.

51. *Chin* 522, but see HJ6063a for a report of a Kung incursion.

52. See Fan Yü-chou, 221; HJ6316; and HJ6317.

53. *Chin* 525, *Ch'ien* 6.30.12.

54. HJ6292. Although undated, HJ24145 indicates that Ch'in was on the verge of destroying them.

55. *Lu* 637.

56. For a basic discussion of the Kuei-fang and their location, see Ch'en Meng-chia, 274–275, and Wang Kuo-wei, "Kuei-fang, K'un-yi, Hsien-yün K'ao." Chao Ch'eng, 2000, 4, believes they inhabited middle Shanxi and western Shanxi.

57. The former is the third line of the sixty-third hexagram entitled "Already Completed" (in the sense of "having passed over," as in fording a river), and the latter the fourth line of the sixty-fourth and last hexagram, "Not Yet Completed."

58. Luo K'un, 1983, 82–87. (*Yi* 865 provides evidence the Kuei were employed against the Ch'iang.) Luo (90–97) successfully deflects any suggestions that some of the strips show the Kuei in an aggressive role. (Ch'en Meng-chia also provides several relevant strips, 1988, 274.)

59. Luo K'un, 85–86.

60. For example, see Fan Yü-chou, 1991, 224, following Li Hsüeh-ch'in.

61. This is essentially Tung Tso-pin's conclusion ("Lun Kung-fang chi Kuei-fang," *Yin-li P'u*, 9:39a–40b). Note that the Kuei are also called the "Chiu Hou" in the *Shih Chi* and other texts.

62. Wang Kuo-wei's "Kuei-fang, K'un-yi, Hsien-yün K'ao" briefly discusses how names and terms vary over time, being their own nominatives and also having Chinese appendages, so that one tribe can come to be referred to by different names. (See also Luo K'un, 1983, 102ff.; Hsü Chung-shu, BIHP 7:2 [1936], 138; and E. G. Pulleyblank, "The Chinese and Their Neighbors.")

63. Luo K'un, 1983, 87ff.

64. Luo K'un, 1983, 99–101.

65. Hsü Chung-shu, BIHP 7:2 (1936), 138.

66. Hsü Chung-shu, 139–140. Hsü notes this clash cannot date to King Wen's time because the Chou had already become too formidable to fear the Kuei. (Hsü's account, dating from the early days of oracle bone studies, never mentions any inscriptions or identifies the Kuei-fang with the Kung.)

67. Luo K'un, 1983, 99–103, citing Hsü's earlier article, also bases his analysis on the premise that two different events are being discussed.

68. For the full explanation with Wang Pi's underlying interpretation, see Luo K'un, 100ff.

69. Luo K'un, 100, also correctly points out that "three" and "nine" function as indefinitely large numbers in ancient Chinese thought; therefore, the actual conflict may not have raged for as long as three years. (However, he fails to note that three and nine are also poignant numbers in *Yi Ching* contemplations, heavily weighted with dynamic implications and metaphysical connotations.)

70. "And then went on to Ching" is sometimes, but incorrectly, read in conjunction with this line. (Hsü, BIHP 7:2 [1936]: 139, sees the latter part as additional grounds for doubting the text's authenticity.)

71. Luo K'un, 1983, 99; Hsü Chung-shu, 139. (Luo points out that the early commentators realized that the term "Kuei-fang" simply referred to peoples populating "distant quarters.") Note that Tung Tso-pin dates the Kung-fang conflict to Wu Ting's twenty-ninth year.

72. For example, Chang Ping-ch'üan classifies the Chou among those on good terms with the Shang during the first two periods at Anyang (encompassing the reigns of Wu Ting and his immediate successors), as well as the fourth period of Wu Yi and Wen Ting, but also notes animosity between them in this same fourth period to which, rather than Wu Ting's era, he ascribes the conflict discussed in the text immediately below. (See Chang, 350, 496, and 512.) Hu Hou-hsüan, Chia-ku T'an-shih-lu, 366, similarly sees significant conflict in the fourth period, whereas Chung Po-sheng, 1991, 95–156, views their relations as interactions between the two poles of a duality.

73. Chang Ping-ch'üan, 432–433. (Inscriptions referring to a Fu Chou include Yi-pien 8894 and HJ22264.)

74. See Ts'ao Ting-yün and Liu Yi-man, KK 2005:9, 60–63, who stress that these inscriptions date to the period when Chou was still at Pin, explaining the unusual reference to "Chung Chou."

75. Ten shells are noted on Yi-pien 5452, another shell on Ping-pien 274.5. Yang K'uan, 1999, 36–37, views the forwarding of women as a deliberate effort at subversion, but the choice was probably less insidious even though there apparently were historical antecedents.

76. Examples of solicitous inquiry include Chia-pien 436, HJ6782 (referring to whether the Fang would "tun" the Chou), and T'ieh 26.1 (whether the Kuan would harm them). In aggregate the king was interested in such developments, though perhaps only because any reduction in potential threats might be welcome. However, he might equally have ordered an assault and been wondering about the results. For example, Lin Hsiao-an (272) interprets HJ6825 as a solicitous inquiry, but Fan Yü-chou (1991, 224) sees the Ch'üan, a satellite of the Shang, acting on the latter's behalf (as on another occasion) and the inquiry being directed to the question of their probable success in harming the Chou. (Some disagreement has plagued the transcription of state names in these inscriptions.)

77. For example, see T'ieh 128.2 and Ch'ien 6.63.1. (Both ming [mandate] and ling [order] appear in the inscriptions.)

78. See Lin Hsiao-an, 246. Examples of the Chou mounting a strike on behalf of the Shang are Yi-pien 7312 and Ping-pien 289, where they seem to be acting as a vanguard for the Ch'üan, who must have reverted to submissive status. (For additional evidence of the Chou acting as a battlefield ally see David N. Keightley, EC 5 [1979–1980]: 25–34, where he notes inscriptions that he translates as "Order Chou to follow Yung's foot," "Order Chou to lead the (to-) tzu-tsu (?) and raid," as well as indirect charges that someone should "order Chou" to do something.)

79. Most analysts, including Lin Hsiao-an, 272, and Fan Yü-chou, 224, date the clash with the Chou to late in Wu Ting's era, but a few, such as P'eng Yü-shang (1994, 153), attribute it to the middle part based on various queries belonging to the Pin diviner group. (Key inscriptions for the conflict are reprised in Ch'en Meng-chia, 1988, 291–292, 492; Chang Ping-ch'üan, 1988, 321, 435, 492; Fan Yü-chou, 224–225; Wang Yü-hsin, 1991, 166ff.; and P'eng Yü-shang, 150–151.)

80. HJ6825 per Fan Yü-chou, 224. Based on numerous inscriptions in which various forces p'u the Chou, Ch'en Meng-chia (291–292) claims that the character "p'u," which apparently signifies a sudden, vehement strike as in HJ6812, appears in oracle bone inscriptions only in reference to the Chou. Although he doesn't draw any particular conclusion from this, it implies that they were targeted with particular ferocity even though other enemies suffered more horrendous fates. However, others have transcribed this character as k'ou, meaning to plunder or invade, a somewhat less severe term, though other strong characters such as shun, tun, and tsai are also found in the inscriptions.

81. HJ6812, HJ6813.

82. HJ6812 *cheng*, and also HJ6813, HJ6814, HJ6815, and *Hsü* 984.

83. HJ6824.

84. See Fan Yü-chou, 1991, 225; HJ6816, HJ6817, HJ6818, HJ6819, HJ6821, HJ6822, and HJ6825. The *tuo yin* were also dispatched on one occasion.

85. Wang Yü-hsin, 1991, 169, cites HJ6782 (which preserves an inquiry about the Fang "severely assaulting" [*shun*] the Chou) as typical of Shang regal concern over the fate of its allies and evidence that the Chou had been subdued early in the last part of Wu Ting's reign. However, being fragmentary, it may simply be another example of the king inquiring about the prospective success of a commander dispatched to inflict just such damage.

86. However, based on attributing several inscriptions to the reigns of Wu Yi and Wen Ting, Chang Ping-ch'üan, 1988, 496, concludes that the Chou was among the states attacked in the relatively active fourth period. This would require ascribing many of the inscriptions already noted to this period.

87. Divination strips supposedly attributable to their stay at Pin have recently been discovered. See Hu Ch'ien-ying, KK 2005:6, 74–86.

88. Wang Yü-hsin, 1991, 166; Fan Yü-chou, 1991, 225.

89. See Fan Yü-chou, 225, based on Yang Shu-ta. However, *Mencius* merely mentions the Ti, so any identification of them as the Ch'üan, however well constructed, is speculative.

90. Yang K'uan, *Hsi-Chou Shih*, 38–40, may well be correct in concluding that this conflict represents another, distinctive event that unfolded when Shang central authority broke down under Wu Yi's dissolute and repressive rule (as indicated in the more nebulous "Hsi-Ch'iang-ch'üan" in the *Hou-Han-shu*), but it seems unlikely that the same sequence of attacks would have been repeated. Rather, attributing the event to Wu Yi's reign is more likely to be erroneous. Conversely, assuming the events occurred late in Wu Ting's reign raises questions about the length of various Chou reigns and whether the period between Chou Tan-fu's action and the conquest of the Shang would not become too long. (For additional discussion see *The Cambridge History of Ancient China*, 299–307, and Yang K'uan, *Hsi-Chou Shih*, 35–45.)

91. See Yen Yi-p'ing, 1980, 159–185, and Luo Hsi-chang and Wang-Chün-hsien, WW 1987:2, 17–26.

92. For a discussion of Shang hunting in the Chou and Chou rulers offering sacrifice to Shang ancestors, see Shaan-hsi Chou-yüan K'ao-ku-tui, WW 1979:10, 38–43, and WW 1981:9, 1–7.

93. In addition to the campaigns discussed in our text, the following inscriptions suggest Wu Ting's scope of activity: HJ6404a (against the Kung); HJ6354a (against the T'u-fang); HJ6417a (with Chih Kuo); HJ6427 (Hsia-wei, T'u-fang); HJ6413 (with Wang Ch'eng); HJ6480; HJ6457 (with Chih Kuo or Hou Kao); HJ6476a (against the Yi, with Chih Kuo against the Pa and Wang Ch'eng against the Hsia-wei); HJ6477a with Hsing against the Pa, or Wang Ch'eng against the Hsia-wei; HJ6482a (with Wang Ch'eng contra the Hsia-wei); HJ6530a (accompanied by the Hsing-fang against Hsia-wei); HJ6542, HJ6543, HJ6552, HJ6553; HJ32a (accompanied by Wang Ch'eng against Hsia-wei, Chih Kuo against the Pa); HJ6607 (against the Ch'ing); *Kuei* 2.15.18 (against the Ma-fang); *K'u* 1094 (against Chi-fang and others); *Ching* 1266 (against Lung and Pa). However, further study is required to determine what motivated him to personally participate rather than depute another commander.

94. As their fates are never specified, they may have been wounded in battle, killed, or simply grown too old to take the field, not to mention fallen out of favor with the king for unrecorded battlefield failures or other reasons, such as becoming too powerful.

95. Lin Hsiao-an, 243.

96. HJ6931.

97. HJ6947a (pursue them), HJ6948, HJ6952 (capture them), HJ6953 (seize them), HJ6954, HJ6958 (pummel them), HJ6959 (capture them), HJ20384, *Ping-pien* 119, *Ping-pien* 249, and *Ping-*

pien 304. Conversely, inquiries about him being endangered by the Hsüan include HJ20383 and HJ20393.

98. HJ6946.

99. See Lin Hsiao-an, 229.

100. *Ping-pien* 119.

101. *Ts'ui* 1167 and *Chia* 2326.

102. HJ6571, HJ6573 (with Prince Shang), HJ13514, and *Ping-pien* 302. (Assuming Fou is the chief of the Chi-fang, HJ6834 inquires whether he will capture Fou and HJ6989 whether Fou will harm Ch'üeh.)

103. HJ6931 and *Kuei* 2.15.11. (See Fan Yü-chou, 1991, 205.)

104. *Ping-pien* 119.

105. HJ53 (whether he will conquer them).

106. HJ6964, HJ6965, *Yi* 5317. See also Fan Yü-chou, 1991, 210.

107. HJ6983, HJ6984.

108. HJ6983 (attacks Yüeh and others), *Yi* 4693 (damages Yüeh). See also Lin Hsiao-an, 230.

109. HJ6931.

110. For example, see HJ6946, HJ6947, HJ6948, HJ6949, HJ6959, HJ6962, HJ6979, HJ6980, HJ6985, HJ6986, HJ6987, HJ20576 (in danger), *Chia* 2902 (the southern campaign), *Ping-pien* 117, and *K'u* 546.

111. *Chia* 2902.

112. Attested by HJ6946. (See Lin Hsiao-an, 232.)

113. HJ4122, HJ4123, HJ4124, HJ6949, HJ32843, HJ32839, and HJ6577 are among the many that inquire if he will suffer misfortune or not. (Others are particularly seen in campaigns against strong enemies.)

114. Lin Hsiao-an, 229. (There is considerable disagreement on how to transcribe and pronounce the second character in his name.)

115. For example, see HJ6087 (with the king), HJ6401 (with the king), HJ6402 (with the king), HJ6404 (with the king), HJ6417a (with the king), HJ6438, HJ6452, and HJ39853 (with the king).

116. For example, see HJ32a, HJ6135, HJ6476 (with the king), and HJ6480 (with Fu Hao).

117. For example, *Ch'ien* 6.60.6, HJ6087 (with the king), HJ6416 (with the king), and HJ6384.

118. HJ22a (with the king), HJ6413, HJ6476, and HJ6482 to HJ6486.

119. For example, HJ6135, *Tun-nan* 81 (with the king), HJ32 (with the king), HJ6476 (with the king), and HJ6583 to HJ6486 (with the king).

120. HJ6480 (with Fu Hao), HJ6461 (with the king), and HJ6476 (with the king).

121. HJ6937 (with the king).

122. Examples of accompanying Fu Hao are HJ6947 and HJ22948; others for the king are HJ6473, HJ7504, HJ33074, and HJ33105 to HJ33108.

123. HJ32 (with the king), HJ326 (with the king), HJ6413 (with the king), HJ6476 (with the king), HJ6477a (with the king), HJ6480, HJ6482 to HJ6486 (with the king), HJ6489 to HJ6493 (with the king), HJ6496, HJ6525, HJ6521, HJ6542 (with officials). Presumably he led his own forces, so the question of his actual role arises, just as with Chih Kuo.

124. HJ6667.

125. For example, HJ6083 (with the king) and HJ6148.

126. For example, HJ6583 and HJ33112.

127. Karlgren, GSR, 1087a, actually pronounced *chou* but generally referred to as "*fu*" because of the later character for woman / wife, "*fu*," which has the additional component of woman, with which it is commonly identified.

128. For a study of the function and significance of the "*fu*" in the Shang, see Chang Cheng-lang, 1986, 103–119; Chao Ch'eng, 2000, 136–156; and Chang Ping-chüan, BIHP 50:1 (1979), 194–199.

129. See Chang Cheng-lang, 1986, 103–119. Also note, for example, HJ924.

130. For example, Fu Ching (HJ6347).

131. Such as Fu Ching, who attacked the Lung (HJ6584, HJ28).

132. HJ6648, HJ6826, HJ18911, and HJ21653 provide examples of them being ordered out or back from external areas. HJ7006 specifically inquires about the prospects for mounting a successful defense.

133. For example, HJ5495.

134. See Chao Ch'eng, 2000, 143.

135. For example, see HJ7006.

136. Space precludes further discussion of the tomb's contents, but for a summary see *The Cambridge History of Ancient China*, 194–202, or Thorp, *China in the Early Bronze Age*, 195ff.

137. Although Shima Kunio provided the first comprehensive collection in his *Inkyo Bokuji Sorui* (139–141), the best compilation remains Yen Yi-p'ing, "Fu Hao Lieh-chuan." Other important discussions include Wang Yü-hsin et al., "Shih-lun Yin-hsü Wu-hao-mu te 'Fu Hao'"; Cheng Hui-sheng, "Kuan-yü Fu Hao te Shen-shih Wen-t'i"; Cheng Chen-hsiang, "A Study of the Bronzes with the 'Ssu T'u Mu' Inscriptions Excavated from the Fu Hao Tomb"; Noel Barnard, "A New Approach to the Study of Clan-sign Inscriptions of the Shang"; Chang Ping-ch'üan, "A Brief Description of the Fu Hao Oracle Bone Inscriptions"; and Ts'ao Ting-yün, "Yin-hsü Fu Hao Mu Ming-wen-chung Jen-wu Kuan-hsi Tsung-k'ao." Yen classifies her as a true "ch'i nü jen" or unorthodox woman, finds it odd that Ssu-ma Ch'ien failed to include her among his exemplary women, and believes that it was through her military achievements that she advanced from the status of an ordinary consort to be one of three prominent wives.

138. Wang Yü-hsin, 1991, 148. Yen Yi-p'ing believes she lived until Tsu Keng's reign, while Chang Ping-ch'üan, 1988, who examines the question of other "Fu Hao" inscriptions and the meaning of *fu*, concludes that the only Fu Hao was the one active in Wu Ting's reign.

139. It is therefore argued that she should be known as Fu Tzu, just as the other *fu* such as Fu Ching are designated by coupling *fu*, not necessarily meaning "wife," and clan names, such as Ching. (See Chang Cheng-lang, "A Brief Discussion of Fu Tzu.") Chang Ping-ch'üan, 129–130, notes other examples of endogenous clan marriage.

140. Yen Yi-p'ing, 1981, 5. For example, HJ6948 inquires about the auspiciousness of her pregnancy and confirms that she gave birth to a daughter. (See also Wang Yü-hsin et al., KKHP 1977:2, 10–17.)

141. See *Ts'ang* 204.3, Hsü 3.1.2 (while on campaign), and Yen Yi-p'ing, 1981, 5–12. Inquiries about the auspiciousness of her coming from and going to external regions are also seen, such as in the series HJ2642, HJ2643, and HJ2645. These coincidentally show that her power is clearly derived, since she can still be ordered about.

142. HJ2638, HJ2672.

143. HJ17380, *Ts'ang* 113.4.

144. See Yen Yi-p'ing, 1981, 18–25, and Wang Yü-hsin et al., KKHP 1977:2, 7–10.

145. See, for example, *Yi* 7782 and Yen Yi-p'ing, 27. Several other *fu* are recorded as having forwarded important materials for divination. (See Chao Ch'eng-chu, 2000, 140–142; for the importance of these turtle shells as tribute in general, see Yeh Hsiang-k'ui and Liu Yi-man, KK 2001:8, 85–92.)

146. Yen Yi-p'ing, 31–32, based on *ts'ang* 244.1.

147. Following the interpretation of *ts'ung* as "to be accompanied by" rather than "accompanied," as in "the king was accompanied by Chih Kuo in an attack on the T'u-fang," even though in later Chinese it would normally be read "the king accompanied Chih Kuo in an attack on the T'u-fang." Although this reasonably presumes that the king would never assume a subordinate role in the field, questions remain.

148. Most analysts stress that she does not appear in the campaigns attributed to the final period, and Wang Yü-hsin even uses her absence as one of his defining chronological criteria. However, Yen Yi-p'ing (1981, 35) believes that in at least one case she attacked the Kung-fang and T'u-fang in a campaign that can be dated to between Wu Ting's twenty-eighth and thirty-second years according to Ch'an T'ang's *Yin-li P'u*.

149. See, for example, HJ6412, *K'u* 237, and Yen Yi-p'ing's enumeration, 32–36.

150. Note HJ6478, HJ6479, and HJ6480.

151. HJ6459. See also HJ6480 (against the Hu, a Yi component) and *Hsü* 4.30.1

152. *K'u* 310. See also Wang Yü-hsin et al., KKHP 1977:2, 2–4.

153. Yen Yi-p'ing, 34–35, notes several inscriptions (such as HJ7283) that order her to first arrange troops for actions against the Lung.

154. For example, HJ6568a enigmatically states, "Fu Hao deputes [*shih*] men to [pound] Mei." (A convenient summary of Fu Hao's campaigns may be found in Wang Yü-hsin, 1991, 149–152.)

155. HJ7283, HJ6347, and *Ying* 150 *cheng*.

156. HJ6480.

157. See HJ2658.

158. Fu Ching, for example, is noted in HJ8035 as offering the military sacrifice before a battle, a role thereafter reserved to men, and in HJ6584 and HJ6585 specifically ordered to inflict heavy damage upon the Lung.

CHAPTER 11

1. Although Chou charges of excessive inebriation were apparently not unfounded, and the Shang had specialized vessels for imbibing alcoholic beverages, much of Shang drinking was associated with feasting, sacrifices, and the ancestral cult (Christopher Fung, JEAA 2:1–2 [2000], 67–92).

2. Sung Hsin-ch'ao, CKSYC 1991:1, 53–63, and Fang Hui, KK 2004:4, 53–67.

3. For example, the predynastic Shang shared certain divinatory practices with the Tung Yi and apparently adopted the practice of prognostication with turtle plastrons from them.

4. The Chü, who seem to have been located in the Shandong area, were a powerful aristocratic family in the Shang and early Chou. Inscriptions (HJ6341) indicate Chü was ordered to instruct 300 archers and campaigned against the Kung-fang (Ho Ching-ch'eng, KK 2008:11, 54–70).

5. Kao Kuang-jen, KKHP 2000:2, 183–198.

6. For example, see HJ6457, HJ6459, HJ6461, HJ6834, and HJ7084.

7. According to the Tung Yi chronicles in the late *Hou Han-shu*.

8. Kao Kuang-jen, KKHP 2000:2, 183.

9. For a site report see Han Wei-lung and Chang Chih-ch'ing, KK 2000:9, 24–29.

10. Kao Kuang-jen, KKHP 2000:2, 190–191.

11. Kao Kuang-jen, KKHP 2000:2, 183–198.

12. Ch'en Hsüeh-hsiang and Chin Han-p'o, KK 2007:5, 86–87; Ch'en Ch'ao-yün, HCCHS 2006:2, 3–8.

13. Kao Kuang-jen, KKHP 2000:2, 184–187.

14. For a seminal study of steppe/sedentary interaction in the first millennium BCE, see Nicola Di Cosmo, *Ancient China and Its Enemies*.

15. Although the climate had been cooling and drying since 3000 BCE, conditions fluctuated during the Shang. From King T'ang to Chung Ting was cooler and drier; from Chung Ting to Wu Ting the temperature rose slightly and rainfall increased, revitalizing the marshes; but from

his reign onward the temperature and rainfall both decreased, resulting in some drying out, harsher conditions, less vegetation, and fewer animals. (Wei Chi-yin, KKWW 2007:6, 44–50. However, as previously noted, it has also been suggested that Wu Ting's reign was actually parched, thereby accounting for his numerous queries about prospects for rainfall.)

16. Luo K'un, 1998, 173. The presence of chariots outside the Shang should be noted.

17. HJ27998.

18. HJ27976.

19. For inscriptions see Chang Ping-chüan, 1988, 496; Ch'en Meng-chia, 1988, 279, and *Hsia Shang Hsi-Chou Chün-shih Shih*, 187ff.

20. Such as HJ27973.

21. Understanding *chih chung* as "call up and put in order" rather than "suffer loss" or "harm." (See, for example, HJ27974, HJ27975, and HJ2972.)

22. See, for example, HJ26887, HJ26896, HJ27978, and HJ27979.

23. For an overview of these border clashes see Lin Huan, HCCHS 2003:3, 57–63.

24. See Yang Sheng-nan, 1982, 359–367, and our discussion in the section on military structure and organization.

25. According to the third-century *Hou Han Shu* chapter "Hsi Ch'iang."

26. Lin Huan, HCCHS 2003:3, 57–63.

27. Oracular inscriptions speak about the five clan troops pursuing and rectifying them. (See Chang Ping-ch'üan, 496.)

28. HJ33049, HJ33050.

29. HJ33059, HJ33060.

30. HJ33039, HJ33040. (Sui is also pronounced Hui by some writers.)

31. See Luo K'un, 1998, 191; HJ27970 and HJ27997.

32. See Ch'en Meng-chia, 1988, 799ff.

33. HJ33019.

34. HJ33213.

35. The account reprised here is based on Luo K'un, 1998, 192ff. The king's concern can be seen in his reports to the ancestors, including HJ33032, HJ33033, HJ33015, and HJ33016.

36. HJ33034–HJ33036.

37. HJ33058. (Not, of course, the Chih Kuo of Wu Ting's era.)

38. HJ31973–HJ31977.

39. HJ33028, HJ33029.

40. HJ33017 and HJ32815, respectively.

41. HJ33031.

42. HJ31978. (Commentators understand the term *yü* as meaning something like "clear out" [or even "extirpate"], but it could simply mean undertaking a prolonged, active defense against them.)

43. HJ33026.

44. See Yang Sheng-nan, 1982, 359–360. Yang transcribes the character as Shao rather than Li, whereas Ch'en Meng-chia (1988, 287ff., followed here) understands it as Li.

45. Based on assigning a number of "nameless" diviner inscriptions to Wen Ting's reign, Li Hsüeh-ch'in, CHSYC 2006:4, 3–7, described a campaign against the Yi capital near Lin-po in Shandong that he attributed to Wen Ting's era. However, in a 2008 reconsideration, CHSYC 2008:1, 15–20, Li concluded that despite the generally accepted convention that nameless inscriptions do not appear post–Wen Ting (and apparently in rejection of his earlier thought that the dates simply do not cohere), they stem from Emperor Hsin's reign and provide crucial battle information for the lengthy campaign reconstructed for Hsin's tenth year (but we have assigned them to Emperor Yi's era in the discussion that follows based on the views of other analysts).

46. Although disagreement continues about their timing and other aspects, it is generally accepted that at least two major expeditionary campaigns were mounted against the Yi during the last two reigns. (For example, see Hsü Chi, STWMYC, 266–268, or Luan Feng-shih, STWMYC, 270–279.)

47. For relevant inscriptions see Tung Tso-pin's "Ti Hsin Jih-p'u"; Ch'en Meng-chia's reconstruction of the campaign, "Yi Hsin Shih-tai Suo-cheng te Jen-fang, Yü-fang," 301–310; or the account provided in Luo K'un, 1998, 195–202, where the campaign objective is uniformly transcribed as the Yi-fang rather than the Jen-fang. Ch'en includes a useful map of the probable route of march that depicts a fairly efficient campaign with limited maneuvering after crossing the Huai, but Tung Tso-pin envisions a rather extensive upward loop prior to the final southern thrust across the Huai River and extensive movement thereafter. (Shima Kunio's reconstruction is also more conservative.)

Ch'en proposed his version as a corrective to Tung Tso-pin's calendar, which he viewed as a meritorious but flawed effort. In addition, he attributes the campaign to Emperor Yi's reign—a view that has garnered general acceptance as evidenced by its adoption in Luo K'un's *Hsia Shang Hsi-Chou Chün-shih Shih*—whereas Tung placed it in Emperor Hsin's era and dates the final query to the seventh month. However, Yen Yi-p'ing, 1989, 317–321, has contributed an overview of the variations in which he concludes that Tung's reconstruction is reliable and the purported variations essentially congruent.

48. Ch'en Meng-chia, 1988, 308.

49. See Li Hsüeh-ch'in, CHSYC 2006:4, 3–7, and CHSYC 2008:1, 15–20. (Li's reconstruction is, however, not without possible problems, such as essentially requiring the campaign to be limited to the Shandong area, around the Hua and Wei rivers, based on his reading of the character for Huai.)

50. Luo K'un, 1998, 200–202. (The mere coincidence of the campaign dates, which are repeatedly attributed to the tenth and fifteenth years, is indicative of fundamental problems.)

51. For a reconstruction of the campaign, see *Hsia Shang Hsi-Chou Chün-shih Shih*, 202–204, or Ch'en Meng-chia, 1988, 309–310.

52. Yen Yi-p'ing, 1989, 321, claims that three campaigns were mounted against the Jen-fang in Hsin's tenth, fifteenth, and twenty-fifth years. Among many, Chang Ping-ch'üan (1988, 433), asserts that the Shang didn't fall because of licentiousness but because Shang military activities in the east wasted its strength, thereby allowing the Chou, which had grown powerful in the neglected west, to easily conquer them. However, for a contrary view, see Kao Kuang-jen, KKHP 2000:2, 183–198.

53. *Tso Chuan*, Chao Kung fourth and eleventh years. The term translated as "martial convocation" is *sou*, which means "to search for" or "to gather/assemble," but also came to designate the annual Chou assembly of forces for the spring hunt, itself an opportunity for military training.

54. However, Ch'en Meng-chia explicitly denies it is the campaign noted late in his reign (1988, 304) or the one noted in bronze inscriptions dating to the emperor's fifteenth year. (The other possibility is the campaign assigned here to Yi's tenth and eleventh years but instead attributed to Hsin.)

55. Wang Yü-hsin, HCCHS 2007:5, 14–20.

56. The identification of the Yi with Yüeh-shih culture in this period—the late Hsia or nineteenth to seventeenth centuries BCE and thereafter—is almost universally accepted. (For example, see Yen Wen-ming, SCKKLC, 306–318; Yen Wen-ming, WW 1989:9, 1–12; and Wang Chen-chung, HCCHS 1988:6, 15–26.) The Shang also incorporated bronze decorative elements from the earlier Liang-chu culture (3200–2100 BCE). (See Jui Kuo-yao and Shen Yüeh-ming, KK 1992:11, 1039–1044.)

57. Chang Kuo-shuo, HCCHS 2002:4, 8–14, and Chu Chün-hsiao and Li Ch'ing-lin, KKHP 2007:3, 295–312, who conclude from an analysis of the ordinary utensils and ceramic vessels from the fourth period at Erh-li-t'ou that the Tung Yi and Shang must have been close allies.

58. Wang Chen-chung, HCCHS 1988:6, 15–26.

59. Fang Yu-sheng, HCCHS 1992:9, 18–20; Tung Ch'i, HYCLC, 1996, 46–53; Sung Yü-ch'in and Li Ya-tung, HYCLC, 1996, 54–59.

60. Ch'en Ch'ao-yün, HCCHS 2006:2, 3–8.

61. For analyses see Yen Wen-ming, WW 1989:9, 1–12, and HCCHS 2002:4, 3–8 and 8–14.

62. Chu Chi-p'ing, KK 2008:3, 53–61.

63. See Hsü 3.13.1 and Ch'en Meng-chia, 1988, 282.

64. See Ch'en Meng-chia, 299.

65. See Ch'en Meng-chia, 298, for inscriptions.

CHAPTER 12

1. Chang Chung-p'ei, WW 2000:9, 55–63.

2. For an enumeration see Chang Ping-ch'üan, 1988, 432–433. (The Shang apparently used marriage relations to cement alliances or emphasize the subjugated status of external clans, but their demand for eligible women may also have caused antagonism and resentment.)

3. For a discussion of these "journeys" see Liu Huan, KK 2005:11, 58–62, and Li Shih-lung, HCCHS 2002:1, 34–40.

4. Most of the inscriptions can be understood in terms of Wu Ting's efforts to impose order (Liu Huan, CKSYC 1995:4, 93–98). However, David Nivison, EC 4 (1977–1978): 52–55, has suggested that the character read as "making a tour of inspection" or "conducting a campaign of rectification" is *Te*, Virtue, and rather than military force refers to making a prominent display of *Te* (which might be defined as theocratic power) to other peoples. (Nevertheless, note that captives were sometimes taken.)

5. Chao Shih-ch'ao, CKKTS 1995:9, 6–18.

6. According to "Wu Yi," *Shang Shu*. Inscriptions related to the hunt continue throughout the Shang and Wu Yi and Wen Ting frequently held them, but they are especially found in Wu Ting's era, another indication of his active lifestyle and warrior values. (See Chang Ping-ch'üan, 1988, 475–487, for relevant inscriptions and a review of earlier findings by Tung Tso-pin and Shima Kunio. See also Meng Shih-k'ai, LSYC 1990:4, 95–104.)

7. Ch'eng Feng, HCCHS 2004:2, 25–26.

8. From Kings Ping Hsin to Wen Ting the term *sheng t'ien*, which has been interpreted as inspecting (*sheng*) the (military activities) of the hunt (*t'ien*), frequently appears (Meng Shih-k'ai, LSYC 1990:4, 101–103).

9. Suggested by Meng Shih-k'ai, 103–104.

10. For example, *Chia-pien* 3939 records an attack on Yü-fang, 3940, one on the Jen-fang.

11. Meng Shih-k'ai, LSYC 1990:4, 98.

12. *Nei-pien* 284.

13. *Nei-pien* 104.

14. *Nei-pien* 433.

15. For the masks interred with military figures, see Ch'ai Shao-ming, KK 1992:12, 1111–1120.

16. For example, the tomb of a high-ranking military commander dated to the third period at Yin-hsü contains a remarkable number of weapons. Of 394 objects, 288 are bronze, the majority being weapons, including three *yüeh* (axes) and 118 *ko* (dagger-axes). There are also four jade *yüeh* and 902 arrowheads (Yang Hsi-chang and Liu Yi-man, KK 1991:5, 390–391). Even victims of the Chou conquest, despite being quickly interred near Anyang, were accompanied by ritual bronze objects. (Tu Chin-p'eng, KK 2007:6, 76–89).

17. Ch'en Ch'ao-yün, HCCHS 2004:6, 30–36.

18. Ch'en Ch'ao-yün, 30–36, especially 35.

19. For example, see Lei Yüan-sheng and Chi Te-yüan's historical overview, CKSYC 1993:4, 3–19; Li Hsin-ta, *Wu-kuan Chih-tu;* or Wang Kuei-min, LSYC 1986:4, 107–119.

20. Even the *Ssu-ma Fa* envisioned the civil and martial having been strictly segregated in earlier times: "In antiquity the form and spirit governing civilian affairs would not be found in the military realm; those appropriate to the military realm would not be found in the civilian sphere. Thus virtue and righteousness did not transgress inappropriate realms" ("Obligations of the Son of Heaven").

21. See, for example, Yang Sheng-nan, CKSYC 1991:4, 47–59. Visibly granting rewards would become a much-practiced means of stimulating martial effort. (For example, see "Stimulating the Officers" in the *Wu-tzu.*)

22. "Obligations of the Son of Heaven."

23. The Greek attitude, especially unwillingness to act in the face of bad omens and prior to the receipt of favorable oracles, is clearly seen throughout *Xenophon*, to the point that Xenophon refused to allow the starving troops to search for provisions for several days.

24. Liu Yi-man, KK 2002:3, 70–72, has suggested that military command and the accompanying titles were hereditarily associated with various clans; certain inscriptions and excavation results appear to at least partially confirm his view. For example, a prominent Chü clan ancestor performed important military functions in Wu Ting's era, was deputed to pursue the Kung-fang (HJ6341) and others, and was ordered to instruct 300 archers (HJ5772). (See Ho Ching-ch'eng, KK 2008:11, 62; for an example of a commander from a clan related by marriage, see Wang Yung-p'o, HCCHS 1992:4, 31–40.)

25. "P'an Keng, Hsia." To the extent that the *Shang Shu* preserves any vestiges of Shang military organization, greater accuracy might be expected for later titles.

26. One sequence shows the Shang receiving intelligence of a Kung-fang incursion, reporting it to the ancestors, holding a meeting, and then appointing Chih Yu as the campaign commander in the temple. (See Li Hsin-ta, *Wu-kuan Chih-tu,* 21–22.)

27. "Appointing the General," which follows two other chapters on the requisite characteristics for generals, an important topic in Warring States military writings. (Essentially the same ceremony is preserved in the "Ping Lüeh" section of the eclectic *Huai-nan-tzu.*) From the Shang onward, if not earlier, the ancestral temple was the site for undertaking military assessments, any decision to engage in military activities being reached before the ancestors.

28. Wang Kuei-min, LSYC 1986:4, 107–119.

29. For a list see Chang Ping-ch'üan, 1988, 439–440. Ch'en Meng-chia, 1988, 508–517, has collated numerous inscriptions with military titles; a (somewhat unsatisfactory) overview may also be found in the volumes of *Chung-kuo Chün-shih Chih-tu-shih.*

30. The Hsia reputedly had an officer in charge of the arrows and quivers, so the *fu* might well have undertaken some sort of responsibility as an armorer for the archers. However, a later term for "bow case," *t'ao,* eventually came to designate "secret plans" as in the *Liu-t'ao* or *Six Secret Teachings,* and to the extent that tactics were formulated, may have already referred to a planning officer. (In addition to *fu* and *tuo fu, fu* appears in conjunction with *ya,* "commander," being generally interpreted as referring to two officials but also possibly to the commander of *fu.* The *tuo fu* is recorded as being assigned defensive responsibilities.)

31. Ch'en Meng-chia's brief overview (1988, 509–511) lists the relevant inscriptions.

32. However, Ko Sheng-hua, HCCHS 1992:11, 13–18, and others believe the Shang's administrative structure included the post of *ssu-ma,* which was responsible not only for the proto-bureaucracy that oversaw military affairs and impositions, but also for the population registers and control of the general populace, making the administrator a sort of minister of state security. Beneath the *ssu-ma* would have been the *tuo ma* and the *ya* and *ta* (or great) *ya,* with the *ya* actually being subordinate to the *tuo ma.*

Ch'en Meng-chia (508) interprets the term *tuo ma ya* as *tuo ma* and *tuo ya*, many *ma* and many commanders, but it seems more likely that an even higher rank, the *tuo-ma-ya* or commander of the horse commanders, existed. If so, it would be a good candidate for evolving into the *ssu-ma* of late times. (Ch'en interprets *tuo ma* as the *ma shih* or horse commander, from which *ssu-ma* evolved.) It also seems that some clauses with *tuo ma ya* might be understood as the *tuo ma* being ordered to take command, *ya* indicating an action rather than a title. (A minor complicating factor is that officials with the surname of Ma—for example, men from the Ma-fang—also had military roles.)

33. For a summary of relevant inscriptions see Ch'en Meng-chia, 1988, 514.

34. Meng Shih-k'ai, LSYC 1990:4, 103–104.

35. For discussion and extensive inscriptional examples, see Ch'en Meng-chia, 1988, 511–514.

36. See Ch'en Meng-chia, 511–512. Ch'en stresses that *"tuo she wei"* should be understood as *"tuo she, tou wei"* or two separate officials, but in many cases it appears that the *tuo she* is being ordered to exercise a protective function, *"wei"* (similar to the use of *"shu"*), rather than two officials being jointly tasked with some effort. (*Tuo she* can also mean the "many archers" [that is, all the archers], especially in a battlefield context.)

37. See Hu Hou-hsüan and Hu Chen-yü, 2003, 107–115. (The character for *shih*, with a person signifier on the left, would subsequently mean "ambassador.")

38. Ch'en Meng-chia, 1988, 516. (Relevant inscriptions will be found on 515–517.)

39. Hsiao Nan, *Ku-wen-tzu Yen-chiu*, 1981, 130, claims they always commanded the *chung* when they were called to action.

40. See Hsiao Nan, 124, or Li Hsüeh-ch'in, CHSYC 2006:4, 3–7, citing *T'un-nan* 2320.

41. See Ch'en Meng-chia, 1988, 517, who classifies the *yin* under internal or *nei-fu* officials.

42. *Ch'ien-pien* 7.1.4. Chao Kuang-hsien, HCCHS 1986:2, 30, believes each *tsu* was led by a *yin*.

43. *Ch'ien-pien* 5.8.2.

44. See Chu Yen-min, CKSYC, 2005:3, 3–13. Throughout history the right flank has tended to be stronger because piercing and crushing weapons are invariably wielded by the right arm.

45. HJ33006 and *Ts'ui* 597.

46. Contrary to Yang Sheng-nan, 1982, 358. (His discussion of *shih* will be found on pages 341–362.)

47. For an overview, see Chin Hsiang-heng's classic *Ts'ung Chia-ku Pu-ts'u Yen-chiu Yin Shang Chün-lü-chung chih Wang-tsu San-hang San-shih*, 1974, 10–16.

48. Numerous strips such as HJ36450, HJ36454, HJ34715, HJ34716, HJ34718, HJ41748, and HJ41750 inquire about the army not being startled or disturbed during the night.

49. Li Hsin-ta, *Wu-kuan Chih-tu*, 1997, 20; Yang Sheng-nan, 1982, 362, attributes the inscription to their reigns, but others (rather unsatisfactorily) claim it dates to Wu Ting's era. Chin Hsiang-heng, 1974, 15–16, claims that the Shang never exceeded three armies; Hsiao Nan, 1981, 124, 128, concludes that the army didn't really attain its final form until Wen Ting's era; Chao Kuang-hsien, HCCHS 1986:2, 31, believes that this represents the first creation of the *shih* and that the *shih* displaced the clan forces.

50. For example, Hsiao Nan (124) confidently ascribes the strip creating the three armies to Wen Ting's era.

51. Chao Kuang-hsien, HCCHS 1986:2, 32, asserts that Wu Ting only had a central army.

52. Liu Chao, 1989, 70–72. Note should be taken of Liu Chao's characterization of the Shang army as largely an imperial bodyguard or defensive force closely tied to the king and his assertion that armies associated with cities performed protective functions, especially at night. However, he also notes that the army occasionally acted as a forward campaign force, something like the elite *hu-pen* ("tiger warriors") of later fame. (Somewhat surprisingly, Liu also cites examples of reconnaissance activity [HJ5605] that would seem to have been intended to initiate so-called

meeting engagements.) In his understanding the army was limited to a (paltry) 100 men, and three components—left, right, and center—were fielded.

53. For example, see Li Hsin-ta, 1997, 20. Hsiao Nan (1981, 128–129) argues for 10,000.

54. Throughout subsequent history Chinese military units were generally organized in multiples of five, ranging from a squad of five through an army of 12,500. However, there were exceptions when a decade-based structure prevailed and a chariot-centered squad of fifteen may have served as a fundamental building block. The Shang seems to have intermixed units based on three and ten, making the reconstruction of the contingents problematic. For example, in an army consisting of 3,000 troops, each of the three component *lü* should number 1,000 men. However, apportioning the 1,000 among three *ta hang* results in the odd number of 333 each, further segmentation yielding the unlikely number of 111 for the *hang*. Although not impossible if the officers or chariot components somehow account for the odd eleven, these figures blatantly contradict the envisioned ideal of 100. Similarly, if the *shih* came to number 10,000 at the end of the dynasty, a *lü* would have to be 3,333. Reconstructing upward rather than downward doesn't really resolve the problem. For example, if the *hang* is 100 and *ta hang* 300, then three *ta hang* comprise a *lü* of 900 and three *lü* an army of only 2,700, far short of the normal levy of 3,000. The problem becomes even more complex if the army were already chariot based, despite the appeal that assigning a ten-man squad to each chariot might have. Three hundred chariots would result in an army of 3,000, but require ignoring the 900 members manning the chariot.

55. Discussions of the *lü* may be found in Chin Hsiang-heng, 1974, 6–8, and Hsiao Nan, 1981, 125–128. The "10,000-man levy" derives from a variant reading of a famous Fu Hao/Ch'iang battle inscription, but according to the *Chou Li*'s reconstruction (as apportioned under the "Hsiao Ssu-t'u") and the number specified by the *Shuo Wen*, it would have been 500 in the Chou. However, Hsiao Nan sets it at 1,000.

56. Hsiao Nan, 130.

57. *T'un-nan* 2350.

58. Li Hsin-ta, *Wu-kuan Chih-tu*.

59. Not everyone agrees that the king's *lü* constituted the middle or that the "right *lu*" occurs in the inscriptions. (See, for example, Chin Hsiang-heng, 1974, 7–8.)

60. See Liu Chao, 1989, 74.

61. Liu Chao, 75ff.

62. Chao Kuang-hsien (31) claims that the three *shih* displaced the *tsu* on the battlefield.

63. According to the *Tso Chuan*, Ting Kung fourth year, the Shang consisted of six or seven clans.

64. See Chin Hsiang-heng, 1974, 2ff.

65. Shen Ch'ang-yün, HCCHS 1998:4, 23–28, believes this even happened with the various clan forces, including the king's.

66. This exercise in hypocrisy is recorded in the *Tso Chuan*, Hsi Kung twenty-eighth year.

67. Note Chin Hsiang-heng's commentary, 1974, 9. (Chin never hazards an estimate of unit size.)

68. As will be discussed in the section on horses, chariots, and cavalry, there is some controversy over whether the term *ma* (horse), generally understood as chariot, doesn't actually refer to the cavalry, with the chariots already being subsumed within the articulated structure.

69. For a convenient summary of relevant inscriptions, see Ch'en Meng-chia, 1988, 512–513.

70. "The Questions of King Wei."

CHAPTER 13

1. Fu Hao's exact role is also questioned: Did she simply represent the king in some ritual fashion, direct the battle, or even participate in some physical way, ranging from acting as an

archer to wielding a shock weapon? (See, for example, Wang Hsiao-wei, ed., *Ping-yi Chih-tu*, 1997, 39–40.)

2. Other clans apart from the Shang royal *tzu* clan and even some foreign peoples such as the Chiang, Chi, Yün, and Jen inhabited the core domain and could be called upon for military personnel (Ch'en Chieh, HCCHS 2003:2, 15–22).

3. Somewhat surprisingly, arguments to the contrary have begun to appear even in PRC publications. For an example based on a revised reading of the "P'an Keng" section in the *Shang Shu*, see Yü Fu-chih, HCCHS 1993:9, 49–55. (See also Li Ch'an, HCCHS 1998:2, 19–24.)

4. For example, in a much-cited appendix to his unpublished PhD dissertation, David Keightley concluded that the Shang did not depend on slaves for productivity.

5. Two useful overviews in the acrimonious debate over the status of the *chung* are Wang Kuei-min, CKSYC 1990:1, 102–114, and Yang Sheng-nan, 1991, 303–352. David Keightley concluded that there is no evidence for the *chung* having been slaves. (*Cambridge History of Ancient China*, 285–286.) Kung Wei-tuan, HCCHS 1986:11, 41–47, among others, concurs.

6. For example, see Ch'ao Fu-lin, CKSYC 2001:4, 3–4.

7. Robin Yates, JEAA 3, nos. 1–2 (2002): 283–331, notes that slaves originated as military captives whose deaths had been temporarily remitted and who had thus become nonpersons.

8. However, Kung Wei-tuan claims that they were not sacrificed (HCCHS 11 [1986]: 41–47).

9. Oracular inquiries about the possibility of them "being lost" (*sang*) have been interpreted in two ways, querying whether they had deserted or had perished. (For an example of the latter see Liu Feng-hua, KKWW 2007:4, 22–26.) They are noted among the king's prognostications about distant agricultural activities, hunting exercises, and military expeditions. (The chaos marking the latter two certainly would have provided ample opportunity to escape.) However, because the *chung* mobilized for these activities seem to have enjoyed a particularly close relationship with the king and the inquiries are far too frequent, this interpretation seems less likely. Furthermore, in the context of the Shang's draconian punishment system, the possibility of successfully escaping may well be doubted. (It should also be noted that there are inquiries regarding the pursuit of defectors or others who betrayed the king, including men of high rank.)

10. Small numbers were employed in farming in Wu Ting's era, only rarely in the hunt or military activities except against a couple of *fang-kuo*. (Ch'ao Fu-lin, CKSYC 2001:4, 3–12.)

11. For an expression of this view see Hsiao Nan, 1981, 129–130.

12. See, for example, *The Cambridge History of Ancient China*, 282–283.

13. However, Ch'ao Fu-lin holds the unusual view that the *chung* were basically members of the royal lineage who undertook various tasks and could even participate in the king's sacrifices, whereas the *jen* were primarily people from other areas (including prisoners) or under the control of other lords. This would account for their greater mobilization in Wu Ting's military campaigns (CKSYC 2001:4, 8–10).

14. At least this is Ch'ao Fu-lin's interpretation (CKSYC 2001:4, 3–12).

15. Ch'ao Fu-lin, 11. Ch'ao believes they formed the basis for the unit known as the *jung*.

16. Chung Po-sheng, 1991, 104–116; Wen Shao-feng and Yüan T'ing-tung, 1983, 286–298. Chung (116) claims that the use of *ts'e* or bound reports (of bamboo strips) shows it was well established.

17. This trepidation would cohere with Chang Tsung-tung's in MS 37 (1986–1987): 5–8 that Wu Ting was paranoid.

18. Claims that horses were already being ridden in the Shang have been advanced by Chang Shih-ju and others, including (for the dedicated purpose of forwarding intelligence to the capital) Wen Shao-feng, Yüan T'ing-tung, and Chung Po-sheng.

19. Wen Shao-feng and Yüan T'ing-tung, *Yin-hsü Pu-ts'u Yen-chiu*, 1983, 292.

20. Wen Shao-feng and Yüan T'ing-tung, *Yin-hsü Pu-ts'u Yen-chiu*, 289–292. The king queried whether he would receive urgent warning by signal drums on more than one occasion.

21. For a discussion see Wen Shao-feng and Yüan T'ing-tung, 286–288. Not all the examples they cite seem to fall into this category, some still perhaps being more correctly interpreted as the king posing (saying) a question as traditionally understood, but certainly those recorded as "*lai yüeh*"—"come to report" or "incoming report"—do.

22. For a discussion of this paradox see Ralph Sawyer, *Sino-Platonic Papers* 157 (2005).

23. For examples see Wen Shao-feng and Yüan T'ing-tung, 1983, 292–297.

24. For example, as already noted in the era summaries, *Ping-pien* 311 preserves apparently simultaneous queries about whether the king should attack the Pa-fang with Hsi or the Hsia-wei with Wang Ch'eng.

25. In HJ6543 and HJ6544 the king seems to be querying which of two alternatives will receive blessings, while HJ6959 preserves two separate inquiries on the same day in which Ch'üeh and Ching are to attack different enemies.

26. For example, HJ6475, whether Chih Kuo or Hou Kao should accompany the king in attacking the Yi, and HJ6477 *cheng*, where the king inquires about having Hsi (?) accompany him in an attack on the Pa or Wang Ch'eng in an attack on the Hsia-wei.

27. For example, HJ7504, choosing between two commanders.

28. For inscriptions, including HJ6476 *cheng*, see Wang Yü-hsin, 1991, 151.

29. For inscriptions, including HJ6480, see Wang Yü-hsin, 150.

30. The classic formulation appears in "Vacuity and Substance" in the *Art of War*, but Wu-tzu and Sun Pin subsequently elaborated the concept.

31. Luo K'un, 1998, 172, based on HJ6480, and Han Feng, 1982.

32. Numerous examples are seen in Wu Ting's period, including HJ27972 and HJ6981. (It should be noted that the character is sometimes understood as meaning "victory.")

33. *Ching* 2.

34. Seen in HJ6667.

35. For example, *Yi* 916 has the king ordering a commander named Mao to slaughter the Wei-fang.

36. For example, HJ5805. *T'un-nan* 2328, as interpreted by Li Hsüeh-ch'in (CKSYC 2006:4, 3–7), indicates that the right and left *lü* were to observe the enemy prior to undertaking a probing attack.

37. This is implied by HJ7888.

38. HJ35345, as cited and interpreted by Li Hsüeh-ch'in, CKSYC 2006:4, 3–7.

39. With an early form of the character *wei* (without the border enclosure) being employed. (For a brief discussion, see Chin Hsiang-heng's famous "San-hang San-shih," 8–9; for examples of two and even three Kung-fang armies surrounding the Shang in Wu Ting's era, see "San-hang San-shih," 15.)

40. "Planning Offensives."

41. Chao Kuang-hsien, HCCHS 1986:2, 30, notes that the various units underwent constant training under officers such as the *Yin*, *She*, and *Shih*.

42. For a brief examination see Han Chiang-su, CKSYC 2008:1, 37–38.

43. Li Hsin-ta, *Wu-kuan Chih-tu*, 1997, 21, believes that Wu Ting particularly valued warriors who had undergone training and assigned positions to them.

44. "Five Instructions."

45. For example, HJ5772. (For further discussion see Ho Ching-ch'eng, KK 2008:11, 54–70, and the archery discussion that follows.)

46. Even Sun-tzu, who stressed the need for training, never discussed the means or measures, and the classic military writings, apart from the *Six Secret Teachings* and *Wei Liao-tzu*, barely

mention the subject. (On night exercises [based on *T'un-nan* 383], see Meng Shih-k'ai, LSYC 1990:4, 103.)

47. For an example see Shang Ch'ing-fu, HCCHS 1999:6, 5–15. According to the *Mo-tzu* ("Ming Kuei"), King T'ang employed the "bird deployment and geese formation" to attack the Hsia, a description interpreted as evidence of a core force with two flanks. Although such projections back into the mists of antiquity are unfounded, not to mention inconsistent with the fundamental nature of society at the time, it should still be remembered that if more than a few men are mobilized, some sort of basic formation or form of deployment, whether circular or square, line or block, is inherently advantageous.

CHAPTER 14

1. More than a dozen stones, including limestone, quartz, sandstone, and jade, were employed for edged weapons in the Hsia and Shang. (For examples recovered from Anyang, see Li Chi, BIHP 23 (1952): 523–526 and 534–535. Wang Chi-huai discusses an early axe fabrication site in KKWW 2000:6, 36–41.)

2. Note that in making a distinction between *metalworking* (defined as limited-scale hammering, forging, etc.) and *metal production*, Ursula Franklin ("On Bronze and Other Metals in Early China," 279–296), among others, has emphasized the importance of scale in metallurgy's role in civilization.

3. See, for example, Yen Wen-ming, 1989, 110–112. In Yünnan, a region of vast copper resources, stone weapons continued to be employed during the early Bronze Age even after primitive axe shapes had appeared, forging and molding were being practiced, and basic alloys were already known. (Yün-nan-sheng Po-wu-kuan, KK 1995:9, 775–787.)

4. Cheng Te-k'un, the chief proponent for indigenous development in his books and articles such as "Metallurgy in Shang China," was seconded by Noel Barnard in an important early review that still retains currency. (In "Review of *Prehistoric China, Shang China, China*," Barnard argues that piece-mold casting techniques were almost unique to China as metals were worked in the West; smithy practices were few; the lost wax process that predominated in the West did not appear until the end of the Warring States period; binary and then ternary alloys were employed early on; and there is essential continuity with the ceramic tradition. In contrast, An Chih-min, KK 1993:12, 1110–1119, has speculated that ancient trade routes could have played an important but unspecified role. For further discussion see Shang Chih-t'an, WW 1990:9, 48–55; Li Shui-ch'eng, KKHP 2005:3, 239–278; and Ch'en Hsü, HSLWC, 171–175.)

It has also long been held that bronze molding techniques evolved to allow the casting of metallic versions of ceramic precursors and that this continuity from ceramic to bronze realizations provides substantial evidence of the indigenous development of metallurgy in China, particularly in the absence of hammering and other smithy techniques. However, a dissenting voice has been raised by John La Plante, EC 13 (1988): 247–273, who claims Chinese molding techniques evolved to facilitate the production of vessels originally fabricated by the hammering and joining of sheet metal.

Finally, it was originally believed (and is still sometimes claimed) that Shang bronze casting relied on the lost wax method, but more recent evidence has clearly shown that it didn't appear until sometime in the Warring States period. Furthermore, in his classic article "Yin-tai T'ung-ch'i," Ch'en Meng-chia (KKHP 1954:7, 36–41) provided an analysis of the evolution of Shang molding techniques that effectively argued that they never employed the lost wax process. His viewpoint was seconded a decade later by Noel Barnard in "Review of *Prehistoric China, Shang China, China*" and more recently updated by T'an Te-jui, KKHP 1999:2, 211–250, who similarly concluded that the lost wax process didn't evolve until well into the Warring States period. The process was also employed in later times out on the periphery of Chinese civilization. (For example, see Chiang Yü, KK 2008:6, 85–90.)

5. For a discussion of these differences, see Miyake Toshihiko, KK 2005:5, 73–88. Metallurgical traditions also evolved in peripheral cultures such as Hsia-chia-tien in Inner Mongolia, Liaoning, and northern Hebei (1700–1200 BCE); Yüeh-shih (1600–1300); and of course San-hsing-tui, whose technological advances and stylistic elements were the result of complex interactions with the core cultural area coupled with indigenous cultural factors and ore characteristics. (For an analysis of metal developments in Hungshan culture in the northeast that date to about 3000 BCE, see Chu Yung-kang, KKHP 1998:2, 133–152.)

6. For example, in the middle Neolithic some areas seem to have specialized in the production of stone implements despite lacking immediately available resources. (For an example, see Li Hsin-wei, KK 2008:6, 58–68.)

7. In recent years several excavations have been carried out at ancient mining sites, leading to a new appreciation not only of their extensiveness and sophistication, but also of the widely varying ore profile. (For example, see Mei Chien-chün et al., KK 2005:4, 78 ff.) Wu Ju-tso, CKKTS 1995:8, 12–20, notes that copper is found naturally intermixed with zinc or lead at numerous small mines around places like Chiao-chou, Kao-mi, An-ch'iu, and Ch'ang-le. Tuan Yü, WW 1996:3, 36–47, describes the low level of tin used in Pa/Shu ritual vessels in comparison with the Shang.

8. An interesting example of a late Shang and early Chou culture that clearly de-emphasized warfare (as evidenced by tools and hunting implements rather than ritual vessels and weapons predominating) is seen on Yü-huan Island, located 1,000 meters off the Zhejiang coast. (See T'ai-chou-shih Wen-kuan-hui, KK 1996:5, 14–20.) The intervening sea not only physically isolated them but also provided a formidable geostrategic barrier.

9. For example, the analysis of bronze containers from a fourth period Shang tomb (1046 BCE) shows that a high proportion of lead (24–27 percent, in comparison with 55–65 percent copper and only 4–6 percent tin) was employed to allow easier casting of more complex shapes (Chao Ch'un-yen et al., WW 2008:1, 92–94).

10. For reports on the phenomenon (but little speculation on the causes), see Kuo Yen-li, KKWW 2006:6, 66–73; Tuan Yü, CKKTS 1994:1, 63–70; and Liu Yi-man, KKHP 1995:4, 395–412. Liu claims that the trend to high-lead-content funerary items and even the use of ceramic versions, reflecting a diminishment in respect for the spirits, is mirrored in a similar shift from sacrificing a large number of victims, up to 1,000 at one time (together with 1,000 cattle and 500 sheep) under Wu Ting to lower amounts under K'ang Ting (200 human victims, 100 cattle, and 100 sheep) and eventually a maximum of 30 human victims under Ti Hsin. (Other explanations are of course possible, including economizing on resources.) More broadly, Yang Chü-hua, HCCHS 1999:4, 28–43, envisions a total reorientation in values, with the Shang esteeming spirits, the Chou valuing ritual, and the Warring States seeing a new human orientation that allowed bronze artifacts to become commodities.

11. Chiu Shih-ching, CKSYC 1992:4, 3–10.

12. Tuan Yü, HCCHS 2008:6, 3–9. For a general discussion of the techniques at Anyang, see Liu Yü et al., KK 2008:12, 80–90.

13. For reports see SHYCS An-yang Kung-tso-tui, KKHP 2006:3, 351–384; Yin-hsü Hsiao-min-t'un K'ao-ku-tui, KK 2007:1, 14–25; Wang Hsüeh-jung and Ho Yü-ling, KK 2007:1, 54–63; and Li Yung-ti et al., KK 2007:3, 52–63.

14. For some of these discoveries see P'eng Ming-han, HCCHS 1996:2, 47–52; Chan K'ai-sun and Liu Lin, WW 1995:7, 18, 27–32; and, for a general discussion, Ch'en Liang-tso, HHYC 2:1 (1984): 135–166 and 2:2 (1984): 363–402.

15. For a discussion of this question see Yen Wen-ming, SCYC 1984:1, 35–44.

16. Huang Sheng-chang, KKHP 1996:2, 143–164.

17. For example, T'ang Yün-ming claims that China was already producing wrought iron in the early Shang. (See WW 1975:3, 57–59, and for further discussion Hsia Mai-ling, HCCHS 1986:6, 68–72. For the history of iron in China, see Donald Wagner, *Ferrous Metallurgy*, or his earlier *Iron and Steel in China*.)

18. See note 7.

19. In addition to any specific references, the discussion that follows is primarily based on Li Shui-ch'eng, KKHP 2005:3, 239–278; Pei-ching Kang-t'ieh Hsüeh-yüan Yeh-chin Shih-tsu, KKHP 1981:3, 287–302; Yen Wen-ming, SCYC 1984:1, 35–44; and An Chih-min, KK 1993:12, 1110–1119.

20. Pei-ching Kang-t'ieh Hsüeh-yüan Yeh-chin Shih-tsu, KKHP 1981:3, 287–302. To date, the most complete overviews of techniques and products are Lu Ti-min and Wang Ta-yeh, 1998, and Hua Chüeh-ming's massive and highly technical *Chung-kuo Ku-tai Chin-shu Chi-shu*, 1999.

21. Shao Wangping, JEAA 2, nos. 1–2 (2000): 195–226.

22. Yen Wen-ming, SCYC 1984:1, 35–44, and WW 1990:12, 21–26.

23. Li Hsüeh-ch'in, CKKTS 1995:12, 6–12, and An Chih-min, KK 1993:12, 1110–1119.

24. Yen Wen-ming, WW 1990:12, 26.

25. Pei-ching Kang-t'ieh Hsüeh-yüan Yeh-chin Shih-tsu, KKHP 1981:3, 287–302.

26. This was pointed out by Sun Shuyun and Han Rubin, EC 9–10 (1983–1985): 260–289.

27. Sun Shu-yün and Han Ju-pin, WW 1997:7, 75–84; Li Hsüeh-ch'in, CKKTS 1995:12, 6–12; and Chang Chih-heng, 1996, 109–112.

28. Sun Shu-yün and Han Ju-pin, WW 1997:7, 75–84; An Chih-min, KKHP 1981:3, 269–285; An Chih-min, KK 1993:12, 1113. (Ch'i-chia culture is more broadly dated as 2200 to 1600 BCE.)

29. Li Shui-ch'eng, 251–254; An Chih-min, KKHP 1981:3, 269–285; An Chih-min, KK 1996:12, 70–78; and Kung Kuo-ch'iang, KK 1997:9, 7–20. (Arsenic alloys are also known in nearby Russia.)

30. Li Shui-ch'eng, 244–245.

31. For critical reports on which this discussion is based, see Sun Shu-yün and Han Ju-pin, WW 1997:7, 75–84; Li Shui-ch'eng, 241–245; and Sun Shu-yün and Han Ju-pin, WW 1997:7, 75–84.

32. Li Shui-ch'eng and Shui T'ao, WW 2000:3, 36–44; Sun Shu-yün et al., WW 2003:8, 86–96; and Li Shui-ch'eng, KKHP 2005:3, 239–278. Copper/arsenic alloys similarly characterize ten of the eleven items found at Tung-hui-shan even though they date to about 1770 BCE. (See Kan-su-sheng Wen-wu K'ao-ku Yen-chiu-suo, KK 1995:12, 1055–1063. Some were forged, some heat treated or cold quenched after forging.)

33. Li Shui-ch'eng, 256–257.

34. Li Shui-ch'eng, 263; Li Hsüeh-ch'in, CKKTS 1995:12, 6–12. Wang Hsün, KKWW 1997:3, 61–68, has suggested that the growing Shang threat prompted the Hsia to develop better bronze weapons.

35. Chin Cheng-yao, WW 2000:1, 56–64, 69. Recognition of lead's properties apparently came last. (Chin also notes a shift in sourcing to Shandong late in the era. Although recognizing that Seima-Turbino and Andronovo developments may have been transmitted through ongoing trade, Chin still argues for radical differentiation.)

36. Sun Shu-yün and Han Ju-pin, WW 1997:7, 75–84.

37. Chin Cheng-yao et al., KK 1994:8, 744–747, 735.

38. Ching Cheng-yao et al., WW 2004:7, 76–88.

39. For Yünnan see Li Shao-ts'en, KKWW 2002:2, 61–67; for the middle Yangtze, Liu Shih-chung et al., KKWW 1994:1, 82–88; for the lower Yangtze, Liu Shih-chung and Lu Pen-shan, KKHP 1998:4, 465–496 and illustrations (discussing a mine that was operated continuously from the middle of the Shang to around the start of the Warring States period); and for Gansu, see Sun Shu-yün and Han Ju-pin, WW 1997:7, 75–84.

40. See note 7.

41. Lu Pen-shan and Liu Shih-chung, WW 1997:3, 33–38.

42. These are the estimates for the mines at Wan-nan down in southern Anhui, where copper, iron, sulfur, and gold are all found. (See Ch'in Ying et al., WW 2002:5, 78–82.) According to Liu Shih-chung et al., KKWW 1994:1, 82–88, the middle and the lower Yangtze were prolific production areas that fed smelters located at Anyang and Wu-ch'eng. The extant slag heap around T'ung-lü-shan amounts to 40,000 tons, and 80,000 tons have been found in the middle Yangtze area.

43. In addition to references in the discussion about the dagger-axe that follows, see Chao Ch'un-yen et al., WW 2008:1, 92–94.

CHAPTER 15

1. "Agricultural Implements," *Liu T'ao*.

2. For significant issues related to bronze artifacts, see An Zhimin, EC 8 (1982–1983): 53–75.

3. For example, numerous specimens of every known dagger-axe style have been recovered from the Anyang area, ranging from tabbed to socketed, from heavily used weapons to funerary replicas (*ming ch'i*). (For one instance see SHYCS An-yang Kung-tso-tui, KKHP 1994:4, 471–497.)

4. An example of a highly unusual weapon is the "wavy" blade sword recovered at Chin-sha, which has six to seven "waves" or bulges along the blade's tapered portion and was produced in both bonze and spectacularly colored jade versions. (For illustrations see pages 13 and 31 of Ch'eng-tu-shih Wen-wu K'ao-ku Yen-chiu-suo, WW 2004:4.)

5. T'ung Chao, *Hou-ch'in Chih-tu*, 1997, 32, notes that bone arrowheads still comprised the majority at the start of the Western Chou and were only gradually displaced by bronze. The stones employed to fabricate Shang weapons include slate, diabase, limestone, quartzite, phylite, sandstone, and jade, with diabase and limestone especially employed for axes and slate for knives.

6. Hayashi Minao's 1972 landmark treatment—*Chugoku In-shu Jidai no Buki*—though it includes considerable material of great importance and frames the discussion, is increasingly outdated. Very little material has been included in the ancient volume of the voluminous *Chung-kuo Chan-shih T'ung-lan*, and the important *Chung-kuo Ku-tai Ping-ch'i T'u-chi*, though extensively illustrated and the only comprehensive overview, provides minimal analytical explanation.

7. Other remnants are found in traditional Chinese opera, but they are deliberately stylized and exaggerated. Over the past three decades numerous experiments conducted with replica weapons and experienced martial artists in Taiwan, Hong Kong, Korea, and Japan suggest that traditional weapons had many limitations and that highly particularized fighting methods had to be scrupulously observed.

8. Traditional martial arts teach the importance of agility and the need to maintain balance in maneuver. Although not applicable to every design, the fundamental principle often cited through the ages is that weapons essentially function as extensions of the human body and therefore must be used in accord with the principles governing all human movement if they are to be successfully and easily handled. Unusual, jerky, or unbalanced movements, although certainly surprising and sometimes effective, generally expose the warrior to danger and death.

9. For one of the few articles that notes both dimensions and weights, see P'ang Wen-lung, KKWW 1994:3, 28–40, 56. For example, one late Shang straight socket *fu* measuring 13 cm. in length and 6.5 cm. in width weighed 600 grams. Two other rectangular examples from the middle Shang (with somewhat indented middle portions and splayed blade tips) are 13 cm. long, 5 cm wide, and weigh 500 grams, and 13 cm. long, 4 cm. wide, but weigh only 300 grams. Two tabbed *ch'i* similarly dating to the middle Shang are recorded as 16.3 cm. long, 5.5 cm.

wide, and 300 grams in weight, and 17.5 cm. long, 7.4 cm. wide at the top of the blade, with a 3.8-cm.-long tab, and a weight of 400 grams. (The former is rectangular with a fairly wide tab, somewhat rounded blade, and slight indentation along the blade's length, while the latter is marked by a somewhat hourglass shape, the blade being the same width at the top and bottom.)

Another flanged *yüeh* dating to the Shang with a somewhat hourglass-shaped blade, two binding slots, and a hole in the tab, with dimensions of 17.6 cm. by 8.8 cm. and thickness of 0.6 m, weighs only 350 grams. (See Fan Chün-ch'eng, KKWW 1995:5, 91.)

10. Surprisingly, Ch'ien Yao-p'eng, KKHP 2009.1: 1–34, has recently claimed that the axe was the most effective weapon in ancient China.

11. Fighters always had to be aware of potential defects and avoid movements that might fatally damage the weapon (such as directly striking a hard surface) when stone was the primary material.

12. For example, a primitive-looking, rectangular specimen with an iron blade that simply tapers upward to a tab area defined by two prominent protruding flanges has, as already noted, given rise to claims that China had commenced making iron and steel by the early Shang. However, other analysts have concluded that the high nickel content suggests a meteoric origination. (For a report, originally written in 1975 but one of many that failed to be brought out because of the Cultural Revolution, see Chang Hsien-te and Chang Hsien-lu, WW 1990:7, 66–71.) For another assertion that China was already producing wrought iron by the early Shang, see T'ang Yün-ming, WW 3 (1975): 57–59; for the history of ferrous metals in China see Donald Wagner, 1996 and 2008.

13. Numerous examples from Anyang have been documented and their composition analyzed by Li Chi in his now classic article, "Yin-hsü Yu-jen Shih-ch'i T'u-shuo," 1952. Additional examples continued to be recovered; for example, see SHYCS, *Yin-hsü Fa-chüeh Pao-kao*, 1958–1961, 171–173. (For an example of a site with stone *fu* intermixed with advanced bronze arrowheads, see Ho-pei-sheng Wen-wu Yen-chiu-suo, KK 1999:7, 1–7.)

14. However, Ch'en Hsü has concluded that it was primarily a symbol of authority rather than an actual weapon or executioner's axe. (See HSLWC, 233–239.)

15. A study of the weapons found in the four periods of Yin-hsü concluded that commanders always had *yüeh*, *ko*, spears, and arrowheads in large numbers, sometimes hundreds, but no other ranks possessed them. Although (as would be expected given disparities in rank) the quality, size, and number varied, sometimes two or three up to a dozen were recovered from a single interment. (See Liu Yi-man, KK 2002:3, especially 66–70.)

16. "Yin Pen-chi."

17. "Chou Pen-chi."

18. *Kuo Yü*, "Lu Yü, Shang."

19. Inscribed *yüeh* recovered from Su-fu-t'un (which has been tentatively identified as the site of the former state of P'u-ku made famous by the Shang conquest) indicate that a "commander Ch'ou" (Ya Ch'ou) governed as ruler of P'u-ku (Ch'en Hsüeh-hsiang and Chin Han-p'o, KK 2007:5, 87).

20. SHYCS Ho-nan Yi-tui, KK 1992:10, 865–874. Some fifty-one *fu* were recovered. (For additional examples of early forms see Chang Chen-hung, WW 1993:9, 32–39, especially the illustrations on 38.) Hsing-lung-wa cultural sites in Inner Mongolia have yielded comparable artifacts. (For example, see Nei-Meng-ku Tzu-chih-ch'ü WWKK YCS, KK 1993:7, 577–586.)

21. For examples see SHYCS Hu-pei Kung-tso-tui, KK 1991:6, 481–494. All the *fu* recovered, though comparatively primitive, rough, and small, were well-used tools. Two sizes reported for *fu* without lashing holes are 6.1 cm. high, 4.9 cm. wide, and 1.6 cm. thick, and 10.2 by 7.3 cm., 3.5 cm. thick. Somewhat more developed styles, almost indistinguishable from the *yüeh* recovered at the site but in a narrower rectangular shape marked by a large binding hole, include

specimens with dimensions of 15.8 by 10 cm. and 1.0 cm. thick; 13 by 8.6, 0.9 cm. thick; and 13.7 by 7.6, 1.3 cm. thick.

22. Although Li Chi's article notes the dimensions and material composition of numerous stone *fu*, some examples that were recovered from sites discussed below in conjunction with the more martially focused *yüeh* might be cited. A stone *fu* found at the Lung-nan site dating to about 3300 BCE is basically rectangular, has a large hole in the upper quarter, tapers gradually outward, and has a slight curvature to the blade (Kao Meng-ho, KK 1 [2000]: 58).

At San-hsing-ts'un twenty-one *fu* were found, mostly double-edged and heavily used, in four distinct styles. (See Chiang-su-sheng San-hsing-ts'un Lien-ho K'ao-ku-tui, WW 2004:2, 10 and the diagrams on 18). Typical dimensions are about 15 cm. long by 8 to 9 cm. wide; all the specimens have a fairly large hole; one style is a clearly tapered but somewhat bulgy rectangular with a large tapered hole; another rectangular design is marked by essentially square tops and blades.

At the Liang-chu site northwest of T'ai-hu the well-used stone *fu* has a somewhat rounded top, expanding downward taper, and curved blade with total dimensions of 10.4 long, 5 cm. wide at the top and 7.4 at the edge, and a thickness of 1.6 cm. (Chiang-su Kao-ch'eng-tun Lien-ho K'ao-ku-tui, WW 2001:5, 4–21.) The *fu* found at the Liang-chu site near Shanghai include one with double tapered edge, slightly rounded top and blade, a large hole, and dimensions of 12 cm. long, 14 cm. wide, and 0.8 cm. thick; 16.8 cm. long, 11.2 cm. wide, 0.85 cm. thick; and two very rounded specimens that splay outward near the blade edge with dimensions of 11.8 cm long and 9.7 cm. wide, another 13 cm. long, 7.3 cm. wide, and 0.6 cm. thick (Shanghai Po-wu-kuan K'ao-ku Yen-chiu-pu, KK 2002:10, 49–63).

Twenty-nine marble, stone, and jade *yüeh* of various sizes have been recovered at Lin-fen in Shanxi that supposedly date to the era of Yao and Shun (2600–2400 BCE). (See Shan-hsi-sheng Lin-fen Hsing-shu Wen-hua-chü, KKHP 1999:4, 470–472, including illustrations on 471 and some appended photos.) Mostly elongated, more like *fu* than the broad *yüeh* of later times, they were produced by grinding rather than flaking and are therefore generally well defined and polished. They all have single moderate lashing holes and a few are exceptionally thin. Some of the sizes listed (length by width by thickness, in centimeters) include 11 x 5 x 0.8; 13 x 7 x 1.3; 15 x 6.2 x 0.6; 12.4 x 6.6 x 0.8; 8.4 x 5.6 x 0.5; 21.7 x 8 x 0.8; 15.4 x 4.9 x 0.8; 12.4 x 4.3 x 0.9; 17.3 x 4.8 x 0.7; 16.7 x 7.4 x 1.2; 25.3 x 12.6 x 1.2; and an astonishingly thin 8.6 x 7.2 x 0.4.

Finally, the *fu* and *yüeh* recovered from a Fujian site are basically similar in style, the *fu* simply being somewhat more rectangular. All have a slight outward taper, have a moderate-size hole in the blade, and show slight rounding of the blade near the edge. Two typical sizes are 10 cm long, 5.6 to 6.0 cm. wide, and 0.9 cm. thick and a very small 4.6 cm. high, 2.8 to 3.2 cm. wide, and 0.5 cm. thick (Fu-chien-sheng Po-wu-kuan, KKHP 1996:2, 165–197).

23. Wang Tsung-yao, KK 1992:7, 663–665. Generally smooth, they reportedly do not show any signs of use and range in thickness from a useless 0.5 cm. to 1.0 cm. thick. Examples include the largest, with dimensions of 15.5 cm. high, 17.2 expanding to 19.4 cm. wide, and 0.8 cm. thick with a large hole in the upper blade, and the smallest, at 11 cm. high, 9.2 to 9.5 cm. wide, and 0.7 cm. thick. The more rectangular style includes one of 15.7 by 9 cm. wide and 0.8 cm. thick with a 3 cm. hole, while a third style with pinched waist and curved blade includes specimens of 22 cm. high, 14.8 to 17.8 cm. wide, and 0.8 cm. thick; 18 cm. high, 13.5 to 17.5 cm. wide, and 1 cm. thick; and 16.8 by 15 to 16 cm. wide, and 0.5 cm. in thickness.

24. See the example described in Fan Chün-ch'eng, KKWW 1995:5, 91.

25. See SHYCS Hu-pei Kung-tso-tui, KK 1991:6, 481–494. (The stone *yüeh* are described on 484–486.) Although the width increases down the body to the blade edge, the *yüeh* recovered here are fairly rectangular. The blades are slightly rounded and have large holes in the upper third (though one has two such holes slightly off-center, perhaps an experiment or mistake). One is 16.7 cm. high by 12.5 cm. wide and 0.9 cm. thick; the second 12.9 cm. high, 10.3 cm.

wide, and 0.8 cm. thick; the third, 17.9 cm. high, 13.6 cm. wide, and 0.9 cm. thick with two holes.

26. Su-chou Po-wu-kuan and Ch'ang-shu Po-wu-kuan, WW 1999:7, 16–30 (description 23–24, diagrams 26, photos 27). Although a corrected radiocarbon date of 5885 BP was obtained, the authors note that classic Liang-chu culture is said by others to be centered about 5,000 BP or 3,000 BCE.

27. All the *yüeh* were secured by a large lashing hole in the upper third of the blade; most show a slight divergence from purely symmetrical shapes; a few are tapered, one sort of rounded on the top and bottom while others are just slightly rounded at the top, have a slight indentation, and widen a little, with the blades marked by various degrees of curvature. Typical sizes include 13.8 cm. high by 10.4 cm. wide, widening to 11.6 cm. but only 0.5 cm. thick; 12.6 by 9.2 to 11.2 cm. wide, 0.9 cm. thick; 13.6 cm. high, 13–14 cm. wide, and 0.9 cm. thick; 11.4 by 10.4 to 11.2 cm., but only 0.6 cm. thick; and a somewhat larger specimen, 15.6 cm. high by 10 to 11 cm. wide and 1.5 cm thick.

28. Chiang-su-sheng San-hsing-ts'un Lien-ho K'ao-ku-tui, WW 2004:2, 4–26. Ten stone *yüeh* are discussed on 8–9; photos with a reconstructed handle appear on 12, diagrams on 17.

29. The jade specimen, which has dimensions of about 12.2 cm. high by 10 cm. wide and a thickness of 1.4 cm., is essentially an oval with a flattened top and one large lashing hole. The stone variants have been divided into two types differentiated by the degree of rectangularity. The example with the cap, shaft filial, and short 53 cm. handle has dimensions of 13.2 by 10.2 by 1.8 cm.; a second example runs 12.4 by 9.4 cm. wide and 1.9 cm thick; and a third, which is thinner at 0.75 cm., is 10.5 by a relatively narrow 5 cm. wide. The second type, said to be squarer and generally marked by a larger hole in the blade, includes one with 45 cm. shaft remnants and dimensions of 11.2 by 11 cm. and a thickness of 1.2 cm., and another of 14.8 by 12.7 and 1.25 cm. thick, which mounted a cap above the blade of 12.3 cm.

30. Chiang-su Kao-ch'eng-tun Lien-ho K'ao-ku-tui, WW 2001:5, 4–21.

31. The stone versions appear in two styles: one gradually tapers down toward the shaft, is slightly rounded at top, has a large hole in the upper portion, is somewhat rough where the shaft would be affixed, and is unsharpened. Sizes include 14.3 cm. high, 9.9 widening to 12.4 cm. at the blade, and 0.65 cm thick, and 15.2 cm. high, 14 to 15.7 cm. wide, and a thickness of 0.6 cm. The second style, which has a slightly pinched middle blade, includes examples of 14.6 cm. high, 9.3 to 13.5 cm. wide, 1.0 cm. thick, and 14 cm. high, 9.4 to 11.4 cm. wide, and 1.2 cm. thick; and a variant form that is rougher and thinner, 11.8 cm. high and 11.3 to 13 cm. wide. (See Chiang-su Kao-ch'eng-tun 15–16 for details.) The jade *yüeh* (described on 16, 18, and 19) basically exhibit a gradual taper in the blade and have large holes near the top. Examples include 16.2 cm. high, 8.2 to 10.5 cm. wide; 15.2 cm. high, 9.4 to 11.5 cm. wide, and 0.6 cm. thick; and 14.8 cm. high and 9.8 to 11.7 cm. wide.

32. Liao-ning-sheng WWKK YCS, WW 2008:10, 15–33. (A photo will be found on 25, and a dimensional sketch on 29.)

33. Fu-chien-sheng Po-wu-kuan, KKHP 1996:2, 165–197. The *fu* and *yüeh* recovered from the site are basically similar in style, the *fu* simply being somewhat more rectangular. Both types display a slight outward taper, rounding of the blade, and moderate-size holes. Typical dimensions for the *yüeh* are 11 by 5.6 to 8.0 cm., 0.8 cm. thick, and 12.6 by 9.2 to 9.9 cm., roughly 0.9 cm. thick.

34. Shanghai Po-wu-kuan K'ao-ku Yen-chiu-pu, KK 2002:10, 49–63 plus six pages of photos. A radiocarbon date of 1690 BCE ± 150 has been reported.

35. Although the grave's occupants have been termed "ordinary people," they may have been reburials necessitated by warfare and the accompanying items reportedly show considerable quality and polish. The five *yüeh* basically fall into relatively square and rectangular versions. An example of the former has dimensions of 20.8 cm. high by 18.1 cm. wide and 0.85 cm. thick;

two of the latter are 23.8 cm. high by 15.2 cm. wide and a very thin 0.55 cm. thick, and 24 cm. high by 16.0 cm. wide and 0.8 cm. thick.

36. The only significant overview article has been Yang Hsi-chang and Yang Pao-ch'eng, "Shang-tai te Ch'ing-t'ung Yüeh," 1986. (Hiyashi Minao's discussion, 132–166, though raising useful issues, has unfortunately become outdated.)

37. Based on specimens recovered prior to 1985, Yang and Yang identified five types in their article "Shang-tai te Ch'ing-t'ung Yüeh": laddered (tapered), square, rectangular, pinched waist, and tongue. (For type examples see 129 and 132. New discoveries from Erh-li-kang and Erh-li-t'ou sites have disproven a few conclusions about origination and evolution that need not be discussed here.)

38. SHYCS Erh-li-t'ou Kung-tso-tui, KK 2002:11, 31–34.

39. Two socketed *fu* (described as tools) were also recovered. (Sizes and descriptions for the *yüeh* are based on photographs and dimensions given in Li T'ao-yüan et al., *P'an-lung-ch'eng Ch'ing-t'ung Wen-hua*, 2002, 48, 49, 128, 129, and Hu-pei-sheng Po-wu-kuan, WW 1976:2, 26–41.)

40. A number of *fu* and unusual *yüeh* dating to this period have been found at the famous site of Lao-niu-p'o. One in the classic tab style is decorated with three large downward-pointing triangles and *t'ao-t'ie* patterns on the upper blade and tab. Overall dimensions are 23 cm. high and 17.2 cm. wide, while the tab is 7.6 cm. long by 7 cm. wide. The blade's thickness tapers down somewhat in the tab, but overall averages an immense 2 cm., allowing for the decorative aspects. The blade was secured by lashings running through two binding slots and a hole in the tab. Another intriguing variant is 19 cm. high and 9.2 cm. wide and has a 7 cm. centered tab with a width of 5 cm. and an odd triangular slot. The usual two binding slots appear on the blade's shoulders, the edge is highly rounded (suggesting it may be a *ch'i*), and three stick figures with smiling heads provide unique decoration. Five *fu* ranging from 12 to 16 cm. in length were also recovered. (For the excavation report see Hsi-pei Ta-hsüeh Li-shih-hsi K'ao-ku Chuan-yeh, WW 1988:6, 1–27. The *yüeh* are described on 9–11.)

Another example from this period recovered in Shandong is also in the classic tab style but decorated by a rather simpler abstract motif on the upper portion of the blade and also the back portion of the tab. (The tab is slightly offset from center and the blade is not symmetrical.) Incorporating the usual two lashing slots and a singe tab hole, the blade is 15.8 cm. high and 9.5 cm. wide, tapering to 8.7 cm., while the tab is 6 cm. long by 4.7 cm. wide. However, being only 0.4 cm. thick, it must have been a purely symbolic weapon (Shan-tung Ta-hsüeh Li-shih-hsi and Hsü Chi, WW 1995:6, 86–87).

41. Sung Hsin-ch'ao, CKSYC 1991:1, 55.

42. For a discussion of the tiger motif see Shih Ching-Sung, KK 1998:3, 56–63.

43. Sizes taken from Yang Hsi-chang and Yang Pao-ch'eng, 1986, 130 and 132. The dimensions vary slightly in other reports.

44. For the report, already noted in the context of commanders having *yüeh* included in their tombs, see SHYCS An-yang Kung-tso-tui, KK 2004:1, 7–19.

45. For example, a *yüeh* marked by a somewhat indented blade shape, fairly wide tab, and comparatively simple abstract decorative patterns dating to the late Shang—possibly Hsin's reign—is only 0.4 cm. thick. (See An-yang Wen-wu Kung-tso-tui, KK 1991:10 906.) Recovered with various ritual bronzes from a commander's grave, it has dimensions of 17 by 13.4 cm. by at the blade (but only 9.9 cm. at the shoulders) and a tab of 6.5 by 4.7 cm. Although there is a slight upper flange extension, no binding slots have been included. (Wooden handle remnants are, however, noticeable.)

46. A *yüeh* found at Anyang that dates to the fourth period has three triangular shapes on the blade similar to the one noted as being found earlier but a *t'ao-t'ie* motif in the upper portion

of the blade. (See illustration 884 in SHYCS An-yang-tui, "1991-nien An-yang Hou-kang Yin-mu te Fa-chüeh," 1993:10, 880–903.)

47. For example, an odd-shaped *yüeh* found at Kuo-chia-chuang at Yin-hsü has two lashing holes near the top of the blade, a tab offset to one side, and a very asymmetrical cutting edge. (See SHYCS An-yang Kung-tso-tui, KK 1998:10, 44–45.) An elaborately decorated, somewhat asymmetrical *yüeh* marked by a tab offset from center with a single lashing hole, slots in the upper blade, and dimensions of 22.4 by 16.8 cm. wide maximum (but 13.6 cm. at the top), and a tab 7 cm. long by 7.6 cm. wide, dating to the late second period at Yin-hsü, has been recovered at Ta-ssu-k'ung (SHYCS An-yang Kung-tso-tui, KK 1992:6, 513).

48. Chang Wen-li and Lin Yün, KK 2004:5, 65–73. Though incorporating indigenous motifs, the *fu* and *yüeh* recovered in the Ch'un-hua district of Shaanxi also reflect Ch'ing-hai, Central Asian, and Shang influences, resulting in highly localized versions that in turn reportedly spread northward and eastward. Socketed versions predominate in the northwest, with socket mounting being employed for both the unusual semicircular blade depicted and simple versions that look like hatchets. (These sockets are often tapered along the length and slightly oval, no doubt requiring considerable effort to achieve a tight fit.) However, more traditional versions that employ straight, centered tabs were also found in three variants, including a tongue-shaped version with pinched sides marked by a very large hole in the center. Most of these have two binding slots at the top and a hole in the tab for lashing. One specimen dated to about the Yin-hsü third period is decorated with an abstract motif that resembles a series of bent cotter pins that are said to have originated in Ch'ing-hai.

49. For the example, see Wang Yung-kang et al., KKWW 2007:3, 11–22. This moderately sized *yüeh* measures 21.5 cm. high by 12.8 cm. wide and was mounted with a narrow 7.2 cm.–long tab. Apart from the sheep's heads protruding at the top of the slightly nonsymmetrical blade, the axe has flanges on both edges, a binding hole in the tab, additional *t'ao-t'ieh* decorations on both the blade and tab, and a 1.5-cm.–wide raised perimeter on the blade itself, thickening the profile. The blade's two faces are identical, ensuring that its awesomeness would still be projected no matter how they were viewed.

CHAPTER 16

1. In pre-imperial times numerous materials were employed to produce knives, even bamboo, which can produce a highly lethal edge if properly prepared.

2. A few exceptions in stone and bronze have been recovered. For example, the weapon set found in the grave of a foreign nobleman that dates to Yin-hsü's second period includes a surprisingly long knife of 32.7 cm. It has a straight tab continuous with the upper edge; the blade turns upward near the point; the front and rear portions are wider than the middle, ranging from 5 to 8.2 cm.; the top is thicker than the edge; and it weighs 335 grams. (See An-yang-shih WWKK YCS, KK 2008:8, 28–29.) A bronze knife recovered from Hsiao-t'un is 27.3 cm. long and 3.2 cm. wide (Shih Chang-ju, BIHP 40 [1969]: 660). Finally, a particularly long *tao* with a turned-up edge almost like seen on elf shoes, a decorated band near the top, a straight bottom edge, a blunt back with a short tab at the top and "Ya Ch'ang" on it, 44.4 cm. long (and thus somewhat functional), and a width of about 5.6 cm. has been recovered from "Ya Ch'ang's" tomb. (See SHYCS An-yang Kung-tso-tui, KK 2004:1, 14.)

3. For a concise overview see Liu Yi-man, KK 1993:2, 150–166.

4. The numerous knives and daggers found in China's peripheral cultures, particularly the Northern complex, have prompted several studies that, though in part charting their evolution, lie outside the scope of *Ancient Chinese Warfare*, as does the more comprehensive question of

the nature and degree of influence from more distant steppe and Eurasian cultures with bronze precursors, including Sintashta-Petrova.

5. Based on early discoveries (BIHP 22 [1950]: 19–79) and references in the *Shang Shu*, Shih Chang-ju concluded that weapons were normally grouped into functional sets. Therefore, after exchanging archery fire at a distance and employing spears at intermediate range, knives were supposedly used (76). Apart from being completely unrealistic—the possibility of anyone still standing after engaging with spears (or dagger-axes) being minuscule—a sample of six sets is too limited to justify such sweeping conclusions. In fact, Ch'en Meng-chia, KKHP 1954.7: 15–59, has challenged the conclusion that the knives were weapons for close combat, notes that they are found with whetstones in sets for maintaining shafted weapons, and cites evidence from other areas where they are found in combination with an axe (*ch'i*) rather than in elaborate sets (52). His analysis of about 100 knives from Yin-hsü (43–57) not only criticizes Li Chi's earlier classifications but also finds that most were tools shaped to specific purposes such as woodworking, kitchen work, and separating hides (46–47). Nevertheless, he terms one category of longer variants that range from 27.5 to 52 cm. a *"chin"* (the basic axe radical). Designed for horizontal mounting along a handle, until being enlarged and mounted horizontally on or atop a long shaft, such as the later Kuan Tao, it could only have been used for executions. (Executions and combat require significantly different modes of action.)

6. Shih Chang-ju (74–76) believes that the knives were hierarchically differentiated by weight and decorative motif, with the horse head being the most prestigious, the ox next, and finally the ram. (The weights range from 382 grams down through 379, 301, 137, 131, and finally an ultralight 62.) Five of the six curve downward; those with animal heads are wider and have a flange, though one has two rings. The knives were carried in two different types of sheaths, with the fancier one being hung from a jade ring.

7. Several early stories are preserved in the *Wu-Yüeh Ch'un-ch'iu* and *Yüeh Chüeh-shu*. A certain mystique came to be associated with the dagger, the preferred weapon of notable assassins.

8. "Waging War."

9. For example see Chou Wei, 1988, 88–98; Yang Hung, 1985, 126.

10. Yang Hung, 1985, 125–26.

11. See Chang Kwang-chih, 1980, 45, and 1982, 13.

12. The development of longer daggers in Wu and Yüeh, where watery terrain precluded the extensive use of chariots, is often, though without justification, cited as evidence that chariot warfare and swords were somehow mutually exclusive. (Similarly, Chiang Chang-hua, KK 1996:9, 78–80, who believes that swords are indigenous, thinks they first evolved in Shu because of Sichuan's wetness.)

13. Whatever the ratio per chariot may have been—10, 25, 72, 75, or even 100 to 1—foot soldiers always comprised the majority of warriors.

14. See Max Loehr's classic article, "The Earliest Chinese Swords," 132–142.

15. See, for example, Chou Wei, 1981, 112–116. (Early dagger-axes, spearheads, and daggers are so similar in appearance as to be virtually interchangeable.)

16. See Yang Hung, 1985, 129.

17. Swords would eventually become too long and unwieldy for the battlefield, as the example of the King of Ch'in being unable to draw his weapon when attacked by the famous assassin Ching K'o shows. For personal defense in normal situations, daggers of moderate length were always more effective.

18. Hayashi Minao, 1972, 199–236, and Chou Wei, 1988, 109–157. Although still the most extensive treatment, Minao's work suffers from being based on materials published prior to 1970; therefore, the only Western Chou "sword" he considers comes from Chang-chia-p'o in Ch'ang-an. Moreover, he rarely provides actual measurements, only figures with rough proportions.

Fortunately, a number of articles have studied the dagger's complex history in considerable detail, including Yang Hung, 1985, 115–130; T'ung En-cheng, KK 1977: 2, 35–55; Lin Shou-chin, KKHP 1962:2, 75–84 and WW 1963:11, 50–55; Hsiao Meng-lung, KKWW 1996:6, 14–27; Sung Chih-min, KK 1997:12, 50–58; Chung Shao-yi, KK 1994:4, 358–362; Chu Yung-kang, WW 1992:12, 65–72; Ho Kang, KK 1991:3, 252–263; Li Po-ch'ien, WW 1982:1, 44–48; Ch'en P'ing, KK 1995:4, 361–375; Chang T'ien-en, WW 1993:10, 20–27, and KK 1995:9, 841–853; and Kuo Pao-chün, KK 1961:2, 114–115. (Essential illustrations may also be found in the *Chung-kuo Ku-tai Ping-ch'i T'u-chi.*)

19. This may be seen from the mix of older and newer weapons found in many graves across the ages. Han tombs even include both bronze and iron swords, even though only the latter may have been used in actual combat. (For one example, see Chung Shao-yi, KK 1994:4, 359.)

20. This sort of knowledge apparently died out after the Han, by which time the *tao* had replaced the sword as a military weapon and even ceremonial versions had been largely replaced by wooden ones (Chung Shao-yi, KK 1994:4, 358–362).

21. One Spring and Autumn burial contains a 50 cm. bronze sword and a long spear with a handle of 120 cm. and a 26.8 cm. head (Ch'ing-yang-hsien Wen-wu Kuan-li-suo, KK 1998:2, 18–24).

22. Considerable insight can be gleaned from modern techniques, even though the latter cannot be simply projected backward. For example, depending on the fighting style, there are several methods for grasping the handle, but it is unlikely that the ancients ever held the blade horizontally or laid back against the wrist and forearm, both of which dramatically alter the type of movements that prove effective.

23. In connection with Pa and Shu see T'ung En-cheng, KK 1977: 2, 40, and for a recent report of a similar double scabbard recovered in Inner Mongolia, Hsiang Ch'un-sung and Li Yi, WW 1995:5. (Daggers from the latter site, which dates to the late Western Chou or early Ch'un Ch'iu, are fairly lengthy, roughly 40 cm., and have blades at least twice the length of the hilt.)

24. Chung Shao-yi, KKHP 1992:2, 130–131. (For representative examples see *Chung-kuo Ku-tai Ping-ch'i T'u-chi*, 13.)

25. Chung Shao-yi, KKHP 1992:2, 130–131.

26. For example, see Max Loehr, 1948, 132–142, and Lin Yün, 1986, 237–273.

27. Prominent archaeologists who have argued for, or simply asserted, indigenous origins include Lin Shou-chin and T'ung En-cheng. (See Lin Shou-chin, KKHP 1962:2, 75–84, and WW 1963:11, 50–55, and T'ung En-cheng, KK 1977: 2, 35–55.)

28. However, Chai Te-fang, KKHP 1988:3, 277–299, divides the north into three regions.

29. For a representative view, see Lin Yün, 1986, 237–241. For a study of the complex interactions marking the northern zone and northern China during the Shang and early Chou (including the effects visible in knives, daggers, and other bronze artifacts), see Yang Chien-hua, KKHP 2002:2, 157–174. The northwest also served as a conduit for Eurasian styles and developments.

30. Significant reports include Chu Yung-kang, WW 1992:12; Pei-ching-shih Wen-wu Kuan-li-ch'u, KK 1976:4, 251–253; Wu En, KKWW 2003:1, 21–30; and Chai Te-fang, KKHP 1988:3, 277–299.

31. Important articles on these daggers and their evolution include Chin Feng-yi, Pt. 1, KKHP 1982:4, 387–426, and Pt. 2, KKHP 1983:1, 39–54, as well as WW 1989:11, 24–35; Lin Yün, KKHP 1980:2, 139–161; Wang Ssu-chou, KK 1998:2, 53–63; Liao-ning-sheng Hsi-feng-hsien Wen-wu-kuan-li-suo, 1995:2, 118–123; Chao Chen-sheng and Chi Lan, KK 1994:11, 1047–1049; Tung Hsin-lin, KKHP 2000:1, 1–30; Chung Shao-yi, KKHP 1992:2, 129–145; Wang Ch'eng, KK 1996:9, 94 (which shows that these styles were sometimes found as far west as Inner Mongolia); and Chai Te-fang, 1988.3: 280–285. Guo Da-shu, 1995, 182–205, includes a well-illustrated examination of knives, daggers, and axes, especially those with ring-type handles. (Not all Liaoning swords

are so distinctive. For example, a damaged stone "sword" dating to about 2500 BCE, about 20 cm. long, was probably a thrusting weapon because it is distinguished by a rhomboidal cross-section and sharpened edges and tip. See Hsü Yü-lin and Yang Yung-fang, KK 1992:5, 395.)

32. Chin Feng-yi, WW 1989:11, 24–35. However, Wang Ssu-chou, KK 1998:2, 53–63, dates their inception to the late Shang or early Western Chou.

33. Chin Feng-yi, WW 1989:11, 24–27, and KKHP 1983:1, 46–48; Wang Ssu-chou, KK 1998:2, 53–63; and Chao Chen-sheng and Chi Lan, KK 1994:11, 1047–1049.

34. Fan Chün-ch'eng, KKWW 1995:5, 91. Even discounting the surpassingly long handle of 23 cm., the blade is still remarkably long for the Shang. However, the dating is dubious.

35. Ma Hsi-lun and Kung Fan-kang, WW 1989:11, 95–96. It has been termed the first Shang bronze sword discovered in this area.

36. Yang Hung, "Chien ho Tao," 116; SHYCS, "Feng-hsi Fa-chüeh Pao-kao," WW 1963. (Other daggers from the Beijing area are even shorter, at 17.5 cm.)

37. Based on calibrating the original radiocarbon dating of 1120 RC ± 90. (See Lin Shou-chin, 1962.2: 80–81. Lin thus significantly revises an original date of late Western Chou, as seen in Pei-ching-shih Wen-wu Kuan-li-ch'u, KK 1976:4, 251–253.) Two rather poor-quality bronze daggers with slightly raised spines about 21 cm. in length, tentatively dated to the mid- or late Shang, have also been found at Ch'eng-tu Shih-erh-ch'iao (Chiang Chang-hua, KK 1996:9, 78–79).

CHAPTER 17

1. The combination of the characters for shield and dagger-axe, *kan* and *ko*, would eventually become a common term meaning "warfare."

2. The *Shuo-wen Chieh-tzu* and its greatly expanded variant the *Shuo-wen Chieh-tzu Ku-lin* long provided the classic interpretations, but the discovery of oracular inscriptions produced an entirely new genre of literature in which interpretations of seemingly incomprehensible characters have been advanced and heatedly debated by Kuo Mo-juo, T'ang Lan, Yen Yi-p'ing, and others.

3. The idea entailed by this pronouncement (and accompanying, fabricated dialogue) probably dates to the middle Warring States period, if not later, when the *Tso Chuan* was compiled. For example, in "Benevolence the Foundation" the *Ssu-ma Fa* states: "Authority comes from warfare, not from harmony among men. For this reason if one must kill men to give peace to the people, then killing is permissible. If one must attack a state out of love for their people, then attacking it is permissible. If one must stop war with war, although it is war it is permissible."

4. *Tso Chuan*, Hsüan Kung, twelfth year.

5. Examples of stone and other dagger-axes may be conveniently found in the chapter on Shang weapons in *Chung-kuo Ku-tai Ping-ch'i T'u-chi*. The authors postulate (on 12) that the first stone *ko* appeared in Guangdong and may have originated as an agricultural implement.

6. For a succinct example of the process and complexity of mutual interaction in China's western area, see Liu Chün-she, KKWW 1994:4, 48–59. Oddities and variations have also been recovered, including one with a stone blade but bronze haft. (See Li Chi, KKHP 4 [1949]: 40–42.) Examples from the late Shang recently discovered in Shaanxi that strongly reflect the process of cultural interaction (with the Northern complex) include one *ko* with a single slot in the middle of the blade, a definite upward tilt, partial downward blade extension, flange extending both up and down, and a tab positioned to extend straight back at the top. A full crescent blade with three binding holes was also found at the site. (See Wang Yung-kang et al., KKWW 2007:3, 11–22 plus back interior cover.)

7. Recently, analysts such as Ching Chung-wei, KK 2008:5, 78–87, have begun to study the major types in order to understand the process of localized variation and stylistic evolution.

Although their conclusions are highly useful, many of the detailed changes did not impact the nature of the weapon or its utilization and, insofar as they would require extensive digressions to reproduce, are not included here. (Key reports for the Shang include Li Chi's well-illustrated classic article, BIHP 22 [1950]: 1–17, and his KKHP 4 [1949]: 38–51; Sun Hua, JEAA5 [2006]: 306–311; and Ching Chung-wei, KK 2008:5, 78–87. However, hundreds of individual reports with only one or two examples that were evaluated for size, evolution, and materials are too numerous to list.)

8. Although every style of *ko* has been recovered at Anyang, replica weapons or *ming ch'i* rapidly multiplied in the last two periods. (For example, see SHYCS An-yang Kung-tso-tui, KKHP 1994:4, 471–497. For an analysis of the increasing lead content and its effects, see Chao Ch'un-yen et al., WW 2008:1, 92–94.)

9. Liu Yi-man, KK 2002:3, 70; SHYCS An-yang Kung-tso-tui, KK 1991:5, 390–391.

10. Liu Yi-man, KK 2002:3, 63–75. Spears tended to proliferate at the very end of the Shang and sometimes exceeded the dagger-axes found in a single grave or locality, but overall the *ko* dominates. (For example, the tomb of a high-ranking Ma-wei nobleman in the Anyang area contains thirty *ko* and thirty-eight spears [SHYCS An-yang Kung-tso-tui, KK 1996:2, 17–35].)

11. For example, the tomb of the progenitor of the Ch'ang clan, an obvious military commander (SHYCS An-yang Kung-tso-tui, KK 2004:1, 7–19).

12. A certain amount of circularity marks these determinations, because rank and prestige are decided by the presence of such weapons, as well as ritual vessels and other minor factors.

13. Yang Hung, "Chan-ch'e yü Ch'e-chan Erh-lun," HCCHS 2000:5, 8.

14. See Huang Jan-wei, 1995, 188–190. *Ko* were sometimes given in combination with other weapons, including *kan* (a staff, but generally interpreted as a shield), *yüeh*, and bows and arrows.

15. Bernhard Karlgren made this point long ago in BMFEA 17 (1945): 101–144. *Fu* and *yüeh* both predate the *ko*.

16. Note that Shen Jung claims that rectangular openings were first seen at Yin-hsü, yet the tabs had long grown too large for simple push-through mounting (Shen Jung, KK 1992:1, 70).

17. Shih Chang-ju, BIHP 22 (1950): 59–65, rather imaginatively calculated a probable length of 1.128 m. by analyzing the characters that appear in the oracle inscriptions. However, remnants recorded in numerous archaeological reports suggest a shorter length of 0.85 to 1.0 meter was common. (For example, even though the blade lengths differ slightly, being 23 and 26 cm., at an early Western Chou chariot site the length for each of a matched pair was 82.5 cm. The shafts expand somewhat toward the butt; the upper 55.5 cm. are wrapped with black lacquered thread but the lower 27 cm. of the wooden shafts are just lacquered red [SHYCS Feng-hsi Fa-chüeh-tui, KK 1990:6, 504–510].)

18. Among many, see Ma Hsi-lun, WW 1995:7, 72–73; Shen Jung, KK 1992:1, 70–71; Li Chien-min, KK 2001:5, 62–64; and Ching Chung-wei, KK 2008:5, 79.

19. For examples, see Li Chi, KKHP 4 (1949): 38–40.

20. Three rather blunt *ko* styles have been recovered from the early dynastic Shang site of Lao-niu-p'o: basically straight *ko* with slightly curved blades, models with curved handles and elaborate patterns on the tab, and a socketed version. (Liu Shih-o, WW 1988:6, 1–22, WW 1988:6, 23–27. See also Kuo Pao-chün, KK 1961:2, 111–118.)

21. Socketed weapons have long been primarily associated with the so-called Northern complex; however, due to the discovery of certain indigenous examples, this identification needs to be reexamined.

22. A recently excavated Anyang tomb that dates to Wu Ting's era contains good examples of socketed, curved, straight-tabbed, and even triangular *ko*. (See An-yang-shih WWKK YCS, KK 2008:8, 22–33.) Moreover, even the excavations of the late 1950s turned up socketed, straight-tabbed, and curved-tab bronze *ko*, as well as a couple of crescent-shaped blades. (See SHYCS Pien-chi, *Yin-hsü Fa-chüeh Pao-kao*, 1958–1961, 245–249. The *ko* and other weapons recovered

were apparently produced in a specialized foundry.) Finally, one early version found in Fu Hao's tomb lacks vertical flanges but displays a slight downward turn in the blade's lower edge, even though the top edge and the top of the tab are essentially continuous.

23. Although Kuo Pao-chün early on claimed that the socketed version never achieved a high percentage in the Shang and rapidly diminished in the Western Chou, most of the 118 *ko* recovered from the tomb of a high-ranking Shang commander dating to the third period of Yin-hsü are socketed. (See SHYCS An-yang Kung-tso-tui, KK 1991:5, 390–391.) Separately, the 71 *ko* found in the tomb of an obvious military commander, perhaps the progenitor of the Ch'ang clan, dated to late in the second period at Yin-hsü, include variants with the full crescent elongation, simple straight-tab versions with rhomboidal blades, curved-tab models, and socketed versions. Lengths average 25 to 26 cm., with the curved tabs being slightly longer at 28.5 cm. (See SHYCS An-yang Kung-tso-tui, KK 2004:1, 14–16.)

24. A socketed *ko* with a stubby profile and short, crescent-style elongation has been recovered from remains dating to Yin-hsü's third period. (See Yin-hsü Hsiao-min-t'un K'ao-ku-tui, KK 2007:1, 41.) Ten of the thirty *ko* recovered from the tomb of the Ma-wei nobleman previously noted, dating to the fourth period at Yin-hsü, are crescent shaped with sockets that have an average blade length of 24 to 25 cm., an 8 cm. curved portion, and the usual 1:4 ratio of tab to blade. One with slots in the crescent still had leather binding remnants (SHYCS An-yang Kung-tso-tui, KK 2004:1, 14–16).

25. Li Chi, KKHP 4 (1949): 42–44; Kuo Pao-chün, KK 1961:2, 111; Ching Chung-wei, KK 2008:5, 78, 82. For examples see Ch'eng Yung-chien, WW 2009:2, 81–82 (two late Shang curved-tab variants and one straight tab), and illustrations 123 to 144 in Karlgren, BMFEA 17 (1945).

26. For the motif of tigers eating people, see Shih Ching-Sung, KK 1998:3, 56–63. As already discussed, the Hu people, with whom this motif is closely associated, were targeted by Wu Ting's campaigns, and an axe head found in Fu Hao's tomb, presumably imported from the south, depicts two tigers jumping at a man's stylized head.

27. The surprising effects of symmetrically weighted and balanced weapons have recently been the subject of Western technological articles. (For example, see Ronald Jager, *Technology and Culture* 4 [1999]: 833–860, particularly the discussion of balance on 838–840.) Although Shang dynasty *ko* never achieved true balance, some of the dynamic advantages (such as less tendency to rotate while being swung through an arc) should have been realized to some degree.

28. For a set of illustrations see Ching Chung-wei, KK 2008:5, 79–82.

29. This was first discussed by Li Chi in his classic article, BIHP 22 (1950): 1–17.

30. For some interesting examples from widely scattered sites, see (for Shandong) Ma Hsi-lun, WW 1995:7, 72–73; (for Anhui) Yang Te-piao, WW 1992:5, 92–93; and (from Wu Ting's era in Henan), Ning Ching-t'ung, WW 1993:6, 61–64. Several *ko* about 27 to 28 cm. in length recovered from P'an-lung-ch'eng show the degree to which *ko* in the Erh-li-kang had already evolved. Three have rhomboidal blades with straight tabs and slight downward curves, projecting flanges, and single tab holes; another has a straight tab but no curve in the blade; one 27 cm. long has a curved handle with a very open stylized motif, a short tab 4.8 cm. long and a blade about 4 cm. wide; two others, 26.8 and 29.2 cm. long, have hooked tabs with decorative designs, flanges, and rhomboidal blades. Finally, two termed *k'uei*, one 29.8 cm. and the other 28.4, though more triangular, are not really fat and bulgy; however, they have no flanges, just reduced tabs with binding holes and the usual lashing holes on the upper portion of the blade. (See Li T'ao-yüan, *Pan-lung-ch'eng Ch'ing-t'ung Wen-hua*, 2002, 130–135.)

31. For an example of a 45-degree blade from the Western Chou, see Pei-ching-shih Wen-wu Kuan-li-ch'u, KK 1976:4, 252.

32. See the references in note 23.

33. Although many articles include a few speculative statements, the classic discussion of dagger-axe fighting techniques remains Shen Jung's KK 1992:1, 69–75. Though fully cognizant

of his analysis, our conclusions are based primarily on experiments with replica weapons employed by experienced martial artists.

34. Wen Kung, eleventh year.

35. For example, see Yang Po-chün, *Ch'un-ch'iu Tso-chuan-chu*, 1990, Vol. 2, 582, or Chung Shao-yi, WW 1995:11, 55–56, who believes it was a cross-shaped *ko*. (It is possible to punctuate the sentence somewhat differently and read it as "struck his throat and used a *ko* to slay him.")

36. See Ronan O'Flaherty, *Antiquity* 81 (2007): 425–426.

37. Terming it a *"k'uei"* or *"chü"* based on later texts describing ritual weapons, although said to be incorrect, is common. (See Shen Jung, WW 1993:3, 78–84.)

38. Ch'eng-ku has been identified with the Pa culture (Huang Shang-ming, KKWW 2002:5, 40–45).

39. Shen Jung, WW 1993:3, 78–84; Chang Wen-hsiang, KKWW 1996:2, 44–49; Li Hsüeh-ch'in and Ai Lan, WW 1991:1, 20–25. However, Sun Hua, JEAA 5 (2006): 310, asserts that the *k'uei*-style *ko* originated in Sichuan rather than Ch'eng-ku. In contradiction, Li Chien-min, KK 2001:5, 60–61, sees the basic type originating in the Hsia but being modified at Hsin-kan. Chen Fangmei also regards the leaf-shaped *ko* at Hsin-kan as a distinctive type. (For context, see also Zhang Changshou, JEAA 2, nos. 1–2 [2000]: 251–272.) For an example of a very simple *k'uei* with a long tab and a single hole, 18 cm. long by 7 cm. wide, that dates approximately to Wu Ting's era, see Ning Ching-t'ung, WW 1993:6, 61–64.

40. See Shen Jung, WW 1993:3, 78–84. (Ch'eng-ku has an early crescent-shaped *ko* with four lashing holes.)

41. For an example with this exact size from Ch'eng-ku, see Kou Pao-p'ing, KK 1996:5, 50. Unlike the original form, there tends to be much less variation in size and extremely long versions are rare, no doubt because of the large amount of copper required and the greater weight.

42. Recent experiments to determine the lethalness of the much-maligned Irish halberd—a piercing rather than crushing weapon—conducted on sheep heads led to the conclusion that the "thought alone" assessments that had disparaged it as a clumsy, ineffective, or perhaps purely symbolic or ritual weapon due to weak hafting, blade fragility, and other factors were wrong. It was further determined that the somewhat rounded tip (similar to the relatively blunt-shaped triangular *ko*) provided strength on impact, was designed to strike bone rather than muscle, and easily penetrated the skull. (See Ronan O'Flaherty, *Antiquity* 81 [2007]: 423–434.)

43. Shen Jung, WW 1993:3, 80–82.

44. The character, which is said not to have existed prior to the Eastern Chou, was written in several ways during the Warring States period, including with *ko* on the right and a chariot pennant on the left (as discovered on the two famous *chi* bearing Shang Yang's and Lü Pu-wei's names), but always included the *ko*. (See Chung Shao-yi, WW 1995:11, 58.)

As Yang Hung pointed out in his "Chung-kuo Ku-tai te Chi," prior to the discovery of actual specimens, the *K'ao-kung Chi*'s discussion of the *chi*'s design prompted some strange misconceptions. Only a few other dedicated articles have pondered the *chi* to date, including three classics: Ma Heng, "Ko Chi chih Yen-chiu," 1929.5: 745–753; Kuo Muo-jo, "Shuo Chi," 1931 (1954 reprint), 172–186; and Kuo Pao-chün, BIHP 5.3 (1935); Chung Shao-yi's recent study, WW 1995:11, 54–60, provides a useful review of the literature and changing opinions about its design, nature, and nomenclature.

45. P'eng Shih-fan and Yang Jih-hsin, WW 1993:7, 14, note that the so-called hooked *chi* (called a *k'uei* by Tsou Hung), which combines a *tao* with a *ko* rather than a spear (seen in a singular example from Pao-chi Yü-ch'üan), is not Shang in origin. It should be remembered that (at least in contemporary martial arts practice) Chinese spears are not just employed for thrusting, but are also frequently used in a slashing and cutting mode.

46. The earliest example has been found at Kao-ch'eng T'ai-hsi in Hebei. (See Ho-pei-sheng Po-wu-kuan, WW 1974:8.) The discovery was quite fortuitous because the warrior's body had

a socketed bronze *ko* on his right side and the combined dagger-axe/spear on his left. In the absence of shaft remnants or impressions, the discovery of the two parts would normally not have led to the conclusion that they comprised parts of a combined weapon. (Yang Hung [157] remarks that the *ko*'s shaft was 87 cm. but the *chi*'s only 64. However, the addition of the spearhead would result in an equivalent weapon length.)

47. Chung Shao-yi, WW 1995:11, 54–60, especially 55–56, asserts that the *chi* that resulted from extending the top portion of the *ko* blade upward in a curve, producing a point, was a distinctively Western Chou weapon that proliferated in the Kuan-chung and central plains areas in the early and middle periods and disappeared in the Spring and Autumn. Moreover, according to evidence found at Liu-li-ho, it should be called a *ko* and therefore named a "cross-shaped *ko*" to distinguish it from the latter unitary *chi* of Spring and Autumn invention.

48. Li Chi, BIHP 22 (1950): 15. Kuo Muo-jo, "Shuo Chi," 179, 182, once claimed that the *ko* had been discontinued by the Han, only *chi* with one or two heads being employed.

49. Yang Hung, "Chung-kuo Ku-tai te Chi," 161; Kuo Muo-jo, "Shuo Chi," 177.

50. For further information on the multiple-*ko* variant of the *chi*, see Sun Chi, WW 1980:12, 83–85, and Chung Shao-yi, WW 1995:11, 59. Considerable controversy exists over the exact term for these multiple-*ko* weapons and whether they require a spear point to be termed a "*chi*" or not, with claims for the latter being based on the discovery of thirty *chi* at Tseng Hou-yi. (See, for example, Chung Shao-yi, 60, or T'an Wei-ssu, *Tseng Hou-yi Mu*, 2001, especially 52–56.)

51. Sun Chi, WW 1980:12, 84.

52. Chung Shao-yi, WW 1995:11, 59 and Sun Chi, 84.

53. In numerous experiments conducted with replicas, the short Shang-style *chi* was felt to be awkward in either of its modes. Effectively exploiting this dual capability requires learning new rotational hand and arm movements, as well as developing the ability to deliver a thrust with the hand turned to disadvantageous positions of lower leverage. (Longer two-handed Spring and Autumn versions do not suffer from this defect because short arc attacks no longer exist, only long-range hooking, piercing, or thrusting movements.) Moreover, the fighter must essentially precommit to either a slicing or thrusting attack because the two possibilities do not equally present themselves in combat situations.

CHAPTER 18

1. Twentieth-century military experience in Southeast Asia attests to the lethality of sharpened sticks—essentially mini-spears—concealed in rice paddies and jungle underbrush. Guerilla actions also involved employing roughly fashioned spears against a variety of forces, and sharply cut bamboo culms continue to be used in street melees and against the police, causing frequent deaths and numerous wounds.

2. Claims that their numbers decreased in the Western Chou with the ascension of chariot combat are common. (For example, see Shen Jung, KKHP 1998:4, 456–458.) However, Western battle experience contradicts arguments for the spear's inapplicability that tend to be made with reference to the effective hooking power of the crescent-shaped *ko*. (Because they are used by thrusting rather than swinging, it can equally be argued that they would be more effective against horses—the usual target—and chariot riders, especially in the congested confusion of a melee when the chariots would have virtually been brought to a halt.) Moreover, in comparison with the masses of fighters clogging the battlefield, even the largest chariot force represented only a fraction of the total combatants.

3. As discussed below, the *p'i* featured a cast bronze spearhead with a tab that was inserted into a shaft until socket mounting was adopted late in the Spring and Autumn period.

4. *Chung-kuo Ku-tai Ping-ch'i T'u-chi*, 26 (based on traditional explanations).

5. Hayashi Miano, *Chugoku Inshu Jidai no Buki*, 1972, 99.

6. Shen Jung, KKHP 1998:4, 452.

7. Li Chien-min, KK 2001:5, 68, notes a somewhat abstract, elongated type found at San-chia-chuang east that has been attributed to Yin-hsü's first period.

8. This unusually long, basically rhomboidal stone spear has a willowy blade that merges into a somewhat rectangular lower (shaft) portion. Dated to about 1000 BCE, it is the largest ritual weapon found in China, the next being an unimaginably oversized stone *ko* of 1.0 m. Fabricated from one piece of smoothly worked sandstone, it has a total length of 120.8 cm. apportioned as a head section of 48.8 cm. and a shaft of 72 cm.; a maximum blade width of 6 cm.; and a very moderate thickness about of 1.2 cm. (Wang Ssu-chou, KK 2006:2, 95–96).

9. Short protrusions that lie in the same dimension as the blade just below, separated by a slight gap, almost like a detached flange, appear on the two *pi* specimens from P'an-lung-ch'eng (shown in *P'an-lung-ch'eng Wen-hua*, 138). However, rather than having lashing holes, the ends are unexpectedly solid. One has a length of 24 cm., the other 23. (*P'i* are rarely discussed in archaeological reports, and there seems to be some disagreement over their defining characteristics. Comments in *P'an-lung-ch'eng Wen-hua* distinguish it from the *mao* solely by the crosslike protrusions and solid shaft, requiring insertion into the shaft rather than mounting over the handle. However, weapons identified as *p'i* that were commonly employed in the Warring States tended to have socket-mounted, elongated, willow-like blades and lengths up to 1.62 m., and some from Ch'in Shih-huang's tomb with a pike- or javelin-like appearance exceed 3.5 m., including their bronze heads of 35–36 cm. None of these later variants have a crosspiece, only a sort of rim at the base. The *p'i* reportedly died out as a battlefield weapon after the Western Han.)

10. Li Chi's "Yin-hsü Yu-jen Shih-ch'i T'u-shuo" (BIHP 23b [1952]) does not show any spears.

11. For example, while Shen Jung sees a major northern contribution, Li Chien-min KK 2001:5, 65, claims the Shang spear is simply an amalgamation of the southern model plus a few Shang elements.

12. Even though Hayashi Minao's comments are not without value, two seminal studies have appeared in the last decade: Shen Jung, KK 1998:4, 447–464, which contains an evolutionary chart that may be overly systematic, and Li Chien-min, KK 2001:5, with a discussion of spears on 65–68.

13. Descriptions of the three will be found in Ho-pei-sheng Po-wu-kuan, WW 1976:2, 33; the clearest photographs appear in *P'an-lung-ch'eng Ch'ing-t'ung Wen-hua*, 136 and 137.

14. Based on binding being the most common method for securing axe and dagger-axe heads, Shen Jung, KKHP 1998:4, 458–460, concluded that the ring holes were simply employed for lashing. (When pegging subsequently dominates, the ears will be employed for pennants and streamers.)

15. Of the two with "ears," the one with the less pronounced rim is 23.6 cm. long, slightly oval, and has a significantly raised spine that extends right to the tip; the other is slightly shorter at 20.2 cm. The third, with upwardly oriented triangles molded onto the shaft, is 22 cm. long with a blade to socket ratio of 2:1; has an essentially rectangular spine with a somewhat rhomboidal cast that runs down most of the length; and includes two protruding upward-hooked flanges at the base which, in a sword, would be intended to stop enemy blades from sliding down onto the warrior's hand.

16. One example discovered in Hubei that lacks ears is particularly interesting because the blade's multihued patina suggests it underwent some form of secondary heat treating. Extremely basic in design, it has a long shaft with a small symmetrical leaf blade, relatively rounded core with raised spine, and slightly raised smaller leaf section on the blades that serves for the relatively simple *t'ao-t'ieh* decorative pattern that is repeated on the lower mounting portion. The length

is 28.7 cm., blade width 7 cm., socket diameter 3.4 cm., and weight 0.55 kg. (See Ho Nu, WW 1994:9, 90 and 91.)

17. Li Chien-min, KK 2001:5, 65–68. Although the specimens clearly reflect the southern spear tradition, local elements loom large, the strongest influence perhaps being from Wu-ch'eng. (The spearheads at Wu-ch'eng and Hsin-kan are said to date to approximately the second period at Yin-hsü.) Twenty-seven can be broken down into subtypes depending on the relative length of the shaft and blade, although they range from 11.8 to a maximum of 30.5 cm., as one cluster falls around 15 cm., another 24–25 cm. The ones with ears are generally short to moderate in length, those without ears often very elongated. Many have blood grooves, some have decorations, and at least one widens out, then the blade edges continue in a sort of indentation down to the rim, as in the late Shang. Although not called *p'i*, two variants have crosspieces with holes in the ends rather than ears. No weights are given. (For further discussion of the merged influences seen at Hsin-kan, see Chan K'ai-sun, CKKTS 1994:5, 34–40.)

18. Li Chien-min, 67.

19. The integrated view has been particularly espoused by Shen Jung, KKHP 1998.4: 452, whereas Liu Yi-man, KK 2002:3, 65, and Li Chien-min (65) assert that the Shang spear is derived from southern precursors. Others, while generally accepting the latter viewpoint, see indigenous Shang stone and bone spears as having exerted a residual effect.

20. For example, three spears dating to the fourth period at Yin-hsü that have somewhat pudgy but moderately long triangular blades and a gradually expanding rhomboidal shaft with pronounced ears lack pegging holes and should therefore be considered purely southern-style embodiments. (For a report see SHYCS An-yang Kung-tso-tui, KK 1993:10, 884.)

21. For examples that date to the fourth period at Yin-hsü, see Yin-hsü Hsiao-min-t'un K'ao-ku-tui, "Ho-nan An-yang-shih Hsiao-min-t'un Shang-tai Mu-tsang 2003–2004-nien Fa-chüeh Chien-pao," KK 2007:1, 32, which depicts two stretched rhomboidal spearheads that are vividly decorated with facelike deigns on the very base and upwardly pointing triangles along the spine. (The shaft openings are somewhat distorted circles.) Comparable in size, one is 25 cm. long with a maximum 5.8 cm.–wide upper blade, the other 26.3 cm. with an identical 5.8 cm. width.

22. Li Chien-min, KK 2001:5, 65–68. Spears recovered from Commander Ch'ang's tomb include southern specimens with ears; a roughly circular spine; but short, stubby, flat leaf blades about 24 cm. in length, and variants with rhomboidal spines and blades with extended lower edges that reach down to the ear holes before cutting sharply inward; they are slightly longer at around 28 cm. (See KK 2004:1, 16.)

23. Hayashi Minao, 1992, 97, and Liu Yi-man, KK 2002:3, 65, claim that the character *mao* does not appear in the oracular inscriptions. However, Karlgren (KGSR1109) interprets the character as a kind of lance and provides a Yin dynasty bone version whose existence, if attested, would contradict assertions of nonexistence.

24. Two spears were among the weapons recently found in a set apparently belonging to a foreign nobleman who had settled in the Anyang area. The leaf-type blades are short but marked by a smooth curve at bottom, while the shafts, which have "ears" on the bottom, are comparatively longer. In one the socket opening is relatively circular, in the other slightly oval. The lengths are given as 21.3 and 21.6 cm., with outer socket diameters of about 3.5 and 3.6 cm. Only one weight is given, 295 grams (An-yang-shih WWKK YCS, KK 2008:8, 22–33). The one found together with a *yüeh* at Ta-ssu-k'ung is marked by *t'ao-t'ieh* decorations at the bottom, protruding ears, a fairly long but narrow blade that quickly curves inward, and a length of 22.2 cm. (SHYCS An-yang Kung-tso-tui, KK 1992:6, 513–514).

25. Yen Yi-p'ing, 1983, 7–10. Tomb 1004 at Hou-chia-chuang contains a significant quantity of armor, shields, and helmets. Some of the short shafts of the 370 upper-layer bronze spears show red coloring.

26. The spears can be divided into two types: a short, stubby, leaf-shaped blade with a roughly circular spine and separate ears at bottom about 23–24 cm. in length, and a longer, flat, leaf-type blade whose edges extend down to the ear holes before cutting inward, has a rhomboidal spine, and averages about 28 cm. in length (SHYCS An-yang Kung-tso-tui, KK 2004:1, 16–17).

27. The spears can again be divided into two styles. The first has a rather broad blade that curves somewhat inward before flaring outward again on the lower third of the shaft, has two lashing holes in the blade itself, and has a pronounced oval socket/spine that extends halfway up to the tip. In the second the blade suddenly flares out then cuts inward, turning slightly upward; shafts are generally rhomboidal or extended ovals, but there are no ears or lashing holes (SHYCS An-yang Kung-tso-tui, KK 1996:2, 380–381).

28. Few articles have pondered the earliest forms of Chinese armor: Yang Hung's "Chung-kuo Ku-tai te Chia Chou" and "Chung-kuo Ku-tai te Chia Chou te Hsin Fa-hsien ho Yu-kuan Wen-t'i" and Albert E. Dien's "A Study of Early Chinese Armor." The *Chung-kuo Ku-tai Ping-ch'i T'u-chi* provides essential illustrations, while an odd, highly colorful reconstruction of a Shang warrior's garb may be found in *Chung-kuo Ku-tai Chün-jung Fu-shih*, 1. (However, it should be noted that the colors on some recently excavated late Warring States tomb figures are quite vibrant, indicating that the Chinese martial realm did not confine itself to somber embellishments.)

29. Yang Hung, 1985, 3; Yen Yi-p'ing, "Yin Shang Ping-chih," 1983, 3.

30. For an oversized illustration that is also reproduced in the *Chung-kuo Ku-tai Ping-ch'i T'u-chi*, 39, see Yen Yi-p'ing, 3.

31. The arrangement of the shields is shown on Yen Yi-p'ing's "Yin Shang Ping-chih," 1983, 4.

32. For a depiction see *Chung-kuo Ku-tai Ping-ch'i T'u-chi*, 39. An additional shield in dark red lacquer has been found in Shandong at T'eng-chou Ch'ien-chang. (See Yang Hung, HCCHS 2000:5, 8.)

33. About 80 cm. high, it was 65 cm. wide at the top but tapered outward to 70 cm. at the bottom and therefore did not form a perfect rectangle. Shih Chang-ju's famous reconstruction of this shield, BIHP 22 (1950): 65–69, upon which all discussion to date has been based, is also cited by Yen Yi-p'ing, 7, and included in the *Chung-kuo Ku-tai Ping-ch'i T'u-chi*, 38.

34. The only large catch yet discovered has been at Hou-chia-chuang's M1004, where 141 helmets in seven discernible styles were recovered. (Unfortunately, little attention has been paid to these helmets subsequent to the initial report, but for a brief discussion see "Chung-kuo Ku-tai te Chia Chou," 8–9, and *Chung-kuo Tu-tai Ping-ch'i T'u-chi*, 39–41, for photos.) Though none have been found at P'an-lung-ch'eng, a similar helmet with an even more exaggerated appearance has been recovered at Hsin-kan. (See P'eng Shih-fan and Yang Jih-hsin, WW 1993:7, 11 plus a separate photo. Yen Yi-p'ing also provides an illustration of a helmet from M1004 in his "Yin Shang Ping-chih," 8. The *Chung-kuo Ku-tai Chün-jung Fu-shih* shows a short version on page 1 and other Shang helmets on page 2.)

35. Even though China has a lengthy tradition of employing masks in theater and street display, bronze face masks seem to have been extremely rare in antiquity, as well as virtually unknown in pre-imperial written materials. (The *Chung-kuo Ku-tai Ping-ch'i T'u-chi* [38] has a photo of a round bronze plate in the shape of a human face with large eye holes that was intended as a shield decoration, and a picture of a rare bronze face mask similarly characterized by large eyeholes appears on 41.)

CHAPTER 19

1. Various dates have been given for the earliest Chinese arrowhead discovered at Shuo-hsien in Shanxi, including Yang Hung's 28,945 BP (1985, 190, based on Chia Lan-p'o KKHP 1972:1, 39–58) and subsequent radiocarbon dating (SHYCS Yen-shih, KK 1997:3).

2. Two important reports among the many that have described ancient finds are Wang Chien et al., KKHP, 1978:3, which discusses black flint arrowheads from a site dated roughly to 23,900–16,400 BP, and Hsi K'o-ting, KK 1994:8, 702–709, which reports on a Paleolithic site dated to 18,000–12,000 BP.

3. For example, see K. C. Chang, *Archaeology of Ancient China*, 93, for a summary of P'ei-li-kang sites dating to about 6,500 to 5,000 BCE where, despite extensive agricultural development and the raising of domesticated animals, bone arrowheads and spearheads show that hunting was still important.

4. One Ma-chia-pang site in northern Zhejiang dated to 5,000 BCE contains seven wooden arrowheads as well as others fabricated from bone and antlers (Chang, *Archaeology of Ancient China*, 201).

5. Rather than a surpassing effort to save civilization from some sort of environmental disaster, this legend is interpreted as a vestigial memory of the Yi people having vanquished nine enemy tribes. (The character for Yi depicts a man bearing a bow on his back, suggesting that they were particularly distinguished for their archery skills.)

6. See, for example, Jim Bradbury, *The Medieval Archer*, especially chapters 1 and 9.

7. "Chou Pen-chi," *Shih Chi*. He similarly shot Chou's two concubines with three arrows each.

8. See Yang K'uan, *Hsi Chou Shih*, 1999, 701–704.

9. "She Yi," *Li Chi*, translated by James Legge, 448, used here because of the sonorous quality of his translations of the ritual writings. (A late Warring States compilation, the *Li Chi* heavily reflects Confucian theory and idealizations. Although an unreliable guide to historical practice, the contents reveal archery's deep psycho-emotional importance.)

10. Li Hsüeh-ch'in, WW 1998:11, 67–70. (Others attribute the vessel to King Mu's reign.)

11. Huang Jan-wei, 1995, 189–191. Although generally awarded in combination, bows and arrows were also granted separately.

12. *Tso Chuan*, Duke Wen, fourth year.

13. *Tso Chuan*, Duke Hsiang, eighth year.

14. *Tso Chuan*, Duke Ting, eighth year.

15. *Tso Chuan*, Duke Chao, first year.

16. *Tso Chuan*, Duke Chao, twenty-fifth year.

17. The issue, first raised by Kelly DeVries (1997, 454–470), prompted a rebuttal by Cliff Rogers (1998, 233–242), who cited extensive evidence of knights and other armored soldiers having been slain by arrows in the medieval period. (For further discussion of the bow's efficacy see, for example, Jim Bradbury, *The Medieval Archer*, and Robert Drews, *The Coming of the Greeks*, especially 113–129.)

18. If 8 inches are taken as the equivalent of one Chou foot, based on the average bow size of 6 feet 3 Chou inches (per the *K'ao-kung Chi*), the range (computed at 50 modern inches for a bow) would have been slightly less than 210 feet. In comparison, the great archery competition at Sanjusangendo in Kyoto took place on the great wooden temple's verandah over a total distance of some 375 feet. Because the overhanging roof precluded arcing the shot more than about 15 degrees, great initial velocity and thus a very powerful bow were required. However, because the archers shot all day and the victor's hits numbered in the thousands, it was also a test of endurance.

19. At 50 paces the large center square probably reached one *chang* or 10 Chinese feet per side or roughly 80 inches. (Based on a range of fifty bow lengths and a bow length of 6 feet, Steele, *Yi Li*, 120, concludes the range was 300 Chou feet and the complete target an enormous 40 Chou feet in width. Some commentators, including the compilers of the *K'ao-kung Chi*, believe the target's shape approximated that of a man with his arms and feet stretched out and was therefore 8 feet at the top and 6 feet at the bottom, but this seems unlikely in a strongly Confucianized ritual competition.)

20. "What can kill men beyond a hundred paces are bows and arrows." ("Discussion of Regulations," in the *Wei Liao-tzu*, a Warring States work.) Throughout Chinese history bows with a maximum effective range of just 100 paces were considered inferior. (Note "Ch'i-hsieh" in the *Wu-pei Chih*.)

21. See Wang Ching-fu, *K'ung Meng Yüeh-k'an* 23: 4 (1984): 56.

22. In another display of power and accuracy, while hunting the king killed a "rhinocerous"—probably some sort of wild buffalo—that attacked his chariot with one shot ("Ch'i-yü, 1," *Kuo-yü*).

23. Duke Ai, second year.

24. Duke Ch'eng, sixteenth year.

25. The archer had even had a prophetic dream about the incident and his own dire fate. Eye wounds seem to have been fairly common. (See, for example, *Tso Chuan*, Duke Chao, fifth year.)

26. Duke Hsüan, fourth year.

27. Duke Ch'eng, sixteenth year. (Although the incident may be a later fabrication inserted by antiwar editors during the *Tso Chuan*'s compilation, it may well reflect prebattle bow tests. It is also recounted in the *Lieh-nü Chuan*, showing that it struck the imagination.)

28. Duke Hsiang, eighteenth year.

29. Duke Ai, fourth year.

30. For example, see Duke Chao, twenty-eighth year, and Duke Ai, sixteenth year.

31. For example, in the "Hsi Chou" section of the *Chan-kuo Ts'e* it is pointed out to an archer who successively hit a willow leaf at a hundred paces that his achievement stems from combining strength and concentration, so if either fails his accomplishments will be dashed. (This insight would subsequently be adapted as a Taoist-flavored persuasion on true skill and not being skillful.)

32. Noticeable in the "Ch'i-yü" section of the *Chan-kuo Ts'e*.

33. Consecutive queries during Wu Ting's era ask whether an ancestor of the Chü clan or if another man named Pi should be ordered to instruct 300 archers (HJ5772; for further discussion see Ho Ching-ch'eng, KK 2008:11, 54–70).

34. "Determining Rank."

35. "Encouraging the Army," *Liu-t'ao*.

36. "Tactical Balance of Power in Defense," *Wei Liao-tzu*.

37. "Controlling the Army."

38. "Martial Cavalry Warriors," *Liu-t'ao*.

39. "Preparation of Strategic Power," *Sun Pin Military Methods*. When China developed the crossbow in the Warring States period, it immediately gained a decided, if temporary, advantage in range over mounted steppe peoples even though the crossbow had a much slower rate of fire.

40. The most important include Shih Chang-ju's pioneering reconstruction of Shang bows and arrows in "Hsiao-tun Yin-tai te Ch'eng-t'ao Ping-ch'i"; Yang Hung's "Kung ho Nu" (1985), especially 190–206; and *Military Technology: Missiles and Sieges*, 101–119. Unfortunately, dramatically in contrast to the masterful section on the crossbow, the discussion is surprisingly cursory.

41. The *K'ao-kung Chi*, an enigmatic work filled with archaic terms but often cited in technological histories and increasingly deemed well founded and informative (Kao Chih-hsi, WW 1964:6, 44), is thought to be a second-century BCE compilation. (Although it concisely ponders the issues and techniques of bow and arrow making, as a seventeenth-century BCE effort the *T'ien-kung K'ai-wu* is of limited use for the Shang.) A number of Western works on traditional bow and arrow making are also useful, including Flemming Arlune's *Bow Builder's Book* and Tim Baker's *Traditional Bowyer's Bible*.

42. See Shih Chang-ju, BIHP 22 (1950): 33–35. (Shih's analysis of the characters for bow and shooting, his reconstruction of the bow's shape and dimensions, and other fundamental conclusions have become so generally accepted that their tentative basis has largely been forgotten.)

43. Shih Chang-ju, 35.

44. Shih Chang-ju, 35. Based on proportions shown in oracular and bronze inscriptions, Shih Chang-ju reached this conclusion in his 1950 article. Nevertheless, *K'ao-kung Chi* materials apparently confirm his conclusions, however imaginatively based, and early Western bows had similar dimensions.

45. Shih Chang-ju, 35.

46. "Ta Lüeh," *Hsün-tzu.*

47. Commenting on these bows, the *Chih* notes that although the names might have been different, the bows were actually the same.

48. These gradations refer to martial capabilities rather than rank.

49. "Ch'i-hsieh, 1," *Chih.*

50. See, for example, Yang Hung's overview in "Kung ho Nu," 1985, 203–206.

51. Translation of these wood species follows *Military Technology*, 110.

52. China has well over 200 species of bamboo, ranging from 6-inch grasses to monstrous eighty-foot culms with 8-inch diameters, many of which provide materials suitable for applications as diverse as weaving and timbering. Larger species such as *Meng-chung* can be split into resilient laminates nearly 0.5 inch thick and 1.5 inches wide that have high tensile strength and well resist fracturing and chipping. (The World War II production facilities at Ch'eng-tu were still using bamboo cores for their compound bows.)

53. As pointed out in *Military Technology*, 111–112.

54. The *T'ian-kung K'ai-wu* similarly emphasizes seasonality in making bows and weapons.

55. The relative strength and flex of the arms had to be adjusted. (However, it is said that the upper limb of a bow should be slightly stronger than the lower one so that it will lift the arrow upon release. [See Alrune, "Bows and Arrows 6000 Years Ago," 18, or Jorge Zschieschang, "A Simple Bow," 39–45.])

56. Alrune, *Bow Builder's Book*, 18; Zschieschang, 39–45.

57. A scientific study of the tensile and sheering strength of the various materials in archery application, as well as the impact of seasonal specificity, is still awaited.

58. T'an Tan-chiung's study, BIHP 23 (1951): 199–243, of essentially the last traditional bow shop in 1942, preserves valuable information on how the materials were selected, processed, and assembled. It also describes the steps involved in arrow fabrication and adds a short note on shooting techniques, numerous useful diagrams, and a table of terminology. Traditional European bow makers similarly advise cutting the timber between late November and mid February, even coordinating with the waning phases of the moon when "the wood is drier and more resistant" (Konrad Vögele, "Woods for Bow Building," 102). Natural drying can take up to three years, showing that early Chinese methods were neither extreme nor unfounded.

59. "Ch'i-hsieh, 1," *Wu-pei Chih.*

60. The episode is found in the records for both Chao and Wei in the *Chan-kuo Ts'e*, as well as in various forms in the *Shih Chi* ("Han, Chao, and Wei Shih-chia"), the *Huai-nan Tzu* ("Chien Hsün"), *Shuo Yüan* ("Ch'üan Mou"), and other parts of the *Han Fei-tzu* ("Shuo Lin, Shang" and "Nan San"). However, the two main versions appear in the *Chan-kuo Ts'e*, right at the beginning of the "Ch'ao Ts'e," and in the *Han Fei-tzu* as part of the "Shih Kuo" or "Ten Excesses" chapter. It ranks among the most famous Warring States stories, well-known throughout the centuries, and has been translated by Burton Watson ("The Ten Faults," *Han Fei-tzu Basic Writings*, 56–62) and J. I. Crump (*Chan-kuo Ts'e*, #229 and #230, 278–283.)

61. Jürgen Junkmanns, "Prehistoric Arrows," 58.

62. For a useful discussion of parameters and practice, see Wulf Hein, "Shaft Material: Wayfaringtree Viburnum," 75–80.

63. Although cracking and crazing are less of a problem for small-diameter bamboos that can be employed as simple shafts, drying methods for bamboo tend to be relatively complex and time-consuming. The high sugar content attracts insects and sustains bacterial growth,

while the culms have a tendency to split if the moisture content is reduced too quickly, particularly in the timber bamboos.

64. Shih Chang-ju, BIHP 22 (1950): 45.

65. Calculated by Shih Chang-ju and reported in his "Hsiao-t'un Yin-tai te Ch'eng-t'ao Ping-ch'i." Early Western arrows were frequently this length and had a maximum diameter of about 0.9 cm.

66. For a discussion see Flemming Alrune, "Bows and Arrows 6000 Years Ago," 30–32.

67. According to the T'ien-kung K'ai-wu, the best feathers come from eagles, next hawks, lesser owls, wild geese, and swans. Because southern birds were not considered as powerful as northern ones, it was believed that southern arrows would not fly as true as northern variants.

68. To date the key article has been Yang Hung, "Kung ho Nu," 1985, 190–232, to which may be added Chang Hung-yen, KK 1998:3, 41–55, 75. Illustrations of the major types may be found in the Chung-kuo Ku-tai Ping-ch'i T'u-chi, Ch'en Hsü and Yang Hsin-p'ing, 2000, 218–232, and the articles noted below.

69. See Li Shui-ch'eng, KKHP 2005:3, 260. However, the late Hsia and the Shang were considerably warmer, possibly comparable to Hang-chou today, an area where bamboo proliferates.

70. Just 2.8 cm. long and dated to 28,845 BP, as already noted the earliest arrowhead yet found was discovered at Shuo-hsien in Shanxi (Yang Hung, "Kung ho Nu," 190).

71. See "Kung ho Nu," 193, reporting a find at Shanxi Ch'in-shui Hsia-ch'uan whose artifacts date to between 23,900 and 16,400 BP. Nine of the thirteen arrowheads have a sort of elongated oval shape; the remainder have the classic triangular profile and sharply tapered bottom edges for inserting into split shafts, a clear advance over earlier forms. However, the number of arrowheads, 13 out of 1,800 recovered objects, is statistically insignificant.

72. For example, for the Paleolithic see T'ao Fu-hai, KK 1991:1, 1–7, or Chang Hung-yen, KK 1998:3, 41–55, 75. (For the production of flint arrowheads, see Hein, "Arrowheads of Flint," 81–95.)

73. Yang Hung, "Kung ho Nu," 192. For a report on the earliest bone arrowhead yet recovered (dating to the Paleolithic, 18,000–12,000 BCE) see Hsi K'o-ting, KK 1994:8, 702–709. Every variety of bone ranging from large animal to human was employed.

74. Yang Hung, "Kung ho Nu," 196, notes that a bone arrowhead discovered at Jiangsu P'ei-hsien Ta-tun-tzu dating to roughly 4500 BCE had penetrated 2.7 cm. into the victim's thigh bone.

75. For example, most of the arrowheads cited by Chang Hung-ch'an, KK 1998:3, 41–55, 75, dating to the eighth and seventh millennia BP from the Liaoning area are small and triangular; relatively few elongated variants are included. (However, others from Hsin-leh in Liaoning that date to 7300–6800 BP, half of which were produced by grinding, are quite elongated.) Samples from a Hung-shan site dated to 5485 BP continue to be small triangles, though some show a slight upward turn in the inner middle at the bottom.

Most of the well-made arrowheads recovered from Hai-la-erh-shih in Inner Mongolia dating to 6000–5500 BP have a slightly elongated triangular shape and are relatively flat with slightly raised spines. Fabricated by pressure flaking, chipping, and polishing, they continued to be only 4.1 to 4.5 cm. long by 1.2 to 1.3 cm. wide. (See SHY Nei-Meng-ku Kung-tso-tui, KK 2001:5, 3–17.)

76. For example, artifacts from Ta-wen-k'ou culture located in Shandong, roughly dated as 3835 to 2240 BCE, include all three. (See Yang Hung, "Kung ho Nu," 193.)

77. Yang Hung (193) believes it would have increased the arrow's power.

78. "Kung ho Nu," 193, and visible in numerous other reports. However, this is a somewhat unexpected result since stone, unlike bone and shell, is impervious to decay.

79. Examples discovered in Ta-wen-k'ou culture out in Shandong (dating to about 3000–2500 BCE) are noted in Shan-tung-sheng WWKK YCS, KK 2000:10, 38–39. Well-formed bone arrowheads characterized by good consistent angles have also been recovered from Han-tan. (See Ho-pei-sheng Wen-hua-chü Wen-wu Kung-tso-tui, KK 1961:4, 197–202.)

80. For a few examples see Shan-tung-sheng WWKK YCS, KK 2000:10, 27; Fu-chien-sheng Po-wu-kuan, KKHP 1996:2, 182–183.

81. For example, as late Liang-chu was transitioning to Kuang-fu-lin around 2000 BCE or slightly later, their stone arrowheads increasingly assumed a somewhat lengthened triangular shape with a t'ing that is formed simply by curving in and extending downward rather than being clearly circular or rhomboidal. Kuang-fu-lin variants then increase in length and include a few very elongated specimens. Average lengths vary from about 4.8 to 5.5 cm.; not all of them include a t'ing. (For artifacts see Shang-hai Po-wu-kuan KK YCS, KK 2008:8, especially 9–10 and 18–19.)

82. Useful reports include Pen-hsi-shih Po-wu-kuan, KK 1992:6, 506, whose specimens range from 4.7 to 6.3 cm. in length, including t'ing of a very short 0.7 to 2.0 cm. and body widths of nearly 2 cm., with some even narrower variants; Liao-ning-sheng WWKK YCS et al., KK 1992:2, 107–121, dated to about 3000 to 2500 BCE, whose 4 to 5 cm. stone specimens are primarily noteworthy for their upward curving and thickness reduction in the bottom portion (for insertion into the shaft); Hsü Yü-lin and Yang Yung-fang, KK 1992:5, 389–398, whose 144 stone arrowheads, produced by grinding, assume a variety of shapes (illustrated on 396), including somewhat more elongated forms, but are mostly triangular with blunt bases, some of which show some upward indentation; Chang Shao-ch'ing and Hsü Chih-kuo, KK 1992:1, 1–10, reporting on the earlier Hung-shan culture in which the t'ing is yet to appear, including some unusual examples with thicker blade edges that taper to give the appearance of double diamonds stuck together, and a number that display upward indentation or notching.

83. Claims that bronze increased an arrow's killing power lack experimental substantiation. Moreover, although they were subject to breakage and chipping and possibly encountered greater penetration resistance due to surface roughness, flint, shell, and materials such as obsidian could be sharpened to a razorlike edge. The Shang, which could have equipped its warriors entirely with bronze arrowheads, still employed large numbers of stone and bone versions.

84. Liu Yi-man, KK 2002:3, 64–65, notes that although some are found in Fu Hao's tomb, 906 grouped into 15 bundles have been discovered at Kuo-chia-chuang, a site generally dated to the third period.

85. Li Chi's report, BIHP 23 (1952): 523–619, includes a few primitive-looking, triangular stone arrowheads with short stubs (on 616), which are characterized by a flat profile rather than the pronounced rhomboidal shape seen in Lungshan manifestations. (See also Li Chi, KKHP 4 [1949]: 54–58.)

86. For examples dating to the mid- to late Yin-hsü taken from among twenty-six specimens that can be described as fairly stubby but elongated triangles with somewhat rhomboidal heads, short projecting downward points, and slightly tapered t'ing that come directly down, see SHYCS, ed., Yin-hsü Fa-chüeh Pao-kao, 1987, 168–171. (Typical sizes are 5.1 to 6.2 cm. long, including a t'ing of 2.2 to 3.5 cm., and a relatively narrow body width of 1.8 to 2.3 cm.) Additional examples dating to late Yin-hsü appear in SHY An-yang Kung-tso-tui, KK 1991:2, 134. (Typical length is 7 cm.)

87. Examples of these basic styles may be found in the Chung-kuo Ku-tai Ping-ch'i T'u-chi, 36, as well as numerous articles published over the years, including SHYCS An-yang Kung-tso-tui, KK 1991:2, 134 (for twenty-two arrowheads in the basic style, with an average 6.2 cm. for the stubbier form); Fu-chien-sheng Po-wu-kuan, KKHP 1996:2,183; and SHYCS An-yang Kung-tso-tui, KK 1991:2, 134 (elongated triangles dating to late Yin-hsü with long but close reverse points, t'ing that taper down slightly, sharply defined blades, and an average length of about 6 cm.).

88. For clear examples taken from among twenty-seven specimens that display a visible rhomboidal core, staggered kuan, and then circular t'ing, see SHYCS, Yin-hsü Fa-chüeh Pao-kao, 1987, 168–171. Sizes range from 5.4 cm. long, including a 2.1 cm. t'ing and width of 2.0 cm., to 6.1 cm. long, including a 3 cm. t'ing and width of 1.7 cm. (For additional examples about 6.2 cm. long dating to late Yin-hsü, see SHYCS An-yang Kung-tso-tui, KK 1991:2, 134.)

89. Shih Chang-ju describes two bundles of ten bronze arrowheads each in BIHP 40 (1969): 659. In one they average a fairly uniform 7.4 cm. long, but in the other bundle they vary from 6.9 to 7.9 cm. Another fifty of highly uniform shape that were found piled together at Ta-ssu-k'ung average 7 to 7.5 cm. in length and about 3.8 cm. wide. However, the only difference is in the length of the *t'ing*, not the shape or size of the body itself (SHYCS An-yang Kung-tso-tui, KK 1992:6, 514).

90. For brief studies of shooting with cords, see Shih Yen, KKWW 2007:2, 38–41, or Hsü Chung-shu, BIHP 4:4 (1934): 433–435.

CHAPTER 20

1. *T'ai-p'ing Yü-lan*, *chüan* 772. (Hsüan refers to a curved pole, Yüan a double pole.) His name is also said to have been derived from his birthplace, Hsüan-yüan-ch'iu.

2. *T'ai-p'ing Yü-lan*, *chüan* 772. Commentaries to the *Chou Li* claim that Kai was the first to harness oxen to vehicles.

3. For a brief summary of legendary views, see Ku Chieh-kang and Yang Hsiang-k'uei, 1937, 39–41.

4. *Kuan-tzu*, "Hsing Shih," dated to the late third century BCE.

5. *Kuan-tzu*, "Hsing Shih Chieh," reportedly composed about the first century BCE.

6. "Hsiao Ch'eng."

7. For example, Hu-pei-sheng Wen-wu KK YCS et al., KK 2000:8, 55–64. Remnants of seven different-sized chariots, harnessing both two and four horses, were found in this mid– to late Warring States Ch'u tomb. (Although changes in the various components, reinforcements against wear, improvements in stability, and efforts to reduce vibration in the Chou have significant implications for the chariot's capability, they fall beyond the compass of this volume. Only fundamental aspects necessary for understanding the chariot's employment in battle during the Shang can be pondered here.)

8. See, for example, E. L. Shaughnessy, "Historical Perspectives," 217, and for H. G. Creel's doubts about barbarians employing chariots, see his (somewhat outdated) *Origins of Statecraft in China*, 266.

9. HJ36481f.

10. For a report see SHYCS Shandong Kung-tso-tui, KK 2000:7, 13–28. The chariot had two horses; unusually large wheels about 1.6 m. in diameter; an axle length of 3.09 m.; and a well-preserved rectangular box (slightly rounded on the left front) about 0.34 m. high, 1.17 m. wide at the front, 1.34 m. wide at the back where there was a narrow opening, and a depth of 1.02 m. The sacrificial victim was accompanied by a bronze *ko*, some arrowheads, and two horses. (Whether the remains belong to an independent state, as suggested, or a Shang outpost in a subjugated state, as well as whether the chariot was imported or locally constructed, are still unanswered questions.)

11. Sun Pin ("Eight Formations") cited this number, and it is frequently mentioned in other Warring States texts. However, 10,000 was often simply used to indicate an indefinite, large number or myriad.

12. *Tso Chuan*, Chao Kung, thirteenth year. If 10 men accompanied each chariot, the infantry component would have amounted to 40,000; if 25, it soars to an astonishing 100,000, exactly the number Sun-tzu speaks about.

13. The evolution of siege warfare would see the introduction of several wheeled devices, some simply large crossbows mounted on chariots, others innovative combinations of rams, ladders, and similar equipment that merely employed wheels to facilitate their movement. (Two chapters in the *Liu-t'ao*, "The Army's Equipment" and "Planning for the Army," describe several

types of specialized Warring States vehicles, including chariots.) Having been displaced, according to "Occupying Enemy Territory," ordinary chariots and the cavalry were "kept at a distance when attacking cities and besieging towns."

14. Whether these wagons were horse or oxen powered (as generally claimed) remains unknown. However, the existence of oxen-pulled wagons is generally assumed in several traditional writings, and a *ta ch'e* is sometimes mentioned (equally without substantiation) as having been employed in the Shang. Wang Hai-ch'eng, *Ou-ya Hsüeh-k'an* 3 (2002): 41, has suggested that the narrow-gauge chariots recovered from Shang sites may have been intended for transporting heavy goods. (Wang also notes that a solid-wheeled cart that might also have been adopted in China has been found in Xinjiang.)

15. At the battle of Pi some vehicles called *t'un-ch'e*, variously glossed as defensive *ch'e* (chariots) but more likely transport wagons, were pressed into service (*Tso Chuan*, Duke Hsüan, twelfth year). In 493 BCE, right at the end of the Spring and Autumn, 1,000 cartloads of grain were captured.

16. Our discussion is based on several articles that have appeared over the decades, some somewhat outdated but others current and highly informative, if often argumentative and contradictory: Shih Chang-ju, BIHP 40:1 (1969), 625–668, and BIHP 58:2 (1987): 253–280; Yang Pao-ch'eng, KK 1984:6, 546–555; Yang Hung, WW 1977:5, 82–90, WW 1984:9, 45–54, and HCCHS 2000:5, 2–18; and Chang Ch'ang-shou and Chang Hsiao-kuang, 1986, 139–162. Others include Kuo Pao-chün, 1997; Hayashi Minao, *Toho Gakuho* 29 (1959), 155–284; and Kawamata Masanori, "Higashi Ajia no Kodai Sensha to Nishi-Ajia," *Koshi Shunju* 4 (1987): 38–58. In English, E. L. Shaughnessy's seminal "Historical Perspectives on the Introduction of the Chariot into China," HJAS 48:1 (1988), 189–237, remains fundamental despite further reports from Sintashta-Petrova, but other useful analyses include Lu Liancheng, *Antiquity* 67 (1993): 824–838; Stuart Piggot, *Antiquity* 48 (1974): 16–24; and, although somewhat outdated, Joseph Needham, 1965, 73–82 and 246–253. Wang Hai-ch'eng's expansive "Chung-kuo te Ma-ch'e Ch'i-yüan," 2002, provides the most comprehensive overview of critical Chinese and Western data to date, as well as highly useful tables and an extensive bibliography.

A very few of the more important chariot excavations over the decades include Ma Te-chih et al., KKHP 9 (1955): 6–67; SHYCS An-yang Fa-chüeh-tui, KK 1977:1, 69–70; SHY An-yang Kung-tso-tui, KK 1972:4, 24–28; SHYCS An-yang Kung-tso-tui, KK 1998:10, 48–65; and SHY Shan-tung Kung-tso-tui, KK 2000:7, 3–28.

17. Damage needs to be minimized because roads have enormous military implications and gauge differences frustrate movement. (For a discussion of roads and the effects of load, see M. G. Lay, 1992.)

18. As the *K'ao-kung Chi* states, "Heaven has its seasons, Earth has its *ch'i*, materials have their excellence, and labor has its skills. Only when these four are brought together can excellence be produced." Just as with the bow, wood selection was sensitive to seasonal and local variation.

19. The full range for Shang dynasty wheels appears to be 120 to 156 cm., with several reported in the 145 cm. range. (For a convenient summary see the table compiled by Yang Hung, "Chan-ch'e yü Ch'e-chan Erh-lun," HCCHS 2000:5, 5; the earlier table [which does not include Mei-yüan-chuang] in "Yin-tai Ch'e-tzu te Fa-hsien yü Fu-yüan," 1985, 555, or Lu Liancheng, *Antiquity* 67 [1993]: 828 and 829.) Yang Hung, "Erh-lun," HCCHS 2000:5, 5, concludes that the wheels averaged 136.8 cm. in diameter. The chariot discovered at T'eng-chou dating to the interstice between the Shang and Chou already shows slightly larger wheels at roughly 157 to 160 cm. (SHYCS Shan-tung Kung-tso-tui, KK 2000:7, 23). Estimates of horse height, generally thought to be less than 130 cm., vary considerably: Shih Chang-ju, BIHP 40, no. 1 (1969): 665, puts them at only 100 to 115 cm., in which case the wheel rims would have stood well above their bodies, whereas Yang Pao-ch'eng, KK 1984:6, 550, reports a height of 140 to 150 cm. for the horses at M7.

20. See SHYCS An-yang Fa-chüeh-tui, KK 1977:1, 69–70. Yang Hung, "Erh-lun," HCCHS 2000:5, 5, and others have questioned the number of twenty-six. However, the T'eng-chou chariot has twenty-two spokes.

21. At Ta-ssu-k'ung the diameter runs between 3.0 and 4.5 cm., at Hsiao-min-t'un about 4 cm., but at Mei-yüan-chuang a rather narrow 2 cm.

22. The T'eng-chou spokes are already reduced to about 2.0 to 4.0 cm. in diameter.

23. For example, see Hsiang Kung, thirty-first year, and Ai Kung, fourth year. Moving parts, even from conformable metals like brass, present extremely complex lubrication problems. Among the complications are temperature, viscosity, contamination, oxidation, and uneven wear, all of which produce increased friction and hot spots, leading to distortion, adhesion, and burnout.

24. Shang fabrication techniques remain uncertain. However, even after steaming or heat soaking, wood bending requires considerable force. ("Ch'i Fa" in the Yi Ching shows that the felloes were definitely being bent in the Western Chou, if not earlier.) Moreover, stresses are invariably induced because the wood is compressed on the interior and stretched on the exterior, producing cracks, crazing, and severe fiber separation. Although most wood species can be bent after sufficient conditioning, only a few prove suitable in useful thicknesses and capable of retaining their form without major fissures or fractures.

25. For some examples see Yang Pao-ch'eng, KK 1984:6, 548. The most frequently noted sizes are 7.5 cm. thick by 5.5 cm. wide, 8 by 5 cm., and 8 by 6 cm.

26. Yang Hung, "Erh-lun," HCCHS 2000:5, 5, gives an average axle estimate of about 300 cm., diameter of 8–10 cm., wheel gauge of about 226 cm., and end caps of 14 cm. (The T'eng-chou chariot has a 232 cm. gauge.)

27. See Wang Hai-ch'eng, 2002, 25.

28. Sun Chi, KK 1980:5, 448, claims that Shang chariots carried only two men.

29. For a convenient summary of the number of figures per chariot, see the comprehensive tables provided by Wang Hai-ch'eng, 2002, 50–52, which show only three cases of three warriors (for the classic discoveries at Hsiao-t'un), ten with one, and two with two out of a total of thirty-eight entries. (Yang Hung, "Erh-lun," HCCHS [2000]:5, 4, also includes a brief summary of accompanying warriors.)

30. According to Yang Hung, "Erh-lun," HCCHS 2000:5, 5. Yang Pao-ch'eng, KK 1984:6, 549, suggests 129–133 cm. by 74 by 45 cm. The compartment at M41 at Mei-yüan-chuang has dimensions of 128 cm. in the front, 144 cm. at the back, a depth of 70 to 75 cm., and a height of 44 cm. (17.32 inches); the southern chariot at M40 is 134 cm. in the front, 146 cm. in the back, and 96 and 82 cm. deep on the sides, with heights of 39 cm. in the front and 50 cm. in the back with an additional crossbar to the front, while the badly damaged northern chariot is 105 cm. wide at the front and 132 cm. at the back with heights of 30 and roughly 40 cm. respectively ("Mei-yüan-chuang," KK 1998:10, 50 and 57).

31. Shih Chang-ju's placement of the opening at the front for the chariot at M40 (in "Hsiao-t'un Ti-ssu-shih Mu te Cheng-li") has been rejected by every other analyst, including Yang Pao-ch'eng (551), who asserts there were not any openings at the front until the late Warring States. (Other aspects of Shih's reconstruction have also been criticized by, for example, Chang Ch'ang-shou and Chang Hsiao-kuang, 1986, 155.) However, in a rebuttal Shih has argued that despite most chariots having the opening in the rear, the chariot reconstruction for M40 correctly places it at the front. (See Shih's "Yin-ch'e Fu-yüan Shuo-ming," BIHP 58:2, 266–268.) Although a fully open back would facilitate ascending and descending, it would not offer any protection to the highly vulnerable rear.

32. Dowels as thin as 2.5 to 3.0 cm. were occasionally employed for smaller chariots, but Warring States versions sometimes soared over 7 cm. (according to the K'ao-kung Chi), but most were 4 to 5 cm., next in frequency being about 6 cm. (Wang Hai-ch'eng, 2002, 23–25; for an example of 3 cm. see "An-yang Yin-hsü Hsiao-min-t'un te Liang-tso Ch'e-ma-hang," KK 1971:1, 70).

33. It should be noted that considerable strides have been made in recovery and reconstruction techniques, enabling dimensional estimates that were highly inaccurate or totally impossible a half century ago. Data from the various reports are therefore not uniformly compatible or reliable.

34. "Mei-yüan-chuang," 64–65.

35. Examples are seen in the reconstructions from M40 at Hsiao-t'un, M40 at Mei-yüan-chuang, and at T'eng-chou. (Note the diagram for the latter, "T'eng-chou-shih," KK 2000:7, clearly showing a much-reduced chariot compartment depth, implying two occupants rather than three.)

36. "Planning for the State," Wu-tzu, mentions leather armored chariots with covered wheels and protected hubs. Wang Hai-ch'eng, 2002, 26, believes that the Shang also used walls fabricated from interlaced leather and that leather thongs were employed as upper rails to reduce weight.

37. The large number of bronze fittings (including highly functional axle caps) recovered with every chariot indicates their penchant to decorate every protruding surface. (Of particular interest, the large number of small, somewhat amorphous bronze dragons found at Hsiao-t'un's M40 apparently were arrayed to form a decorative border around the outside edges of the compartment. See Shih Chang-ju, "Hsiao-t'un Ti-ssu-shih Mu te Cheng-li," 641–643.)

38. Yang Pao-ch'eng, KK 1984:6, 547–548.

39. One of the chariots at M40 also seems to place the axle somewhat to the rear of center.

40. Shih Chang-ju's original reconstruction of the chariot at Hsiao-t'un's M40 (663–665) shows them being employed on the axle. However, this aspect was universally rejected because it was generally believed that the "crouching rabbit" was a Chou dynasty innovation. (For example, see Yang Pao-ch'eng, 555.) Chu Ssu-hung and Sung Yüan-ju, KKWW 2002:3, 85, also claim that there is absolutely no evidence for the "fu-t'u" prior to the Chou, and in his subsequent "Yin-ch'e Fu-yüan Shou-ming," 269, Shih acknowledged his error. However, the chariots recovered from Mei-yüan-chuang, especially from M41 (KKWW 2003:5, 38–41), are said to already employ it, in which case the incipient beginnings can be traced back to innovations at the end of the Shang.

41. For example, at Ta-ssu-k'ung the shaft and axle each have 15 cm.–wide grooves about 6 cm. deep that result in the shaft projecting just 3 cm. above the axle (KK 1974:2, 25).

42. "Mei-yüan-chuang," KK 1998:10, 41 and 57.

43. See Yang Pao-ch'eng, 549.

44. See, for example, Joseph Needham's concerns in Physics and Physical Technology: Mechanical Engineering, 303–305.

45. As seen on Ch'in bronze models; however, their use cannot be projected back into the Shang. (For a discussion of harnessing methods see Wang Hai-ch'eng, 2002, 33–37, who concludes that the same yoke saddle method was used in Egypt as in the Shang, or Sun Chi, KK 1980:5, 448–460.)

46. Bridles and harnessing are discussed by Yang Hung in "Erh-lun," HCCHS 2000:5, 6, and his "Ma-chü te Fa-chan," WW 1984:9, 45, as well as by Sun Chi.

47. Yang Hung, "Ma-chü te Fa-chan," 45. Fully formed snaffle bits have been recovered from Han dynasty sites, suggesting that they may have appeared in the late Warring States period. However, a bit formed from two figure-eight pieces is already visible in remains from Mei-yüan-chuang ("Mei-yüan-chuang," 51–52). For additional discussion also see Wang Hai-ch'eng, 2002, 27–28.

48. Numerous articles and a few books by Stuart Piggot, Robert Drews, and David W. Anthony have discussed, even vehemently argued, the horse and chariot's history in the West.

49. Four-wheeled war wagons are not unknown, but two-wheeled vehicles induce less drag and are easier to maneuver under identical loads, though they obviously have less carrying capacity.

50. Robert Drews's theory about the demise of chariot warfare in the West around 1200 BCE, as reprised in his *End of the Bronze Age*, tends to dominate current reconstructions.

51. For a brief summary of these positions see Wang Hai-ch'eng, 2002, 2–3 and 45–46.

52. Decades ago Hayashi Minao, *Toho Gakuho*, 225, confidently asserted that the Shang had chariots by 1300 BCE and that they were used in hunting, whereas Edward L. Shaughnessy, HJAS 48, no. 1 (1988): 190, holds that the chariot's introduction should be dated to 1200 BCE.

53. Based on a detailed reexamination of Chinese and Western artifacts, Wang Hai-ch'eng's recent overview concludes the chariot was imported from the steppe. However, for an example of the ongoing arguments for indigenous origination, see Wang Hsüeh-jung, 1999, 239–247. (Wang opts for indigenous origination because all the necessary elements—tools, bronze, and technology—were all advancing at the same time, and some archaeological indications of earlier use, such as small hub caps, have been ignored. He also claims that other animals were employed, including sheep and oxen.)

54. See David W. Anthony, *The Horse, the Wheel, and Language*, 200 ff. (Anthony's views, being the best substantiated and argued, are adopted herein. However, other theories have been proposed, particularly by Marija Gimbutas.)

55. Anthony, 216–222. In an earlier article Anthony's conclusions were somewhat more conservative, 3500 to 3000 BCE. (See Anthony and Brown, *Antiquity* 74 [2000]: 76; see also Anthony and Brown, *Antiquity* 65 [1991]: 22–38, now outdated.) However, opposing views about the origins of riding range from outright rejection of horses ever being ridden before they were harnessed to chariots to a grudging recognition of the possibility while dismissing any military activity or significance until nearly 1000 BCE. (For one overview, see Robert Drews, *Early Riders*.)

56. One question that might well arise from this initial use of horses for riding and raiding is why chariots ever developed as a war vehicle, particularly since they seem to have functioned solely as command and archery platforms. Cavalry can also fight dismounted, just as many chariot warriors historically did in the West.

57. Anthony, *The Horse, the Wheel, and Language*, 66–72.

58. One argument against a steppe origination is the reputedly small size of these chariots, making them unstable at speed and incapable of carrying more than a single warrior. However, Anthony (399–403) points out that numerous javelin points have been found with several chariots that have a gauge of 1.4 to 1.6 m., suggesting the driver-warrior employed a javelin as his primary means of combat.

59. Although many aspects remain nebulous because of the inherent difficulty posed by reconstruction efforts, the priorities of the excavators, and problems of access, enough information is available in secondary publications to tentatively establish the basic features and dimensions.

60. See Li Shui-cheng, 2002, 171–182; Mei Jianjun, 2003, 1–39; and Mei Jianjun, BMFEA 75 (2003): 31–54. See also Stuart Piggott, *Antiquity* 48 (1974): 16–24.

61. The question of whether key discoveries ranging from metallurgy to stirrups are repeated as an artifact of human experience or, because they are inevitably singular, must be transmitted, underpins any debate on the chariot's origination in China.

62. For reports see SHYCS Erh-li-t'ou Kung-tso-tui, KK 2004:11, 3–13, and Wang Hsün, KKWW 1997:3, 61–68.

63. For reports see SHYCS Ho-nan Ti-erh Kung-tso-tui, KK 1998:6, 3; Tu Chin-p'eng et al., KK 1998:6, 13–14; and Robin Yates, "The Horse in Early Chinese Military History," 26–27.

64. Wang Hsüeh-jung, 1999, 239–247, reflecting on the discovery of narrow cart tracks at Yen-shih Shang-ch'eng, notes that they are beneath the inner protective wall; their gauge is 1.2 m.; the ruts are 20 cm. wide but only 3 to 5 cm. deep (implying a brief period of use); and they lie only 20 to 30 cm. from the core wall, so they must have been used to haul soil for it.

65. Feng Hao, KKWW 2003:5, 38–41.

66. For a brief summary of the individual aspects, see Wang Hai-ch'eng, 2002, 7–9. (Wang concludes importation to be fully proven, a conclusion that Yang Hung also came to accept in his "Erh Lun.")

67. For an early form of this conclusion see Cheng Te-k'un, *Archaeology in China: Chou China*, 265–272, or Kawamata Masanori, *Koshi Shunju* 4 (1987): 35–58.

CHAPTER 21

1. Minimal numbers of bones have been discovered at the Yangshao cultural site of Pan-p'o and the Lungshan site of Pao-t'ou Chuan-lung-ts'ang, creating a sort of thread through time. (See Wang Hsün, KKWW 1997:3, 61–68.) The early (Lungshan) Shang site of T'ang-yang Pai-chü also shows evidence of horse raising, as does Ch'eng-tzu-yai. The Anyang area shows a distinct lack of bones except when deliberately interred in graves or horse-and-chariot pits. (See Yüan Ching and T'ang Chi-ken, KK 2000:11, 75–81, who, however, conclude that horses were being imported and rather than eaten, interred with the deceased. Wang Hai-ch'eng, 2002, 38–40 and 47–52, provides a summary of current knowledge.)

2. For example, see Shih Chang-ju, KKHP 2 (1947): 21–22. As already noted, it has been claimed that Shang intelligence efforts relied on mounted riders and that small contingents of "horse" sometimes preceded the army into the field. Although the complete lack of evidence strongly implies that early riding simply did not exist, insofar as the absence of evidence doesn't prove nonexistence and it seems unlikely that people working closely with horses would not have begun riding, if only for herding, training, and control purposes, dogmatic assertions of impossibility are hardly justified. Moreover, if Anthony's view that the horse was ridden even before the chariot appeared is correct, this knowledge would certainly have accompanied the chariot's introduction into China.

3. Wang Yü-hsin, CKSYC 1980:1, 99–108; Liu Yi-man and Ts'ao Ting-yün, HCCHS 2005:5, 24–32. (For examples of horse-focused inquiries see HJ22247 and HJ22347.) The "Hsi-ts'u" in the *Yi Ching* also suggests horses were classified by their distinctive features, including colors.

4. Wang Yü-hsin, 99–105.

5. Prognosticatory inquiries reflect this importance insofar as humans and oxen were the most likely sacrificial offerings, followed by sheep, pigs, dogs, and finally horses. (Liu and Ts'ao, 29, estimate that queries about oxen or men number somewhere between 300 and 1,000; those for sheep, pigs, or dogs about 100; and only a few refer to horses. However, horses captured from steppe enemies seem to have sometimes been sacrificed. [See Wang Yü-hsin, 106.])

6. *Tso Chuan*, Chuang Kung, eighteenth year, records that a king gave three horses each to several feudal lords to preserve their loyalty. (Even the refugee Ch'ung Erh was twice given twenty teams of four horses as a hedge against his enmity should he become powerful [*Tso Chuan*, Hsi Kung, twenty-third year].)

7. *Tso Chuan*, Hsüan Kung, second year. (He managed to escape when only half had been paid.)

8. *Tso Chuan*, Hsiang Kung, second year. (Ch'i had mounted a generally successful invasion of Lai.)

9. Found in the *Tso Chuan*, *Kung-yang*, and *Ku-liang* for Hsi Kung's second year, the episode is reprised in numerous early Warring States and Han texts, including the *Han Fei-tzu* ("Shih Kuo"), *Shuo Yüan*, and *Ch'un-ch'iu Fan-lu*, generally being employed to illustrate shortsightedness. It is also retold in the *Chan-kuo Ts'e* and *Shih Chi*; is included in the *Thirty-six Strategies;* and numbers as one of the important examples of unorthodox techniques in the military writings (such as the *Wu-ching Tsung-yao's* "Ch'i Ping"). (For further discussion see Sawyer, *The Tao of Deception*.)

10. Or, according to the *Ku-liang*, their teeth had grown longer.

11. Note, for example, "Shan Kuo Kuei," *Kuan-tzu*. China's "horse problem" and its ill-fated attempts to rectify the shortage in later centuries have been extensively discussed in recent articles.

12. *Kung-tzu Chia-yü*.

13. "Horses' Hooves," *Chuang-tzu*.

14. An incident in the *Chan-kuo Ts'e* ("Chao," 4) affirms the general recognition that specialized knowledge is required to select horses, neither a well-known administrator nor one of the king's concubines being considered capable of buying a horse. This belief is further reflected in two tales about Po Le. In the first, a merchant whose superlative horse had drawn no interest despite having been displayed in the market for three days paid him to simply look at it intently for a few moments and then glance at it over his shoulder as he walked away, manifesting an appearance of interest. Being observed, his behavior increased the horse's value tenfold (*Chan-kuo Ts'e*, "Yen," 2). In the second, Po recognized the great stallion Chi even though he had become decrepit and reduced to hauling salt wagons (*Chan-kuo Ts'e*, "Ch'u," 4).

15. See Liu Yi-man and Ts'ao Ting-yün, HCCHS 2005:5, 28; Wang Yü-hsin, CKSYC 1980:1, 101.

16. The last four characters comprise a well-known Chinese aphorism generally translated as "eject the bit and gnaw the reins," which is generally understood as meaning "the more you force a horse, the more it resists." Although the dynamic tension of most bridles, coupled with their "oppressive" cheek pieces, makes it difficult to spit out a well-fitted bit, horses reportedly have other ways of shifting it onto their back teeth or chomping down on it, thwarting its effects. (The terms found in this passage have widely differing interpretations; the translation is, at best, an approximation of Chuang-tzu's intent.)

17. The "Hsing-chün Hsü-chih" section of the Sung dynasty *Wu-ching Tsung-yao* contains a section on "Selecting Horses" ("Hsüan Ma") for the cavalry that emphasizes the same ideas of measure and constraint. The text notes that horses are sensitive, riders and horses need to know each other, and training is required; "if the horses and men have not been trained, they cannot engage in battle." The *Wu-tzu* states: "Only after the men and horses have become attached to each other can they be employed." Even the trainer's suitability was the subject of Shang divinatory inquiry (Wang Yü-hsin, 101).

18. "Hsing Shih" and "Hsing Shih Chieh," *Kuan-tzu*.

19. "Hsiao Ch'eng," *Kuan-tzu*.

20. "Yen Hui," *Kung-tzu Chia-yü*.

21. In 549 BCE, Chin attacked Ch'i, prompting Ch'u to strike Chin's ally of Cheng in order to draw off the invaders. After both sides had deployed, the Duke of Chin selected two men to ride forth and pique Ch'u. Because they were fighting within Cheng, they in turn asked Cheng to provide a chariot driver (who would be familiar with the terrain). However, being hot tempered, the driver didn't react well to being forced to wait outside their tent while they ate. Therefore, when they had embarked on their mission and the two were riding in the chariot, nonchalantly playing their instruments, he suddenly rode into the enemy, compelling them to dismount and fight. However, the driver then turned about and started to depart, forcing them to hastily jump on before they cut down their pursuers with arrows. After the mission's completion they severely berated him because all men in a chariot are supposedly brothers. At the Battle of Ta-chi another experienced warrior, disgruntled that he had been bypassed when the pre-battle feast of lamb had been portioned out, deliberately drove the commander's chariot into the enemy, resulting in his capture (*Tso Chuan*, Hsüan Kung 2, 607 BCE). Incidents such as these suggest that penetrating the enemy's ranks was not a primary chariot function.

22. *Tso Chuan*, Ch'eng Kung, sixteenth year.

23. *Tso Chuan*, Hsi Kung, fifteenth year. Although the dialogue is certainly a late reconstruction and may be completely fictional, it doesn't lessen the validity of the insights or recognition of the need for horses to be well trained prior to combat employment.

24. Not being Spring and Autumn concepts, *ch'i* and *yin* are both anachronistic.

25. "Ma Chiang" ("Horse Generals," synonymous with "cavalry generals").

26. *Tso Chuan*, Hsüan Kung, twelfth year.

27. As does Xenophon in *The Cavalry Commander*.

28. A similar statement (which will be cited below) is found in Sun Pin's *Military Methods*.

29. "Obligations of the Son of Heaven."

30. *Wu-tzu* 3, "Controlling the Army."

31. Because the dialogue is prefaced by Marquis Wu asking whether "there are methods for taking care of the chariots and cavalry" and cavalry appears as a referent in the second paragraph, some analysts have dated the work to the late Warring States or early Han. However, most of the contents reflect chariot practices; references to cavalry may simply be later editorial accretions.

32. The *Hu-ch'ien Ching* includes a chapter titled "Cheng Ma" ("Expeditionary Horses") that specifies the regulations that should govern the care and use of army horses and emphasizes the need to find grass and water. In addition to citing a number of measures from the *Wu-tzu*, it discusses several steps for securing the camp, including putting donkeys on the perimeter to thwart raiders.

33. Cited in *Wu-pei Chih*, *chüan* 141.

34. For a discussion of the horse's inherent symbolism and felt power, see Elizabeth A. Lawrence, *Hoofbeats and Society*.

35. *Gallic War*, Book 4.

36. *Tso Chuan*, Chao Kung, first year. This episode, historically considered an example of unorthodox innovation, will be more fully reprised in the next section. (For further discussion of the unorthodox aspects, see Sawyer, *Tao of Deception*.) Somewhat more than two centuries later King Chao Wu-ling's forceful adoption of barbarian dress to facilitate the cavalry's development evoked similar opposition.

37. The *Ku-chin T'u-shu Chi-ch'eng* (*chüan* 34 of "Shen Yi Tien") also preserves information on the seasonal horse sacrifice in its section "Ma Shen" ("Horse Spirits").

38. The selections are taken from "Ma Chan," which itself contains sections entitled "Horse Divination" (*ma chan*) from the *Sung Shu* and "Horse Oddities" (*ma yi*) collected from the Chou onward.

39. In many cases a single chariot and a pair of horses have been found together, though there are also instances of multiple chariots with correspondingly larger numbers of horses. Whether buried intact or in sections, Shang chariots exist mainly as vestiges, just impressions in the sand. The horses are normally aligned along the shaft and the deceased is sometimes accompanied by dogs or grooms.

40. *Ch'ien-pien* 2.19.1 refers to 20 pairs, while HJ21777 and HJ11459 each note 50 pairs. According to Li Hsüeh-ch'in, HCCHS 2005:4, 38–40, a Chou fragment has 200 plus 50 for 250 pairs.

41. Based on finds at Hsiao-t'un Kung-tien-ch'ü, M20, Shih Chang-ju concluded that a Shang chariot that employed four horses had been discovered, but his conclusions have generally been challenged and Shih subsequently recanted his conclusion in "Shuo-ming." However, it continues to be cited as evidence that four-horse chariots existed in the Shang. (Based on the excavation of at least two late tombs in which four horses had been offered in sacrifice, Yang Pao-ch'eng, KK 1984:6, 547, similarly concluded that the Shang started using teams of four. However, in the absence of definitive evidence that the four were actually harnessed together, they might represent two teams of two.)

42. This is the view long held by a number of analysts in varying degree, including Kawamata Masanori, *Koshi Shunju* 4 (1987): 38–58.

43. Some interments contain astonishing numbers of horses. For example, at Lin-tzu in Shandong, site of the ancient state of Ch'i, one assemblage dating to Duke Chin's era (547–489 BCE) contains 600 skeletons, of which 228 have been excavated. "Kuei Tu" in the *Kuan-tzu* speaks of

an army of a million men that includes 10,000 chariots and 40,000 horses, clearly a ratio of four horses to one vehicle.

44. For example, although a Kuo noblewoman named Liang Chi was accompanied by nineteen chariots and thirty-eight horses, thus meeting the rule for a single pair, she far exceeded the allowable five chariots and ten horses that her status allowed. Conspicuous consumption became so ostentatious that critics not only decried the trend and abuses of privilege, but also concluded that they indicated character flaws. Even a high polish could indicate weakness of character and presage ill results (*Tso Chuan*, Hsiang Kung, twenty-eighth year). Similarly, *Kuan-tzu* ("Li Cheng," 4) notes that opulent chariots are a sign of misplaced priorities; the *Yi Ching*'s "Chieh" observes that inappropriately riding in a chariot only attracts robbers.

45. See Chou Hsin-fang, CKSYC 2007:1, 41–57. Chou concludes that privileges were not really systematized until Ch'in Shih-huang's reign; six were employed by both the king and feudal lords, and the question of the emperor having six stemmed from an ongoing debate between old and new text schools.

46. Tombs from the earls of Wei in the Western Chou at Hsin-ts'un in Chün-hsien have yielded twelve chariots and seventy-two horses—a very early, quite substantial representation of six horses per chariot, including one described as a war chariot. (See Hu-pei-sheng K'ao-ku-suo, KK 2003:7, 51–52.)

CHAPTER 22

1. For representative viewpoints see Ku Chieh-kang and Yang Hsiang-k'uei, 1937, 39–54; E. L. Shaughnessy, "Historical Perspectives," 199, 213–221; Herrlee G. Creel, 1970; or Hayashi Minao, *Toho Gakuho* 29 (1959): 278.

2. "Obligations of the Son of Heaven."

3. See Shaughnessy, "Historical Perspectives," 220–224.

4. See Shaughnessy, 216. As noted, the term for horses, *ma*, is generally understood as referring to chariots. Ironically, the inscription refers to the king (rather inauspiciously) falling out of his chariot.

5. "Determining Rank," *Ssu-ma Fa*.

6. For example, see "The Army's Equipment," *Liu-t'ao*.

7. "The Army's Equipment."

8. "Determining Rank," *Ssu-ma Fa*.

9. "Equivalent Forces," *Liu-t'ao*.

10. "Equivalent Forces," *Liu-t'ao*.

11. "Planning for the Army."

12. "The Army's Equipment."

13. "The Army's Equipment."

14. "Equivalent Forces." The text includes ratios for cavalry as well, noting that when "not engaged in battle one cavalryman is unable to equal one foot soldier" and considers one chariot to be equivalent to ten cavalrymen on easy terrain and six on difficult ground.

15. *Questions and Replies*.

16. *Questions and Replies*. This aspect continued to be emphasized in the chariot warfare section of the *Ts'ao-lü Ching-lüeh*, which quotes Li Ching in this regard.

17. For an overview see "Yung Ch'e" in the *Wu-ching Tsung-yao*. (Li Ching also discusses this aspect with regard to his own campaign against the Turks.) In "Military Instructions II" the *Wei Liao-tzu* mentions employing "a wall of chariots to create a solid defense in order to oppress the enemy and stop them" and "'arraying the chariots' refers to making the formations tight with the spears deployed to the front and putting blinders on the horses' eyes."

18. *Wu-ching Tsung-yao, Ch'ien-chi, chüan* 4.

19. The *Tso Chuan* contains accounts (such as Hsiang Kung, eighteenth year) in which *ta ch'e* (great vehicles) are connected together to block a defile.

20. "Employing Chariots," *Wu-ching Tsung-yao.*

21. For an example of four occupants, see *Tso Chuan*, Chao Kung, twentieth year. At Kuo-chia-chuang only two people were buried with the chariot; many other interments have only one.

22. "Five Instructions," *Military Methods.*

23. "Martial Chariot Warriors," *Liu-t'ao.* The Chinese foot at the time of the *Liu-t'ao*'s compilation was about eight inches. (The passage reflects the inception of cavalry.)

24. For an example see *Tso Chuan*, Hsiang Kung, thirty-first year.

25. *Tso Chuan*, Ch'eng Kung, second year. Prior to the battle Han dreamt his father told him not to stand to the side.

26. In 251 BCE, Yen, despite being a peripheral state, attacked Chao with 600,000 men and 2,000 chariots, one of the rare instances of a 300:1 ratio.

27. A *Tso Chuan* passage on Ch'u's organization (known as the "double battalion of King Chuang") gave rise to considerable confusion over the centuries and prompted a pointed discussion in the *Questions and Replies.* (See Sawyer, *Seven Military Classics*, 331.)

28. "Ta Ch'en."

29. "Ta Ch'en," "Hsiao K'uang."

30. "Sheng Ma." "Shan Chih Shu" also refers to a chariot having twenty-eight men.

31. "Ta Ch'en."

32. In his discussion of the *Hsin Shu*'s chariot methods Li Ching concluded that troops were attached only to the attack chariots.

33. See "The Army's Equipment," which allocates fixed numbers of infantry to the roughly 600 specialized chariots integrated into the ideal 10,000-man army.

34. The reconstruction discussed above, advanced by Yen Yi-p'ing, NS 7 (1983): 16–28, is based on extensive reports from Yin-hsü and suggestions made by Shih Chang-ju. (See KKHP 2 [1947]: 1–81, especially 15–24 on chariots, weapons, and personnel, and BIHP 40 [1969:11]: 630–634, as well as Shih's response to various criticisms, BIHP 58:2 [1987:6]: 273–276.)

35. Shih Chang-ju arbitrarily explained away one potential problem—the apparent existence of two chariots in each of the two graves at the top—by deeming the second one an auxiliary vehicle because it lacked any "occupants." However, other explanations are possible, including that the fundamental unit should be seven chariots rather than five or that the other two were scout or reconnaissance vehicles.

36. The issue of five—whether the base of five included the respective unit-level leaders or they were additional—plagues historical reconstructions of Chinese military organization. If the rule of five is rigorously carried out, 5 men comprise a squad, 5 squads a company of 25, 5 companies a battalion of 125, 5 battalions a regiment of 625, 5 regiments a *shih* or army of 3,125, and 5 armies a division or *chün* of 15,625, numbers that do not cohere with the commonly discussed 2,500 for an army or *shih* and 12,500 for a division or *chün.* Moreover, there is always the question whether the officers constitute additional personnel or are to be subsumed within the respective units, posing insurmountable problems at the highest level because squad members suddenly have multiple ranks. (One ad hoc explanation envisions the lower leaders coming from within the unit but the higher ones being additional.)

37. Yen Yi-p'ing, NS 7 (1983): 28.

38. See E. L. Shaughnessy's comments, HJAS 48, no. 1 (1988): 194–199. Shaughnessy notes that half the graves have been ignored.

39. *Tso Chuan*, Chao Kung, twenty-first year. According to "Ming Kuei" in the *Mo-tzu*, King T'ang employed the goose formation when attacking Chieh, the Hsia's last tyrant.

40. Texts such as the *Wei Liao-tzu* ("Offices, 1") state: "The Whirlwind Formation and swift chariots are the means by which to pursue a fleeing enemy."

41. "Equivalent Forces." Further discussion of how the vastly increased number of chariots and infantry seen in the Warring States period actually functioned must be deferred. However, larger numbers simply exacerbate problems of coordination and the overall congestion, particularly if the chariots and infantry are not segregated and employed in distinctly different modes.

42. "Ten Questions," *Military Methods*.

43. *Tso Chuan*, Hsi Kung, twenty-eighth year. The Battle of Ch'eng-p'u, included among the examples in the *Wu-ching Tsung-yao*'s "Ch'üan Ch'i," has been the subject of innumerable articles over the years and is extensively discussed in the two major Chinese military histories. Further explication in English may also be found in Frank A. Kierman Jr., "Phases and Modes of Combat in Early China."

44. "Waging War."

45. "Stimulating the Officers," *Wu-tzu*. The passage states: "Marquis Wu assented to his plan, granting him another 500 strong chariots and 3,000 cavalry. They destroyed Ch'in's 500,000 man army as a result of his policy to encourage the officers." However, it should be noted that the mention of cavalry, adamantly said not to exist in Wu Ch'i's era, has raised doubts about the passage's veracity.

46. For a theoretical example see "Responding to Change" in the *Wu-tzu*, where 1,000 chariots and 10,000 cavalry are to be divided into five operating groups supported by infantry. Similarly, when encountering the enemy in a confined valley, the chariots are to be divided into operational groups, four of which should conceal themselves on the sides to constrain the enemy's options and mount ambushes. In "Eight Formations" Sun Pin also stressed dividing the chariots into discrete operational contingents (though without mentioning infantry) and suiting their numbers to the terrain's characteristics.

47. "When the Three Armies are united as one man they will conquer. There are drums (directing the deployment of) the flags and pennants; drums for the chariots; drums for the horses (cavalry); drums for the infantry; drums for the different types of troops; drums for the head; and drums for the feet. All seven should be properly prepared and ordered" ("Strict Positions," *Ssu-ma Fa*). Sun-tzu also speaks about multiplying the drums to ensure strong control.

CHAPTER 23

1. Herrlee G. Creel, 1970, 262–282, was among the first to question the chariot's capabilities. Studies of the chariot's history and impact in the West by noted historians such as John Keegan and others have similarly debated its real combat role.

2. For a discussion of poisoning water supplies in Chinese warfare, see Sawyer, *Fire and Water*.

3. "Wen" in the *Kuan-tzu* discusses the importance of enumerating the state's resources, including the artisans who can be employed on expeditionary campaigns.

4. In addition to various grooms and ordinary stable hands, designated personnel were responsible for lubricating the axles in the Spring and Autumn. (See *Tso Chuan*, Hsiang Kung, thirty-first year and Ai Kung, third year.) A few oracular inscriptions suggest the Shang experienced some of these problems.

5. *Tso Chuan*, Ai Kung, second year.

6. *Tso Chuan*, Hsi Kung, fifteenth year.

7. For example, see the incident preserved in *Tso Chuan*, Chao Kung, twenty-first year.

8. *Tso Chuan*, Ch'eng Kung, sixteenth year.

9. "Waging War." He also states that seven-tenths of the people's resources will be consumed.

10. "Waging War."

11. *Tso Chuan*, Ch'eng Kung, second year.

12. *Tso Chuan*, Hsiang Kung, twenty-third year.

13. *Tso Chuan*, Ch'eng Kung, second year.

14. *Tso Chuan*, Hsüan Kung, twelfth year.

15. "Responding to Change," *Wu-tzu*.

16. As recounted in *Judges* 4 and 5. (Since Barak's troops came down from the mountain, it wasn't the hilly terrain that proved inimical, but the rain, noted only in the poeticized account in *Judges* 5.)

17. "Ten Deployments," *Military Methods*.

18. For example, as when the state of Chin attacked Cheng at T'ung-ch'iu in 468 BCE.

19. Ch'eng Kung, sixteenth year. In the incident already discussed in which horses unfamiliar with the terrain were employed, the chariot turned into a mire and was halted. (*Tso Chuan*, Hsi Kung, fifteeth year.)

20. The identification of terrain-imposed limitations certainly dates back to the Western Chou, but the first articulation is found in the *Art of War*.

21. "Ti T'u," presumably a late Warring States chapter.

22. "Hsiao K'uang," *Kuan-tzu*.

23. "Ten Questions," *Sun Pin Military Methods*.

24. For a brief retelling of the incident, see Sawyer, *Tao of Deception*, 189–191.

25. *Tso Chuan*, Ch'eng Kung, seventh year. The Sichuan area similarly lagged behind in their employment. (The *Pei-cheng Lü* [*chüan* 7] makes the point that environment shapes natural tendencies and that skills in riding or using boats best derive from familiarity from an early age rather than from instruction. Thus Wu naturally inclined to boats and Chin to cavalry.)

26. "Battle Chariots" states: "The infantry values knowing changes and movement; the chariots value knowing the terrain's configuration; and the cavalry values knowing the side roads and the Tao of the unorthodox."

27. "Battle Chariots."

28. They essentially replicate a series found in the *Wu-tzu*'s "Responding to Change" in which victory is inevitable. (Generally speaking, the *Wu-tzu* is less concerned with tactics than with the essential principles governing chariot operations that have already been discussed for the horses, and it is only in the *Liu-t'ao* that their battlefield exploitation becomes apparent.)

29. Sun Pin, for example, asserted that in a dispersed deployment the "chariots do not race, the infantry does not run" ("Ten Deployments," *Military Methods*).

30. If a chariot with wheels about 3 feet in diameter (and therefore with a circumference of nearly 9.5 feet) was moving at the still-significant speed of 5 miles an hour or about 440 feet per minute, the wheel would be turning at about 46 rpm. One revolution per second would have been slow enough for the most unskilled warrior to insert a spear near the outer rim between the spokes. Larger wheels would have been even slower, but higher speeds of 10 mph would still have been feasible.

31. *Tso Chuan*, Yin Kung, ninth year. Even if a late fabrication, it no doubt reflects concepts common at the time of compilation in the Warring States period.

32. For further discussion see Sawyer, *Tao of Deception*, 23–24.

33. *Tso Chuan*, Chao Kung, first year. Li Ching cites this episode as an example of the unorthodox in *Questions and Replies*. However, Li Ching thought they still represented chariot tactics even though they were deployed as infantrymen. (For further discussion, see Sawyer, *Tao of Deception*, 38–40.)

34. Duke Ai, second year, records that a battle commander riding in a chariot was brought down by a spear.

35. For example, see *Tso Chuan*, Chao Kung, twenty-sixth year.

36. For an example, see *Tso Chuan*, Ai Kung, second year.

37. *Tso Chuan*, Ch'eng Kung, sixteenth year. Incidents of pursuit and seizure are recorded in Homeric warfare, and several well-publicized experiments with replica chariots in the West have shown that foot soldiers could have easily surrounded and overtaken chariots in the melee's chaos.

38. "The Infantry in Battle," *Liu-t'ao*.

39. T'ai Kung's assertion attracted T'ang T'ai-tsung's attention because it contradicted Sun-tzu's admonitions. (See Book III of *Questions and Replies*.)

40. Examples are found in the *Liu-t'ao* chapter "Certain Escape," including (in "Incendiary Warfare") employing the chariots to thwart incendiary attacks.

41. For examples, see "Crow and Cloud Formation in the Mountains," *Liu-t'ao*.

42. Caltrops, which might be conceived of as primitive land mines, repeatedly proved highly effective in stopping enemy advances and shaping the battlefield. (For examples of the types of caltrops and their modes of use, see "The Infantry in Battle.")

43. In an odd illustration, Shih Chang-ju, BIHP 40 (1969:11), 666, places the archer on the right side of the chariot. (In the illustration both the chariot's occupants and the horses are too small in comparison with the diameter of the wheels. However, the occupants are still tightly confined.)

44. For further analysis see Yang Hung's classic discussion in "Chan-ch'e yü Ch'e-chan."

45. In the *Tso Chuan* (Hsiang Kung, twenty-third year) a fighter who asks permission to act as the warrior on the right leaps aboard the chariot and, wielding a "sword" in his right hand, grasps the strap (or possibly the traces holding the outer horses) with his left.

Chapter 24

1. See Domicio Proenca Jr. and E. E. Duarte, *Journal of Strategic Studies* 28, no. 4 (2005): 645–677. Useful discussions of logistics include Martin Van Crevald, *Supplying War*; John Lynn, ed., *Feeding Mars*; and Kenneth Macksey, *For Want of a Nail*.

2. The classic study is Donald W. Engels, *Alexander the Great and the Logistic of the Macedonian Army*.

3. The early sixth-century BCE Chou vessel known as the *Tao Ting Ming* records men being locally sent out to confiscate rice and millet.

4. "King's Wings," *Liu-t'ao*. This is also the first known description of a general staff.

5. For the nature and history of water denial measures, see Sawyer, *Fire and Water*.

6. The *Art of War* mentions thirst and the *Ssu-ma Fa* includes water among its seven administrative affairs in "Determining Rank."

7. "Encouraging the Army" in the *Six Secret Teachings* emphasizes that the true general shares "hunger and satiety with the men."

8. Studies of ancient Chinese logistics are extremely rare, basically limited to Yang Sheng-nan's fiscally oriented work, LSYC 1992:5, 81–94, and part of the *Hou-ch'in Chih-tu*, edited by T'ung Chao (1997).

9. "Waging War." A chapter titled "Ch'ing-chung Chia" in the *Kuan-tzu* generally dated to the second century BCE similarly notes that an army of 100,000 *chi* bearers will exhaust all the firewood and grass for ten *li* and a single day's combat will cost 1,000 *chin* (units of gold).

10. "Employing Spies." The "Pa Kuan" ("Eight Observations") section of the *Kuan-tzu* similarly speaks about the dire impact of having just one-tenth of the populace serve in the military for extended terms.

11. "Waging War."

12. Under Duke Ai's second year the *Tso Chuan* records the capture of an astonishing 1,000 wagons (*ch'e*) filled with grain that was being transported to another area.

13. "Orders for the Vanguard."

14. "Nine Terrains," *Art of War*.

15. See, among several, T'ung Chao, *Hou-ch'in Chih-tu*, 25.

16. The *Ssu-ma Fa* counsels restraint, but "Martial Plans" and "Military Instructions II" in the *Wei Liao-tzu* also advocate not disturbing the farmers. Probably the earliest prohibition against disturbing farmers and their animals is embedded in the "Fei Shih" section of the *Shang Shu*.

17. "Controlling the Army," *Wu-tzu*.

18. "Crow and Cloud Formation in the Marshes." "Cavalry in Battle" also refers to the problems posed by an enemy having cut off the supply lines.

19. "Secret Tallies."

20. "Evaluating the Enemy." Victory is so certain that divination, then a fundamental pre-battle practice, was deemed unnecessary.

21. "Fatal terrain," a well-developed concept in China's sophisticated martial psychology, is first articulated in "Nine Terrains" in *Art of War*, but "Certain Escape" in the *Six Secret Teachings* advises burning the supply wagons to elicit this sort of unshakable commitment.

22. "Tactical Balance of Power in Attacks," *Wei Liao-tzu*.

23. "Tactical Balance of Power in Defense," *Wei Liao-tzu*.

24. The *Wei Liao-tzu* stressed the connection in "Discussion of Regulations." Shang Yang's reforms are credited with significantly shaping Ch'in's martial character.

25. "Maneuvering the Army," *Art of War*. The chapter adds: "If they kill their horses and eat the meat, the army lacks grain." (For a general discussion of assessment and deception in field reconnaissance, see "Field Intelligence" in Sawyer, *Tao of Spycraft*.)

26. "Military Instructions II," *Wei Liao-tzu*.

27. *"Military Pronouncements."* The last section parallels a passage in the *Kuan-tzu's* "Pa Kuan."

28. His ingenuity is cited to illustrate the topic of "The Hungry" in the *Hundred Unorthodox Strategies*.

29. For a brief discussion of the growth of animal husbandry, see Yen Wen-ming, SCKKLC, 351–361.

30. The *Art of War* admonishes commanders not to target fortified cities for siege or assault.

31. Ch'en Chien-hua, CKSYC 4 (2004): 3–14.

32. For relevant inscriptions see Chin Hsiang-heng, "San-hang San-shih," 1974, 7–8.

33. "Waging War." "Pa Kuan" ("Eight Observations") in the *Kuan-tzu* notes that having to transport provisions about the countryside due to production and distribution problems will quickly deplete a state's reserves and ultimately cause famine.

34. See, for example, T'ung Chao, *Hou-ch'in Chih-tu*, 1997, 23.

35. T'ung Chao, *Hou-ch'in Chih-tu*, 20–21.

36. T'ung Chao (25) sees them as a sort of early version of the *t'un-t'ien* system.

37. Unfortunately none of the volumes on China's transport history, including Wang Chan-yi's massive *Chung-kuo Ku-tai Tao-lu Chiao-t'ung-shih*, does more than speculate on the pre-Chou period.

CHAPTER 25

1. Several books have recently tried to debunk the ardently held view of early societies being tranquil and cooperative, untarnished by conflict and warfare.

2. The tendency to idealize prehistoric societies even within the late Neolithic horizon as matriarchal and thus egalitarian can, for example, be seen in Yen Wen-ming's 1988 article on Pan-p'o. Pan-p'o already shows strong defensive characteristics and evidence of warfare, yet the article asserts production and consumption were undertaken in common and the society was marked by equality.

3. Conflict is thus thought to have beset the central Hua-Hsia and Yi cultures even in the archaic period, reflecting the basic dichotomization of east and west.

4. "Wei 2." *Chan-kuo Ts'e*, states that the "Tung Yi populace did not arise." (For further discussion see Wang Yü-ch'eng, CKSYC 1986:3, 71–84.)

5. To take just one example, P'ei An-p'ing, WW 2007:7, 75–80, 96, traces the origin of (provocatory) privileges and private property to about 4500 BCE, about the time of Ch'eng-t'ou-shan.

6. In addition, as attested by a large number of artifacts but sparse locally available materials, stone weapons production became specialized. (See, for example, Li Hsin-wei, KK 2008:6, 58–68.)

7. *T'ai-p'ing Yü-lan, chüan* 193.

8. Analysts such as Hsü Hung, STWMYC, 286–295, have recently been scrutinizing the ancient period in an attempt to formulate some defining characteristics for different types of sites.

9. Two examples are the prominent but discontinuous 500-meter-long, 10-meter-wide ditch in the northeastern corner of Erh-li-t'ou and a 110-meter-long, 14-meter-wide remnant just north of the Shang capital of Yen-shih. (For a report on the former, see Hsü Hung et al., KK 2004:11, 23–31; for the latter, see SHYCS Ho-nan Erh-tui, KK 2000:7, 1–12.)

10. "Military Disposition."

11. Some upon being abandoned, though for unknown reasons.

12. "Shih Chih Chieh," *Yi Chou-shu*. This incident makes up a pair with one previously cited from the *Yi Chou-shu* to illustrate the belief that "one who practices warfare ceaselessly will perish." An idea that probably originated in the Spring and Autumn period, it is manifestly expressed in the *Ssu-ma Fa*, which states that "even though a state may be vast, those who love warfare will inevitably perish. Even though calm may prevail under Heaven, those who forget warfare will certainly be endangered." In "Audience with King Wei" Sun Pin similarly said: "Victory in warfare is the means by which to preserve vanquished states and continue severed generations. Not being victorious in warfare is the means by which to diminish territory and endanger the altars of state. For this reason military affairs must be investigated. Yet one who takes pleasure in the military will perish and one who finds profit in victory will be insulted. The military is not something to take pleasure in, victory not something through which to profit." The *Art of War* is of course particularly known for its assertion that "warfare is the greatest affair of state, the Tao to survival or extinction."

13. "Audience with King Wei."

14. For productive distribution see Chen Fangmei, JEAA 2, nos. 1–2 (2000): 228. (The Shang's fundamentally strong martial orientation is well attested by the high proportion of weapons found in the Anyang area.)

15. See Yen Wen-ming, KKWW 1982:2, 38–41, or Shao Wangping, JEAA 2, nos. 1–2 (2000): 199.

16. Western battle scenes from Sumer and Egypt dating to the third millennium BCE are well-known, so finding evidence for warfare in China of comparable age should not be unexpected. Yet it is frequently explained away as post-death treatment of the body or other benign activity, including accidental death or reinterment. Moreover, one of the main problems in discriminating between combat casualties and sacrificial victims is the tendency to presuppose an absence of conflict and view any skeletons showing the effects of deliberately inflicted violence as "sacrificial victims," as if the sacrificial character somehow obviates the slaying aspect. (For a discussion see Mark Golitko and Lawrence H. Keeley, *Antiquity* 81 [2007]: 332–342, who emphasize that the presence of fortifications should argue for combat rather than sacrifice.) For the purpose

of our investigation we have assumed that conflict has always existed and have interpreted the evidence accordingly rather than dismissing or transmorphing it.

Among the innumerable reports of graves with shattered skeletons, various crushing blows, dismemberment, and other untoward acts committed against the deceased, one puzzling one from Yün-men (radiocarbon dated to 1260 BCE, ± 90 or somewhere around the early Shang when corrected) where 172 stone arrowheads were found is particularly dramatic. Several of the interred had been shot multiple times, two more than ten times, with arrows in their chests, stomachs, and heads, and one woman was bound, suggesting that this was indeed a form of execution or that they had been used for target practice. There is also evidence of turning skulls into vessels and dishonoring the deceased, as well as scalping, in the Lungshan, though the latter has surprisingly not yet been identified as a practice of ancient Chinese warfare. (Scalping has been discussed in two articles: Yen Wen-ming, KKWW 1982:2, 38–41, and Ch'en Hsing-ts'an, WW 2000:1, 48–55.)

17. Sarah Allen, JAS 66, no. 2 (2007): 461–496, has recently advanced a concept of cultural hegemony derived from the appearance and ascendancy of Erh-li-t'ou culture.

18. A typical overview arguing for multiple origins is Yen Wen-ming, WW 1987:3, 38–50.

19. For examples of this position see Fan Yü-chou, HCCHS 2006:5, 11–15; Hsüeh Jui-che, HCCHS 2006:4, 13–22; and Tai Hsiang-ming, KKHP 1998:4, 389–418.

20. For an overview of the Yi-Luo area's contribution, see Ch'en Hsing-ts'an et al., KKHP 2003:2, 161–218. Other topographical divisions are possible, including segmentation by river systems. (For example, see Ch'en Hsü, HSLWC [reprint of 1996], 282–292.)

21. According to Ch'en Sheng-po and others, HCCHS 2005:4, 7–8) the lower Yangtze River area was comparatively free of coercion and large-scale warfare.

22. Both Hung-shan and Liang-chu, two cultures that esteemed jade, may have perished because they were subverted by their religious beliefs. (See Li Po-ch'ien, WW 2009:3, 47–56.) More generally, devastating floods may have had an irrecoverable impact. (See Chin Sung-an and Chao Hsin-p'ing, CKKTS 1994:10, 14–20.) The Yüeh-shih lithic industry was more productive and their ceramics were also more advanced (Yen Wen-ming, SCKKLC, 306–318; Feng Chen-kuo, LSYC 1987:3, 54–65).

23. Yen Wen-ming, 312–313, stresses this lack of overarching self-identity in his analysis of the Tung Yi collapse.

24. The question of cultural determinism has recently been discussed by John Keegan, John Lynn, Jeremy Black, and Victor Davis Hanson. It is tempting to speculate on the psychological impact that the "citadel mentality" may have had on the Chinese Tao of warfare from its inception in the ancient period.

25. Yüan Ching, WW 2001:5, 51–57; Liu Chin-hsiang and Tung Hsin-lin, KK 1996:2, 61–64. Rice cultivation commenced in the lower Yangtze around 4000 BCE, presumably after a millennium of gathering naturally occurring variants. Although millet's deliberate cultivation in the north has been identified with Hsing-lung-wa culture beginning around 6000 BCE and millet penetrated the Yellow river basin around 5500, rice did not become intermixed until about 2500 to 2000. (For an overview see Dorian Q. Fuller et al., Antiquity 81 [2007]: 316–331.) Other archaeologists discern even earlier inception points. (For example, see Yen Wen-ming's four articles, SCKKLC, 351–361, 362–384, 385–399, and 400–406.)

26. For an excellent example see Yen Wen-ming, SCKKLC, 262–266.

27. Warfare compels people to adapt whatever might be at hand as well as to innovate to meet the challenges. The Shang certainly did not need to employ chariots in warfare because their opponents rarely had any, yet they were willing to risk these high-prestige vehicles in combat.

✦ Integrated Bibliography ✦

Allen, Sarah. "Erh-li-t'ou and the Formation of Chinese Civilization: Toward a New Paradigm." *Journal of Asian Studies* 66, no. 2 (2007):461–496.

Alrune, Flemming. "Bows and Arrows 6000 Years Ago." In *The Bow Builders Book,* edited by Flemming Alrune, 13–38.

Alrune, Flemming, ed. *The Bow Builders Book: European Bow Building from the Stone Age to Today.* Atglen, PA: Schiffer, 2007.

An Chih-min. 安志敏。關於鄭州 "商城" 的幾個問題。考古 1961.8:448–450.

———. 試論中國的早期銅器。考古 1993.12:1110–1119.

———. 塔里木盆地及其周圍的青銅文化遺存。考古 1996.12:70–76.

An Chin-huai. 安金槐。試論鄭州商代城址—隞都。文物 1961.5:73–80.

———. 試論豫西地區龍山文化類型中晚期與夏代文化早期的關係。夏文化研究論集 (1996): 3–10.

———. "The Shang City at Cheng-chou and Related Problems." In *Studies of Shang Archaeology,* edited by Chang Kwang-chih, 15–48.

———. 再論鄭州商代城址—隞都。先秦、秦漢史 1993.11:33–38.

———. 豫西潁河上游在探索夏文化遺存中的重要地位。考古與文物 1997.3:54–60.

An Chin-huai and Yang Yü-pin. 安金槐，楊育彬。偃師商城若干問題的再探討。考古 1998.6:14–19.

An-hui-sheng Wen-wu K'ao-ku Yen-chiu-suo. 安徽省文物考古研究所。安徽省濉溪縣石山子遺址動物骨骼鑒定與研究。考古 1992.3:253–262.

An-yang Wen-wu Kung-tso-tui. 安陽文物工作隊。河南安陽郭莊村北發現一座殷墓。考古 1991.10:903–909.

An-yang-shih Wen-wu K'ao-ku Yen-chiu-suo. 安陽市文物考古研究所。河南安陽市殷墟郭家莊東南五號商代墓葬。考古 2008.8:22–33.

Anthony, David W. *The Horse, the Wheel, and Language: How Bronze-age Riders from the Eurasian Steppes Shaped the Modern World.* Princeton, N.J.: Princeton University Press, 2007.

Anthony, David W., and Dorcas R. Brown. "Enolithic Horse Exploitation in the Eurasian Steppes: Diet, Ritual and Riding." *Antiquity* 74 (2000): 75–86.

———. "The Origin of Horseback Riding." *Antiquity* 65 (1991): 22–38.

Arkush, Elizabeth N., and Mark W. Allen, eds. *The Archaeology of Warfare: Prehistories of Raiding and Conquest.* Gainesville: University Press of Florida, 2006.

Bagley, Robert. "Shang Archaeology." In *The Cambridge History of Ancient China*, edited by M. Loewe and E. L. Shaughnessy, 124–231.

Bagley, Robert, ed. *Ancient Sichuan: Treasures from a Lost Civilization.* Seattle: Seattle Art Museum, 2001.

Barnard, Noel. "A New Approach to the Study of Clan-sign Inscriptions of the Shang." In *Studies of Shang Archaeology*, edited by Chang Kwang-chih, 141–206.

———. "Review of *Prehistoric China, Shang China, China.*" *Monumenta Serica* XXII.1963: 213–255.

Beijing Ta-hsüeh K'ao-ku-his. 北京大學考古系。鄭州市岔河遺址 1988 年試掘簡報。考古 2005.6:17–31.

Birrell, Anne. "The Four Flood Myth Traditions of Classical China." *TP* 83 (1997): 213–259.

Black, Jeremy. *Rethinking Military History.* New York: Routledge, 2004.

———. *War: Past, Present, and Future.* New York: St. Martin's Press, 2000.

Bradbury, Jim. *The Medieval Archer.* Woodbridge, UK: Boydell Press, 1985.

Carman, John, and Anthony Harding, eds. *Ancient Warfare: Archaeological Perspectives.* Gloucestershire, UK: Sutton Publishing Limited, 2004.

Chai Te-fang. 翟德芳。中國北方地區青銅短劍分群研究。考古學報 1988.3:277–299.

Ch'ai Shao-ming. 柴曉明。論商周時期的青銅面飾。考古 1992.12:1111–1120.

Chan K'ai-sun. 詹開遜。試論新干青銅器的裝飾特點。考古 1995.1:63–74, 36.

———. 從新干青銅器的造型看商代中原文化對南方的影響。中國古代史（一）1994.5:34–40.

Chan K'ai-sun and Liu Lin. 詹開遜，劉林。談新干商墓出土的青銅農具。文物 1993.7:27–32, 18.

Chang Ch'ang-shou and Chang Hsiao-kuang. 張長壽，張孝光。殷周車制略說。中國考古學研究夏鼐先生考古五十年紀念論文集, 1986, 139–162.

Chang Ch'ang-shou and Chang Kuang-chih. 張長壽，張光直。河南商丘地區殷商文明調查發掘初步報告。考古 1997.4:24–31.

Chang Chen-hung. 張鎮洪。1986–1987 年西樵山發掘簡報。文物 1993.9:32–39.

Chang Cheng-lang. "A Brief Discussion of Fu Tzu." In *Studies of Shang Archaeology*, edited by K. C. Chang, 103–111.

Chang Chih-heng. 張之恒。夏代都城的變遷。夏文化研究論集, 1996, 109–112.

Chang Ch'i-yün. 張其昀。遠古史. In 中華五千年史。台北，中國文化大學出版部, 1961.

Chang Chung-p'ei. 張忠培。中國古代文明形成的考古學研究。先秦、秦漢史 2000. 4:2–24.

———. 窺探凌家灘墓地。文物 2000.9:55–63.

Chang Fu-hsiang. 張富祥。"殷"名號起源考。先秦、秦漢史 2001.5:57–60.

Chang Ho-jung and Luo Erh-hu. 張合榮，羅二虎。試論雞公山文化。考古 2006.8:57–66.

Chang Hsien-te and Chang Hsien-lu. 張先得，張先祿。北京平谷劉家河商代銅鉞鐵刃的分析鑒定。文物 1990.7:66–71.

Chang Hsiu-p'ing et al. 張秀平，毛元佑，王朴民。影響中國的100次戰爭。廣西人民出版社, 1993.

Chang Hsü-ch'iu. 張緒球。屈家嶺文化古城的發現和初步研究。考古 1994.7:629–634.

Chang Hsüeh-hai. 張學海。城起源研究的重要突破。考古與文物 1999.1:36–43.

———. 試論山東地區的龍山文化城。文物 1996.12:40–52.

Chang Hsüeh-lien and Ch'iu Shih-hua. 張雪蓮，仇士華。夏商周斷代工程中應用的系列樣品方法測年及相關問題。考古 2006.2:81–89.

Chang Hsüeh-lien et al. 張雪蓮，仇士華，蔡蓮珍，薄官成，王金霞，鐘建。新砦—二里頭—二里岡文化考古年代序列的建立與完善。考古 2007.8:74–89.

Chang Hung-yen. 張宏彥。關於仰韶文化時空范圍的界定問題。考古與文物 2006.5:66–70.

——. 東亞地區史前石鏃的初步研究。考古 1998.3:41–55, 75.

——. 渭水流域仰韶文化分期問題。文物 2006.9:62–69, 78.

Chang Kuang-chih. 張光直. *See* Chang Kwang-chih.

Chang Kung. 張弓。中國古代的治水與水利農業文明—評魏特夫的 "治水專制主義" 論。中國古代史（一）1994.2:4–18.

Chang Kuo-shuo. 張國碩。鄭州商城與偃師商城并為亳都說。考古與文物 1996.1: 32–38, 31.

——. 關於殷墟的幾個問題。考古與文物 2000.1: 42–48.

——. 夏商時代都城制度研究。鄭州，河南人民出版社，2001.

——. 論夏末早商的商夷聯盟。先秦、秦漢史 2002.4: 8–14.

——. 論夏商周三族的起源。三代文明研究（一），1999, 280–285.

Chang Kwang-chih. *Archaeology of Ancient China*. 4th ed. New Haven, Conn.: Yale University Press, 1986.

——. 張光直。中國青銅時代—中國考古藝術研究中心集刊（二）。香港，中文大學出版社，1982.

——. "The Chinese Bronze Age: A Modern Synthesis." In *The Great Bronze Age of China*, edited by Wen Fong, 35–50.

——. "On the Meaning of Shang in Shang Dynasty." *Early China* 20 (1995): 69–78.

——. *Shang Civilization*. New Haven, Conn.: Yale University Press, 1980.

——. 張光直。殷周關係的再檢討。中央研究院歷史語言研究所集刊 51.2, 1980.6: 197–216.

Chang Kwang-chih, ed. *Studies of Shang Archaeology*. New Haven, Conn.: Yale University Press, 1986.

Chang Kwang-chih and Xu Pingfang, eds. *The Formation of Chinese Civilization*. New Haven, Conn.: Yale University Press, 2005.

Chang Li and Wu Chien-p'ing. 張力，吳健平。浙江余杭瓶窯，良渚古城結構的遙感考古。文物 2007.2: 74–80.

Chang Li-tung. 張立東。夏都與夏文化。夏文化研究論集，1996, 113–118.

Chang Mao-jung. 張懋鎔。高家堡出土青銅器研究。考古與文物 1997.4: 38–41, 49.

——. 早期金文和遠古陶器刻繪符號。古文字與青銅器論集。北京。科學出版社，2002, 20–23.

Chang P'ei-yü. 張培瑜。武丁、殷商的可能年代。考古與文物 1999.4: 62–65.

Chang Ping-ch'üan. "A Brief Description of the Fu Hao Oracle Bone Inscriptions." In *Studies of Shang Archaeology*, edited by K. C. Chang, 121–140.

——. 張秉權。甲骨文與甲骨學，國立編譯館主編，台北，1988.

——. 略論婦好卜辭。漢學研究 1983.6:27–40.

——. 卜辭中所見殷商政治統一的力量及其達到的範圍。中央研究院歷史語言研究所集刊 50.1, 1979.3:175–229.

——. 殷代的農業與氣象。中央研究院歷史語言研究所集刊 42.2, 1970.12:267–336.

Chang Shao-ch'ing and Hsü Chih-kuo. 張少青，許志國。遼寧康平縣趙家店村古遺址及墓地調查。考古 1992.1: 1–10.

Chang Sung-lin et al. 張松林。(河南省文物管理局南水北調文物保護辦公室，鄭州市文物考古研究院)。河南新鄭市唐戶遺址裴李崗文化遺存發掘簡報。考古 2008.5: 3–20.

Chang Te-shui. 張德水。夏國家形成的地理因素。夏文化研究論集，1996, 170–175.

Chang T'ieh-niu and Kao Hsiao-hsing. 張鐵牛，高曉星。中國古代海軍史。北京，八一出版社，1993.

Chang T'ien-en. 張天恩。秦器三論—益門春秋墓幾個問題淺談。文物 1993.10: 20–27.

——. 關中西部夏代文化遺存的探索。考古與文物 2000.3: 44–50, 84.

——. (北京大學考古文博院)。陝西彬縣、淳化等縣商時期遺址調查。考古 2001.9: 13–21.

——. 巴蜀文化與中原文化的關係試探。考古與文物 1998.5: 68–77.

———. 再論秦式短劍。考古 1995.9: 841–853.

Chang Ts'ui-lien. 張翠蓮。試論豫東東部地區的岳石文化遺存。考古與文物 2001.2: 36–46.

Chang Tsung-tung. "A New View of King Wuding." *Monumenta Serica* 37 (1986–1987): 1–12.

Chang Wei-hua. 張維華。湯都四遷芻議。先秦、秦漢史 1993.11: 48–56.

Chang Wei-lien. 張渭蓮。氣候變遷與商人南下。先秦、秦漢史 2006.4: 23–32.

Chang Wen-hsiang. 張文祥。寶雞弓 | 魚？國墓地淵源的初步探討—兼論蜀文化與城固銅器群的關係。考古 與文物 1996.2: 44–49.

Chang Wen-li and Lin Yün. 張文立, 林沄。黑豆嘴類型青銅器中的西來因素。考古 2004.5: 65–73.

Ch'ang Yao-hua. 常耀華。關於夏代文字的一點思考—兼論中國文字的起源。夏文化研究論集,1996, 252–265.

Chang Yü-shih et al. (國家文物局考古領隊培訓班)。鄭州西山仰韶時代城址的發掘。文物 1999.7: 4–15.

Chao Chen-sheng and Chi Lan. 趙振生, 紀蘭。遼寧阜新近年來出土一批青銅短劍及短劍加重器。考古 1994.11: 1047–1049.

Chao Ch'eng. 趙 誠。甲骨文與商代文化—漢字與文化叢書。瀋陽。遼寧人民出版社, 2000, 136–156.

Chao Chih-ch'üan. 趙芝荃。論夏、商文化的更替問題—為紀念二里頭遺址發掘 40 周年而作。考古與文物 1999.2: 23–29.

———. 論夏文化起、止年代的問題。夏文化研究論集, 1996, 277–283.

———. 夏商分界界標之研究。考古與文物 2000.3: 28–32.

———. 夏社與桐宮。考古與文物 2001.4: 36–40.

———. 夏代前期文化綜論。考古學報 2003.4: 459–482.

———. 再論偃師商城的始建年代。先秦、秦漢史 2000.1: 18–27.

Chao Ch'un-ch'ing. 禹貢五服的考古學觀查。先秦、秦漢史 2007.1: 9–19.

Chao Ch'un-yen et al. 趙春燕, 岳占偉, 徐廣德。安陽殷墟劉家莊北 1046 號墓出土銅器的化學組成分析。文物 2008.1: 92–94.

Chao En-yü. 趙恩語。夏初年代的勘定。先秦、秦漢史 1985.11: 17–19.

Ch'ao Fu-lin. 晁福林。補釋甲骨文 "眾?" 字并論其社會身份的變化。中國史研究 2001.4: 3–12.

———. 試論夏代社會結構的若干問題。夏文化研究論集, 1996, 136–142.

———. 從盤庚遷殷說到 "尚書 盤庚" 三篇的次序問題。中國史研究 1989.1: 57–67.

———. 我國文明時代初期社會發展道路及夏代社會性質研究。先秦、秦漢史 1996.6: 23–32.

Chao Kuang-hsien. 趙光賢。古代漢苗二族關係史辨誤。歷史研究 1989.5: 24–34.

———. 從古文獻上證明夏代的存在。夏文化研究論集, 1996, 122–123.

———. 殷代兵制述略。先秦、秦漢史 1986.2: 29–36.

Chao Shih-ch'ao. 趙世超。巡守制度試探。中國古代史 1995.9: 6–18.

———. 炎帝與炎帝傳說的南遷。先秦、秦漢史 1999.2: 43–45.

Chao Tien-tseng. 趙殿增。巴蜀文化幾個問題的探討。文物 1987.10: 18–23.

Chen Fang-mei. "Some Thoughts on the Dating of Late Shang Bronze Weaponry." *Journal of East Asian Archaeology* 2, nos. 1–2 (2000): 227–250.

Ch'en Ch'ang-yüan. 陳昌遠。商族起源地望發微—兼論山西垣曲商城發現的意義。歷史研究 1987.1: 136–144.

Ch'en Ch'ang-yüan and Ch'en Lung-wen. 陳昌遠, 陳隆文。論先商文化淵源及其殷先公遷徙之歷史地理考察。先秦、秦漢史 2002.4: 15–27.

Ch'en Chao-yün. 陳朝雲。夏商周中原文明對淮河流域古代社會文明化進程的影響。先秦、秦漢史 2006.2: 3–8.

———. 商代聚落模式及其所體現的政治經濟景觀。先秦、秦漢史 2004.6: 30–36.

Ch'en Chieh. 陳絜。試論殷墟聚落居民的族系問題。先秦、秦漢史 2003.2: 15–22.

———. 從商金文的"寢某"稱名形式看殷人的稱名習俗。先秦、秦漢史 2001.4: 46–52.

Ch'en Chih. 陳致。夷夏新辨。中國史研究 2004.1: 3–22.

Ch'en Ch'un and Kung Hsin. 陳淳, 龔辛。二里頭、夏與中國早期國家研究。先秦、秦漢史 2004.6: 3–12.

Ch'en Hsing-ts'an. 陳星燦。中國古代的剝頭皮風俗及其他。文物 2000.1: 48–55.

Ch'en Hsing-ts'an et al. 陳星燦, 劉莉, 李潤權, 華翰維, 艾琳。中國文明腹地的社會復雜化進程—伊洛河地區的聚落形態研究。考古學報 2003.2: 161–218.

Ch'en Hsü. 陳旭。炎黃歷史傳說與中華文明。夏商文化論集 (reprint of 1998), 293–302.

———. 鄭州小雙橋商代遺址即隞都說。夏商文化論集 (reprint of 1997), 163–170.

———. 鄭州小雙橋商代遺址的年代和性質。夏商文化論集 (reprint of 1995), 145–154.

———. 鄭州商城宮殿基址的年代及其相關問題。夏商文化論集 (reprint of 1985), 36–44.

———. 鄭州商代鑄銅遺址的年代及相關問題。夏商文化論集 (reprint of 1992), 45–54.

———. 鄭州商代王都的興與廢。夏商文化論集 (reprint of 1987), 64–72.

———. 鄭州商文化的發現與研究。夏商文化論集 (reprint of 1983), 23–35.

———. 鄭州商文化淵源試析。夏商文化論集 (reprint of 1990), 96–103.

———. 中國文明起源多元論與有中心論。夏商文化論集 (reprint of 1996), 284–292.

———. 二里頭遺址是商都還是夏都。夏商文化論集 (reprint of 1985), 8–15.

———. 河南古代青銅冶鑄業的興起。夏商文化論集 (reprint of 1985), 171–175.

———. 夏商文化論集。北京, 科學出版社, 2000.

———. 小雙橋遺址的發掘與隞都問題。夏商文化論集 (reprint of 1996), 159–162.

———. 關於鄭州商城湯都亳的爭議。夏商文化論集 (reprint of 1993), 73–84.

———. 關於鄭州商文化分期問題的討論。夏商文化論集 (reprint of 1988), 85–95.

———. 關於偃師商城和鄭州商城的年代問題。夏商文化論集 (reprint of 1985), 119–128.

———. 論河南早商都邑遺址的年代及相關問題。考古與文物 2000.1: 33–38.

———. 1995 年鄭州小雙橋遺址的發掘結語。夏商文化論集 (reprint of 1996), 155–158.

———. 商周青銅鉞。夏商文化論集 (reprint of 1984), 233–239.

———. 試論商代青銅武器的分期。夏商文化論集 (reprint of 1983), 218–232.

———. 商代隞都探尋。夏商文化論集 (reprint of 1991), 137–144.

———. 商代戰爭的性質及其歷史意義。夏商文化論集 (reprint of 1988), 240–248.

———. 商代第一都—鄭州商城。夏商文化論集 (reprint of 1993), 115–118.

———. 豫東石文化與鄭州商文化的關係。夏商文化論集 (reprint of 1994), 104–110.

Ch'en Hsüeh-hsiang and Chin Han-p'o. 陳雪香、金漢波。"2006 年商文明國際學術研討會" 紀要。考古 2007.5: 84–88.

Ch'en Kao-hua and Ch'ien Hai-hao. 陳高華, 錢海皓。中國軍事制度史。鄭州。大象出版社, 1997.

Ch'en Kuo-ch'ing and Chang Ch'uan-chao. 陳國慶, 張全超。(吉林大學邊疆考古研究中心, 內蒙古自治區文 物考古研究所)。內蒙古赤峰市上機房營子遺址發掘簡報。考古 2008.1: 46–55.

Ch'en Ku-ying. 陳鼓應。堯禹在先秦諸子中的意義與問題。先秦、秦漢史 1985.7: 4–16.

Ch'en Liang-tso. 陳良佐。我國古代的青銅農具—兼論農具的演變。漢學研究 1984.6: 135–166, 1984.12: 363–402.

Ch'en Li-chu. 陳立柱。夏文化北播及其與匈奴關係的初步考察。歷史研究 1997.4: 18–35.

———. 陳立柱。亳在大伾說。先秦、秦漢史 2004.4: 28–37.

Ch'en Lien-ch'ing. 陳連慶。《禹貢》研究。陳連慶教授學術論文集。中國古代史研究。長春, 吉林文史出版 社, 1991, 863–891.

Ch'en Meng-chia. 陳夢家。夏史及商史說。古史辨。Vol. 7B, 330–333.

————. 殷虛卜辭綜述。中華書局, 北京, 1988.

————. 殷代銅器。考古學報 1954.7:15–59.

Ch'en P'ing. 陳平。試論寶雞益門二號墓短劍及有關問題。考古 1995.4: 361–375.

Ch'en Sheng-po. 論良渚文化中心聚落的特殊性。先秦、秦漢史 2005.4: 7–8.

Ch'en Sheng-yung. 陳剩勇。東南地區: 夏文化的萌生與崛起—從中國新石器時代晚期主要文化圈的比較 研究探尋夏文化。先秦、秦漢史 1991.5: 15–36.

Cheng Chen-hsiang. "A Study of the Bronzes with the 'Ssu T'u Mu' Inscriptions Excavated from the Fu Hao Tomb." In *Studies of Shang Archaeology*, edited by K. C. Chang, 81–102.

Cheng Hui-sheng. 鄭慧生。甲骨卜辭研究。開封, 河南大學出版社, 1998.

Cheng Kuang. 鄭光。夏商文化是二元還是一元—探索夏文化的關鍵之二。考古與文物 2000.3: 33–43.

Cheng Shao-tsung. 鄭紹宗。夏商時期河北古代文化的關係問題。三代文明研究 (一), 1999, 423–428.

Cheng Te-k'un. *Archaeology in China: Chou China*. Cambridge, UK: Heffer, 1963.

————. *Archaeology in China: Shang China*. Cambridge, UK: Heffer, 1960.

————. "Metallurgy in Shang China." TP (1974): 209–229.

Cheng-chou-shih Wen-wu Kung-tso-tui. 鄭州市文物工作隊。鄭州市大河村遺址博物館。鄭州大河村遺址, 1983, 1987 年發掘報告。考古 1996.1: 111–142.

Cheng-chou-shih WWKK YCS. 鄭州市文物考古研究所, 滎陽市文物保護管理所。河南滎陽大師姑遺址 2002 年度發掘簡報。文物 2004.11: 4–18.

————. 鄭州市文物考古研究所, 北京大學考古文博學院。河南鞏義市花地嘴遺址 "新砦期" 遺存。考古 2005.6: 3–6.

Ch'eng Feng. 程峰。焦作府城早期商城發現的史學價值。先秦、秦漢史 2004.2: 24–26, 35.

Ch'eng P'ing-shan. 程平山。夏代紀年考。先秦、秦漢史 2004.5: 10–21.

————. 論新砦古城的性質與啟時期的夏文化。考古與文物 2007.3: 59–63.

Ch'eng-tu-shih WWKK Kung-tso-tui. 成都市文物考古工作隊。四川崇州市雙河史前城址試掘簡報。考古 2002.11: 3–19.

Ch'eng-tu-shih WWKK Kung-tso-tui et al. 成都市文物考古工作隊, 郫縣博物館。四川省郫縣古城遺址調 查與試掘。文物 1999.1: 32–42.

————. 成都市文物考古工作隊, 四川聯合大學歷史系考古教研室, 溫江縣 文管所。四川省溫江縣魚鳧村遺址調查與試掘。文物 1998.12: 38–56.

Ch'eng-tu-shih WWKK YCS. 成都市文物考古研究所。成都金沙遺址I區 "梅苑" 地點發掘一期簡報。文物 2004.4: 4–65.

Ch'eng Tung and Chung Shao-yi, eds. 成東, 鍾少異。中國古代兵器圖集。解放軍出版社, 1990.

Ch'eng Yung-chien. 程永建。介紹幾件商代青銅、玉器。文物 2009.2: 79–82, 96.

Ch'i Wu-yün. 齊烏雲。山東沭河上游史前自然環境變化對文化演進的影響。考古 2006.12: 78–84.

Chia Chin-piao et al. 賈金標, 朱永剛, 任亞珊, 李伊萍。關於葛家莊遺址北區遺存的幾點認識。考古 2005.2: 71–78.

Chia Lan-p'o et al. 賈蘭波。山西峙峪舊石器時代遺址發掘報告。考古學報 1972.1: 39–58.

Chiang Chang-hua. 江章華。巴蜀柳葉形劍研究。考古 1996.9: 74–80.

Chiang Chang-hua et al. 江章華, 王毅, 張擎。成都平原先秦文化初論。考古學報 2002.1: 1–22.

Chiang Ch'ung-yao. 蔣重躍。"歷數" 和 "尚賢" 與禪讓說的興起。先秦、秦漢史 2007.1: 41–46.

Chiang Kang. 蔣剛。盤龍城遺址群出土商代遺存的幾個問題。考古與文物 2008.1: 35–46.

Chiang Lin-ch'ang. 江林昌。摒棄中國古文明研究中的兩種誤解。先秦、秦漢史 2006.4: 3–12.

Chiang Yü. 江瑜。中國南方和東南亞古代銅鼓鑄造技術探討。考古 2008.6: 85–90.

Chiang-hsi-sheng Po-wu-kuan. 江西省博物館, 北京大學歷史系考古專業, 清江縣博物館。江西清江吳城商代 遺址發掘簡報。文物 1975.7: 51–71.

Chiang-su-sheng Kao-ch'eng-tun Lien-ho K'ao-ku-tui. 江蘇省高城墩聯合考古隊。江陰高城墩遺址發掘簡報。文物 2001.5: 4–21.

Chiang-su-sheng San-hsing-ts'un Lien-ho K'ao-ku-tui. 江蘇省三星村聯合考古隊。江蘇金壇三星村新石器時代 遺址。文物 2004.2: 4–26.

Ch'iao Teng-yün and Chang Yüan. 喬登雲, 張沅。邯鄲境內的先商文化及其相關問題。三代文明研究 (一), 1999, 162–174.

Ch'ien Yao-p'eng. 錢耀鵬。中國古代斧鉞制度的初步研究。考古學報 2009.1: 1–34.

———. 關於西山城址的特點和歷史地位。文物 1999.7: 41–45.

———. 關於半坡聚落及其形態演變的考察。考古 1999.6: 69–77.

———. 關於半坡遺址的環壕與哨所—半坡聚落形態考察之一。考古 1998.2: 45–52.

———. 堯舜禪讓的時代契機與歷史真實—中國古代國家形成與發展的重要線索。先秦、秦漢史 2001.1: 32–42.

Chin Cheng-yao. 金正耀。二里頭青銅器的自然科學研究與夏文明探索。文物 2000.1: 56–64, 69.

Chin Cheng-yao et al. 金正耀, 平尾良光, 彭適凡, 馬淵久夫, 三輪嘉六, 詹開遜。江西新干大洋洲商墓青 銅器的鉛同位素比值研究。考古 1994.8: 744–747, 735.

———. 金正耀, 朱炳泉, 常向陽, 許之咏, 張擎, 唐飛。成都金沙遺址銅器研究。文物 2004.7: 76–88.

Chin Ching-fang and Lü Shao-kang. 金景芳, 呂紹綱。"甘誓" 淺說。先秦、秦漢史 1993.5: 13–17.

Chin Feng-yi. 靳楓毅。論中國東北地區含曲刃青銅短劍的文化遺存。考古學報 (上) 1982.4: 387–426; (下) 1 (1983): 39–54.

———. 大凌河流域出土的青銅時代遺物。文物 1988.11: 24–35.

Chin Hsiang-heng. 金祥恆。從甲骨卜辭研究殷商軍旅中之王族三行三師。中國文字 52, 1974, 1–26.

Chin Kuei-yün. 靳桂云。燕山南北長城地帶中全新世氣候環境的演化及影響。考古學報 2004.4: 485–505.

Chin Sung-an and Chao Hsin-p'ing. 靳松安, 趙新平。試論山東龍山文化的歷史地位及其衰落原因。中國古代史 1994.10: 14–20.

Ch'in Hsiao-li. 秦小麗。二里頭文化的地域間交流—以山西省西南部的陶器動態為中心。考古與文物 2000.4: 43–57.

Ch'in Wen-sheng. 秦文生。祖乙遷邢考。三代文明研究 (一), 1999, 133–136.

Ch'in Ying et al. 秦穎, 王昌燧, 張國茂, 楊立新, 汪景輝。皖南古銅礦冶煉產物的輸出路線。文物 2002.5: 78–82.

Ching-chou Po-wu-kuan and Fu-kang Chiao-yü Wei-yüan-hui. 荊州博物館, 福岡教育委員會。湖北荊州市 陰湘城遺址東城牆發掘簡報。考古 1997.5: 1–10, 24.

Ching-chou Po-wu-kuan and Chia Han-ch'ing. 荊州博物館, 賈漢清。湖北公安雞鳴城遺址的調查。文物 1998.6: 25–30.

Ching-chou-shih Po-wu-kuan et al. 荊州市博物館, 石首市博物館, 武漢大學歷史系考古專業。湖北石首市 走馬嶺新石器時代遺址發掘簡報。考古 1998.4: 16–38.

Ching Chung-wei. 井中偉。由曲內戈形制辨祖父兄三戈的真偽。考古 2008.5: 78–87.

Ching San-lin. 荊三林。試論殷商源流。先秦、秦漢史 1986.5: 37–46.

Ch'ing-yang-hsien Wen-wu Kuan-li-suo. 青陽縣文物管理所。安徽青陽縣龍崗春秋墓的發掘。考古 1998.2: 18–24.

Ch'iu Shih-ching. 裘士京。江南銅材和 "金道錫行" 初探。中國史研究 1992.4: 3–10.

Chou Chi-hsü. 周及徐。華夏古 "帝" 考—黃河文明探源之一。中國文化研究 2007.3: 93–104.

Chou Hsin-fang. 周新芳。"天子駕六" 問題考辨。中國史研究 2007.1: 41–57.

Chou Tzu-ch'iang. 周自強。從古代中國看（東方專制主義）的謬誤。中國古代史
（一）1994.2: 19–30.

Chou Wei. 周緯。中國兵器史稿。台北。 明文書局，1981.

Chu Chen. 朱楨。商代後期都城研究綜述。先秦、秦漢史 1989.8: 3–10.

Chu Chi-p'ing. 朱繼平。從商代東土的人文地理格局談東夷族群的流動與分化。考古
2008.3: 53–61.

Chu Chün-hsiao and Li Ch'ing-lin. 朱君孝，李清臨。二里頭晚期外來陶器因素試析。
考古學報 2007.3: 295–312.

Chu Kuang-hua. 朱光華。洹北商城與小屯殷墟。考古與文物 2006.2: 31–35.

———. 早夏國家形成時期的聚落形態考察。考古與文物 2002.4: 19–25.

Chu Ssu-hung and Sung Yüan-chu. 朱思紅，宋遠茹。伏兔、當兔與古代車的減震。考
古與文物 2002.3: 85–88.

Chu Yen-min. 朱彥民。卜辭所見 "殷人尚右" 觀念考。中國史研究 2005.3: 3–13.

———. 商族的遷徙與冀中南之亳。三代文明研究（一），1999, 296–299.

———. 殷墟都城探論。南開大學出版社，1999.

Chu Yung-kang. 朱永剛。試論我國北方地區銎柄式柱脊短劍。文物 1992.12: 65–72.

———. 朱永剛。東北青銅文化的發展階段與文化區系。考古學報 1998.2: 133–152.

Ch'ü Hsiao-ch'iang. 屈小強。三星堆傳奇—古蜀王國的發祥。香港, 中天出版社，1999.

Ch'ü Wan-li. 屈萬里。史記殷本紀及其他紀錄中所載殷商時代史事。台灣大學文史
哲學報 14 (1965.11): 87–118.

Ch'ü Ying-chieh. 曲英杰。古代城市—20世紀中國文物考古發現與研究叢書。北京,
文物出版社，2003.

Chung-kuo K'e-hsüeh-yüan K'ao-ku Yen-chiu-suo. 中國科學院考古研究所。灃西發掘報
告。北京, 文物出版社，1963.

Chung-kuo Ku-tu She-hui. 中國古都學會。中國古都研究。太原, 山西人民出版社，
1994.

Chung Po-sheng. 鍾柏生。卜辭中所見殷代軍政之一：戰爭啟動過程及其準備工作。
中國文字 NS 14 (1991): 95–156.

Chung Shao-yi. 鍾少異。古相劍術芻論。考古 1994.4: 358–362.

———. 試論戟的幾個問題。文物 1995.11: 54–60.

———. 試論扁莖劍。考古學報 1992.2: 129–145.

Chung-kuo She-hui K'o-hsüeh-yüan K'ao-ku Yen-chiu-suo [SHYCS]. 中國社會科學院考古
研究所: SHYCS 殷墟發掘報告 (1958–1961). 文物出版社，1987.

SHYCS An-yang Fa-chüeh-tui. 安陽發掘隊。安陽殷墟孝民屯的兩座車馬坑。考古
1977.1: 69–70, 72.

SHYCS An-yang Kung-tso-tui. 安陽工作隊。安陽新發現的殷代車馬坑。考古 1972.4:
24–28.

———. 安陽郭家莊 160 號墓。考古 1991.5: 390–391.

———. 河南安陽市花園莊54號商代墓葬。考古 2004.1: 7–19.

———. 河南安陽市洹北商城的勘察與試掘。考古 2003.5: 3–16.

———. 河南安陽市郭家莊東南 26 號墓。考古 1998.10: 36–47.

———. 河南安陽市梅園莊東南的殷代車馬 坑。考古 1998.10: 48–65.

———. 1980 年河南安陽大司空村M 539 發掘 簡報。考古 1992.6: 509–517.

———. 1984–1988 年安陽大司空村北地殷代 墓葬發掘報告。考古學報 1994.4: 471–497.

———. 1987 年秋安陽梅園莊南地殷墓 的發掘。考古 1996.2: 125–142.

———. 1991 年安陽後岡殷墓的發掘。考古 1993.10: 880–903.

———. 2000–2001 年安陽孝民屯東南地殷代 鑄銅遺址發掘報告。考古學報 2006.3:
351–384.

———. 殷墟大司空 M303 發掘報告。考古學報 2008.3: 353–394.

————. 1991. 年安陽後岡殷墓的發掘。考古1993.10:880–903.

SHYCS Erh-li-t'ou Kung-tso-tui. 二里頭工作隊。河南偃師二里頭遺址中心區的考古新發現。考古 2005.7: 15–20.

————. 河南偃師市二里頭遺址發現一件 青銅鉞。考古 2002.11: 31–34.

————. 河南偃師二里頭遺址宮城及宮殿 區外圍道路的勘察與發掘。考古 2004.11:3–13.

SHYCS Feng-hsi Fa-chüeh-tui. 灃西發掘隊。陝西長安張家坡 M170 號井叔墓發掘簡報。考古 1990.6: 504–510.

SHYCS Ho-nan Erh-tui. 河南二隊。河南偃師商城宮城北部 "大灰溝" 發掘簡報。考古 2000.7: 1–12.

SHYCS Ho-nan Hsin-chai-tui. 河南新砦隊。河南新密市新砦遺址 2002 年發掘簡報。考古 2009.2:3–15.

————. 河南新密市新砦遺址東城牆發掘簡報。考古 2009.2: 16–31.

SHYCS Honan Ti-erh Kung-tso-tui. 河南第二工作隊。河南偃師商城宮城池苑遺址。考古 2006.6:13–31.

————. 南偃師商城 IV 區 1999 年 發掘簡報。考古 2006.6:32–42.

————. 河南偃師商城東北隅發掘簡報。考古 1998.6:1–8.

SHYCS Ho-nan Yi-tui. 河南一隊。河南郟縣水泉新石器時代遺址發掘簡報。考古 1992.10:865–874.

SHYCS Hu-pei Kung-tso-tui. 湖北工作隊。湖北黃梅陸墩新石器時代墓葬。考古 1991.6: 481–495.

SHYCS Nei-Meng-ku Kung-tso-tui. 內蒙古工作隊, 呼倫貝爾盟民族博物館。內蒙古海拉爾市團結遺址的調 查。考古 2001.5:3–17.

SHYCS Nei Meng-ku Ti-yi Kung-tso-tui. 內蒙古第一工作隊。內蒙古赤峰市興隆溝聚落遺址2002–2003年的發 掘。考古 2004.7:3–8.

SHYCS Shan-tung Kung-tso-tui. 山東工作隊。山東滕州市前掌大商周墓地 1998 年 發簡報。考古 2000.7: 13–28.

SHYCS Shih-yen-shih. 實驗室。放射性碳素測定年代報告 (14). 考古 1987.7:653–659.

SHYCS Tung-hsia-feng K'ao-ku-tui et al. 東下馮考古隊。山西夏縣東下馮龍山文化遺址。考古學報 1983.1: 55–92.

Creel, Herrlee G. *The Origins of Statecraft in China*. Chicago: University of Chicago Press, 1970.

Crump, J. I. *Chan-kuo Ts'e*. London: Oxford, 1970.

DeVries, Kelly. "Catapults Are Not Atomic Bombs: Towards a Redefinition of 'Effectiveness' in Pre-modern Military Technology." *War in History* 4, no. 4 (1997): 454–470.

Di Cosmo, Nicola. *Ancient China and Its Enemies: The Rise of Nomadic Power in East Asian History*. Cambridge, UK: Cambridge University Press, 2002.

Dien, Albert E. "A Study of Early Chinese Armor." *Artibus Asiae* 43 (1981–1982): 5–66.

Drews, Robert. *The Coming of the Greeks*. Princeton, NJ: Princeton University Press, 1988.

————. *Early Riders: The Beginnings of Mounted Warfare in Asia and Europe*. New York: Routledge, 2004.

————. *The End of the Bronze Age: Changes in Warfare and the Catastrophe ca. 1200 B.C.* Princeton, N.J.: Princeton University Press, 1993.

Engels, Donald W. *Alexander the Great and the Logistic of the Macedonian Army*. Berkeley: University of California, 1978.

Eno, Robert. "Was There a High God *Ti* in Shang Religion?" *Early China* 15 (1990):1–26.

Fan Chün-ch'eng. 樊俊成。延川縣出土的幾件青銅器。考古與文物 1995.5:91.

Fan Li. 樊力。論石家河文化青龍泉三類型。考古與文物 1999.4: 50–61.

Fan Yü-chou. 范毓周。中原文化在中國文明形成進程中的地位與作用。先秦、秦漢史 2006.5:11–15.

————. "Military Campaign Inscriptions from YH127." *Bulletin of the School of Oriental and African Studies* 52, no. 3 (1989): 533–548.

——. 范毓周。殷代武丁時期的戰爭。甲骨文與殷商史第三輯, 上海古籍出版社, 1991, 175–239.

Fang-ch'eng K'ao-ku Kung-tso-tui. 防城考古工作隊。山東費縣防故城遺址的試掘。考古 2005.10: 25–36.

Fang Chieh. 方介。韓愈（對禹問）析義—兼論韓愈與孟子政治理念之歧異。漢學研究 11: 1 (1993): 15–28.

Fang Hsiao-lien. 方孝廉。夏代及其文化。夏文化研究論集 1996, 266–273.

Fang Hui. 方輝。商周時期魯北地區海鹽業的考古學研究。考古 2004.4: 53–67.

Fang Li-chung. 房立中。戰爭的起源及其原始形態。先秦、秦漢史 1989.3: 18–26.

Fang Yen-ming. 方燕明。登封王城崗城址的年代及相關問題探討。考古 2006.9: 16–23.

Fang Yu-sheng. 方酉生。夏王朝中心在伊洛和汾澮河流域考析—兼與（夏后氏居于古河濟之間考）一文商榷。先秦、秦漢史 1996.6: 33–39.

——. 夏與東夷族關係新探。先秦、秦漢史 1992.9: 18–20.

——. 小雙橋遺址為仲丁隞都說商榷。考古與文物 2000.1: 39–41.

——. 論登封告成王城崗遺址為禹都陽城說—兼與（禹都陽城即濮陽說）一文商榷。考古與文物 2001.4: 29–35.

——. 略論新砦期二里頭文化—兼評《來自"新砦期"論證的幾點困惑》。先秦、秦漢史 2003.1: 35–39.

——. 試論小雙橋遺址非仲丁所遷之隞都。考古 2002.8: 81–86.

——. 從鄭州白家莊期商文化說到仲丁都隞。先秦、秦漢史 1998.1: 58–63.

——. 從三處窖藏坑看鄭州商城為何王都。考古與文物 1999.3: 39–42.

——. 偃師二里頭遺址第三期遺存與桀都斟鄩。考古 1995.2: 160–169, 185.

Feng Chen-kuo. 逢振鎬。東夷及其史前文化試論。歷史研究 1987.3: 54–65.

Feng Hao. 馮好。關於商代車制的幾個問題。考古與文物 2003.5: 38–41.

Feng Shih. 馮時。山東丁公龍山時代文字解讀。考古 1994.1: 37–54.

——. "文邑" 考。考古學報 2008.3: 273–290.

Feng T'ien-yü. 馮天瑜。從神話傳說透視上古歷史—上古史研究方法的一種探索。先秦、秦漢史 1984.11: 5–14.

Ferrill, Arther. *The Origins of War: From the Stone Age to Alexander the Great*. New York: Thames and Hudson, 1985.

Franklin, Ursula. "On Bronze and Other Metals in Early China." In *The Origins of Chinese Civilization,* edited by David N. Keightley, 279–296. Berkeley: University of California, 1983.

Fu-chien-sheng Po-wu-kuan. 福建省博物館。福建浦城縣牛鼻山新石器時代遺址第一、第二次發掘。考古 學報 1996.2: 165–197.

Fu Pei-jung. "On Religious Ideas of the Pre-Chou China." *Chinese Culture* 26, no. 3 (1985): 23–39.

Fuller, Dorian, Emma Harvey, and Ling Qin. "Presumed Domestication? Evidence for Wild Rice Cultivation and Domestication in the Fifth Millennium BC of the Lower Yangtze Region." *Antiquity* 81 (2007): 316–331.

Fung, Christopher. "The Drinks Are on Us: Ritual, Social Status, and Practice in Dawenkou Burials, North China." *JEAA* 2, nos. 1–2 (2000): 67–92.

Golitko, Mark, and Lawrence H. Keeley. "Beating Ploughshares back into Swords: Warfare in the *Linearbandkeramik*." *Antiquity* 81 (2007): 332–342.

Guilaine, Jean, and Jean Zammit. *The Origins of War: Violence in Prehistory*. Translated by Melanie Hersey. Malden, Mass.: Blackwell Publishing, 2005.

Guo Da-shu. "'Northern-type' Bronze Artifacts Unearthed in the Liaoning Region and Related Issues." In *The Archaeology of Northeast China*, edited by Sarah M. Nelson, 182–205.

Han Chia-ku. 韓嘉谷。河北平原兩側新石器文化關係變化和傳說中的洪水。考古 2000.5: 57–67.

Han Chiang-su. 韓江蘇。從殷墟花東 H3 卜辭排譜看商代舞樂。中國史研究 2008.1: 21–40.

Han Chien-yeh. 韓建業。夏文化的起源與發展階段。先秦、秦漢史 1998.1: 44–49.

Han Feng. 寒峰。甲骨文所見商代軍制數則。甲骨探史錄。1982.

Han K'ang-hsin and P'an Ch'i-feng. 韓康信, 潘其風。殷代人種問題考察。歷史研究 1980.2: 89–98.

Han Wei-lung and Chang Chih-ch'ing. 韓維龍, 張志清。長子口墓的時代特徵及墓主。考古 2000.9: 24–29.

Hanson, Victor Davis. *Hoplites: The Classical Greek Battle Experience.* London: Routledge, 1993.

———. *A War Like No Other: How the Athenians and Spartans Fought the Peloponnesian War.* New York: Random House, 2005.

———. *The Western Way of War: Infantry Battle in Classical Greece.* New York: Alfred A. Knopf, 1989.

Hayashi Minao. 林巳奈夫。中國先秦時代的馬車。東方學報 29(1959):155–284.

———. 中國殷周時代的武器。京都。京都大學人文科學研究所, 1972.

Hein, Wulf. "Arrowheads of Flint." In *The Bow Builder's Book: European Bow Building from the Stone Age to Today,* edited by Flemming Alrune, 81–95.

———. "Shaft Material: Wayfaringtree Viburnum." In *The Bow Builder's Book: European Bow Building from the Stone Age to Today,* edited by Flemming Alrune, 75–80.

Ho Chien-an. 何建安。從王灣類型、二里頭文化與陶寺類型的關係試論夏文化。先秦、秦漢史 1987.1: 33–46.

Ho Ching-ch'eng. 何景成。商末周初的舉族研究。考古 2008.11: 54–70.

Ho Kang. 賀剛。先秦百越地區出土銅劍初論。考古 1991.3: 252–262.

Ho Kao. 何浩。顓頊傳說中的神話與史實。歷史研究 1992.3: 69–84.

Ho Nu. 何駑。湖北江陵江北農場出土商周青銅器。文物 1994.9: 86–91.

Ho-nan-sheng Kung-yi-shih Wen-wu Pao-hu Kuan-li-suo. 河南省鞏義市文物保護管理所。河南省鞏義市里泃遺址調查。考古 1995.4: 297–304.

Ho-nan-sheng Po-wu-kuan. 河南省博物館, 鄭州市博物館。鄭州商代城址試掘簡報。文物 1977.1: 21–31.

Ho-nan-sheng Wen-wu K'ao-ku Yen-chiu-suo. 河南省文物考古研究所。鄭州商城外郭城的調查與試掘。考古 2004.3: 40–50.

———. 河南鄭州商城宮殿區夯土牆 1998 年的發掘。考古 2000.2: 40–60.

———. 河南輝縣市孟莊龍山文化遺址發掘簡報。考古 2000.3: 1–20.

———. 河南淮陽平糧台龍山文化城址試掘簡報。文物 1983.3: 21–36.

———. 登封王城崗遺址的發掘。文物 1983.3: 8–20.

Ho-pei-sheng Po-wu-kuan et al. 河北省博物館。河北槁城縣台西村商代遺址 1973 年的重要發現。文物 1974: 8.

Ho-pei-sheng Wen-hua-chü Wen-wu Kung-tso-tui. 河北省文化局文物工作隊。河北邯鄲澗泃村古遺址發掘簡報。考古 1961.4: 197–202.

Ho-pei-sheng Wen-wu Yen-chiu-suo. 河北省文物研究所, 石家莊市文物研究所, 正定縣文物保護管理所。河北正定縣曹村商周遺址發掘簡報。考古 2007.11: 26–35.

———. 河北省文物研究所, 保定市文物管理處, 容城縣文物保管所。河北容城縣上坡遺址發掘簡報。考古 1999.7: 1–7.

Hsi K'o-ting. 席克定。貴州的石器時代考古。考古 1994.8: 702–709.

Hsi-pei Ta-hsüeh Li-shih-hsi K'ao-ku Chuan-yeh. 西北大學歷史系考古專業。西安老牛坡商代基地的發掘。文物 1988.6: 1–22.

Hsia Mai-ling. 夏麥陵。殷代能冶鐵嗎？先秦、秦漢史 1986. 6:68–72.

Hsiang Ch'un-sung and Li Yi. 項春松, 李義。寧城小黑石泃石槨墓調查清理報告。文物 1995.5: 4–22.

Hsiang-kang Ku-wu Ku-chi Pan-shih-ch'u. 香港古物古跡辦事處。香港涌浪新石器時代遺址發掘簡報。考古 1997.6: 35–53.

Hsiao Meng-lung. 肖夢龍。試論吳越青銅兵器。考古與文物 1996. 6:14–27.

Hsiao Nan. 肖楠。試論卜辭中的師和旅。古文字研究第六輯（四川大學歷史系古文字研究室編），1981, 123–131.

Hsiao Ping. 蕭兵。蚩尤是南中國的英雄祖先—兼論蚩尤族與蛇牛圖騰、割體葬儀、戰鼓靈力、獵頭之風、楓木信仰等的民俗關係。中國古代史（一）1994.11: 7–12.

Hsü Chao-ch'ang. 許兆昌。胤征義和事實考。先秦、秦漢史 2004.4: 22–27.

Hsü Chao-feng. 徐昭峰。從 湯始居亳"說到湯都鄭亳。考古與文物 1999.3: 43–48.

Hsü Chao-feng and Yang Yüan. 徐昭峰，楊遠。鄭州大師姑發現的早商文化與商湯滅夏。考古與文物 2008.5:26–30.

Hsü Chi. 徐基。濟南大辛莊遺址發現商代青銅兵器。文物 1995. 6:86–87.

———. 山東商代考古研究的新進展。三代文明研究（一），1999, 257–269.

Hsü Chung-shu. 徐中舒。《羌族史稿》序。歷史研究 1983.1: 60–61.

———. 殷周之際史蹟之檢討。中央研究院歷史語言研究所集刊 7.2 (1936):137–164.

Hsü Hung. 許宏。二里頭遺址發掘和研究的回顧與思考。考古 2004.11:32–38.

———. 論夏商西周三代城市之特質。三代文明研究（一），1999, 286–295.

Hsü Hung et al. 許宏，陳國梁，趙海濤。二里頭遺址聚落形態的初步考察。考古 2004. 11:23–31.

Hsü Po-hung. 徐伯鴻。盤庚遷殷年代考及大龜四版歷法復原。先秦、秦漢史 1998. 4:29–36.

Hsü Shun-chan. 許順湛。尋找夏啟之居。先秦、秦漢史 2004. 6:13–17.

———. 再論夏王朝前夕的社會形態。夏文化研究論集，1996, 128–135.

———. 再論黃帝時代是中國文明的源頭。考古與文物 1997. 4:19–26.

Hsü Yü-lin and Yang Yung-fang. 許玉林，楊永芳。遼寧岫岩北溝西山遺址發掘簡報。考古 1992. 5:389–398.

Hsüeh Jui-che. 薛瑞澤。論河洛文化的濫觴期。先秦、秦漢史 2006. 4:13–22.

Hu Chia-ts'ung. 胡家聰。從 《管子》看田氏齊國崇奉黃帝—兼論 "百家言黃帝" 的時代思潮。先秦、秦漢史 1991. 1:19–26.

Hu Hou-hsüan. 胡厚宣。甲骨文土方為夏民族考。先秦、秦漢史 1991. 2:13–20.

———, ed. 甲骨探史錄。北京，三聯書店，1982.

Hu Hou-hsüan and Hu Chen-yü. 胡厚宣，胡振宇。甲骨文師為武官。殷商史。上海，上海人民出版社，2003, 107–115.

Hu-pei-sheng Po-wu-kuan. 湖北省博物館。盤龍城商代二里岡期的青銅器。文物 1976. 2:26–41.

———. 盤龍城一九七四年度田野考古紀要。文物 1976. 2:5–15.

Hu-pei-sheng Wen-wu K'ao-ku Yen-chiu-suo. 湖北省文物考古研究所。湖北隨州市黃土崗遺址新石器時代 環壕的發掘。考古 2008. 11:3–14.

Hu-pei-sheng Wen-wu K'ao-ku Yen-chiu-suo et al. 湖北省文物考古研究所，十堰市博物館，丹江口市博物 館。湖北丹江口市吉家院墓地的清理。考古 2000.8:55–64.

———. 湖北棗陽市九連墩楚墓。考古 2003.7: 10–14.

Hua Chüeh-ming. 華覺明。中國古代金屬技術—銅和鐵造就的文明。鄭州。大象出版社，1999.

Huang Chien-hua. 黃劍華。三星堆文明與中原文明的關係。先秦、秦漢史 2001. 6:21–27.

Huang Hsin-chia. 黃炘佳。三皇五帝及華夏文化探源—中國上古神話譜系的文化人類學研究。先秦、秦漢史 1993. 11:25–32.

Huang Huai-hsin. 黃懷信。仰韶文化與原始華夏族—炎、黃部族。考古與文物 1997. 4:33–37.

Huang Jan-wei. 黃然偉。殷周史料論集。三聯書店（香港）有限公司，1995.

Huang Shang-ming. 黃尚明。城固洋縣商代青銅器群族屬再探。考古與文物 2002. 5:40–45.

Huang Sheng-chang. 黃盛璋。論中國早期（銅鐵以外）的金屬工藝。考古學報 1996. 2:143–164.

Huang Shih-lin. 黃石林。三論夏文化問題—夏文化探索與中原龍山文化。夏文化研究論集, 1996, 16–26.

Huber, Louisa G. F. "The Bo Capital and Questions Concerning Xia and Early Shang." *Early China* 13 (1988): 46–77.

Hunan-sheng Wen-wu K'ao-ku Yen-chiu-suo. 湖南省文物考古研究所。灃縣城頭山古城址, 1997–1998 年度發掘 簡報。文物 1999.6: 4–17.

Hunan-sheng Wen-wu K'ao-ku Yen-chiu-suo and Hunan Sheng Li-hsien Wen-wu Kuan-li-suo. 湖南省文物考古 研究所, 湖南省灃縣文物管理所。灃縣城頭山屈家嶺文化城址調查與試掘。文物 1993.12: 19–30.

Ivanhoe, Philip J., ed. *Chinese Language, Thought, and Culture: Nivison and His Critics*. Peru: Open Court, 1996.

Jager, Ronald. "Tool and Symbol: The Success of the Double-Bitted Axe in North America." *Technology and Culture* 4 (1999): 833–860.

Jen Shih-nan. 任式楠。中國史前城址考察。考古 1998. 1:1–16.

———. 公元前五千年前中國新石器文化的幾項主要成就。考古 1995. 1:37–49.

Jui Kuo-yao and Shen Yüeh-ming. 芮國耀, 沈岳明。良渚文化與商文化關係三例。考古 1992. 11:1039–1044.

Junkmanns, Jurgen. "Early Stone Age Bows." In *The Bow Builders Book*, edited by Flemming Alrune, 47–55.

———. "Prehistoric Arrows." In *The Bow Builders Book*, edited by Flemming Alrune, 57–73.

Kan Chih-keng, Li Tien-fu, Ch'en Lien-k'ai. 干志耿, 李殿福, 陳連開。商先起源於幽燕說。歷史研究 1985.5:21–34.

Kan-su-sheng Wen-wu K'ao-ku Yen-chiu-suo. 甘肅省文物考古研究所, 吉林大學考古學系。甘肅民樂縣東灰山 遺址發掘紀要。考古 1995. 12:1057–1063.

———. 肅省文物考古研究所。甘肅秦安縣大地灣遺址仰韶文化早期聚落 發掘簡報。考古 2003. 6:19–31.

Kao Chih-hsi. 高至喜。記長沙、常德出土弩機的戰國基—兼談有關弩機、弓矢的幾個問題。文物 1964. 6:33–45.

Kao Kuang-jen. 高廣仁。海岱區的商代文化遺存。考古學報 2000. 2:183–198.

Kao Meng-ho. 高蒙河。從江蘇龍南遺址論良渚文化的聚落形態。考古 2000. 1:54–60.

Kao T'ien-lin. 高天麟。候馬東呈王新石器時代遺址發掘的重要意義。考古 1992. 1:62–68.

Kao Wei et al. 高煒, 楊錫璋, 王巍, 杜金鵬。偃師商城與夏商文化分界。考古 1998. 10:66–79.

Karlgren, Bernhard. "The Book of Documents." *BMFEA* 22 (1950): 1–81.

———. "Glosses on The Book of Documents." *BMFEA* 20 (1948): 39–315.

———. *Grammata Serica Recensa*. *BMFEA* 29 (1957): 1964 (reprint).

———. "Some Weapons and Tools of the Yin Dynasty." *BMFEA* 17 (1945): 101–144 and 40 plates.

Kawamata Masanori. "Higashi Ajia no Kodai Sensha to Nishi-Ajia." *Koshi Shunju* 4 (1987): 38–58.

Keegan, John. *A History of Warfare*. New York: Alfred A. Knopf, 1993.

Keightley, David N. *The Ancestral Landscape: Time, Space, and Community in Late Shang China (ca. 1200–1045 B.C.)*. Berkeley: University of California Press, 2000.

———. "The Cradle of the East: Supplementary Comments." *Early China* 3 (1977): 55–61.

———. "Graphs, Words, and Meanings: Three Reference Works for Shang Oracle-bone Studies, with an Excursus on the Religious Role of the Day or Sun." *JAOS* 117, no. 3 (1997): 507–524.

———. "Public Work in Ancient China: A Study of Forced Labor in the Shang and Western Chou." PhD dissertation, Columbia University, 1969.

———. "The Religious Commitment: Shang Theology and the Genesis of Chinese Political Culture." *History of Religions* 17 (1978): 211–224.

――――. "Shang Divination and Metaphysics." *Philosophy East and West* 38 (1988): 367–397.

――――. "The Shang State as Seen in the Oracle Bone Inscriptions." *Early China* 5 (1979–1980): 25–34.

――――. *Sources of Shang History: The Oracle-Bone Inscriptions of Bronze Age China*. Berkeley: University of California Press, 1978.

――――. "Theology and the Writing of History: Truth and the Ancestors in the Wu Ding Divination Records." *JEAA* 1, nos. 1–4 (1999): 207–230.

――――. "Religion and the Rise of Urbanism." *JAOS* 93, no. 4 (1973): 527–538.

Keightley, David N., ed. *The Origins of Chinese Civilization*. Berkeley: University of California Press, 1983.

Kierman, Frank A., Jr. "Phases and Modes of Combat in Early China." In *Chinese Ways in Warfare*, edited by Frank A. Kierman Jr. and John Fairbank, 27–66.

Kierman, Frank A., Jr., and John Fairbank, eds. *Chinese Ways in Warfare*. Cambridge, MA: Harvard University Press, 1974.

Ko Sheng-hua. 葛生華。我國夏商時期的官吏制度。先秦、秦漢史 1992.11: 13–18.

Kou Pao-p'ing. 苟寶平。陝西城固縣徵集的商代銅戈。考古 1996.5: 50.

Ku Chieh-kang. 顧頡剛。古史辨。 7 vols. 上海。上海古籍出版社, 1981 (reprint of 1941 ed.).

Ku Chieh-kang and Yang Hsiang-k'uei. 顧頡剛, 楊向奎。中國古代車戰考略。東方雜誌 34:1 (1937): 39–54.

Kuei-chou-sheng WWKK YCS et al. 貴州省文物考古研究所, 貴州威寧縣雞公山遺址, 2004 年發掘簡報。考古 2006.8: 11–27.

――――. 貴州威寧縣吳家大坪 商周遺址。考古 2006.8: 28–39.

Kung Kuo-ch'iang. 龔國強。新疆地區早期銅器略論。考古 1997.9: 7–20.

Kung Wei-tuan. 宮為端。商代原始社會考。先秦、秦漢史 1986.11: 41–47.

Kung Wei-ying. 龔維英。上古南中國外來和土著各族關係淺探—兼論楚文化對南華諸族的影響。先秦、秦漢 史 1988.9: 40–48.

Kuo Hung-chi. 郭洪紀。從"武經七書" 看儒家對傳統兵學的整合。中國哲學史 1994.10: 65–71.

Kuo Mo-juo, ed. 郭沫若。甲骨文合集。北京。中華書局, 13 vols., 1978–1982.

Kuo Muo-jo. 郭沫若。說戟。殷周青銅器銘文研究卷二。上海, 1931 (1954 reprint), 172–186.

Kuo Pao-chün. 郭寶鈞。中國青銅器時代。台北, 駱駝出版社, 1976.

――――. 戈戟餘論。中央研究院歷史語言研究所集刊 5.3 (1935): 313–326.

――――. 殷周車器研究。北京, 文物出版社, 1998.

――――. 殷周的青銅武器。考古 1961.2: 111–118.

Kuo Yen-li. 郭妍利。論商代青銅兵器的明器化現象。考古與文物 2006.6: 66–73.

Lang Shu-te. 郎樹德。甘肅秦安縣大地灣遺址聚落形態及其演變。考古 2003.6: 83–89.

Lawrence, Elizabeth A. *Hoofbeats and Society: Studies of Human-Horse Interactions*. Bloomington: Indiana University Press, 1985.

Lay, M. G. *Ways of the World: A History of the World's Roads and of the Vehicles That Used Them*. New Brunswick, N.J.: Rutgers, 1992.

Le Blanc, Charles. "A Re-examination of the Myth of Huang-ti." *Journal of Chinese Religions* 13/14 (1985/1986): 45–63.

LeBlanc, Steven A. *Constant Battles: The Myth of the Peaceful, Noble Savage*. New York: St. Martin's, 2003.

Legge, James, trans. *Li Chi*. London: Oxford University Press, 1885.

――――. *The She King*. London: Oxford University Press, 1871.

――――. *The Shoo King or Book of Historical Documents*. London: Oxford University Press, 1894.

Lei Hsing-shan. 雷興山。對關中地區商文化的幾點認識。考古與文物 2000.2: 28–34.

Lei Yüan-sheng and Chi Te-yüan. 雷淵深，季德源。中國歷代軍事職官制度。中國史研究 1993. 4:3–19.

Li Ch'iao. 李喬。再次証明半坡陶文是古彝文始祖。先秦、秦漢史 1992. 5:21–26.

Li Chi. *Anyang*. Seattle: University of Washington, 1977.

——. 李濟。記小屯出土之青銅器。中國考古學報 1949. 4:1–69.

——. 殷墟有刃石器圖說。中央研究院歷史語言研究所集刊 23b (1952):523–619.

——. 豫北出土青銅句兵分類圖解。中央研究院歷史語言研究所集刊 1950. 22:1–17.

Li Chien-min. 李健民。論新干商代大墓出土的青銅戈、矛及其相關問題。考古 2001. 5:60–69.

Li Feng. 李鋒。鄭州大師姑城址商湯滅夏前所居亳說新論。華夏考古 2006. 2:67–72.

——. 鄭州大師姑城址商湯韋亳之我見。考古與文物 2007. 1:62–66.

Li Hai-jung. 李海榮。關中地區出土商時期青銅器文化因素分析。考古與文物 2000. 2:35–47.

Li Hsien-teng. 李先登。再論關於探索夏文化的若干問題。夏文化研究論集, 1996, 27–34.

Li Hsien-teng and Yang Ying. 李先登，楊英。論五帝時代。先秦、秦漢史 2000.3: 9–19.

Li Hsin. 李鑫。西山古城與中原地區早期城市的起源。考古 2008.1: 72–80.

Li Hsin-ta. 李新達。武官制度卷。中國軍事制度史。鄭州。大象出版社, 1997.

Li Hsin-wei. 李新偉。地理信息系統支持的興隆洼文化手工業生產專業化研究。考古 2008. 6:58–68.

Li Hsüeh-ch'in. 李學勤。周公廟卜甲四片試釋。先秦、秦漢史 2005. 4:38–40.

——. 中國青銅器及其最新發現。中國古代史 1995. 12:6–12.

——. ed. (中國先秦史學會, 洛陽市第二文物工作隊)。夏文化研究論集。北京: 中華書局, 1996.

——. 商代夷方的名號和地望。中國史研究 2006. 4:3–7.

——. 帝辛征夷方卜辭的擴大。中國史研究 2008. 1:15–20.

——. 柞伯簋銘考釋。文物 1998. 11:67–70.

——. 有關古史的十個新發現。先秦、秦漢史 2005. 5:3–7.

Li Hsüeh-ch'in and Ai Lan. 李學勤，艾蘭。針刻紋三角援戈及相關問題。文物 1991. 1:20–25.

Li Hsüeh-shan. 李雪山。卜辭所見商代晚期封國分佈考。先秦、秦漢史 2005.1: 29–34.

Li Hu. 黎虎。夏商周史話。北京, 北京出版社, 1984, 130–136.

Li Liu. *The Chinese Neolithic: Trajectories to Early States*. Cambridge, UK: Cambridge University Press, 2004.

Li Liu and Chen Xingcan. *State Formation in Early China*. London: Duckworth, 2003.

Li Liu and Hung Xu. 許宏，劉莉。關於二里頭遺址的省思。文物 2008. 1:43–52.

——. "Rethinking Erlitou: Legend, History, and Chinese Archaeology." *Antiquity* 81 (2007): 886–901.

Li Min. 李民。中原古代文明進程中的 "萬邦" 時期。先秦、秦漢史 2005. 3:6–8, 13.

——. 關於盤庚遷殷後的都城問題。先秦、秦漢史 1988. 4:41–48.

——. 《尚書 甘誓》 "三正" 考辨。中國史研究 1980. 2:157–161.

——. 殷墟的生態環境與盤庚遷殷。歷史研究 1991. 1:111–120.

——. "区"與殷末周初之管地。先秦、秦漢史 1996. 2:44–46.

Li Nai-sheng et al. 李乃勝。安徽蒙城縣尉遲寺遺址紅燒土排房建築工藝的初步研究。考古 2005. 10:76–82.

Li Po-ch'ien. 李伯謙。中國古代文明演進的兩種模式—紅山、良渚、仰韶大基隨葬玉器觀察隨想。文物 2009. 3:47–56.

——. 中原地區東周銅劍淵源試探。文物 1982. 1:44–48.

——. 關於早期夏文化—從夏商周王朝更迭與考古學文化變遷的關係談起。先秦、秦漢史 2000. 3:20–23.

Li Shao-lien. 李紹連。關於商王國的政體問題—王國疆域的考古佐證。三代文明研究
　　（一）, 1999, 304–312.

Li Shao-ts'en. 李曉岑。從鉛同位素比值試析商周時期青銅器的礦料來源。考古與文
　　物 2002. 2:61–67.

Li Sheng. 厲聲。先秦國家形態與疆域、四土芻見。中國邊疆史地研究 2006. 9:1–8.

Li Shih-lung. 李世龍。中國古代帝王巡游活動述論。先秦、秦漢史 2002. 1:34–40.

Li Shui-ch'eng. 李水城。西北與中原早期冶銅業的區域特徵及交互作用。考古學報
　　2005. 3:239–278.

Li Shui-ch'eng and Shui T'ao. 李水城，水濤。四壩文化銅器研究。文物 2000. 3:36–44.

Li Shuicheng. "The Interaction between Northwest China and Cantral Asia during the Second
　　Millennium BC: An Archaeological Perspective." In *Ancient Interactions: East and West
　　in Eurasia*, edited by Katie Boyle et al., 171–182. Cambridge, UK: McDonald Institute
　　for Archaeological Research, 2002.

Li T'ao-yüan et. al. 李桃元，何昌義，張漢軍。盤龍城青銅文化。湖北美術出版社,
　　2002.

Li Tsung-t'ung. 李宗侗。炎帝與黃帝的新解釋。中央研究院歷史語言研究所集刊
　　39:27–39.

Li Tzu-chih. 李自智。中國古代都城布局的中軸線問題。考古與文物 2004. 4:33–42.

———. 先秦陪都初論。考古與文物 2002. 6:43–50.

Li Wei-ming. 李維明。20 世紀夏史與夏文化探索綜論。先秦、秦漢史 2000. 5:40–45.

———. 先商文化淵源與播化。考古與文物 2000. 3:51–55.

Li Yen. 李埏。夏、商、周—中國古代的一個歷史發展階段。先秦、秦漢史 1998. 2:19–24.

Li Yu-mou. 李友謀。炎黃文化與裴李崗文化。中國古代史 1994. 2:39–45.

Li Yung-hsien. 李永先。也談伏羲氏的地域和族系。先秦、秦漢史 1988. 10:13–20.

Li Yung-ti et al. 李永迪，岳占偉，劉煜。從孝民屯東南地出土陶范談對殷墟青銅器的
　　幾點新認識。考古 2007. 3:52–63.

Liao-ning-sheng Hsi-feng-hsien Wen-wu-kuan-li-suo. 遼寧省西豐縣文物管理所。遼寧西
　　豐縣新發現的幾座 石棺墓。考古 1995. 2:118–123.

Liao-ning-sheng Wen-wu K'ao-ku Yen-chiu-suo. 遼寧省文物考古研究所。牛河梁紅山文
　　化第二地點一號冢 石棺墓的發掘。文物 2008. 10:15–33.

Liao-ning-sheng Wen-wu K'ao-ku Yen-chiu-suo et al. 遼寧省文物考古研究所, 吉林大學
　　考古學系, 旅順博物 館。遼寧省瓦房店市長興島三堂村新石器時代遺址。考
　　古 1992.2: 107–121.

Liao-ning-sheng Wen-wu K'ao-ku Yen-chiu-suo and Chi-lin Ta-hsüeh K'ao-ku-hsüeh-hsi. 遼寧
　　省文物考古研究 所, 吉林大學考古學系。遼寧阜新平頂山石城址發掘報告。
　　考古 1992. 5:399–417.

Lien Shao-ming. 連劭名。關於商代稱謂的幾個問題。先秦、秦漢史 2000. 1:27–31.

Lin Hsiang-keng. 林祥庚。中華民族的象徵—黃帝及其傳說之試釋。先秦、秦漢史
　　1984. 1:3–10.

Lin Hsiao-an. 林小安。殷武丁臣屬征伐與行祭考。甲骨文與殷商史第二輯, 223–302.

Lin Huan. 林歡。晚商時期晉豫交界地帶的軍事駐地及相關地理問題。先秦、秦漢
　　史 2003. 3:57–63.

Lin Shou-chin. 林壽晉。論周代銅劍的淵源。文物 1963. 11:50–55.

———. 東周式銅劍初論。考古學報 1962. 2:75–84.

Lin Yün. "A Reexamination of the Relationship between Bronzes of the Shang Culture and of
　　the Northern Zone." In *Studies of Shang Archaeology*, edited by Chang, 237–273.

———. 林沄。中國東北系銅劍初論。考古學報 1980. 2:139–161.

Liu Chan, ed. 劉展。中國古代軍制史—國家哲學社會科學 "75" 規劃軍事學課題。軍
　　事科學出版社, 1992.

Liu Chao. 劉釗。卜辭所見殷代的軍事活動。古文字研究16, 中華書局, 1989, 67–139.

Liu Chao-ying and P'ei Shu-lan. 劉超英，裴淑蘭。河北商代帶銘銅器綜述。三代文明研究（一），1999, 365–372.

Liu Ch'i-yi. 劉啟益。"隞都" 質疑。文物 1961. 10:39–40.

Liu Chin-hsiang and Tung Hsin-lin. 劉晉祥，董新林。淺論趙寶溝文化的農業經濟。考古 1996. 2:61–64.

Liu Ch'ing-chu. 劉慶柱。中國古代都城考古學研究的幾個問題。考古 2000. 7:60–69.

———. 中國古代都城遺址布局形制的考古發現所反映的社會形態變化研究。考古學報 2006. 3:281–312.

Liu Chün-she. 劉軍社。試論先周文化與相鄰諸文化的關係。考古與文物 1994. 4:48–59.

Liu Chung-hui. 劉忠惠。對 "堯典" 背景的哲學思考。中國哲學與哲學史 1997. 1: 11–15.

Liu Fan-ti. 劉范弟。善卷傳說及其與蚩尤的關係考論。中國典籍與文化 1999, 70–74.

Liu Feng-hua. 劉風華。甲骨新綴八例。考古與文物 2007. 4:22–26.

Liu Hsiu-ming. 劉修 明。"治水社會" 和中國的歷史道路。中國史研究 1994. 2:10–18.

Liu Hsü. 劉緒。關於西亳說的幾個問題。夏文化研究論集，1996, 38–41.

———. 從夏代各部族的分佈和相互關係看商族的起源地。先秦、秦漢史 1989. 7: 21–26.

Liu Huan. 劉桓。釋甲骨文 ⊠ 字—兼說"王 ⊠ 于（某地）卜辭的性質。考古 2005. 11:58–62.

———. 說武丁時的一次南征。中國史研究 2002. 4:3–9.

———. 殷代 "德方" 說。中國史研究 1995. 4:93–98.

Liu Kuo-hsiang. 劉國祥。興隆洼文化聚落形態初探。考古與文物 2001. 6:58–67.

Liu Shih-chung et al. 劉詩中，曹柯平，唐舒龍。長江中游地區的古銅礦。考古與文物 1994. 1:82–88.

Liu Shih-chung and Lu Pen-shan. 劉詩中，盧本珊。江西銅嶺銅礦遺址的發掘與研究。考古學報 1998. 4:465–496.

Liu Shih-o. 劉士莪。西安老牛坡商代墓地初論。文物 1988. 6:23–27.

Liu Shih-o and Yüeh Lien-chien. 劉士莪，岳連建。西安老牛坡遺址第二階段發掘的主要收穫。先秦、秦漢史 1991. 10:15–19.

Liu Yi-man. 劉一曼。安陽小屯西地的先商文化遺存—兼論 "梅園莊一期" 文化的時代。三代文明研究（一），1999, 148–161.

———. 安陽殷墓青銅禮器組合的幾個問題。考古學報 1995. 4:395–412.

———. 論安陽殷墟墓葬青銅武器的組合。考古 2002. 3:63–75.

———. 殷墟青銅刀。考古 1993. 2:150–166.

Liu Yi-man and Ts'ao Ting-yün. 劉一曼，曹定云。殷墟花東H3卜辭中的馬—兼論商代馬匹的使用。先秦、秦漢史 2004. 5:24–29.

Liu Yung-hua, ed. 劉永華。中國古代軍戎服飾。上海古籍出版社，1995.

Liu Yü et al. 劉煜，岳占偉，何毓靈，唐錦瓊。殷墟出土青銅禮器鑄型的制作工藝。考古 2008. 12:80–90.

Liu Yü-t'ang. 劉玉堂。夏商王朝對江漢地區的鎮撫。先秦、秦漢史 2001. 4:53–60.

Loehr, Max. "The Earliest Chinese Swords and the Akinakes." *Oriental Art* I (1948):132–142.

Loewe, Michael, ed. *Early Chinese Texts: A Bibliographical Guide.* Berkeley, Calif.: Society for the Study of Ancient China, 1993.

Loewe, M., and E. L. Shaughnessy, eds. *The Cambridge History of Ancient China: From the Origin of Civilization to 221 B.C.* Cambridge, UK: Cambridge University Press, 1999.

Lu Liancheng. "Chariot and Horse Burials in Ancient China." *Antiquity* 67 (1993):824–838.

Lu Lien-ch'eng. 盧連成。商代社會疆域地理的政治架構與周邊地區青銅文化。中國古代史 1995. 4:30–56.

Lu Pen-shan and Liu Shih-chung. 盧本珊，劉詩中。銅嶺商周銅礦開采技術初步研究。文物 1993. 7:s33–38.

Lu Ti-min and Wang Ta-yeh, eds., 路迪民, 王大業 中國古代冶金與金屬文物。西安, 陝西科學技術, 1998.

Lü Wen-yü. 呂文郁。論堯舜禹時代的部族聯合體。先秦、秦漢史 2000. 1:10–17.

Luan Feng-shih. 欒豐實。商時期魯北地區的夷人遺存。三代文明研究（一），1999, 270–279.

———. 試論仰韶時代東方與中原的關係。考古 1996. 4:45–58.

Luo Hsi-chang and Wang-Chün-hsien. 羅西章, 王均顯。周原扶風地區出土西周甲骨的初步認識。文物 1987. 2:17–26.

Luo K'un. 羅琨。二里頭文化南漸與伐三苗史跡索隱。夏文化研究論集, 1996, 197–204.

———.『高宗伐鬼方』史蹟考辨。甲骨文與殷商史, 胡厚宣編, 上海古籍出版社, 1983, 83–125.

———. 殷商時期的羌和羌方。甲骨文與殷商史第三輯, 上海古籍出版社, 1991, 405–426.

Luo K'un and Chang Yung-shan. 羅琨, 張永山。夏商西周軍事史 (中國軍事通史 Vol. 1)。軍事科學出版社, 1998.

Lynn, John A. *Battle: A History of Combat and Culture.* Boulder, Colo.: Westview Press, 2003.

Lynn, John, ed. *Feeding Mars: Logistics in Western Warfare from the Middle Ages to the Present.* Boulder, Colo.: Westview Press, 1993.

Ma Heng. 馬衡。戈戟之研究。燕京學報 1929. 5:745–753.

Ma Hsi-lun. 馬璽倫。山東沂水新發現一件帶鳥形象形文字的銅戈。文物 1995. 7:72–73.

Ma Hsi-lun and Kung Fan-kang. 馬璽倫, 孔凡剛。山東沂水發現商代青銅器。文物 1989. 11:95–96.

Ma Shih-chih. 馬世之。中原龍山文化城址與華夏文明的形成。夏文化研究論集, 1996, 103–108.

———. 新砦城址與啟都夏邑問題探索。考古與文物 2007. 3:54–58.

Ma Te-chih et al. 馬得志, 周永珍, 張雲鵬。一九五三年安陽大司空村發掘報告。考古學報 1955. 9:60–79.

Macksey, Kenneth. *For Want of a Nail.* London: Brassey's, 1989.

Mei Chien-chün et al. 新疆出工銅斧的初步科學分析。考古 2005. 4:78 ff.

Mei Jianjun. "Cultural Interaction between China and Central Asia during the Bronze Age." *Proceedings of the British Academy* 121 (2003): 1–39.

———. "Qijia and Seima-Turbino: The Question of Early Contacts between Northwest China and the Eurasian Steppe." BMFEA 75 (2003):31–54.

Meng Shih-k'ai. 孟世凱。《夏商史話》。北京, 中國青年出版社, 1986, 200–285.

———. 殷商時代田獵活動的性質與作用。歷史研究 1990. 4:95–104.

Miao Ya-chüan. 繆雅娟。關於中國文明形成的思考。先秦、秦漢史 2004. 3:13–19, 26.

Miyake Toshihiko. 三宅俊彥。卡約文化青銅器初步研究。考古 2005. 5:73–88.

Nan P'u-heng et al. 南普恆, 秦穎, 李桃元, 董亞巍。湖北盤龍城出土部分商代青銅器鑄造地的分析。文物 2008.8: 77–82.

National Palace Museum. 國立故宮博物院。故宮青銅兵器圖錄。台北, 故宮, 1995.

Needham, Joseph. *Civil Engineering and Nautics, Science and Civilisation in China.* Vol. 4, Pt. 3 of *Science and Civilisation in China.* Cambridge, UK: Cambridge University Press, 1971.

———. *Physics and Physical Technology: Mechanical Engineering.* Vol. 4, Pt. 2 of *Science and Civilisation in China.* Cambridge, UK: Cambridge University Press, 1965, 73–82, 246–253.

Nei-Meng-ku Tzu-chih-ch'ü WWKK YCS. 內蒙古自治區文物考古研究所。內蒙古林西縣白音長汗新石器時代 遺址發掘簡報。考古 1993. 7:577–586.

Nei-Meng-ku WWKK YCS. 內蒙古文物考古研究所。內蒙古赤峰市三座店夏家店下層文化石城遺址。考古 2007. 7:17–27.

Nelson, Sarah, ed. *The Archaeology of Northeast China.* London: Routledge, 1995.

Ning Ching-t'ung. 寧景通。河南伊川縣發現商墓。文物 1993. 6:61–64.

Ning Yüeh-ming et al. 寧越敏，張務棟，錢今昔。中國城市發展史。合肥，安徽科學技術出版社，1994.

Nivison, David. "The 'Question' Question with Additional Comments." *Early China* 14 (1989): 115–126.

———. "Royal 'Virtue' in Shang Oracle Inscriptions." *Early China* 4 (1977–1978):52–55.

Nivison, David, and Kevin D. Pang. "Astronomical Evidence for the *Bamboo Annals'* Chronicle of Early Xia" [with appended discussion and responses from Huang Yi-long, John S. Major, David W, Pankenier, Zhang Peiyu, Nivison, and Pang]. *Early China* 15 (1990):86–196.

Nylan, Michael. "The *Chin Wen/Ku Wen* Controversy in Han Times." *T'oung Pao* 80, nos. 1–3 (1994):83–145.

———. "The *Ku Wen* Documents in Han Times." *T'oung Pao* 81, nos. 1–3 (1995):22–50.

O'Flaherty, Ronan. "A Weapon of Choice—Experiments with a Replica Irish Early Bronze Age Halberd." *Antiquity* 81 (2007):423–434.

P'an Chi-an. 潘繼安。陶寺遺址為黃帝及帝譽之都考。考古與文物 2007. 1:56–61.

P'ang Ming-chüan. 潘明娟。從鄭州商城和偃師商城的關係看早商的主都和陪都。考古 2008. 2:55–63.

P'ang Wen-lung. 龐文龍。岐山縣博物館藏商周青銅器錄遺。考古與文物 1994.3:28–40, 56.

Pankenier, D. W. "Astronomical Dates in Shang and Western Zhou." *Early China* 7 (1981–1982): 2–37.

———. "Mozi and the Dates of Xia, Shang, and Zhou: A Research Note." *Early China* 9–10 (1983–1985): 175–183.

Pei-ching Kang-t'ieh Hsüeh-yüan Yeh-chin Shih-tsu. 北京鋼鐵學院冶金史組。中國早期銅器的初步研究。考古學報 1981.3: 287–302.

Pei-ching-shih Wen-wu Kuan-li-ch'u. 北京市文物管理處。北京地區的又一重要考古收穫—昌平白浮西周 木槨墓的新啟示。考古 1976. 4:246–258, 228.

Pei-ching Ta-hsüeh K'ao-ku Wen-po Hsüeh-yüan. 北京大學考古文博學院，河南省文物考古研究所。河南 登封市王城崗城址 2002、2004 年發掘簡報。考古 2006. 9:3–15.

P'ei An-p'ing. 裴安平。澧陽平原史前聚落形態的特點與演變。考古 2004. 11:63–76.

———. 史前私有制的起源—湘西北澧陽平原個案的分析与研究。文物 2007. 7:75–80, 96.

P'ei Ming-hsiang. 裴明相。論登封王城崗城堡的性質。夏文化研究論集, 1996, 60–65.

Pen-hsi-shih Po-wu-kuan. 本溪市博物館，桓仁縣文管所。遼寧桓仁縣抱圈沟遺址。考古 1992. 6:505–508.

P'eng Ming-han. 彭朋瀚。江西新干晚商遺存出土青銅農具淺析。先秦、秦漢史 1996. 2:47–52.

———. 商代虎方文化初探。中國史研究 1995. 3:101–108.

P'eng Shih-fan and Yang Jih-hsin. 彭適凡，楊日新。江西新干商代大墓文化性質芻議。文物 1993. 7:10–18.

P'eng Yü-shang. 彭裕商。殷墟甲骨斷代。北京，中國社會科學出版社，1994.

Pi Shuo-pen, P'ei An-p'ing, and Lü Kuo-nien. 畢碩本，裴安平，閻國年。基于空間分析方法的姜寨史前聚落 考古研究。考古與文物 2008. 1:9–17.

Piggot, Stuart. "Chariots in the Caucasus and China." *Antiquity* 48 (1974):16–24.

———. *The Earliest Wheeled Transport: From the Atlantic Coast to the Caspian Sea.* Ithaca, N.Y.: Cornell University Press, 1983.

———. *Wagon, Chariot and Carriage: Symbol and Status in the History of Transport.* New York: Thames & Hudson, 1992.

Proenca, Domicio, Jr., and E. E. Duarte. "The Concept of Logistics Derived from Clausewitz: All That Is Required So That the Fighting Force Can Be Taken as a Given." *Journal of Strategic Studies* 28, no. 4 (2005): 645–677.

Pulleyblank, E. G. "The Chinese and Their Neighbors in Prehistoric and Early Historic Times." In *The Origins of Chinese Civilization*, edited by D. N. Keightley, 411–466. Berkeley: University of California Press, 1983.

Qiu Xigui. "An Examination of Whether the Charges in Shang Oracle-bone Inscriptions Are Questions." *Early China* 14 (1989): 77–114.

Rogers, Clifford J. "The Efficacy of the English Longbow: A Reply to Kelly DeVries." *War in History* 5, no. 2 (1998):233–242.

Sawyer, Ralph D. "Chinese Warfare: The Paradox of the Unlearned Lesson." *American Diplomacy Magazine* (Fall 1998).

———. *Fire and Water: The Art of Incendiary and Aquatic Warfare in China*. Boulder, Colo.: Westview Press, 2004.

———. "Martial *Qi* in China: Courage and Spirit in Thought and Military Practice." *Journal of Military and Strategic Studies* (Winter 2008/2009).

———. *One Hundred Unorthodox Strategies: Battle and Tactics of Chinese Warfare*. Boulder, Colo.: Westview Press, 1996.

———. "Paradoxical Coexistence of Prognostication and Warfare." *Sinoplatonic Papers* 157 (2005).

———. *The Seven Military Classics of Ancient China*. Boulder, Colo.: Westview Press, 1993.

———. *Ssu-ma Fa*. In *The Seven Military Classics of Ancient China,* by Ralph Sawyer. Boulder, Colo.: Westview Press, 1993.

———. *Sun Pin Military Methods*. Boulder, Colo.: Westview Press, 1995.

———. *Sun-tzu Art of War*. Boulder, Colo.: Westview Press, 1994.

———. *T'ai Kung Liu-t'ao (Six Secret Teachings)*. In *Seven Military Classics of Ancient China*. Boulder, Colo.: Westview Press, 1993.

———. *The Tao of Deception: Unorthodox Warfare in Historic and Modern China*. New York: Basic Books, 2007.

———. *The Tao of Spycraft: Intelligence Theory and Practice in Traditional China*. Boulder, Colo.: Westview, 1998.

———. *The Tao of War*. Boulder, Colo.: Westview Press, 1999.

———. "Three Strategies of Huang Shih-kung." In *Seven Military Classics of Ancient China*. Boulder, Colo.: Westview Press, 1993.

Serruys, Paul. "Studies of the Language of the Shang Oracle Inscriptions." *T'oung Pao* 60, nos. 1–3 (1974):12–120.

Shaan-hsi Chou-yüan K'ao-ku-tui. 陝西周原考古隊。扶風縣齊家村西周甲骨發掘簡報。文物 1981.9: 1–7.

———. 陝西岐山鳳雛村發現周初甲骨文。文物 1979.10: 38–43.

Shan-hsi-sheng Lin-fen Hsing-shu Wen-hua-chü. 山西省臨汾行署文化局，中國社會科學院考古研究所山西工作隊。山西臨汾下靳村陶寺文化墓地發掘報告。考古學報 1999. 4:459–486.

Shan-tung Ta-hsüeh Dung-fang K'ao-ku Yen-chiu-suo. 山東大學東方考古研究中心，山東省文物考古研究所，濟南市考古所。濟南市大辛莊遺址出土商代甲骨文。考古 2003. 6:3–6.

Shan-tung-sheng Po-wu-kuan. 山東省博物館。山東益都蘇埠屯第一號奴隸殉葬墓。文物 1972. 8:17–30.

Shan-tung-sheng WWKK YCS. 山東省文物考古研究所。山東章丘市西河新石器時代遺址1997年的發掘。考古 2000. 10:15–28.

———. 山東滕州市西公橋大汶口文化遺址發掘簡報。考古 2000. 10:29–44.

Shan-tung-sheng WWKK YCS Liao-ch'eng ti-ch'ü Wen-hua-chü Wen-wu Yen-chiu-suo. 山東省文物考古研究所, 聊城地區文化局文物研究室。山東陽谷縣景陽崗龍山文化城址調查與試掘。考古 1997. 5:11–24.

Shang-hai Po-wu-kuan K'ao-ku Yen-chiu-suo. 上海博物館考古研究部。上海松江區廣富林遺址, 2001–2005 年發掘簡報。考古 2008. 8:3–21.

Shang Chih-t'an. 商志譚。蘇南地區青銅器合金成份的特色及相關問題。文物 1990.9: 48–55.

Shang Ch'ing-fu. 商慶夫。中國原始陣法"伏羲先天圓陣" 考述。先秦、秦漢史 1999. 6:5–15.

Shanghai Po-wu-kuan K'ao-ku Yen-chiu-pu. 上海博物館考古研究部。上海金山區亭林遺址 1988, 1990 年良渚 文化墓葬的發掘。考古 2002. 10:49–63.

Shao Wangping. "The Interaction Sphere of the Longshan Period." In *The Formation of Chinese Civilization*, edited by Chang Kwang-chih and Xu Pingfang, 85–124.

———. "The Longhsan Period and Incipient Chinese Civilization." *JEAA* 2, nos. 1–2 (2000): 195–226.

Shaughnessy, Edward L. *Before Confucius*. Albany: State University of New York Press, 1997.

———. "The 'Current' *Bamboo Annals* and the Date of the Zhou Conquest of Shang." *Early China* 11–12 (1985–1987): 33–60.

———. "Historical Perspectives on the Introduction of the Chariot into China." *HJAS* 48, no. 1 (1988): 189–237.

———. "Micro-periodization and the Calendar of a Shang Military Campaign." In *Chinese Language, Thought, and Culture: Nivison and His Critics*, edited by Philip J. Ivanhoe, 58–82. Peru: Open Court, 1996.

———. "On the Authenticity of the *Bamboo Annals*." *HJAS* 46, no. 1 (1986): 149–180.

———. "Shang Shu." In *Early Chinese Texts: A Bibliographical Guide*, edited by Michael Loewe, 376–389. Berkeley, Calif.: Society for the Study of Ancient China, 1993.

Shen Ch'ang-yün. 沈長雲。夏代是杜撰的嗎—與陳淳先生商榷。先秦、秦漢史 2005. 5:8–15.

———. 夏后氏居于古河濟之間考。中國史研究 1994. 3:113–122.

———. 說殷墟卜辭中的 "王族"。先秦、秦漢史 1998. 4:23–28.

Shen Chien-hua. 沈建華。卜辭所見商代的封疆與納貢。中國史研究 2004. 4:3–14.

Shen Jung. 沈融。論早期青銅戈的使用法。考古 1992. 1:69–75.

———. 商與西周青銅矛研究。考古學報 1998. 4:447–464.

———. 試論三角援青銅戈。文物 1993. 3:78–84.

Shih Chang-ju. 石璋如。小屯第四十墓的整理與殷代第一類甲種車的初步復原。中央研究院歷史語言研究所 集刊 40.b (1969):625–667.

———. 小屯殷代的成套兵器 (附殷代的策)。中央研究院歷史語言研究所集刊 22 (1950):19–79.

———. 殷車復原說明。中央研究院歷史語言研究所集刊 58.2 (1987):253–280.

———. 殷墟最近之重要發現附論小屯地層。田野考古報告第二冊 (1947):1–81.

Shih Ching-Sung. 施勁松。論帶虎食人母題的商周青銅器。考古 1998. 3:56–63.

Shima Kunio. 島邦男。殷墟卜辭研究。Hirosaki, 1958.

———. 編。殷墟卜辭綜類。東京, 1967, 1971.

Shirakawa Shizuka. 白川靜。金文通釋。56 vols. 神戶, 1962–1984.

SHYCS. *See* Chung-kuo She-hui K'e-hsüeh-yüan K'ao-ku Yen-chiu-suo.

Sleeswyk, Andre Wegeners. "Reconstruction of the South-Pointing Chariots of the Northern Sung Dynasty: Escapement and Differential Gearing in 11th Century China." *Chinese Science* 2 (1977): 4–36.

Ssu-ch'uan-sheng Wen-wu Kuan-li Wei-yüan-hui et al. 四川省文物管理委員會, 四川省文物考古研究所, 廣漢市文化局、文管所。廣漢三星堆遺址二號祭祀坑發掘簡報。文物 1989.5: 1–20.

Stevens, Anthony. *The Roots of War: A Jungian Perspective.* New York: Paragon House, 1989.

Su-chou Po-wu-kuan and Ch'ang-shu Po-wu-kuan. 蘇州博物館，常熟博物館。江蘇常熟羅墩遺址發掘簡報。文物 1999.7: 16–30.

Sukhu, Gopal. "Yao, Shun, and Prefiguration: The Origins and Ideology of the Han Imperial Genealogy." *Early China* 30 (2005–2006): 91–153.

Sun Chi. 孫機。從胸式系駕法到鞍套式系駕法—我國古代車制略說。考古 1980.5: 448–460.

———. 有刃車軎與多戈戟。文物 1980.12: 83–85.

Sun Hua. 孫華。夏文化探索中若干問題的思考。夏文化研究論集, 1996, 35–37.

———. "The Zhuwajie Bronzes." *JEAA* 5 (2006): 306–311.

Sun Sen. 孫淼。夏商史稿。北京，文物出版社，1987.

Sun Shuyun and Han Rubin. "A Preliminary Study of Early Chinese Copper and Bronze Artifacts." Translated by Julia K. Murray. *Early China* 9–10 (1983–1985): 260–289.

Sun Shu-yün and Han Ju-pin. 孫淑雲，韓汝玢。甘肅早期銅器的發現與冶煉、製造技術的研究。文物 1997. 7:75–84.

Sun Shu-yün et al. 孫淑雲 (北京科技大學冶金與材料史研究所，甘肅省文物考古研究所)。火燒溝四壩文化 銅器成分分析及制作技術的研究。文物 2003. 8:86–96.

Sun Ya-ping and Sung Chen-hao. 孫亞冰，宋鎮豪。濟南市大辛莊遺址新出甲骨卜辭探析。考古 2004.2: 66–75.

Sung Chen-hao. 宋鎮豪。夏商人口初探。歷史研究 1991.4: 92–106.

———. 再談殷墟卜用甲骨的來源。三代文明研究（一），1999, 392–398.

Sung Chih-min. 宋治民。三叉格銅柄鐵劍及相關問題的探討。考古 1997.12: 50–58.

———. 試論蜀文化和巴文化。考古學報 1999.2: 123–140.

Sung Hsin-ch'ao. 宋新潮。商代政治疆域與商文化影響範圍。中國史研究 1991.1: 53–63.

Sung Yü-ch'in and Li Ya-tung. 宋豫秦，李亞東。"夷夏東西說" 的考古學觀察。夏文化研究論集, 1996, 54–59.

Sung Yü-ch'in et al. 宋豫秦，鄭光，韓玉玲，吳玉新。河南偃師市二里頭遺址的環境信息。考古 2002.12: 75–79.

Tai Hsiang-ming. 戴向明。黃河流域新石器時代文化格局之演變。考古學報 1998.4: 389–418.

T'ai-chou-shih Wen-kuan-hui. 台州市文管會，玉環縣文管會。浙江玉環島發現的古文化遺存。考古 1996.5: 14–20.

T'an Tan-chiung. 譚旦冏。成都弓箭製作調查報告。中央研究院歷史語言研究所集刊 23a (1951): 199–243.

T'an Te-jui. 譚德睿。中國青銅時代陶范鑄造技術研究。考古學報 1999.2: 211–250.

T'an Wei-ssu. 譚維四。曾侯乙墓—20世紀中國文物考古發現與研究叢書。文物出版社, 2001.

T'ang Chi-ken. 唐際根。中商文化研究。考古學報 1999.4: 393–420.

T'ang Lan. 唐蘭。關於江西吳城文化遺址與文字的初步探索。文物 1975.7: 72–76.

T'ang Yün-ming. 唐雲明。槁⊠城台西商代鐵刃銅鉞問題的探討。文物 1975.3: 57–59.

T'ao Fu-hai. 陶富海。山西襄汾縣大崮堆山史前石器制造場新材料及其再研究。考古 1991.1: 1–7.

T'ao Yang and Chung Hsiu, eds. 陶陽，鍾秀。黃帝戰蚩尤。中國神話。上海文藝出版社, 1990, 504–508.

Teng Shu-p'ing. 鄧淑苹。晉、陝出土東夷系玉器的啟示。考古與文物 1999.5: 15–27.

Thorp, Robert. *China in the Early Bronze Age.* Philadelphia: University of Pennsylvania Press, 2006.

Thorp, Robert L. "Erhlitou and the Search for the Xia." *Early China* 16 (1991): 1–38.

T'ien Ch'ang-wu. 田昌五。夏商文化中的年代問題。先秦、秦漢史 1987.12: 12–16.

T'ien Chi-chou. 田繼周。夏代的民族和民族關係。先秦、秦漢史 1985.9: 25–30.

Ting Shan. 丁山。由陳侯因資錞銘黃帝論五帝。中央研究院歷史語言研究所集刊 32: 517–536.

Trigger, Bruce G. "Shang Political Organization: A Comparative Approach." *JEAA* 1 (1999): 43–62.

Ts'ao Ping-wu. 曹兵武。從垣曲商城看商代考古的幾個問題—（垣曲商城 1985–1986 年度勘查報告）讀後。文物 1997.12: 85–89.

Ts'ao Ting-yün. 曹定雲。夏代文字求證—二里頭文化陶文考。考古 2004.12: 76–83.

———. 殷墟婦好墓銘文中人物關係綜考。考古與文物 1995.5: 44–54.

———. 殷墟花東H3卜辭中的"王"是小乙—從卜辭中的人名"丁"談起。先秦、秦漢史 2007.5: 21–29.

Ts'ao Ting-yün and Liu Yi-man 曹定云，劉一曼。殷墟花園莊東地出土甲骨卜辭中的"中周"與早期殷周關係。考古 2005.9: 60–68.

Tsou Heng. "The Yanshi Shang City: A Secondary Capital of the Early Shang." Translated and edited by Lothar von Von Falkenhausen. *JEAA* 1, nos. 1–4 (1999): 195–205.

———. 鄒衡。鄭州小雙橋商代遺址隞（斁）都說輯補。考古與文物 1998.4: 24–27.

———. 鄭州商城即湯都亳說（摘要）。夏商周考古學論文集（續集）(reprint of 1978), 97–100.

———. 西亳與桐宮考辨。夏商周考古學論文集（續集）(reprint of 1987), 123–158.

———. 夏商周考古學論文集（續集）。北京，科學出版社，1998.

———. 夏文化的研究及其有關問題。夏商周考古學論文集（續集）(reprint of 1978), 3–10.

———. 邢台與先商文化、祖乙遷邢研究。三代文明研究（一），1999, 42–44.

———. 關於夏文化的上限問題—與李伯謙先生商討。考古與文物 1999.5: 50–54.

———. 關於探討夏文化的幾個問題。夏商周考古學論文集（續集）(reprint of 1979), 14–20.

———. 論菏澤（曹州）地區的岳石文化。夏商周考古學論文集（續集）(reprint of undated), 64–83.

———. 漫談商文化與商都。夏商周考古學論文集（續集）(reprint of 1993), 221–226.

———. 偃師商城即太甲桐宮說（摘要）。夏商周考古學論文集（續集）(reprint of 1984), 120–122.

———. 湯都垣亳說考辨。夏商周考古學論文集（續集）(reprint of undated), 204–218.

———. 再論"鄭亳說"—兼答石加先生。夏商周考古學論文集（續集）(reprint of 1981), 101–106.

———. 綜述夏商四都之年代和性質。夏商周考古學論文集（續集）(reprint of 1988), 173–188.

———. 綜述早商亳都之地望。夏商周考古學論文集（續集）(reprint of 1995), 117–119.

———. 桐宮再考辨—與王立新、林沄兩位先生商談。夏商周考古學論文集（續集）(reprint of 1998), 166–168.

Ts'ui Ta-hua. 崔大華。論經學的歷史發展。中國哲學史 1995.1: 55–63.

Tu Cheng-sheng. 杜正勝。夏代考古及其國家發展的探索。考古 1991.1: 43–56.

———. 從考古資料論中原國家的起源及其早期的發展。中央研究院歷史語言研究所集刊 58.1 (1987): 1–81.

Tu Chin-p'eng. 杜金鵬。安陽后岡殷代圓形葬坑及其相關問題。考古 2007.6: 76–89.

———. "鄭亳說"立論前提辨析。考古 2005.4: 69–77.

———. 夏商周始祖起源傳說新析。夏文化研究論集, 1996, 160–164.

———. 試論商代早期王宮池苑考古發現。考古 2006.11: 55–65.

———. 偃師二里頭遺址4號宮殿基址研究。文物 2005.6: 62–71.

———. 偃師商城初探。北京，中國社會科學出版社，2003.

———. 偃師商城考古新發現及其意義。先秦、秦漢史 1999.5: 38–40.

Tu Chin-p'eng and Wang Hsüeh-jung. 杜金鵬，王學榮。偃師商城近年考古工作要覽—
　　紀念偃師商城發現 20 周年。考古 2004.12: 3–12.

Tu Chin-p'eng, Wang Hsüeh-jung, Chang Liang-jen, Ku Fei. 杜金鵬，王學榮，張良仁，谷
　　飛。試論偃師商城東北隅考古新收穫。考古 1998.6: 9–13, 38.

Tu Tsai-chung. 杜在忠。邊線王龍山文化城堡試析—兼述我國早期國家誕生、文化融
　　合等有關問題。中國古代史 1995.8: 5–11.

Tu Yung. 杜勇。關於歷史上是否存在夏朝的問題。先秦、秦漢史 2006.6: 3–7.

T'u Wu-chou. 屠武周。神、炎帝和黃帝的糾葛。先秦、秦漢史 1984.3: 9–14.

Tuan Chen-mei. 段振美。殷墟考古史。河南，中州古籍出版社，1991.

Tuan Hung-chen. 段宏振。關於早商文化的起始年代及形成問題。三代文明研究
　　（一），1999, 214–222.

Tuan Yü. 段渝。略論古蜀與商文明的關係。先秦、秦漢史 2008.6: 3–9.

———. 巴蜀青銅文化的演進。文物 1996.3: 36–47.

———. 巴蜀古代城市的起源、結構和網絡體系。先秦、秦漢史 1993.4: 37–54.

———. 三星堆文化與古蜀文明—關於三星堆文化研究的論爭和前景。中國古代史
　　1994.1: 63–70.

———. 大禹史傳的西部底層。先秦、秦漢史 2005.1: 7–13.

———. 從血緣到地緣: 古蜀酋邦向國家的演化。先秦、秦漢史 2006.5: 16–20.

Tuan Yü and Ch'en Chien. 段渝，陳劍。成都平原史前古城性質初探。先秦、秦漢史
　　2002.2: 57–62.

Tung Ch'i. 董琦。夏代的中原。夏文化研究論集，1996, 46–53.

———. 論早期都邑。文物 2006.6: 56–60, 87.

———. 再談偃師商城年代可定論。考古與文物 1996.1: 27–31.

Tung Hsin-lin. 董新林。魏營子文化初步研究。考古學報 2000.1: 1–30.

Tung Tso-pin. 董作賓。殷曆譜。中國年曆總譜（全二冊）。香港，香港大學，1960.

Tung-hsia-feng K'ao-ku-tui. 東下馮考古隊。山西夏縣東下馮遺址東區、中區發掘簡報。
　　考古 1980.2: 97–107.

Tung-hsien-hsien K'ao-ku-tui. 東先賢考古隊。河北邢台市東先賢遺址1998年的發掘。
　　考古 2003.11: 27–40.

T'ung Chao. 童超。後勤制度卷，中國軍事制度史。鄭州。大象出版社，1997.

T'ung Chu-ch'en. 佟柱臣。中國夏商王國文明與方國文明試論。考古 1991.11: 1003–1018.

———. 從二里頭類型文化試談中國的國家起源問題。文物 1975.6: 29–33, 84.

T'ung En-cheng. 童恩正。我國西南地區青銅劍的研究。考古學報1977.2: 35–55.

Turney-High, H. H. *Primitive War.* Columbia: University of South Carolina Press, 1949. Reprint,
　　1991.

Van Crevald, Martin. *Supplying War: Logistics from Wallenstein to Patton.* 2nd ed. Cambridge, UK:
　　Cambridge University Press, 2004.

Van Ess, Hans. "The Old Text/New Text Controversy: Has the Twentieth Century Got It
　　Wrong？" *T'oung Pao* 80, nos. 1–3 (1994): 149–170.

Vogele, Konrad. "Woods for Bow Building." In *The Bow Builders Book*, edited by Flemming
　　Alrune, 97–103.

Von Falkenhausen, Lothar. "The External Connections of Sanxingdui." *JEAA* 5 (2006): 191–245.

Wagner, Donald. *Ferrous Metallurgy.* Vol. 5, Pt. 2 of *Science and Civilisation in China* (Joseph Need-
　　ham, series founder). Cambridge, UK: Cambridge University Press, 2008.

———. *Iron and Steel in China.* Leiden: E. J. Brille, 1996.

Wang Chan-yi. 王展意。中國古代道路交通史。北京，人民交通出版社，1994.

Wang Chen-chung. 王震中。"中商文化" 概念的意義及其相關問題。考古與文物
　　2006.1: 44–49.

———. 帝嚳并非商之始祖。先秦、秦漢史 2005.1: 3–6, 64.

———. 東夷的史前史及其燦爛文化。先秦、秦漢史 1988.6: 15–26.

Wang Ch'eng. 王成。內蒙古伊敏河煤礦出土曲刃青銅短劍。考古 1996.9: 94.

Wang Chien et al. 王建。下川文化—山西下川遺址調查報告。考古學報, 1978, 3.

Wang Chi-huai. 王吉懷。史前遺存中生產工具與建築工具的比較研究。考古與文物 2000.6: 36–41.

Wang Chih-p'ing. 王志平。一則蚩尤傳說的新解釋—兼論神話傳說中的語源迷誤。中國典籍與文化 1999.4: 95–98.

Wang Ch'ing. 王青。試論華夏與東夷集團文化交流及融合的地理背景。中國史研究 1996.2: 125–132.

Wang Ching-fu. 王經甫。鄉射之禮。孔孟月刊 23: 4 (1984): 52–56.

Wang En-t'ien et al. 王恩田。專家筆談丁公遺址出土陶文。考古 1993.4: 344–354, 375.

Wang Hai-ch'eng. 王海城。中國馬車的起源。歐亞學刊, 2002, 1–75.

Wang Heng-chieh. 王恒杰。從民族學發現的新材料看大汶口文化陶尊的"文字"。考古 1991.12: 1119–1120, 1108.

Wang Hsiao-wei, ed. 王曉衛。兵役制度卷。中國軍事制度史。鄭州，大象出版社，1997, 39–40.

Wang Hsüeh-jung. 王學榮。商代早期車轍與雙輪車在中國的出現。三代文明研究（一），1999, 239–247.

———. 偃師商城與二里頭遺址的幾個問題。考古1996.5: 51–60.

Wang Hsüeh-jung and Ho Yü-ling. 王學榮，何毓靈。安陽殷墟孝民屯遺址的考古新發現及相關認識。考古 2007.1: 54–63.

Wang Hsüeh-jung and Hsü Hung. 王學榮，許宏。"中國。二里頭遺址與二里頭文化國際學術研討會" 紀要。考古 2006.9: 83–90.

Wang Hsün. 王迅。二里頭文化與中國古代文明。考古與文物 1997.3: 61–68.

Wang Hui. 王暉。出土文字資料與五帝新證。考古學報 2007.1: 1–28.

———. 殷商為神本時代說。先秦、秦漢史 2000.6: 36–41.

Wang Hung-hsing. 王紅星。從門板灣城壕聚落看長江中游地區城壕聚落的起源與功用。考古 2003.9: 61–75.

Wang Jui. 王睿。垣曲商城的年代及其相關問題。考古 1998.8: 81–90.

Wang K'e-lin. 王克林。從出土文物看夏遺民的遷徙。考古與文物 2001.2: 48–53.

Wang Kuan-ying. 王冠英。殷周的外服及其演變。歷史研究 1984.5: 80–99.

Wang Kuei-min. 王貴民。商朝官制及其歷史特點。歷史研究 1986.4: 107–119.

———. 商代"眾人"身分為奴隸論。中國史研究 1990.1: 102–114.

Wang Kuo-wei. 王國維。鬼方昆夷玁狁考。觀堂集林。北京，中華書局, 1923, 1959 reprint.

Wang Li-chih. 王力之。商人屢遷中的湯亳。考古與文物 2003.4: 41–42.

Wang Shen-hsing. 王慎行。卜辭所見羌人考。古文字與殷周文明。西安，陝西人民教育出版社 , 1992, 113–142.

Wang Ssu-chou. 王嗣洲。遼寧普蘭店市孫屯村發現一件大型石矛。考古 2006.2: 95–96.

———. 論中國東北地區大石蓋墓。考古 1998.2: 53–63.

Wang Tsung-yao. 汪宗耀。湖北蘄春坳上灣新石器時代遺址。考古 1992.7: 663–665.

Wang Tzu-chin. 王子今。漢代"蚩尤" 崇拜。先秦、秦漢史 2006.6: 70–75.

Wang Wei. 王巍。公元前2000年前後我國大範圍文化變化原因探討。考古 2004.1: 67–77.

Wang Wen-hua et al. 王文華, 陳萬卿, 丁蘭坡。河南滎陽大師姑夏代城址的發掘與研究。文物 2004.11: 61–64, 74.

Wang Wen-kuang and Chai Kuo-ch'iang. 王文光, 翟國強。"五帝"世系與秦漢時期"華夷共祖"思想。中國邊疆 史地研究 2005.9: 1–18.

Wang Yen-chün. 王燕均。黃帝族發祥於東方初証。先秦、秦漢史 1988.6: 11–15.

Wang Yi. "Prehistoric Walled Settlements in the Chengdu Plains." *JEAA* 5 (2006): 109–148.

Wang Yi and Sun Hua. 王毅, 孫華。寶墩村文化的初步認識。考古 1999.8: 60–73.

Wang Yü-ch'eng. 王育成。中國父系氏族時代戰爭問題探索。中國史研究 1986.3: 71–84.

Wang Yü-che. 王玉哲。商族的來源地望試探。歷史研究 1984.1: 61–77.

Wang Yü-hsin. 王宇信。商代的馬和養馬業。中國史研究 1980.1: 99–108.

———. 談上甲至湯滅夏前商族早期國家的形成。先秦、秦漢史 2007.5: 14–20.

———. 武丁期戰爭卜辭分期的嘗試。甲骨文與殷商史第三輯, 上海古籍出版社, 1991, 142–174.

Wang Yü-hsin, ed. 王宇信。甲骨文與殷商史第三輯。上海, 上海古籍出版社, 1991.

Wang Yü-hsin, Chang Yung-shan, and Yang sheng-nan. 王宇信, 張永山, 楊升南。試論殷墟五號墓的"婦好"。考古學報 1977.2: 1–22.

Wang Yüeh-ch'ien and T'ung Wei-hua. 王月前, 佟偉華。垣曲商城遺址的發掘與研究—紀念垣曲商城發現 20 周年。考古 2005.11: 3–18.

Wang Yung-kang et al. 王永剛, 崔風光, 李延麗。陝西甘泉縣出土晚商青銅器。考古與文物 2007.3: 11–22.

Wang Yung-p'o. 王永波。並氏探略—兼論殷比干族屬。先秦、秦漢史 1992.4: 31–40.

Watson, Burton. *Han Fei-tzu: Basic Writings*. New York: Columbia University Press, 1964.

Wei Chi-yin. 魏繼印。殷商時期中原地區氣候變遷探索。考古與文物 2007.6: 44–50.

Wei Chien and Ts'ao Chien-en. 魏堅, 曹建恩。內蒙古中南部新石器時代石城址初步研究。文物 1999.2: 57–62.

Wei Ch'ung-wen. 衛崇文。晉南夏人探蹤。先秦、秦漢史 1991.6: 29–31.

Wen Fong, ed. *The Great Bronze Age of China*. New York: Alfred A. Knopf, 1980.

Wen Shao-feng and Yüan T'ing-tung. 溫少峰, 袁庭棟。殷墟卜辭研究—科學技術篇。四川省社會科學院出版社 , 1983, 258–298.

Wheatley, Paul. *The Pivot of the Four Quarters: A Preliminary Enquiry into the Origins and Character of the Ancient Chinese City*. Edinburgh, UK: Edinburgh University Press, 1971.

Wu En. 烏恩。論蒙古鹿石的年代及相關問題。考古與文物 2003.1: 21–30.

Wu Ju-tso. 吳汝祚。初探海岱地區古代文明的起源。中國古代史 1995.8: 12–20.

Wu Jui. 吳銳, 整理。楊向奎先生論炎帝文明。中國哲學史 1996.3: 4–8.

Xu. Jay. "Defining the Archaeological Cultures at the Sanxingdui Site." *JEAA* 5 (2006): 149–190.

Yang Chien-hua. 楊建華。燕山南北商周之際青銅器遺存的分群研究。考古學報 2002.2: 157–174.

Yang Chü-hua. 楊菊華。中國青銅文化的發展軌跡。先秦、秦漢史 1999.4: 28–43.

Yang Hsi-chang and T'ang Chi-ken. 楊錫璋, 唐際根。豫北冀南地區的中商遺存與盤庚以先的商都遷徙。三代文明研究 (一), 1999, 248–256.

Yang Hsi-chang and Yang Pao-ch'eng. 楊錫璋, 楊寶成。商代的青銅鉞。中國考古學研究夏鼐先生考古五十年紀念論文集, 1986, 128–138.

Yang Hsi-mei. 楊希枚。論商周社會的上帝太陽神。中國史研究 1992.3: 36–40.

Yang Hsin-kai and Han Chien-yeh. 楊新改, 韓建業。苗蠻集團來源與形成的探索。先秦、秦漢史 1997.2: 25–30.

———. 禹征三苗探索。中國古代史 1995.8: 32–40.

Yang Hua. 楊華。從鄂西考古發現談巴文化的起源。考古與文物 1995.1: 30–43.

Yang Hung. 楊泓。戰車與車戰二論。先秦、秦漢史 2000.5: 2–18

———. 中國古兵器論叢 (增訂本)。北京, 文物出版社, 1985.

———. 中國古代馬具的發展和對外影響。文物 1984.9:45–54.

Yang K'uan. 楊寬。中國古代都城制度史研究。上海古籍出版社, 1993.

———. 中國古代冶鐵技術發展史。上海:上海人民出版社, 1982.

———. 中國上古史導論。古史辨, vol. 7, pt. 1 (1941, reprint 1981), 65–404.

———. 楊寬。西周史。台灣商務印書館, 1999.

Yang Pao-ch'eng. 楊寶成。殷代車子的發現與復原。考古1984.6: 546–555.

———. 楊寶成。試論殷墟文化的年代分期。考古 2000.4: 74–80.

Yang Po-chün. 楊伯峻, 編。春秋左傳注（全四冊）修訂本。北京, 中華書局, 1990.

Yang Sheng-nan. 楊升南。夏時期的商人。夏文化研究論集 1996, 143–148.

———. 夏代軍事制度初探。先秦、秦漢史 1991.9: 42–48.

———. 略論商代的軍隊。甲骨探史錄。北京, 三聯書店, 1982, 340–399.

———.《尚書 甘誓》"五行" 說質疑。中國史研究 1980.2: 161–163.

———. 商代人牲身份的再考察。歷史研究 1988.1: 134–146.

———. 商代的財政制度。歷史研究 1992.5: 81–94.

———. 商代的土地制度。中國史研究 1991.4: 47–59.

———. 商代的王權和對王權的神化。中國史研究 1997.4: 16–23.

———.「殷人屢遷」辨析。甲骨文與殷商史第二輯, 185–222.

———. 殷墟卜辭中眾的身分考。甲骨文與殷商史第三輯, 上海古籍出版社 1991, 303–352.

Yang Te-piao. 楊德標。安徽省含山縣出土的商周青銅器。文物 1992.5: 92–93.

Yang Yü-pin. 楊育彬。夏商考古研究的新進展—（偃師商城初探）讀後。考古 2004.9: 87–92.

———. 關於夏文化的幾個問題。夏文化研究論集 1996, 11–15.

Yao Cheng-ch'üan et al. 姚政權, 吳妍, 王昌燧, 趙春青。河南新密市新砦遺址的植硅石分析。考古 2007.3: 90–96.

Yates, Robin D. S. "The Horse in Early Chinese Military History." *Academia Sinica* (2002): 1–78.

———. *Military Technology: Missiles and Sieges.* Vol. 5, Pt. 6 of *Science and Civilisation in China* (Joseph Needham, series founder). Cambridge, UK: Cambridge University Press, 1994.

———. "Slavery in Early China: A Socio-cultural Approach." *JEAA* 3, nos. 1–2 (2002): 283–331.

Yeh Hsiang-k'ui and Liu Yi-man. 葉祥奎, 劉一曼。河南安陽殷墟花園莊東地出土的龜甲研究。考古 2001.8: 85–92.

Yen Wen-ming. "Neolithic Settlements in China: Latest Finds and Research." *JEAA* 1, nos. 1–4 (1999): 143.

———. 嚴文明。澗溝的頭蓋杯和剝頭皮風俗。考古與文物 1982.2: 38–41.

———. "Chiao-tung Yüan-shih Wen-hua Ch'u-t'an," 史前考古論集 (reprint of 1986), 227–247.

———. 近年聚落考古的進展。考古與文物 1997.2: 34–38.

———. 中國農業和養畜業的起源。史前考古論集 (reprint of 1989), 351–361.

———. 中國史前文化的統一性與多樣性。文物 1987.3: 38–50.

———. 中國稻作農業的起源。史前考古論集 (reprint of 1982), 362–384.

———. 夏代的東方。史前考古論集 (reprint of 1985), 306–318.

———. 略論中國文明的起源。文物 1992.1: 40–49, 25.

———. 論中國的銅石并用時代。史前研究 1984.1: 36–44, 35.

———. 論半坡類型和廟底溝類型。仰韶文化研究。北京, 文物出版社 1989, 110–121.

———. 半坡村落及渭河流域的原始部落。史前考古論集 (reprint of 1988), 146–153.

———. 碰撞與征服。史前考古論集(reprint of 1990), 262–266.

———. 史前聚落考古的重要成果—《姜寨》評述。文物 1990.12: 22–26.

———. 史前考古論集。北京, 科學出版社, 1998.

———. 史前稻作農業遺存的新發現。史前考古論集 (reprint of 1990), 400–406.

———. 再論稻作農業的起源。史前考古論集 (reprint of 1989), 385–399.

———. 東夷文化的探索。文物 1989.9: 1–12.

———. 仰韶文化研究。北京, 文物出版社, 1989.

Yen Yi-p'ing. 嚴一萍。校「正人方日譜」。甲骨古文字研究第二輯, 藝文印書館, 1989, 317–321.

———. 周原甲骨。中國文字新 1 (1980): 159–185.

———. 婦好列傳。中國文字 1981.3: 1–104.

———. 夏紀元遺址為般庚遷殷前古都推測。甲骨古文字研究第二輯, 157–173.

———. 殷商兵志。中國文字 NS7 (1983): 1–82.

Yi Mou-yüan. 易謀遠。中國文明源頭問題研究述評。先秦、秦漢史 1991.2: 4–12.

Yin Chün-k'o. 尹鈞科。中國古代都城制度及其在古都學研究中的地位。中國古都研究。太原, 山西人民出版社, 1994, 110–120.

Yin-hsü Hsiao-min-t'un K'ao-ku-tui. 殷墟孝民屯考古隊。河南安陽市孝民屯商代環狀溝。考古 2007.1: 37–40.

———. 河南安陽市孝民屯商代墓葬 2003–2004 年發掘簡報。考古 2007.1: 26–36.

———. 河南安陽市孝民屯商代鑄銅遺址 2003–2004 年的發掘。考古 2007.1: 14–25.

Yu Shen. 尤慎。春秋及其以前舜帝傳說新考。先秦、秦漢史 2006.3: 39–44.

Yü Feng-ch'un. 于逢春。華夷衍變與大一統思想框架的構築—以（史記）有關記述為中心。中國邊疆史地研究 2007.2: 21–34.

Yü Fu-chih. 余福智。(盤庚) 篇人際關係分析—對中國古代社會經濟形態的初步思考。先秦、秦漢史 1993.9: 49–55.

Yü Shu-sheng. 余樹聲。從中國國家起源看魏特夫對歷史的歪曲。中國史研究 1994.2: 3–9.

Yü T'ai-shan. 余太山。說大夏的遷徙—兼考允姓之戎。夏文化研究論集, 1996, 176–196.

Yüan Ching. 袁靖。中國新石器時代家畜起源的問題。文物 2001.5: 51–58.

———. 論中國新石器時代居民獲取肉食資源的方式。考古學報 1999.1: 1–22.

Yüan Ching and T'ang Chi-ken. 袁靖, 唐際根。河南安陽市洹北花園莊遺址出土動物骨骼研究報告。考古 2000.11: 75–81.

Yüan Kuang-k'uo. 袁廣闊。關於孟莊龍山城址毀因的思考。考古 2000.3: 39–44.

———. 孟莊龍山文化遺存研究。考古 2000.3: 21–38.

Yüan Kuang-kuo and Ch'in Hsiao-li. 袁廣闊, 秦小麗。河南焦作府城遺址發掘報告。考古學報 2000.4: 501–536.

Yüan Kuang-k'uo and Hou Yi. 袁廣闊, 侯毅。從城牆夯築技術看早商諸城址的相對年代問題。文物 2007.12: 73–76.

Yüan Kuang-kuo and Tseng Hsiao-min. 袁廣闊, 曾曉敏。論鄭州商城內城和外郭城的關係。考古 2004.3: 59–67.

Yüeh Lien-chien. 岳連建。商代邊遠地區二里岡期文化分析—兼論商代早期的政治疆域。先秦、秦漢史 1993.10: 29–40.

Yün-nan-sheng Po-wu-kuan. 雲南省博物館。雲南劍川海門口青銅時代早期遺址。考古 1995.9: 775–787.

Zhang Changshou. "A Comparative Study of the Ding Bronze Vessels from Xin-gan." *JEAA* 2, nos. 1–2 (2000): 251–272.

Zhu Zhangyi, Zhang Qing, and Wang Fang. "The Jinsha Site: An Introduction." *JEAA* 5 (2006): 247–276.

Zschieschang, Jorge. "A Simple Bow." In *The Bow Builders Book*, edited by Flemming Alrune, 39–45.